STATISTICAL COMPUTING

STATISTICS: Textbooks and Monographs

A SERIES EDITED BY

D. B. OWEN, Coordinating Editor
Department of Statistics
Southern Methodist University
Dallas, Texas

Volume 1: The Generalized Jackknife Statistic, *H. L. Gray and W. R. Schucany*

Volume 2: Multivariate Analysis, *Anant M. Kshirsagar*

Volume 3: Statistics and Society, *Walter T. Federer*

Volume 4: Multivariate Analysis: A Selected and Abstracted Bibliography, 1957-1972, *Kocherlakota Subrahmaniam and Kathleen Subrahmaniam* (out of print)

Volume 5: Design of Experiments: A Realistic Approach, *Virgil L. Anderson and Robert A. McLean*

Volume 6: Statistical and Mathematical Aspects of Pollution Problems, *John W. Pratt*

Volume 7: Introduction to Probability and Statistics (in two parts)
Part I: Probability; Part II: Statistics, *Narayan C. Giri*

Volume 8: Statistical Theory of the Analysis of Experimental Designs, *J. Ogawa*

Volume 9: Statistical Techniques in Simulation (in two parts), *Jack P. C. Kleijnen*

Volume 10: Data Quality Control and Editing, *Joseph I. Naus*

Volume 11: Cost of Living Index Numbers: Practice, Precision, and Theory, *Kali S. Banerjee*

Volume 12: Weighing Designs: For Chemistry, Medicine, Economics, Operations Research, Statistics, *Kali S. Banerjee*

Volume 13: The Search for Oil: Some Statistical Methods and Techniques, *edited by D. B. Owen*

Volume 14: Sample Size Choice: Charts for Experiments with Linear Models, *Robert E. Odeh and Martin Fox*

Volume 15: Statistical Methods for Engineers and Scientists, *Robert M. Bethea, Benjamin S. Duran, and Thomas L. Boullion*

Volume 16: Statistical Quality Control Methods, *Irving W. Burr*

Volume 17: On the History of Statistics and Probability, *edited by D. B. Owen*

Volume 18: Econometrics, *Peter Schmidt*

Volume 19: Sufficient Statistics: Selected Contributions, *Vasant S. Huzurbazar (edited by Anant M. Kshirsagar)*

Volume 20: Handbook of Statistical Distributions, *Jagdish K. Patel, C. H. Kapadia, and D. B. Owen*

Volume 21: Case Studies in Sample Design, *A. C. Rosander*

Volume 22: Pocket Book of Statistical Tables, *compiled by R. E. Odeh, D. B. Owen, Z. W. Birnbaum, and L. Fisher*

Volume 23: The Information in Contingency Tables, *D. V. Gokhale and Solomon Kullback*

Volume 24: Statistical Analysis of Reliability and Life-Testing Models: Theory and Methods, *Lee J. Bain*

Volume 25: Elementary Statistical Quality Control, *Irving W. Burr*

Volume 26: An Introduction to Probability and Statistics Using BASIC, *Richard A. Groeneveld*

Volume 27: Basic Applied Statistics, *B. L. Raktoe and J. J. Hubert*

Volume 28: A Primer in Probability, *Kathleen Subrahmaniam*

Volume 29: Random Processes: A First Look, *R. Syski*

Volume 30: Regression Methods: A Tool for Data Analysis, *Rudolf J. Freund and Paul D. Minton*

Volume 31: Randomization Tests, *Eugene S. Edgington*

Volume 32: Tables for Normal Tolerance Limits, Sampling Plans, and Screening, *Robert E. Odeh and D. B. Owen*

Volume 33: Statistical Computing, *William J. Kennedy, Jr. and James E. Gentle*

OTHER VOLUMES IN PREPARATION

STATISTICAL COMPUTING

William J. Kennedy, Jr.
Department of Statistics
Iowa State University
Ames, Iowa

James E. Gentle
ISML, Inc.
Houston, Texas

MARCEL DEKKER, INC. New York and Basel

Library of Congress Cataloging in Publication Data

Kennedy, William J., Jr., [Date]
Statistical computing.

(Statistics, textbooks and monographs ; v. 33)
Includes index.
1. Mathematical statistics--Data processing.
I. Gentle, James E., [Date] joint author.
II. Title.
QA276.4.K46 519.5'028'54 80-10976
ISBN 0-8247-6898-1

MARCEL DEKKER, INC.
270 Madison Avenue, New York, New York 10016

Current printing (last digit):
10 9 8 7 6 5 4 3 2 1

PRINTED IN THE UNITED STATES OF AMERICA

To our families,
Carole and Eve Kennedy,
and Joyce Gentle

PREFACE

The purpose of this book is to present material that is generally considered to be in the area called *statistical computing*. Many numerical methods and algorithms are discussed from a computational viewpoint, and techniques for implementing given algorithms in a computer are often considered. The diversity of statistical computer applications is reflected in the book through the various chapters, each of which deals with a relatively general but unique application area. This book is designed to serve both as a textbook at the beginning graduate level, and as a reference text for people interested in computer applications of statistics. Exercises are provided at the end of most chapters for the student, and an extensive list of references is given on each subject to benefit all readers.

Much of the material in this text forms the basis for the Statistics 580 and 581 courses that have been presented at Iowa State University for the past several years. This two-quarter course sequence is taken by graduate students in statistics, computer science, mathematics, and related disciplines. Emphasis in these courses, and in this book, is on numerical methods in statistical computing. Seminumerical and nonnumerical topics, such as computer graphics for example, are not discussed in detail.

The authors are indebted to many people for assistance which led to the completion of this text. All cannot be mentioned in this short space, but especially we wish to thank Professors H. O. Hartley, W. J. Hemmerle, and C. E. Gates for major contributions to our education in statistical computing. We also thank Dr. T. A. Bancroft and Dr. H. A. David for their continuing energetic support of statistical computing which has provided the atmosphere and resources needed for growth in this subject area at ISU. Thanks also go to Professors V. A. Sposito, T. J. Boardman, and W. J. Hemmerle who read parts of the manuscript and made valuable suggestions. Finally, we thank Valerie Engeltjes for the excellent typing support that she provided through several revisions of the original manuscript; and Joyce Johnson, Marlene Sposito, and Darlene Wicks for preparation of the final camera-ready copy.

William J. Kennedy, Jr.

James E. Gentle

CONTENTS

1 / INTRODUCTION

1.1 ORIENTATION

During the past few years a tremendous amount of computer software has been developed to support statistical computing requirements. One significant feature of the software which has evolved over the years is its ever-improving quality in terms of capability, efficiency, and reliability. Much of this improvement can be traced to the development of better numerical methods, algorithms, and programming techniques along with an awareness of the characteristics of computer floating-point arithmetic.

Top-quality computer software for statistical applications is not easily produced. Development begins with a careful choice of the best among competing numerical methods. Selected methods are then specified in detail, with an eye toward computer implementation, in the form of algorithms. Each algorithm must subsequently be carefully coded in a computer language to form the nucleus of the finished software package. Additional considerations such as ease of use and flexibility are important to the overall makeup of the software. Decisions in this area will usually lead to the use of other methods and associated algorithms, many of which may be semi- or nonnumerical in nature. These latter algorithms must also be carefully coded to insure optimal performance of the final product. Thus both numerical and nonnumerical methods and associated algorithms form a basis for computer software, and careful implementation of selected algorithms is a vitally important aspect of the developmental process.

The need for new and improved methods and algorithms is never-ending. Research workers in statistics and related sciences are continually producing new results which require computational support. In many cases an algorithm to support a new computing requirement can be formed by modifying and/or specializing some existing algorithms. In other cases an extension of some numerical method may lead to a desirable algorithm. The most difficult situation is one in which a totally new numerical method must be derived to satisfy the computing requirement. Fortunately, this is not the usual case.

1.2 PURPOSE

The purpose of this book is to present selected computational methods, algorithms, and other subject matter which is important to the effective use of the given methods and algorithms in the computer. Most of the computational methods and algorithms contained in this book have been used in one or more software systems. There is usually more than one "good" method, and possibly several algorithms for each method, for handling a given problem, and the reader will find this fact reflected frequently in the chapters of this book.

Since the literature in statistical computing is extensive, and since it is not possible in the space of this text to treat every subject area and all important aspects within the area, the references listed at the end of each chapter are an important part of the chapter. The authors have attempted to include an extensive list of publications (including many that are not cited in the body of the chapter) which will allow the reader to move quickly to an in-depth study in any of the major areas defined in this text. The literature in statistical computing is to be found primarily in journals on statistics, computer science, and numerical analyses. A list of the most frequently referenced journals in the field and the abbreviations used for each is given in Table 1.1. Proceedings of conferences on statistical computing also contain many relevant articles. The annual Symposium on the Interface of Statistics and

Table 1.1 Journal Titles and Abbreviations

Abbreviation	Journal Title
JACM	*Journal of the Association for Computing Machinery (ACM)*
CACM	*Communications of the ACM*
TOMS	*ACM Transactions on Mathematical Software*
Comp. Rev.	*ACM Computing Reviews*
Comp. Sur.	*ACM Computing Surveys*
Numer. Math.	*Numerische Mathematik*
SIAM J.[a]	*Journal of the Society for Industrial and Applied Mathematics (SIAM)*
SIAM J. Appl. Math.	*SIAM Journal on Applied Mathematics*
SIAM Rev.	*SIAM Review*
SIAM J. Num. Anal.	*SIAM Journal on Numerical Analysis*
MTAC[b]	*Mathematical Tables and Other Aids to Computation*
Math. Comp.	*Mathematics of Computation*
JSCS	*Journal of Statistical Computation and Simulation*
Ann. Stat.	*Annals of Statistics*
AMS	*Annals of Mathematical Statistics*
JASA	*Journal of the American Statistical Association*
Amer. Stat.	*The American Statistician*
Techno.	*Technometrics*
Comm. Stat. (A)	*Communications in Statistics, Part A*
Comm. Stat. (B)	*Communications in Statistics, Part B*
JRSS (A)	*Journal of the Royal Statistical Society, Series A*
JRSS (B)	*Journal of the Royal Statistical Society, Series B*
Appl. Stat.	*Applied Statistics (Also called JRSS, Series C)*
Comp. J.	*Computer Journal*

Table 1.1 (Continued)

Abbreviation	Journal Title
BIT	*Nordisk Tidskrift for Information-behandling*
Computing	*Computing*
J.O.T.A.	*Journal of Optimization Theory and Applications*
J. Inst. Maths. Applics.	*Journal of the Institute of Mathematics and its Applications*

[a]The journal title *SIAM Journal* was used until 1966 when the name was changed to *SIAM Journal on Applied Mathematics*.

[b]This title was used from 1943 through 1959. In 1960 the title *Mathematics of Computation* was adopted.

Computer Science, the proceedings of which are published, has been a major factor in determining the growth of the field of statistical computing. The Statistical Computing Section of the American Statistical Association sponsors sessions at the annual meetings of the ASA, and the proceedings of these sessions are also published.

Exercises are given at the end of most chapters. The purpose of each exercise will be either to reinforce or to extend the material given in the chapter. None of the exercises is intended to be extremely difficult to solve.

1.3 PREREQUISITES

This book is designed for use by readers who have background in statistics equivalent to undergraduate training in statistical methods and theory; mathematical training in elementary calculus, matrix algebra, and numerical analysis; and a familiarity with the FORTRAN computer-programming language.

A complete discussion of an algorithm and its implementation on the computer often requires consideration of details which are dependent on computer hardware characteristics, the computer language used, and in some cases the language compiler. In situations in which computer hardware specifications are pertinent to the discussion, this text assumes use of the IBM 370 series computer. Needed hardware specifications for this machine are given in Chapter 2. This reduces what would otherwise be a much larger prerequisite in applied computer science, and it gives the user of machines other than the IBM 370 (or equivalents) a basis for translating the text material to another machine-compatible form. There are only a few cases where the discussion is affected by the specific machine, the major ones being pseudorandom number generation and error analysis.

1.4 PRESENTATION OF ALGORITHMS

Many different computational methods will be described and discussed in succeeding chapters. In some instances an algorithm associated with some given computational method will be provided. When such is the case, and the algorithm is not stated in the form of a FORTRAN program, then the general form of the statement of the algorithm will be as follows. The first line will contain the word "*ALGORITHM*" followed by the name given to the algorithm and a FORTRAN-like argument list, enclosed in parentheses, containing input and output quantities. For example, a first line might be

ALGORITHM RAND(I,J,X)

Following the first line, in some cases, will be a short set of comments designed to facilitate overall understanding of the algorithm. Then one or more numbered steps will be given along with a description of the operations to be performed in each step. These will take the form

Step	*Description*
1.	Set I = I*J
⋮	⋮

Comments will also appear between the steps in the algorithm as often as they are needed to explain the process.

When algorithms are stated in a programming language, such as FORTRAN, the meaning of the "=" symbol as an assignment of value is clear. However, there is some possibility of confusion when the "=" sign is used in the statement of algorithms using other than a programming language. The question may arise, in a specific situation, as to whether the "=" symbol is specifying an assignment of value or whether a statement of equality is intended. Sometimes the symbol "← " or ":=" is used to denote assignment of value. This will not be done in this text. The context in which the symbol "=" is used will serve to define whether an assignment of value or a statement of equality is intended. If there is room for doubt, a comment will be made to clarify the situation.

2 / COMPUTER ORGANIZATION

2.1 INTRODUCTION

The purpose of this chapter is to describe the computer in general terms and consider some of the internal computer operations which are of special interest to the user in scientific applications. The computer is a tool to be used in performing calculations, and a basic understanding of how this tool operates is essential.

2.2 COMPONENTS OF THE DIGITAL COMPUTER SYSTEM

The computer system may be thought of as being made up of five components: (1) the main storage, (2) a control component, (3) an arithmetic and logic component, (4) an input component, and (5) an output component. Figure 2.1 depicts the way in which these components are interconnected.

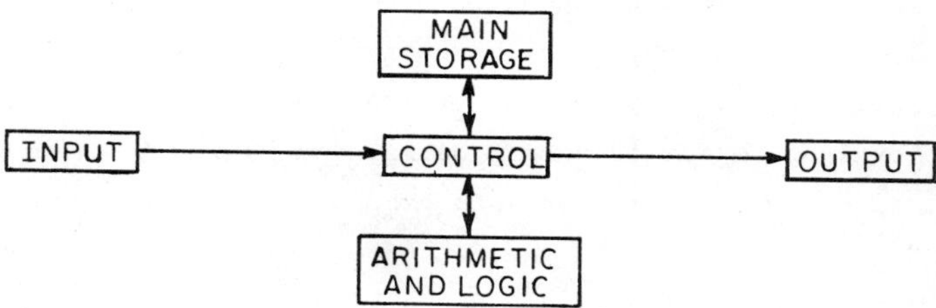

Figure 2.1 Components of the computer system.

The main storage of the computer is sometimes called the *memory* of the machine. This main storage component is used to hold information relative to the computational task (*job*) being performed at the moment. Stored information typically consists of program instructions, numeric constants, input data, intermediate results, and final results of computations.

The main storage is a hardware device made up of small elements, each of which is capable of assuming either of two states at any given time. The computer is constructed so that the state of a memory element can be sensed *(read)* and restored to its original state before sensing; hence the reading of states does not destroy the states. The actual form of the memory elements has been changing over the years with advances in technology. The general trend has been toward smaller physical size and less expense for given storage capacity.

The computer user can think of main storage as a conceptual pool containing information. The transition from physical hardware reality to the abstraction of information is made by using the natural correspondence between state of a memory element and the binary digits zero and one. The smallest unit of information is reflected in the state of an element and is called a *bit*, the term being a contraction of *binary digit*. In IBM 370-type computers the information in main storage is addressed in units called *bytes*. A byte contains 8 bits of information. Main storage byte locations are numbered beginning with 0 and proceeding to the integer 1 less than the number of bytes available in the main storage. This numbering system is used to provide an address for each unit (byte) of information in the main storage. In other computers, bits may be organized into groups of various sizes, called *words*. For example, in the CDC 6600 computers a word is 60 bits and in the Univac 1108 a word is 36 bits. Four bytes, hence 32 bits, constitute a word in the IBM 370. Taking into account the specific word size is important when one speaks of memory size (in bytes or words) or when one considers the precision with which numbers are represented or arithmetic is performed, as we shall do later.

The *control* component, as its name implies, controls the activities of each of the other components. The sequence in which operations are performed; the retrieval, storage, and decoding of instructions from main storage; and the execution of instructions are all functions of the control component.

The *arithmetic and logic* component performs all arithmetic operations, comparisons, and conversion of data. This component accepts operands which are represented in various modes (eg., integer, floating-point, character) and lengths (in terms of bytes), performs operations in areas called *registers*, and returns the results of specified operations. The control and arithmetic and logic components together are often called the *central processing unit* (CPU).

The final two components, input and output, will be discussed concurrently, in order to allow a more complete description of the hardware devices which form a part of each of these components.

One of the main vehicles for initial input is the punched card. Cards are commonly punched using a typewriterlike device called a *card punch*. The punched cards are fed through a card reader which senses the holes and transmits the characters recorded on cards in the form of electrical impulses. The electrical impulses are then used to initialize bytes of main storage in such a way that they represent the characters punched on the cards.

Magnetic tape is a commonly used information storage medium. A single reel of tape can contain several million characters of information. Tape reels are used in place of cards in situations where large amounts of information are being processed or need to be stored. Information is physically recorded on tape in quantities called *records*. Within a record the basic elements of information *(characters)* are recorded by magnetizing spots on the tape. The magnetic tape unit is a hardware device upon which a reel of magnetic tape is mounted. The tape unit has sensors called *reading* and *writing heads* which respectively sense or record magnetic spots on the tape. Information travels between tape unit and main storage as input (to main storage) or output (from main storage) depending upon the direction of transfer.

One characteristic common to input and output operations using punched cards or magnetic tape is that each involves sequential processing of records one by one from first to last. Other information storage devices allow input or output of selected records--for example, the 16th record, the 34th record, and then the 19th record. Such devices are called *direct-access devices*. One direct-access device is the magnetic disk device. This device holds at least one magnetic disk pack, each disk pack being composed of a series of plates. These plates are coated with a magnetic oxide and look much like phonograph records stacked on top of one another with a space between each. When mounted on the disk device, the disk pack revolves at high speed; read/write heads, attached to the disk device and suspended between the plates, sense and record information on the magnetic oxide coating. The main advantage of disk over tape is that disk input and output is faster due to the capability of the disk unit read/write heads to position themselves in a fraction of a second, whereas minutes may elapse before a tape can be positioned prior to execution of an input or output operation.

The *line printer* is a device used to record characters, output from the computer, on paper. Many printers can record up to 1200 lines of 132 characters per minute. Nevertheless, the printer rate of speed in character transfer, as is the case with the card reader, is much slower than that of either the tape or disk device.

One input/output device which is becoming increasingly more important is a typewriterlike device with either a cathode-ray tube display, an attached paper for printing purposes, or both. Such devices are attached to the computer and may allow immediate feedback of results to the user. This allows the user to achieve what is often called an *interactive mode* of computer use. One limitation on such devices is that they are slow because the rate of data input is that of the person typing on the keyboard of the machine.

Other input/output/storage devices (not discussed here), including drums, data cells, optical scanners, bulk core, paper tape, and incremental plotters, are all available for use at many of the larger computer installations.

2.3 REPRESENTATION OF NUMERIC VALUES

In order to make effective use of the computer in scientific applications the computer user must be aware of the ways in which numeric data are stored and how arithmetic operations are carried out. This section contains a description of two modes for representing numerical data. These are the binary integer (fixed-point) and the floating-point representations.

2.3.1 Integer Mode Representation

Integer mode in the IBM 370 uses either 2 or 4 bytes to represent an integer. The bits are thought of as being numbered left to right: 0 through 15 or 0 through 31, as the case may be. The bit numbered 0 designates the sign of the number: 0 is positive, and 1 negative. In the case of positive integers, the remaining bits together form the binary representation of the integer. Negative integers are represented in what is called *two's complement notation.* If $-x$ is a negative integer, then the two's complement representation of $-x$ will be, using 32 bits, $2^{32} - x$. Using 2 bytes (16 bits), the number -13 would be represented as

```
2^16 = 1 0 0 0 0 0 0 0 0 0 0 0 0 0 0 0 0
     - 0 0 0 0 0 0 0 0 0 0 0 0 1 1 0 1
       -------------------------------
       1 1 1 1 1 1 1 1 1 1 1 1 0 0 1 1
```

The representation or interpretation of numbers in two's complement is quite simple using the following rule: Given the representation of x (positive or negative) in this scheme, the representation of $-x$ is obtained by (1) complementing all bits and (2) adding 1 to the result. For example, using 16 bits,

```
2 is represented as 0 0 0 0 0 0 0 0 0 0 0 0 0 0 1 0
                    1 1 1 1 1 1 1 1 1 1 1 1 1 1 0 1   Bit complement
                                                + 1   Add 1
                    -------------------------------
```

```
-2 is represented as 1 1 1 1 1 1 1 1 1 1 1 1 1 1 1 0
                     0 0 0 0 0 0 0 0 0 0 0 0 0 0 0 1 Bit complement
                                                + 1 Add 1
                     -------------------------------
 2 is represented as 0 0 0 0 0 0 0 0 0 0 0 0 0 0 1 0
```

Two special numbers are 0 and -2^{p-1}, where p is the number of bits used. The number 0 is represented (as one would guess) by 0 0 0 ... 0. The smallest number representable in this scheme, -2^{p-1}, is 100 ... 0. The corresponding positive number cannot be represented. In a full word on the IBM 370 the smallest number representable in integer mode is -2^{31} and the largest number is $2^{31} - 1$.

2.3.2 Representation in Floating-Point Mode

The floating-point mode of representation of numeric quantities is ordinarily in either *single precision*, using one computer word, or *double precision*, using two words. On the IBM 370 these occupy 32 and 64 bits, respectively, and in either case the zero-numbered bit is used to designate the sign of the number, just as in integer mode representation. The bits numbered 1 to 7 are collectively called the *characteristic*. Bits numbered 8 to 31 or 8 to 63 make up the *fraction* or *mantissa*.

The fraction is always recorded as a positive binary integer. Two's complementation never applies to floating-point numeric representation. The binary point is taken to be immediately to the left of bit position 8. Bit 8 is the coefficient of 2^{-1}, bit 9 the coefficient of 2^{-2}, and so on. Normally the fraction is looked upon as a hexadecimal (base-16) number, having either six hexadecimal digits when the fraction is 24 bits long or 14 digits in double precision. The radix point is still taken to be immediately to the left of the first (leftmost) hexadecimal digit in the fraction.

A floating-point number is represented by a fraction multiplied by a power of 16, the sign being attached to the result. The exponent denoting the appropriate power of 16 is given by the characteristic in the floating-point representation. The bit positions 1 to 7 containing the characteristic can represent integers in the range 0 to 127 (decimal). A particular notation is used for characteristics, called *excess-64 notation.* In excess-64 notation the convention is to consider the exponent as a number in excess of 64, which means that the characteristic is 64 units larger than the exponent. Thus a characteristic of 67 (decimal) implies that the fraction is to be multiplied by 16^3, and a characteristic value 61 yields a multiplier of 16^{-3}. The excess-64 notation allows the representation of magnitudes as small as 16^{-64} and as large as 16^{63}. In the floating-point representation of nonzero numbers, each number is *normalized* so that the leading digit of the hexadecimal fraction is nonzero.

As an example, consider the floating-point representation of -7.25. The quantity to be represented is $-(2^2 + 2 + 2^0 + 2^{-2}) = -(2^{-2} + 2^{-3} + 2^{-4} + 2^{-6}) \times 16^1$. The floating-point representation is

```
1  1 0 0 0 0 0 1   0 1 1 1 0 1 0 0   0 0 0 0 0 0 0 0   0 0 0 0 0 0 0 0

Sign  Characteristic                    Fraction
```

Consider a second example wherein we represent 0.50. The characteristic in this case is 64 (decimal) = 1 0 0 0 0 0 0 (binary). The floating-point representation is

```
0  1 0 0 0 0 0 0   1 0 0 0 0 0 0 0   0 0 0 0 0 0 0 0   0 0 0 0 0 0 0 0

Sign  Characteristic                    Fraction
```

The floating-point representation, as we have seen, involves the use of hexadecimal (base-16) numeric representation. The IBM 370 can in fact be thought of as a hexadecimal computer. The decimal numbers 0 to 15 are required in general for expressing numbers in base 16. Since single-character quantities are much more convenient to use, the characters A, B, C, D, E, and F are used to represent, respectively, the numbers 10, 11, 12, 13, 14, and 15. Thus, for example, the pattern

0 1 1 1 1 1 0 1   1 0 1 0 0 0 0 1   0 0 0 0 0 0 0 0

will be recorded in hexadecimal as

7DA100

and the fraction

1 1 1 1 0 1 1 1   0 0 0 0 0 0 0 0   0 0 0 0 0 0 0 0

as

F70000

The floating-point representation

1 1 0 0 0 0 0 1   0 1 1 1 0 1 0 0   0 0 0 0 0 0 0 0   0 0 0 0 0 0 0 0

of the number -7.25 will be treated as

C1740000

2.4 FLOATING- AND FIXED-POINT ARITHMETIC

Knowing how the computer represents numeric data in two different modes only gives part of the story. One must also know how the basic arithmetic operations are performed. This knowledge will be used in subsequent chapters where algorithms are given which take advantage of some of the subtle details of the ways in which the computer performs arithmetic operations both on floating-point and fixed-point numbers.

2.4.1 Floating-point Arithmetic Operations

We now describe the steps taken in carrying out each of the floating-point operations of addition, subtraction, multiplication, and division in the IBM 370. The operations to be described are carried out in special areas called *floating-point registers*. These registers are contained in the arithmetic and logical component of the computer. Each register is 64 bits in length.

Floating-point addition and subtraction are performed in the following steps:

Addition and Subtraction

1. Each operand is placed in a separate register
2. The characteristics are compared and the fraction of the number having smaller characteristic is shifted right a number of hexadecimal digits equal to the difference between the characteristics. We call this shifting an *alignment*.
3. Addition or subtraction of fractions takes place.
4. The larger of the two characteristics is taken as the resultant characteristic.
5. The correct sign is inserted.
6. *Normalization* is accomplished, for nonzero fractions, by executing left shifts and characteristic decreases until the leading hexadecimal digit is nonzero. *Cancellation* is said to have occurred when normalization is necessary.

If either operand is not normalized prior to the execution of step 1, then *prenormalization* takes place. In performing step 3 a carry-out of the high-order hexadecimal digit may occur. If this happens, the digits of the fraction are shifted right one hexadecimal position, a hexadecimal 1 is inserted as the highest-order hexadecimal digit of the fraction, and the characteristic is increased by 1 to take account of the shift.

For example, the sum of the fractions C10243 and 612405 is

$$\begin{array}{r} \text{C10243} \\ \underline{\text{612405}} \\ \text{1222648} \end{array}$$

where the 1 in the sum represents a carry. The resultant fraction will be obtained by shifting one digit to the right. Thus, we obtain 122264 as the resultant fraction. Note that rounding does not take place; rather, this result is obtained by truncating to six hexadecimal digits. (Some computers round instead of "chop.")

In every single-precision addition or subtraction the operand having smaller characteristic is furnished a *guard digit* placed immediately to the right of the sixth digit of the fraction. The guard digit is zero at the beginning of each operation and is altered only when a right shift takes place. Digits which are shifted right during alignment go through the guard digit position. The last digit shifted out of the register being used is retained as the guard digit. The guard digit participates in the arithmetic operation and during the normalization step the guard digit may reappear in the fraction of the final result.

Consider the addition of the floating-point numbers 4310574A and C2112097, with the guard digit shown in parentheses.

Step 2	4310574A(0)
	C2011209(7)
Steps 3, 4, 5	430F4540(9)
Step 6	42F45409(0)

The guard digit participates in the intermediate and final result in this example. Note that if no left shift had been necessary, the guard digit would not have appeared in the final result, and if two left shifts had been made during normalization, a zero would have appeared as the last digit in the fraction.

The floating-point operation of multiplication is carried out in the following steps, assuming both operands are prenormalized:

Multiplication

M1. Addition of the two characteristics (less 64) takes place.

M2. Multiplication of the two fractions is accomplished. In single precision only the first six hexadecimal digits participate; however, the 48 bits numbered 8 to 55 of the register are used to hold the product. The last 8 bits in the register are zero. In double precision all 56 bits of each operand take part and the product is truncated to 56 bits.

M3. The resultant sign is determined.

M4. Postnormalization takes place.

Consider the multiplication of the single-precision numbers stored as 41200000 and 42335421. The multiplication is carried out in the following steps:

M1. 41 + 42 - 40 = 43

M2. 200000 × 335421 = 066A842000000

M3. Sign bit = 0

The intermediate result is 43066A8420000000. The final step consists of a normalization, and we obtain

M4. 4266A842 00000000

Note that unless the resultant fraction is zero, no more than one left shift will be necessary in postnormalization for multiplication.

Division is carried out in three steps instead of four because postnormalization is never necessary; however, a right shift may be needed. As an illustration of the floating-point division operation, consider the quotient 41400000 ÷ 41600000 of two single-precision floating-point numbers. The operations are carried out as follows:

Division: 400000/600000

Yields: AAAAAA

Characteristic: 41 - 41 + 40 = 40, allowing for the right shift

Sign: Positive

Quotient: 40AAAAA

As in multiplication, only the first 24 bits in each fraction participate in a single-precision division, but the entire register is used to hold the quotient. When the value in the register is stored in a single word of main storage, the fraction is truncated to six hexadecimal digits.

The ability of digital computers to represent numbers in floating-point mode simplifies the programming required in scientific applications, because otherwise the programmer would need to maintain an associated scale factor with each number, making the programming task much more complicated. Most scientific application programs use floating-point variables in an effort to obtain numeric results which are "close" approximations to the real numbers which form the exact solutions to the problem at hand. In most cases the computer solution is quite satisfactory, as we shall see from the examples given in succeeding chapters.

2.4.2 Fixed-Point Arithmetic Operations

It is seldom possible to carry integers throughout in scientific applications, but certainly integer (fixed-point) arithmetic is used in many ways, such as subscripting, for example. This subsection will briefly describe the integer arithmetic operations and give some relevant facts about each one. The reader can consult appropriate texts (for example, Struble 1969) for a more complete discussion of fixed-point arithmetic. Examples of addition will be given in terms of an 8-bit word, the first bit being the sign bit, instead of our usual 32 bits. This is done without loss of generality.

Addition and Subtraction

The arithmetic operations of addition and subtraction are carried out as addition using the two's complement form for negative integers. In the course of addition a carry-over into the sign bit may occur, and this serves to change the sign bit. That is to say, addition is carried out by binary addition using all the bits in the word. The following examples are given to illustrate this point. Note that the final answers in binary in both columns are incorrect due to carry-over into the sign bit.

Decimal	Binary	Decimal	Binary
57	00111001	-57	11000111
-35	11011101	-35	11011101
22	00010110	-92	10100100
92	01011100	-36	11011100
114	01110010	-128	10000000
36	00100100	-22	11101010
150	10010110	-150	01101010

Multiplication

The product of two fixed-point 32-bit values is formed in the usual way. If the absolute product requires more than 31 bits for its representation, an overspill occurs and only the last 32 bits are taken as the product. Thus as in addition, the sign bit can be inadvertently modified. This overspill or carry-out feature means that we must control the magnitude of integers used in high-level-language programs, such as FORTRAN, else a carry-out may occur. The IBM 370 versions of FORTRAN do not take note of this carry-out.

Division

The sign of the quotient in division is determined algebraically, so we do not have the carry-out problem in this operation as we did in the other three. The quotient is obtained by computing the integer part of the quotient and also a remainder. The remainder is ignored and the integer part of the quotient is taken as the result.

EXERCISES

2.1 Convert the following decimal numbers to the floating-point hexadecimal form of the IBM 370.

(a) -1.2 (c) 8192.0 (e) -0.12

(b) 4.25 (d) -8192.0 (f) 0.0425

2.2 Express the following decimal integers in the fixed-point form in the IBM 370.

(a) -1 (c) 4096 (e) 4354

(b) 23 (d) -128 (f) -4354

2.3 Assume an 8-bit computer word, with sign bit as the leading bit. Show the results which will be obtained in fixed-point computations using FORTRAN for this machine.

(a) 16 - 12 (c) (-109) + (-21) (e) 51 × 4

(b) 109 + 21 (d) -49 × 3 (f) -17 ÷ 3

2.4 Perform the indicated floating-point arithmetic and express the result in a floating-point form. Use the computer to verify results.

(a) 41F06123 - 40A91702

(b) C2111AAA + 422620AA

(c) 41F06123 ÷ 40A91702

(d) 40A09000 ÷ 40017920

(e) C2111AAA × 40A91702

(f) 40A09000 × 40017920

(g) C2111AAA - C21BBAAA

(h) 42BBF001 + 42BBE001

2.5 What is meant by the term *cancellation*? Will cancellation always occur in addition and subtraction operations? Why? Why is cancellation not a problem in multiplication and division?

2.6 The hexadecimal floating-point word uses a six-digit fraction to represent leading significant digits on given numbers. What is the approximate decimal number equivalent to these six hexadecimal digits?

2.7 Let F be the set of IBM 370 single-precision floating-point numbers. For x as below in ordinary base-10 representation, find a and b is an element of F such that $a = \max \{y | y < x\}$ and $b = \min \{z | z > x\}$, where y and z denote floating-point numbers.

(a) x = 0.001

(b) x = 1.00

(c) x = 1000

2.8 A decimal fraction (e.g., 0.5125) can be converted to a hexadecimal fraction by repeated multiplication by 16 with the integer part of the product being a desired hexadecimal digit. The resulting fractional part of the product is used in the next multiplication by 16. For example, to convert 0.5125 to hexadecimal, proceed as follows:

0.5125*16 = 8.2 ⟶ 0.8

0.2*16 = 3.2 ⟶ 0.83

0.2*16 = 3.2 ⟶ 0.8333 etc.

Use this method to convert the base-10 number 0.3758125 to hexadecimal.

2.9 What is the function of the guard digit in floating-point arithmetic operations? Why are guard digits not supplied in single-precision floating-point multiplication and division?

2.10 How many different normalized single-precision floating-point numbers are representable by the computer?

REFERENCES

Gear, C. W. (1974), *Computer Organization and Programming, 2nd ed.*, McGraw-Hill, New York.

IBM Corporation (1970), *IBM System/360 Principles of Operation*, form GA22-6821, P.O. Box 390, Poughkeepsie, N.Y.

Struble, G. (1969), *Assembler Language Programming: The IBM System/360*, Addison-Wesley, Reading, Mass.

3 / ERROR IN FLOATING-POINT COMPUTATION

3.1 INTRODUCTION

The uninitiated computer user often tends to believe that the results obtained through execution of a "logically correct" set of computer instructions contain little or no error. As the user gains experience in application he or she inevitably finds examples where the computer gives almost totally incorrect results which are not due to "bugs" in one's program. These traumatic experiences leave the new user shaken and undermine one's confidence in computed results. This is a natural reaction; moreover, a certain amount of suspicion, to stimulate close checking of results, is certainly desirable. Computed results often contain some error. Usually the amount of error is not large, and computed results are usually sufficiently accurate. The user who is aware of the kinds of inaccuracies that appear in computations and the way in which they are propagated is in a much better position to anticipate the amount and kind of error that is likely to appear in his or her answers. In this chapter we will discuss errors in general and then closely examine one important kind, the so-called rounding error. Specifically, we will define different types of error, discuss the amount of error which can be introduced in certain kinds of computations, describe methods for obtaining a priori bounds for error in computations, and suggest some techniques which can be used in an effort to control certain kinds of errors in floating-point computations.

3.2 TYPES OF ERROR

An *absolute error* is, by definition, the difference between an observed or calculated value and the true value; and a *relative error* is an absolute error divided by the true value (whenever it is nonzero).

Three different general types of error in computations can be defined. These are:

1. Blunders
2. Errors due to use of an approximate computational method
3. Errors due to approximation imposed by the computer

The error, either relative or absolute, in a result obtained in computer application can be the sum total of any combination of these three types.

By a *blunder* we mean a gross error or mistake such as introduction of a "bug" in a program or incorrect recording of data. A blunder, if detected, is readily correctable, whereas the other two types of errors may not be.

The second type of error is introduced when we use one function or process as an approximation to a true function or process. For example, evaluation of the first n terms in the series expansion of a function may give values which are only approximately equal to the true function values even when calculations are carried out exactly. A rather precise bound for this type of error can often be obtained mathematically. We will, in succeeding chapters, study specific examples of this type of error.

Finally, due to the finite word length of a computer, there are errors introduced in the course of computations. The initial representation of many numbers, both rational and irrational, even those in the permissible range, will contain error due to truncation. The arithmetic operations may introduce other errors due to the chopping of fractions in floating-point operations. This type of error is not always discussed in textbooks; however, it is certainly an important factor in scientific computer applications, so we will now consider it in more detail.

3.3 ERROR DUE TO APPROXIMATION IMPOSED BY THE COMPUTER

Consideration of errors, either relative or absolute, due to approximation imposed by the computer dictates that we discuss what we mean by an error when we are dealing with the computer. The definition of *error* must involve the idea of a true value. What do we mean when we say *true value*? To clarify the nature of the question, let us consider two different situations relative to a given problem. First, we begin with *exact* initial values and imagine that we perform *exactly* all calculations to arrive at an *exact* answer. The second situation is that we begin with the computer approximation to the initial values and then perform all calculations *exactly*. This exact value appears to be the best that we can hope to obtain in computer application. If we are unable to represent all the initial values exactly in the computer, then two different values can result in these two situations. They differ because we actually have solved two different problems. Thus the question is: Which of the exact answers should be taken as the true value to be used in defining error? The answer to this question is that in practice both of these exact results are used at different times to determine error. To avoid confusion, we will use the term *true value* to mean the exact value of the first situation (beginning with exact data) and the term *c-true value* to mean the exact result from the second situation described, in which the initial values are the computer approximations.

To illustrate the difference described above, consider the following example in which we assume the use of a computer which performs floating-point decimal arithmetic with three-digit accuracy. The problem is to subtract 1.613 from 1.624. The true value in this case is obviously 0.011. The c-true value is obtained by subtracting 0.161×10 from 0.162×10, which gives 0.1×10^{-1}. Application on the computer, assuming floating-point subtraction is carried out as described in Chapter 2 (except in base 10 here), gives the computer result

$$\begin{array}{r} 0.162 \times 10 \\ -0.161 \times 10 \\ \hline 0.100 \times 10^{-1} \end{array}$$

The absolute error with respect to the true value is

$$0.11 - 0.010 = 0.001$$

and the absolute error with respect to the c-true value is zero.

3.4 ANALYZING ERROR IN A FINITE PROCESS

The magnitude of error due to approximations imposed by the computer and the way in which it is generated are of particular interest in scientific applications. Often a bound for the magnitude of the total error in an arithmetic process can be described mathematically in terms of original data and intermediate errors that can be bounded. The separation of total error into component parts is called an *error analysis*. An error analysis often alerts the prospective user to problem areas in his proposed numeric process, and it may suggest ways in which error can be reduced.

There are two different general techniques for a priori analysis of error which are used in practice. These are called *forward* or *direct* error analysis and *backward* or *inverse* error analysis. It will become obvious as we discuss these techniques that other different, but somewhat similar, techniques can be defined. We will, however, concern ourselves only with the two techniques mentioned above because they are the most commonly used.

A direct error analysis is carried out by obtaining an expression for error at each intermediate step of the computation in terms of true values and of errors made in previous steps. Often, errors considered in a direct analysis are relative errors defined with respect to the true value. Thus at each stage of the direct error analysis we are required to compare our computer result with the true value. This type of error analysis is usually difficult to carry out.

The inverse error analysis makes no reference to the true value until the final step of the analysis. Rather, the inverse analysis defines the computed value at each stage in terms of c-true values. A part of each computed value is error, but the error in this analysis is defined with respect to c-true values. This type of analysis looks upon the final computed results as being the exact solution to a problem which uses a perturbation of the data for the real problem of interest, and it gives bounds on the perturbations. We then decide whether the problem actually solved was "sufficiently" close to the desired problem.

An inverse error analysis, due to the definition of error which is used, is concerned only with the assessment of *rounding error* at each intermediate stage of the computations. Rounding error, as we shall see later, is defined with respect to the c-true value. Due to the fact that the inverse error analysis is most often used, we will concentrate on this type of analysis in subsequent discussion.

3.5 ROUNDING ERROR IN FLOATING-POINT COMPUTATIONS

The theory of rounding error in floating-point computations is given in detail in the texts by Wilkinson (1963) and Sterbenz (1974). The reader who is interested in a detailed treatment of rounding error and error analysis should consult these texts. In this section we consider rounding errors in the basic floating-point arithmetic operations. Application of these results to two different situations which frequently occur in practice will be given in Section 3.6.

The method used in storing normalized nonzero floating-point numbers allows us to express them in the general form

$$\pm\ .\ f_1 f_2 \ldots f_t \times \beta^e$$

where the $f_1, \ldots, f_t$ and e are integers satisfying the inequalities

$$1 \leq f_1 \leq \beta - 1$$

$$0 \leq f_1 \leq \beta - 1 \qquad \text{for } i = 2, 3, \ldots, t$$

$$-m \leq e \leq M$$

The IBM System/370 uses $\beta = 16$, $m = 64$, $M = 63$, and $t = 6$ in single precision.

We use the symbol F to denote the set of all floating-point numbers. This set is a finite subset of the numbers on the real line, and the reader can easily verify that the numbers are not equally spaced on this line (see Exercise 2.7). In computer applications we can deal only with elements of F. Real numbers which fall outside the allowable range cannot be approximated. A real number in this range is approximated by choosing an element from F to represent it. The choice is made, in compatibility with the computer being considered and the FORTRAN language, by representing the real number in normalized exponential form and chopping the fraction after a specified number of significant figures. Thus, for example, if the real number is representable as

411ABC106413...

then the element chosen from F would be (assuming single-precision length)

411ABC10

This chopping can be replaced by a true rounding operation, but this is not usually done in IBM 370 FORTRAN.

The foregoing comments about the set F, coupled with our knowledge of the way in which the computer performs floating-point arithmetic, suggests that the well-known axioms of the real number system may not apply in this situation. That this is indeed true is easily shown by example. Specifically, the failure of the associative law of addition, the associative law of multiplication, the distributive law, and the cancellation law can be demonstrated for the set F coupled with the floating-point arithmetic operators (see Exercise 3.3).

We now turn to the basic arithmetic operations and discuss rounding errors in these operations. Let x and y denote any two elements of F, and o denote any of the operators $-$, $+$, $*$, $/$. The result of a floating-point operation is denoted by $f\ell(\ )$. When $(x o y) \neq 0$, it can be written as

$$f\ell(xoy) = (xoy)(1 + \varepsilon) \tag{3.1}$$

The quantity (xoy) is the *c-true value* defined in Section 3.3, since x and y are in F. We call (xoy)ε the *rounding error* in the operation; however, it might more properly be called the chopping error. The quantity ε is the relative error with respect to the c-true value, since

$$\varepsilon = \frac{f\ell(xoy) - (xoy)}{(xoy)}$$

Forsythe and Moler (1967) have shown that $|\varepsilon| < \beta^{1-t}$, under certain assumptions which are approximately satisfied on the IBM 370. (Kaneko and Liu 1973 show that this bound for addition is actually $\beta^{1-t}/(1 - \beta^{-1})$, or $16^{-4}(15^{-1})$, on machines like the IBM 370. In the following, however, we will find it convenient to ignore this minor adjustment and use β^{1-t} as the bound for floating-point operations on c-true values.)

Finally, we need to describe the error due to entry of data into the computer. This will be accomplished by using the previously described notation, but with respect to the true value. Thus the conversion from true to c-true value will be indicated by $f\ell(x) = x(1 + \varepsilon)$, where $|\varepsilon| < \beta^{1-t}$ and x is the true value.

We now have expressions which bound the relative error due to entering data into the computer and performing any of the four basic arithmetic operations. The relative error ε is usually studied in error analysis. Its absolute counterpart, εx, is more highly data-independent and not as descriptive. Some examples will now be given for illustration. The examples will use $\beta = 10$, $t = 4$, and chopping operations.

$$0.1245 + (0.2167)10^4 \tag{i}$$

The floating-point sum in this case is 2167. because of the shifting which will take place in the register operations. The c-true value in this example is 2167.1245, and the relative error is

$$\frac{-0.1245}{2167.1245} \cong (-0.6)10^{-4}$$

which is smaller than 10^{-3} in absolute value.

$$(0.4316)10^{-2} - (0.4303)10^{-2} \tag{ii}$$

The floating-point difference is $(0.1300)10^{-4}$. In example (ii) the c-true value is also $(0.1300)10^{-4}$; thus the rounding error is zero. Note that the cancellation of a higher-order digit resulted in zero rounding error.

$$(0.4310)10^{-2} \times (0.1614)10^{2} \tag{iii}$$

In (iii) the floating-point product is $(0.6956)10^{-1}$ and the c-true value is $(0.6956634)10^{-1}$. The reader can verify that the relative error in this case is substantially less than 10^{-3} in absolute value.

At this point in our discussion of rounding error we will pause to compare relative error with respect to c-true values and the relative error defined in the direct error analysis, that is, with respect to true values. Consider, for example, the difference $(0.431675)10^{-2} - (0.430344)10^{-2}$, where each number is a true value. The true value is

$$\frac{(0.1331)10^{-4} - (0.1300)10^{-4}}{(0.1331)10^{-4}} = \frac{(0.31)10^{-6}}{(0.1331)10^{-4}}$$

$$= (0.23)10^{-1}$$

Thus, error defined with respect to the true value in this single operation reflects the effect of cancellation, whereas the relative error with respect to c-true values does not. Cancellation effects are highly data dependent, so meaningful a priori error bounds are difficult to obtain when error is defined in terms of true values in intermediate computational stages. The reader should note that cancellation only occurs in addition of quantities having different signs (subtraction) and not in multiplication, division, or addition of quantities having like sign. We will not attempt mathematically to assess the effects of cancellation in the error analyses in this text, but we will recommend certain procedures which minimize cancellation and point out situations where cancellation sometimes becomes a serious problem. Cancellation is not considered to be an error in the usual sense; it is a natural result of the computational method and the data used.

The result of any one of the four floating-point operations can be interpreted, in the inverse error analysis, as the exact solution of a slightly different problem. Consider addition, for example. The floating-point value obtained is the exact result of summing the two perturbed numbers $x(1 + \varepsilon_1)$ and $y(1 + \varepsilon_2)$. Often it is helpful to think of a final floating-point result as being the c-true value obtained by using slightly different (perturbed) data in computations. The possible magnitude of perturbations is indicated in the course of analysis.

3.6 ROUNDING ERROR IN TWO COMMON FLOATING-POINT CALCULATIONS

The formation of finite sums and inner products are two types of computations which occur repeatedly in scientific applications. In this section we will investigate rounding errors in these computations.

Consider first a sequence of additions of floating-point values

$$x_1 + x_2 + x_3 + \dots + x_n$$

The addition takes place recurisvely as

$$\begin{aligned} s_1 &= x_1 \\ s_k &= f\ell(s_{k-1} + x_k) \\ &= (s_{k-1} + x_k)(1 + \varepsilon_k). \end{aligned}$$

We will assume $|\varepsilon| \leq \beta^{1-t}$, as discussed in Section 3.5. The floating-point sum is

$$s_n = x_1(1 + \gamma_1) + x_2(1 + \gamma_2) + \dots + x_n(1 + \gamma_n) \tag{3.2}$$

where

$$1 + \gamma_1 = \prod_{j=2}^{n} (1 + \varepsilon_j)$$

$$1 + \gamma_k = \prod_{j=k}^{n} (1 + \varepsilon_j) \qquad (k = 2, 3, \dots, n)$$

$$(1 - \beta^{1-t})^{n-1} \leq 1 + \gamma_1 \leq (1 + \beta^{1-t})^{n-1}$$

$$(1 - \beta^{1-t})^{n+1-k} \leq 1 + \gamma_k \leq (1 + \beta^{1-t})^{n+1-k}$$

$$(k = 2, 3, \ldots, n)$$

The absolute error with respect to the c-true value is

$$e = \sum_{j=1}^{n} x_j - \sum_{j=1}^{n} x_j(1 + \gamma_j) = -\sum_{j=1}^{n} \gamma_j x_j$$

The absolute value of e is bounded above by

$$|e| \leq \sum_{j=1}^{n} |\gamma_j| \; |x_j| \tag{3.3}$$

Wilkinson (1963) showed that the upper bound given in inequality (3.3) is almost attainable. Note that the smallest *upper bound* is obtained if the x's are ordered in increasing order of magnitude prior to addition. This is shown by inequality (3.3) when we note that $|\gamma_1|$ has the largest bound of the $|\gamma_j|$ and, after ordering, $|x_1|$ is the smallest of the $|x_j|$. This does not, however, mean the smallest error is always obtained using this ordering.

The computation of finite products is another type of arithmetic computation often performed. Consider the product

$$p_n = f\ell\left(\prod_{i=1}^{n} x_i\right) \tag{3.4}$$

where the x_i are floating-point values. Let us define partial products as

$$p_1 = x_1$$

$$p_r = f\ell(p_{r-1}x_r)$$

Then

$$p_n = (1 + E) \prod_{j=1}^{n} x_j$$

where $1 + E = \prod_{j=2}^{n} (1 + \varepsilon_j)$. The magnitude of the absolute error in this case can be shown to have the upper bound (see Forsythe and Moler 1967)

$$|e| \leq 1.01(n - 1)\beta^{1-t}|\theta \prod_{j=1}^{n} x_j| \qquad (3.5)$$

where $|\theta| \leq 1$. In this case the bounds on the error do not depend on the order of multiplication; however, the computed value may depend on order of multiplication. Inspect of the bounds points to the fact that the relative error in a finite product is small unless n is very large.

The computation of vector inner products is frequently required in statistical applications. This computation, involving evaluation of finite sums of products, has proven to be a frequent source of numerical difficulty. This is not surprising in view of the fact that finite sums can have large relative error due to cancellation. Nevertheless, programmers sometimes lose sight of this fact when computing quantities such as variances and covariances. Most numerical analysts recommend that inner products be formed in double precision when single precision is used for the bulk of the other computations. For double-precision programs, inner products should be found using extended precision. This will not ensure a small relative error in the inner product; but if severe cancellation does not occur, a satisfactory level of accuracy can be maintained in taking the inner product.

3.7 CONDITION AND NUMERICAL STABILITY

Theoretical error analyses do not always prepare one for the numerical difficulties which are sometimes encountered using a given computational method. The purpose of this section is to discuss the concepts of data condition and the numerical stability of computational methods and algorithms. These concepts assist in understanding how computational methods interact with numerical data to produce the strange results which are sometimes observed by computer users.

Given two different computational methods for doing the same job, an error analysis may or may not show that one method is generally to be preferred over the other. Some methods quite often give unsatisfactory results in computer application and we say that these methods are numerically unstable. Numerical instability will not be rigorously defined, but we think of a computational method as being *numerically unstable* whenever careful analysis indicates that serious accumulation of rounding error may occur and/or severe cancellation can take place, and practical applications show that for many sets of data the errors introduced in computation are indeed sufficiently large so that the final results are unsatisfactory. The term *numerically stable* is applied to methods which tend to minimize the effect of rounding error and cancellation in computations. We will often use these concepts when comparing methods and say that one method is not as numerically stable as another.

Regardless of the method used, some data can be analyzed more accurately than others when using the computer. Given a set of data and the problem to be solved, we call the data *ill-conditioned* with respect to the problem if the quantities to be computed are sensitive to small changes in the data. For example, consider the matrix

$$A = \begin{bmatrix} 5 & 7 & 6 & 5 \\ 7 & 10 & 8 & 7 \\ 6 & 8 & 10 & 9 \\ 5 & 7 & 9 & 10 \end{bmatrix}$$

which has the inverse

$$A^{-1} = \begin{bmatrix} 68 & -41 & -17 & 10 \\ -41 & 25 & 10 & -6 \\ -17 & 10 & 5 & -3 \\ 10 & -6 & -3 & 2 \end{bmatrix}$$

Now if we alter the (3,3) element of A by subtracting 0.01, the resulting matrix is

$$A_1 = \begin{bmatrix} 5 & 7 & 6 & 5 \\ 7 & 10 & 8 & 7 \\ 6 & 8 & 9.99 & 9 \\ 5 & 7 & 9 & 10 \end{bmatrix}$$

The inverse of A_1, rounded to two decimal places, is

$$A_1^{-1} = \begin{bmatrix} 71.04 & -42.79 & -17.89 & 10.54 \\ -42.79 & 26.05 & 10.53 & -6.32 \\ -17.89 & 10.53 & 5.26 & -3.16 \\ 10.54 & -6.32 & -3.16 & 2.09 \end{bmatrix}$$

Here a perturbation of .1% in one element of A resulted in changes of 5% in several elements of A^{-1}. Thus the elements of A^{-1} are sensitive to small changes in the elements of A, and we say that the matrix A is ill-conditioned with respect to the problem of matrix inversion.

There are various measures of ill-conditioning of certain problems which have been suggested in the literature. Problems involving matrix operations have been studied in considerable detail (e.g., Wilkinson 1963). A commonly used measure of condition of a matrix for a variety of algebraic processes (inversion, for example) is the "condition number," to be discussed in Section 7.2.4.

Ill-conditioned data tend to display the numeric instability of a method, often in a dramatic fashion. This is due to the fact that a computed solution is the exact solution of a problem using some perturbation of the original data. Wampler (1970) used sets of ill-conditioned data in many different regression programs. His results showed a tremendous variation in the calculated coefficient estimates from the various programs. Naturally, the programs which implemented numerically stable methods gave better results.

The concept of numerical instability can be illustrated using an example given by Forsythe (1970). Consider the convergent series for exp (x),

$$\exp(x) = 1 + x + \frac{x^2}{2!} + \frac{x^3}{3!} + \ldots$$

Use of the first 25 terms in this series to approximate $e^{-5.5}$ using five-digit decimal arithmetic yields 0.0026362 as the c-true sum, while the true value is approximately 0.0040867. Theoretically, the first 25 terms should yield a good approximation to $e^{-5.5}$ unless cancellation occurs. The reason for this relatively large difference between c-true and true value is that cancellation has indeed eliminated significant digits in intermediate results to a degree which makes the final result useless. The first few terms in the alternating series are as follows:

± Term Number	Positive Term	Negative Term	Cumulative Sum
1	1.000	-5.5000	-4.5000
2	15.125	-27.730	-17.105
3	38.129	-41.942	-20.918
4	38.446	-30.208	-12.680
5	20.768	-12.692	-4.6040
6	6.9803	-3.4902	-1.1139

Cancellation in this case in subtle but nevertheless present. If we compare the magnitude of the true value with terms like 20.768, we can readily see that cancellation is inevitable. These larger values cannot make a significant contribution to the final result unless we carry *more digits* in the computations. We can, for this problem, obtain much better results if we use the series expansion of e^{-x} to compute $e^{-5.5}$ even though this method also has drawbacks. A moment's reflection will convince the reader that cancellation is not the basic cause of the poor result which was obtained. The choice of method for the five-digit computations was the basic cause in that such choice actually forced catastrophic cancellation to be inevitable in the given situation.

Since real-life ill-conditioned data sets will present themselves from time to time, there is little that we can do to avoid them. We can, however, utilize numerically stable algorithms and use available techniques for uncovering ill-condition. Once alerted to possible numerical inaccuracies in computer results, we will take special care in checking answers and, if needed, we will recompute using greater precision. These actions, in themselves, will not suffice in every conceivable situation, but they will prove useful in the majority of cases.

3.8 OTHER METHODS OF ASSESSING ERROR IN COMPUTATION

For a given real-life computation problem, it is rarely feasible to do a full-scale error analysis as outlined in previous sections. Assessing the numerical accuracy of the computations for a problem with unknown solution is an inherently difficult task. Two general techniques that are sometimes useful in practice are (1) checking the computed solution for internal consistency, and (2) computing solutions of perturbations of the problem that have known effects.

It is sometimes the case that the solution to a problem must satisfy extraneous conditions, that is, conditions that were not employed in obtaining the solution. For example, in a least-squares regression with a model containing a constant term, the residuals must sum to zero. After computing the least-squares solutions, the residuals could be computed and their sum could be compared to zero. A sum which differs from zero by more than some small tolerance would indicate numerical inaccuracies somewhere. (Of course, it could be in the performance of the consistency check itself.) Kennedy and Gentle (1977) apply this type of technique to other problems in statistical estimation.

Internal consistency checks are made with respect to c-true values; hence they will not show errors made in the entry of data into the computer. It should be noted further that internal consistency checks are frequently only negative indicators; that is, incorrect solutions may sometimes pass the tests.

The second general technique mentioned above consists of perturbing the data so as to produce a known effect, solving the problem with the new data set, and comparing this solution to the solution of the original problem. Vignes and LaPorte (1974) discussed this techniques in general computing applications. Mullet and Murray (1971) and Kennedy and Gentle (1977) considered applications in computations for statistical analyses.

Perturbation techniques also have the disadvantages mentioned above for the internal consistency checks-namely, that they are done with respect to c-true values, not the real data. They may indicate inaccuracies when the solution is in fact accurate, since the inaccuracies may occur only in the computations on the perturbed problem; and they may fail to indicate inaccuracies that do occur, since corresponding errors may occur in the computations on the perturbed problem. Nevertheless, both of the techniques can be judiciously applied to alert the user to possible errors in the computations and to indicate the need for the use of more stable (and possibly more expensive) algorithms than would be employed on ordinary problems.

3.9 SUMMARY

This chapter has presented a brief discussion of the sources of error in computer-generated floating-point results. There is a large amount of literature on this subject, and several different general viewpoints on the most desirable methods to use in analyses are reflected in this literature. Most error analyses are done a priori and are inverse analyses. Such analyses do provide some information about the computational method under consideration, and many times they form the basis for comparison of competing methods. The limitation of such analyses, from an applied viewpoint, is that they are not extremely helpful when attempting to assess the potential for error for a given set of data in a specific scientific application. Some work on computing error bounds has been done. Chartres and Geuder (1967) and Moore (1966) are examples. Bounds obtained in practice, however, are often very conservative and provide little help.

Experience coupled with information provided by error analyses, however, has further the cause of statistical computing in the following ways:

1. By identifying certain programming techniques which work well and others which must be avoided.
2. By pointing out the best computational methods and algorithms in the various application areas

3. By publicizing the need for caution in scientific computer use to avoid, as much as possible, inaccuracies in computed results

Chapter 4 deals, in part, with programming techniques which are "good" from the standpoint of numerical accuracy. The remaining chapters describe many computational methods and associated algorithms which are considered to be the most stable among possible choices. We will not, in subsequent chapters, attempt to provide error analyses. References to existing analyses will, however, be given in many cases.

EXERCISES

3.1 Assume a computer with $\beta = 10$, $t = 4$, and a rounding rule which calls for chopping of fractions. Find the relative error in each of the following computations, described in terms of true values.

(a) 0.67423 + 1.71423

(b) 26.215 - 5.3261

(c) 1.23 + 4.52

(d) 1.167 - 1.162

(e) (4.23) × (5.42)

(f) 4.23 - 5.42

3.2 Let a, b, and c denote single-precision floating-point values. Analyze the error in computing $f\ell(ab + c)$ and discuss your results. Will your results be true if cancellation occurs?

3.3 Give an example which shows that

$$f\ell[f\ell(a + b) + f\ell(c)] \neq f\ell[f\ell(a) + f\ell(b + c)]$$

In other words, show that floating-point addition is not associative.

3.4 Assume that the computer rounding rule is to "chop" and retain only the first t digits. Use decimal arithmetic with $t = 4$ and find the rounding error and relative error introduced through conversion of the following numbers to floating-point numbers.

(a) 1.16702

(b) -6.66

(c) 1742.52

(d) 16.6299

(e) -16.6299

(f) 0.00067418

3.5 Give an example which shows that the sum of $n > 2$ floating-point numbers can have a large relative error.

3.6 Consider the true value $y = f\beta^e$ in normalized floating-point form. Chop the fraction f after the first t digits, giving f_1 as the first t digits. Now y can be expressed as

$$y = f_1\beta^e + f_2\beta^{e-t}$$

where f_2 is a normalized fraction containing digits $t + 1$, $t + 2$, ... from f. With y in this form,

(a) What is the numeric least upper bound for the absolute value of f_2, that is, $|f_2|$?

(b) What is the greatest lower bound of $|f_1|$?

(c) What is the least upper bound for the value of the following?

$$\left| \frac{f_2\beta^{e-t}}{f_1\beta^e} \right|$$

3.7 Compare the maximum rounding and relative error in $x + x + x$ with that of 3x.

3.8 Let x be a real number within the range of floating-point numbers. Let e be the unique integer such that

$$\beta^{e-1} \leq x \leq \beta^e$$

where β is the number base for floating-point numbers.

(a) Show that in the interval $[\beta^{e-1}, \beta^e]$ the floating-point numbers are equally spaced and give the spacing in terms of the number t of digits in a floating-point fraction.

(b) Assuming that rounding instead of chopping takes place in conversion, find an upper bound for $|f\ell(x) - x|$. How does this upper bound compare with that obtained when the chopping operation is carried out?

REFERENCES

Aird, T. J., and R. E. Lynch (1975), Computable Accurate Upper and Lower Error Bounds for Approximate Solutions of Linear Algebraic Systems, *TOMS 1*, 217-231.

Amble, O. (1961), On the Accuracy of Floating-Point Computers, *BIT 1*, 220-221.

Ashenhurst, R. L. (1964), Function Evaluation in Unnormalized Arithmetic, *JACM 11*, 168-187.

Babuska, I. (1972), Numerical Stability in Problems in Linear Algebra, *SIAM J. Num. Anal. 9*, 53-77.

Bauer, F. L. (1974), Computational Graphs and Rounding Error, *SIAM J. Num. Anal. 11*, 87-96.

Brown, W. S., and P. L. Richman (1969), The Choice of Base, *CACM 12*, 560-561.

Burrill, D. F. (1974), Computational Precision in Statistics: Variances and Covariances, *JSCS 3*, 105-115.

Caprani, O. (1975), Round Off Errors in Floating-Point Summation, *BIT 15*, 5-9.

Chartres, B. A., and J. C. Geuder (1967), Computable Error Bounds for Direct Solution of Linear Equations, *JACM 14*, 63-71.

Cody, W. J. (1971), Desirable Hardware Characteristics for Scientific Computation, *SIGNUM Newsletter 6*, 16-31.

Descloux, J. (1963), Note on the Round-Off Errors in Iterative Processes, *Math. Comp. 17*, 18-27.

Dorn, W. S. (1970), Numerical Instability, *Proceedings of Share XXXIV, vol. 111.*

Flehinger, B. J. (1966), On the Probability that a Random Integer Has Initial Digit A, *Amer. Math. Monthly 73*, 1056-1061.

Forsythe, G. E. (1970), Pitfalls in Computation, or Why a Math Book Isn't Enough, *Amer. Math. Monthly* 77, 931-955.

Forsythe, G. E. and C. B. Moler (1967), *Computer Solution of Linear Algebraic Systems*, Prentice-Hall, Englewood Cliffs, N.J.

Goldberg, I. B. (1967), 27 Bits Are Not Enough for 8 Digit Accuracy, *CACM 10*, 105-106.

Goldstein, M. (1963), Significance Arithmetic on a Digital Computer, *CACM 6*, 111-117.

Good, D. I., and R. L. London (1970), Computer Interval Arithmetic: Definition and Proof of Correct Implementation, *JACM 17*, 603-612.

Goodman, R., and A. Feldstein (1977), Effect of Guard Digits and Normalization Options on Floating Point Multiplication, *Computing 18*, 93-106.

Householder, A. S. (1958), Generated Error in Rotational Tridiagonalization, *JACM 5*, 335-338.

Howell, K. M. (1967), Multiple Precision Arithmetic Techniques, *Comp. J. 9*, 383-387.

Kahan, W. (1964), Further Remarks on Reducing Truncation Error, *CACM 8*, 40.

Kaneko, T., and B. Liu (1970), Accumulation of Round-off Error in Fast Fourier Transforms, *JACM 17*, 637-654.

Kaneko, T., and B. Liu (1973), On Local Round-off Errors in Floating-Point Arithmetic, *JACM 20*, 391-398.

Kennedy, W. J., and J. E. Gentle (1977), Examining Rounding Error in LAV Regression Computations, *Comm. Stat. (B)6*, 415-420.

Kennedy, W. J., J. E. Gentle, and V. A. Sposito (1977), A Computer Oriented Method for Generating Test Problems for L_1 Regression, *Comm. Stat. (B)6*, 21-27.

Knuth, D. E. (1969), *The Art of Computer Programming, vol. II*, Addison-Wesley, Reading, Mass.

Kuki, H. (1971), Mathematical Function Subprograms for Basic System Libraries: Objectives, Constraints, and Trade-Off, in *Mathematical Software*, edited by John Rice, Academic Press, New York, 187-200.

Kuki, H., and J. Ascoly (1971), FORTRAN Extended Precision Library, *IBM Sys. J. 10*, 39-61.

Kuki, H., and W. J. Cody (1973), A Statistical Study of the Accuracy of Floating-Point Number Systems, *CACM 16*, 223-230.

Linnainmaa, S. (1975), Towards Accurate Statistical Estimation of Rounding Errors in Floating-Point Computations, *BIT 15*, 165-173.

Linz, P. (1970), Accurate Floating-Point Summation, *CACM 13*, 361-362.

McCracken, D. D., and W. S. Dorn (1964), *Numerical Analysis and FORTRAN Programming*, Wiley, New York.

Miller, W. (1972), Automatic A Priori Roundoff Analysis, *Computing 10*, 213-219.

Miller, W. (1974), Remarks on the Complexity of Roundoff Analysis, *Computing 12*, 149-161.

Miller, W. (1975a), Software for Roundoff Analysis, *TOMS 1*, 108-128.

Miller, W. (1975b), Computer Search for Numerical Instability, *JACM 22*, 512-521.

Miller, W. (1976), Roundoff Analysis by Direct Comparison of Two Algorithms, *SIAM J. Num. Anal. 13*, 382-392.

Moore, R. (1966), *Interval Analysis*, Prentice-Hall, Englewood Cliffs, N.J.

Mullet, G. M. and T. W. Murray (1971), A New Method for Examining Rounding Error in Least-Squares Regression Computer Programs, *JASA 66*, 496-498.

Phillips, G. M. and B. N. Sahney (1975), An Error Estimate for Least Squares Approximation, *BIT 15*, 426-430.

Pinkham, R. S. (1961), On the Distribution of First Significant Digits, *AMS 32*, 1223-1230.

Ralston, A. (1965), *A First Course in Numerical Analysis*, McGraw-Hill, New York.

Sterbenz, P. H. (1974), *Floating-Point Computation*, Prentice-Hall, Englewood Cliffs, N.J.

Stewart, G. W. (1973), *Introduction to Matrix Computations*, Academic Press, New York.

Stewart, G. W. (1977), On the Perturbation of Pseudo-Inverses, Projections, and Linear Least Squares Problems, *SIAM Rev. 19*, 634-662.

Stoutemyer, D. R. (1977), Automatic Error Analysis Using Computer Algebraic Manipulation, *TOMS 3*, 26-43.

Sweeney, D. W. (1965), An Analysis of Floating-Point Addition, *IBM Syst. J. 4*, 31-42.

Vignes, J., and M. LaPorte (1974), Error Analysis in Computing, *Information Processing 74* (IFIP, Stockholm), 610-614.

Von Neumann, J., and H. H. Goldstine (1947), Numerical Inverting of Matrices of High Order, *Bull. Amer. Math. Soc. 53*, 1021-1099.

Wampler, R. (1970), A Report on the Accuracy of Some Widely Used Least Squares Computer Programs, *JASA 65*, 549-565.

Wilkinson, J. H. (1961), Error Analysis of Direct Methods of Matrix Inversion, *JACM 8*, 281-330.

Wilkinson, J. H. (1962), Error Analysis of Eigenvalue Techniques Based on Orthogonal Transformations, *SIAM J. Appl. Math. 10*, 162-195.

Wilkinson, J. H. (1963), *Rounding Errors in Algebraic Processes*, Prentice-Hall, Englewood Cliffs, N.J.

Wilkinson, J. H. (1965), *The Algebraic Eigenvalue Problem*, Oxford University Press, London.

Wilkinson, J. H. (1971), Modern Error Analysis, *SIAM Rev. 13*, 548-568.

Wolfe, J. M. (1964), Reducing Truncation Errors by Programming, *CACM 7*, 355-356.

4 / PROGRAMMING AND STATISTICAL SOFTWARE

The ultimate goal of research workers in the area of statistical computing is to produce dependable, efficient, and useful computer programs. It is not sufficient to have good numerical methods and associated algorithms, because the algorithms must finally be properly expressed in a computer language if they are to be useful. In other words, the act of programming is an important part of the overall process of providing computational support. Programming is considered by some people to be a rather elementary and less important activity when compared to other required actions in the developmental process. However, this is definitely not the case. Many examples can be given wherein excellent algorithms perform poorly in the computer because the program that implemented these algorithms was poorly written (but was "bug-free" in that poor results were not due to errors in the code). It is difficult to produce "good" programs, and a substantial amount of knowledge and attention to detail is needed.

This chapter consists of a relatively brief introduction to the programming aspect of statistical computing. Some relatively general topics related to programming languages and program development will be given. Then we will mention some of the better programs that are available for statistical computing applications. The scope of this book precludes undertaking a detailed discussion of the many small specific items that the programmer must carefully consider in order to write a "good" program. Also, many things that can be said about coding are very situation-dependent, so it is extremely difficult, if not impossible, to generalize in this area. In subsequent chapters some of the discussion will supplement the material in this chapter, and the authors hope that the total effect of this book will be to provide the reader with a firm foundation upon which he can build research-level expertise in all aspects of statistical computing.

4.1 PROGRAMMING LANGUAGES: INTRODUCTION

There exist a multitude of languages for communicating instructions to the computer. For the earliest computers, the user supplied instructions by arranging electrical circuits to effect the execution of the desired operation. Later the circuits were fixed ("hardwired"); and codes, usually in binary, were supplied by the user to indicate which operations were to be performed (i.e., which circuits were to be turned on). Such a set of codes, in which codes correspond in essentially a one-to-one fashion with the single operations performed by the circuitry of the computer, is called a "machine language." To facilitate the writing of programs in machine language, mnemonic codes were developed, called "assembly" or "assembler" languages. The instructions of an assembly language also correspond generally in a one-to-one fashion with operations of hardware components.

During the 1950s and later, there emerged languages in which one sentence corresponded to a sequence of many machine instructions. Such languages are referred to as high-level or algorithmic languages. Examples of commonly used algorithmic languages are FORTRAN, COBOL, PL/1, and ALGOL. A name such as *FORTRAN* generally refers to a family of similar languages or dialects. It reality, a specific language may be defined by the set of rules that transform the statements of the language into machine instructions. Programs implementing these transformation rules are called *compilers*. Compilers, which are usually written by the manufacturers of the machine, such as IBM, Burroughs, and so on, therefore define languages or dialects; thus IBM FORTRAN G may be slightly different from IBM FORTRAN H or WATFIV, although each is a FORTRAN language. (Adding to this diversity is the fact that a specific compiler will be in use for a limited time, as new releases or versions are continually being made.)

Standards bureaus, such as the American National Standards Institute (ANSI) or the International Standards Organization (ISO), have attempted to mollify the language diversity by promulgating standard versions of languages.

To attempt to define languages without reference to specific hardware devices, syntactic "metalanguages" may be used to specify the *syntax* or form of instructions of the languages. By means of metalanguages, such as the Vienna Definition Language (see, e.g., Wegner 1972) and the Backus-Naur Form (see, e.g., Donovan 1972), programming languages may be somewhat standardized by format definition. The *semantics* or meaning of a language, however, is more difficult to specify.

Another class of programming languages commonly used in certain applications is referred to as *very-high-level languages* or *special-purpose languages*. One statement in these languages may be equivalent to several statements of an algorithmic language; in fact, one way of constructing a very-high-level language is by specifying rules for translating sentences in the language into FORTRAN subprograms. Special-purpose languages are frequently used in statistical analyses (SAS, SPSS, etc.) and simulation (GPSS, SIMSCRIPT, etc.).

The higher-level languages are generally easier to learn, write, read, and debug than the lower-level languages. The higher-level languages are to some extent independent of the machine. A FORTRAN program will produce generally the same results on an IBM 370 as it would on a CDC 7600. Some algorithms, such as for random number generation, by their very nature are coded differently for different machines. The results of the computations specified by the same program on different machines may also be different due to differences in rounding done by the two machines. Aside from these two considerations, however, good high-level language programs are fairly "portable" from one machine to another.

An additional advantage of high-level languages results from the effort expended to develop compilers that will generate efficient code. For a given problem a good compiler will generally produce more efficient machine code from a reasonably good high-level source program than a run-of-the-mill machine language programmer would produce.

For a while in the late 1960s a number of people envisioned achieving greater flexibility in programming by means of compiler-generators (compiler-compilers or parser generators), which would allow the programmer essentially to define his own language and then proceed to write programs in that language. Johnson (1976) gives an optimistic report on the state of theoretical progress in the area. Development of compiler-generators lagged, however, as interest among data analysts never really flourished. Standard programming languages and the special-purpose statistical languages do not appear to be overly restrictive for the needs of statistical computing. The primary areas of activity have been in development of algorithms and programs useful to a large number of data analysts.

A history and perspective of most of the important programming languages is given by Sammet (1969). Chambers (1977) gives some useful comparisons of the common algorithmic languages.

4.2 COMPONENTS OF PROGRAMMING LANGUAGES

The basic purpose of computer programs is to operate on data: to store, move, display, and transform (arithmetically or otherwise) data. The features of a general-purpose, higher-level language all relate in some way to these activities with data.

4.2.1 Data Types

Higher-level programming languages generally make provision for working with various types of data. The types of data handled by these languages may be conveniently classified primarily as numeric and nonnumeric. Within each of these classes further distinctions exist in most languages. The data type determines how the values are represented internally and how operations are performed on the data. In most FORTRAN languages numeric data can be of integer (fixed-point), short floating-point, long floating-point, or complex types. PL/1 provides for all of these numeric data types as well as a decimal fixed-point type.

The commonly used types of nonnumeric data are *logical data* and *character data*. Logical data types carry 0-1, or true-false, information. A value of this type frequently occupies more than one bit of memory, however. In IBM 370 FORTRAN a LOGICAL variable occupies a byte or a word. Logical data in PL/1 are assigned as bit-strings, which can be of varying lengths and on which a variety of operations are available, providing much greater flexibility than FORTRAN.

Older versions of FORTRAN provide no character type of data storage and operations; character data must be treated as numeric data in these languages. PL/1 provides for variable-length character-strings with a special storage type and special operations for character manipulation.

4.2.2 Data Structures

The relationship of the various items of data, called *records*, with one another is referred to as the *data structure*. Programming languages, to varying extents, provide for specification of the data structure by rules for naming data. Most languages provide for scalar variables, which show no structure (no relationship to other variables), and array variables, which show an orthogonal structure in as many directions as the dimension of the array.

The natural structure of much of the data processed in statistical analyses is a one- or two-dimensional orthogonal array, as the variables in the linear model $\underline{y} = X\underline{\beta}$, for example. (Note, *orthogonal* here does not mean the matrix is orthogonal in the usual mathematical sense. In a structural sense *orthogonal* means "rectangular," or the higher-dimensional analogue of rectangular.) In FORTRAN and PL/1 orthogonal data structures are formed and referenced by subscription (indexing) variables. The number of dimensions possible varies with the version of the language. The compilers will generally store the actual data as a linear array and provide addressing of the linear array in conformity with the storage conventions. (Matrices, for example, are stored by columns in FORTRAN and by rows in PL/1.)

Other kinds of data structures occur with less frequency in statistical application. One special case of a rectangular array that is occasionally useful is the "factorial array." Array elements may be assigned values on the basis of the values of two or more other variables, without explicitly providing storage for this interaction. Some special-purpose statistical programming languages provide for this type of data structure, which occurs often in the analysis of certain design models.

Another useful structure is obtained by linking linear arrays into a sequence, forming the "linked-list" structure. The individual linear arrays can be of various lengths. Each array contains a pointer to the next, and possibly to the previous, array in the sequence.

Triangular data arrays also occur occasionally, but most programming languages do not provide explicitly for this type of structure.

Structures expressing hierarchial relationships are available in some languages, such as PL/1 and COBOL. Naming rules are provided to refer to the whole data set, or to some particular subset, in a convenient fashion. Hierarchically structured files are becoming more important in statistical analyses since the U.S. Bureau of the Census uses this kind of file in its public use samples, arranging neighborhood, household, and person records in hierarchical fashion. Some of the special-purpose statistical software packages now handle files with a hierarchical structure.

4.2.3 Syntax

The rules for arrangement of the various symbols of a language to form sentences to name sentences and the rules governing the arrangement of the sentences to form programs determine the ease of programming in the language and also, to some extent, the power of the language. These rules constitute the *syntax* of the language. The basic symbols include statement labels; variable names; constants, such as numeric literals and string literals; operators; and punctuation marks.

4.2.4 Control Structures

Groups of instructions having some unifying logical relationship are referred to as *control structures*. These program modules may be formed by only two or three statements and be an integral part of a larger program. Some particularly common and useful modules are referred to as *basic control structures*, which we will discuss in more detail in Section 4.3.

Modules that can be separately compiled, that is, in which variable names and statement labels may have meanings independent of their meaning in other modules, are called *control sections* or external routines. FORTRAN provides for four types of control sections: main programs, subroutine subprograms, function subprograms, and COMMON blocks. The locality of names and labels to control sections and the ability to invoke the routines from various places in other routines greatly enhance the power and ease of use of a language.

4.3 PROGRAM DEVELOPMENT

Programming has traditionally been a fairly individualistic activity. Programs frequently have been designed for a very specific problem, and almost unusable on a quite similar problem. In recent years, however, there has been a growing awareness of the need for a programming methodology that would facilitate the development of ad hoc programs, while at the same time making the programs less ad hoc, and enable persons other than the original programmer to use, and perhaps modify, the programs. In the following subsections we will discuss some of the qualities of good programs and methods to enhance the likelihood of producing good programs.

The basic objective of programming is to do the programming and computing *correctly* and *efficiently*. The correctness aspect relates to which computations are done and how resistant the computations are to rounding errors. Obviously, to be of any value, the program must perform the correct computations; that is, the method being programmed must be implemented correctly. Even when the correct computations are attempted, however, the results may be incorrect due to rounding errors (see Chapter 3). The program must properly implement algorithms that maintain good numerical accuracy.

The second objective mentioned above, efficiency, relates to a number of computing resources: programmers' time (development and maintenance), data storage media, and computing units. There are tradeoffs among the various aspects of efficiency. The relative importance of these is determined by the costs associated with the various resources. Programmers' time is becoming more costly relative to the hardware resources; hence more effort is currently being expended to increase efficiency in this area. The methods of *structured programming* (see, e.g., Dahl, Dijkstra, and Hoare 1972) show promise for improving the efficiency of the programmer or programming team. The central idea of these methods is to restrict flow of program control to a very simple progression through the statements comprising the program. Application of the ideas of structured programming produces code that can be read from top to bottom without having to trace through complex branching maneuvers in the process. Program testing, debugging, maintenance, and modification are thus more easily accomplished.

Structured programming methods begin with a decomposition of the flowchart or model of the program into a set of *basic control logic structures* (BCLS). The results of Bohm and Jacopini (1966) show that every flowchart can be represented equivalently by means of only three basic structures, shown in Fig. 4.1.

Note that each BCLS has only one entry and one exit. Following program control flow is thus simplified. By nesting and proper

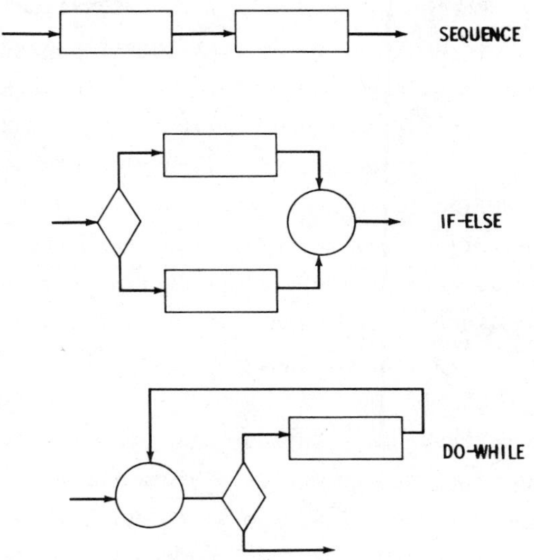

Figure 4.1 Basic control logic structures.

sequencing of BCLSs it is possible to avoid uncondition branches, which tend to obscure the readability of programs (see Dijkstra 1968).

In addition to the three BCLSs of Fig. 4.1, two other control logic structures, shown in Fig. 4.2, are useful.

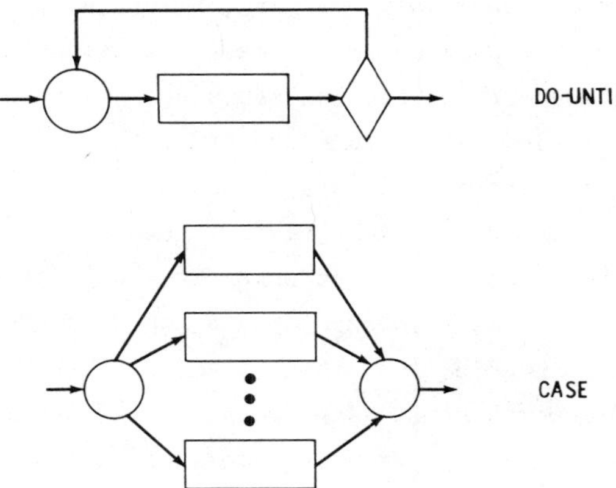

Figure 4.2 Other control logic structures.

Long programs generally are harder to understand than short ones. It is therefore useful to segment longer programs, making successive divisions of the problem into segments or small programs. This may be done by a succession of refinements of the large program. At the highest level the overall task is divided into a few major subtasks, which define segments of the program. Within each segment successive levels of subtask definition and segmentation are undertaken until each segment subtask is defined in terms of small blocks of logic that will result in a small program. In accord with the precepts of structured programming, each segment has only one entry and one exit.

Segmenting subtasks this way may also result in the consolidation of portions of the program that perform the same task. A single subroutine may be called from different points.

Other features of programs that contribute to their readability, and hence to the overall efficiency of the activity of programming, are achieved by conforming to coding conventions specifying such things as indentation and labeling of statements and naming of variables. The coding shown in Fig. 4.3 illustrates a form that is easy to read. Anyone knowledgeable in FORTRAN could use the program, having only the source listing. When uniform conventions are followed in the coding, large sets of programs are made much easier to use.

Consideration should also be given to efficient use of computing resources other than programmers' time. The two main concerns here would be efficiency in storage and CPU usage. The basic principle in storage efficiency is: do not use memory needlessly. Special attention needs to be paid to array usage. For example, although the identity matrix frequently appears in statistical analysis, there is probably no situation in which it should be stored. If it is being computed as a check on other computations, only one element at a time need be inspected. Many statistical analyses permit "out-of-memory" processing; for example, in regression computations on the model $\underline{y} = X\underline{\beta}$, it may not be necessary to store

```
C**STAT*************************************************************
C
C  FUNCTION          - COMPUTE SAMPLE MEAN AND VARIANCE
C
C  USAGE             - CALL STAT(X,N,XMEAN,XVAR,IER)
C
C  PARAMETERS      X - INPUT, REAL, N-VECTOR CONTAINING SAMPLE
C                  N - INPUT, INTEGER, SAMPLE SIZE
C              XMEAN - OUTPUT, REAL, SAMPLE MEAN
C               XVAR - OUTPUT, REAL, SAMPLE VARIANCE
C                IER - OUTPUT, INTEGER, ZERO UNLESS ERROR IN
C                      VALUE INPUT FOR N
C
C  PRECISION         - SINGLE
C
C  REQ'D ROUTINES    - NONE
C
C  PROGRAMMER        - J. C. DEBUGGER
C
C  LATEST REVISION 6/27/78
C
C*******************************************************************
C
      SUBROUTINE STAT(X,N,XMEAN,XVAR,IER)
      DIMENSION X(N)
      DATA ZERO,ONE/0.0,1.0/
      IER=0
      IF(N.LT.1) GO TO 20
C                                        INITIALIZE
      XMEAN=X(1)
      XVAR=ZERO
      CT=ONE
C                                        IF N=1, RETURN; OTHERWISE
C                                        ACCUMULATE SUMS
      IF(N.EQ.1) RETURN
      DO 10 I=2,N
         CT=CT+ONE
         XI=X(I)
         XMEAN=XMEAN+XI
         ADJX=CT*XI-XMEAN
         XVAR=XVAR+ADJX*ADJX/(CT*(CT-ONE))
   10 CONTINUE
C                                        DIVIDE TO GET MEAN AND
C                                        VARIANCE
      XMEAN=XMEAN/CT
      XVAR=XVAR/(CT-ONE)
      RETURN
C                                        SET ERROR CODE FOR N LESS
C                                        THAN ONE
   20 IER=1
      RETURN
      END
```

Figure 4.3 Example program listing.

the X matrix. The savings can be significant if there are many observations. Even in the modern-day virtual memory systems, which allow use of combinations of main memory and auxiliary storage, the need to conserve memory is still often present.

Efficiency in CPU usage can be improved by attention to the details of the computations. For example, addition of a fixed-point and a floating-point number requires first of all a type conversion. Care in programming may reduce the number of type conversions required. As a second example, consider the alternate FORTRAN expressions for the square of the floating-point variable X;

```
X*X
X**2
X**2.
```

On most compilers the first two expressions yield the same machine language code, which results in a simple floating-point multiplication. The third expression results in taking a logarithm, multiplying two floating-point numbers, and exponentiation. Furthermore, the third expression may give an error condition if X has a negative value when the statement containing the expression is executed. Larson (1971) and Kernighan and Plauger (1974) give other useful suggestions for achieving higher efficiency in programming.

4.4 STATISTICAL SOFTWARE

Most of the computing done in statistical analyses is now done by programs resident in a library of the computer installation, rather than by a program written specifically for the problem at hand. The availability of such programs for statistical analyses has been a boon to applied statisticians. The selection and use of such "canned" programs or "program packages," however, requires careful consideration.

Statistical program packages can be classified in a number of ways. Schucany, Minton, and Shannon (1972) gives a useful classification and discussion. Formerly much of the statistical computing was done using stand-alone programs designed for a specific type of analysis. The tendency in recent years has been toward the use of more general-purpose systems. For the software packages currently in common use, a basic distinction can be made between the multipurpose systems which essentially constitute very-high-level programming languages and the packages of subprograms which are to be invoked by a user program in a general algorithm language. The latter type of package provides maximum flexibility. The package is easy to maintain and additions to the system present little problem. Two widely distributed packages of this type are the IBM Scientific Subroutine Package (IBM Corporation 1968) and the International Mathematical and Statistical Libraries (IMSL Corporation 1977). Both of these software packages consist of many subroutines (approximately 250 for the IBM and 400 for IMSL) which can be invoked easily by FORTRAN programs.

The multipurpose systems are capable of performing one or more of several different kinds of analyses upon one entry to the system. This type of package generally provides a simple and somewhat flexible command language that is recognized and translated by the system. Advantages of this type of package include potential use by a wide range of users, some degree of flexibility, good data-handling capabilities, and the ability to perform more than one type of analysis in a single computer run. Widely distributed packages of this type include SAS (Barr et al. 1976), SPSS (Nie et al. 1975), and BMDP (Dixon and Brown 1977). Moore (1978) gives an introduction to the use of these three systems.

As the complexity of the software systems for statistical analysis has increased, the need has arisen for objective evaluation and comparison of these systems by those who are knowledgeable in the area and will to devote the necessary time to the task. A committee for this purpose was established by the Statistical Computing Section of the American Statistical Association in 1974. This committee has since issued periodic reports of its activities. The first report (Francis, Heiberger, and Velleman 1975) described the plan of action and discussed the criteria to be applied in the evaluations. The criteria take into consideration the intended kinds of uses and users of the system, and include such things as the user interface, statistical effectiveness, and implementation. Details of the criteria relate to the documentation, control language, data structures, printed output, graphics, versatility, accuracy, extensibility, and portability of the program packages.

REFERENCES AND FURTHER READINGS

Aho, A. V., J. E. Hopcroft, and J. D. Ullman (1974), *The Design and Analysis of Computer Algorithms*, Addison-Wesley, Reading, Mass.

Baker, F. T. (1972), Chief Programmer Team Management of Production Programming, *IBM Systems J. 11*, 56-73.

Baker, F. T., and H. D. Mills (1973), Chief Programmer Teams, *Datamation 19* (December), 58-61.

Barr, A. J., J. H. Goodnight, J. P. Sall, and J. T. Helwig (1976), *A User's Guide to SAS 76*, SAS Institute, Raleigh, N.C.

Bohm, C., and G. Jacopini, (1966), Flow Diagrams Using Machines and Languages with Only Two Formation Rules, *CACM 9*, 366-371.

Chambers, J. M. (1977), *Computational Methods for Data Analysis*, Wiley, New York.

Clark, R. L. (1973), A Linguistic Contribution to GO TO-less Programming, *Datamation 19*, no. 12 (December), 62-63.

Dahl, O. J., E. W. Dijkstra, and C. A. R. Hoare (1972), *Structured Programming*, Academic Press, New York.

Dijkstra, E. W. (1968), GO TO Statement Considered Harmful, *CACM 11*, 147-148.

Dijkstra, E. W. (1972), The Humble Programmer, *CACM 18*, 859-866.

Dixon, W. J. and M. B. Brown (1977), *BMDP: Biomedical Computer Programs, P-Series*, University of California Press, Berkeley.

Donaldson, J. R. (1973), Structured Programming, *Datamation 19*, no. 12 (December), 52-54.

Donovan, J. J. (1972), *Systems Programming*, McGraw-Hill, New York.

Francis, I., R. M. Heiberger and P. F. Velleman (1975), Criteria and Considerations in the Evaluation of Statistical Program Packages, *Amer. Stat. 29*, 52-56.

IBM Corporation (1968), *IBM System/360 Scientific Subroutine Package*, IBM Systems Reference Library, White Plains, N.Y.

IMSL Corporation (1977), *IMSL Reference Manual*, Houston.

Johnson, S. C. (1976), Compiler-Compilers: Where Have We Been and Where Are We Going? *Proceedings of the Ninth Interface Symposium on Computer Science and Statistics*, edited by D. C. Hoaglin and R. E. Welsch, Prindle, Weber & Schmidt, Boston, 56-58.

Kernighan, B. W., and P. J. Plauger (1974), *Elements of Programming Style*, McGraw-Hill, New York.

Kernighan, B. W., and P. J. Plauger (1976), *Software Tools*, Addison-Wesley, Reading, Mass.

Knuth, D. E. (1968), *The Art of Computer Programming, vol. 1: Fundamental Algorithms*, Addison-Wesley, Reading, Mass.

Knuth, D. E. (1974), Structured Programming with GOTO Statements, *Comp. Sur. 6*, 261-302.

Larson, C. (1971), The Efficient Use of FORTRAN, *Datamation 17*, no. 8 (August), 24-31.

Manna, Z., and E. A. Ashcraft (1970), The Translation of GO TO Programs to WHILE Programs, Stanford University Report CS-188, Palo Alto, Calif.

McCracken, D. D., and C. M. Weinberg (1972), How to Write a Readable FORTRAN Program, *Datamation 18*, no. 10 (October), 73-77.

Mills, H. (1971a), Top Down Programming in Large Systems, *Debugging Techniques in Large Systems*, edited by R. Rustin, Prentice-Hall, Englewood Cliffs, N.J.

Mills, H. (1971b), *Mathematical Foundations for Structured Programming*, IBM Corp., Federal Systems Division, Report no. RSC 71-5108, San Jose, Calif.

Moore, R. W. (1978), *Introduction to the Use of Computer Packages for Statistical Analysis*, Prentice-Hall, Englewood Cliffs, N.J.

Nie, N. H., C. H. Hull, J. G. Jenkins, K. Steinbrenner , and D. H. Bent (1975), *SPSS: Statistical Package for the Social Sciences*, McGraw-Hill, New York.

Sammet, J. E. (1969), *Programming Languages: History and Fundamentals*, Prentice-Hall, Englewood Cliffs, N.J.

Wegner, P. (1972), The Vienna Definition Language, *Comp. Sur. 4*, 5-63.

5 / APPROXIMATING PROBABILITIES AND PERCENTAGE POINTS IN SELECTED PROBABILITY DISTRIBUTIONS

The purpose of this chapter is to present computational methods that have proven to be useful for computing probabilities and percentage points for several of the best-known univariate continuous probability distributions. There is a very large amount of literature on this subject and many different methods have been proposed. As we shall see, it is often difficult to compute desired probabilities and percentage points to a specified level of accuracy, and the best of existing algorithms are not entirely satisfactory in many cases.

Many applications in statistical computing require the evaluation of a cumulative distribution function (cdf), $F(x)$, to determine the probability p corresponding to a given percentage point x_p. The first of two major problem types, considered in this chapter, is to compute p as

$$p = F(x_p)$$

using given F and x_p. In most of the cases we will consider, $F(x)$ is not expressible in closed form, so we must somehow approximate an integral

$$F(x) = \int_{-\infty}^{x} f(t)\, dt \tag{5.1}$$

The second major type of problem is that of evaluating the inverse of a given cdf. This problem occurs whenever we want to determine the percentage point x_p which produces a given probability p. Thus the problem is to compute x_p as

$$x_p = F^{-1}(p) \tag{5.2}$$

It is seldom possible to determine a simple closed form for $F^{-1}(p)$, so the usual situation is that we have only $F(x)$ and p, and must use these quantities to find x_p.

Since neither of these problems generally has a closed form solution, some type of mathematical approximation must be derived to provide a usable computational method. The same situation prevails in the case of standard mathematical functions such as sin(x), log(x), and exp(x). Ideally, an algorithm based on a given mathmatical approximation will provide accuracy essentially to "machine precision." By this we mean that the combined effect of error introduced by use of the mathematical approximation *and* rounding error in computations is negligible, and the computed value is correct to within (say) at most 2 bits in the mantissa of the floating-point result. This level of accuracy over a wide range of arguments in a function is often very difficult to achieve. The typical situation calls for a considerable amount of "tuning" of software which implements a numerical method that is capable of producing a desired level of accuracy. Efficiency also is a consideration, and sometimes it is necessary to balance level of accuracy against the cost of computation. The situation that we strive to achieve is, of course, to have an algorithm which gives results to machine precision for very small cost. This situation prevails for many of the standard mathematical functions, but it is not always the case for probability functions.

In the first section of this chapter we will establish the notation and conventions to be used in the remainder of the chapter. Section 5.2 contains a description of some methods in numerical analysis which are especially useful in deriving algorithms for the application area that is the subject of this chapter. Subsequent sections will consider specific families of probability distributions and describe some of the available methods and algorithms specific to each family.

The scope of this chapter precludes consideration of details relative to computational methods for noncentral distributions such as the chi-square, t, and F. These distributions do, however, often arise in statistical computing problems, and there are several approximations available for each distribution. A large portion of the literature on this subject is cited in the References and Further Readings at the end of this chapter.

5.1 NOTATION AND GENERAL CONSIDERATIONS

5.1.1 Probability Distributions

As was stated earlier, we will limit discussion of probability distributions to a few of the best-known and most often used continuous distributions. The purpose of this subsection is to facilitate subsequent discussion by defining each cdf which will be referenced later in the chapter. We begin with the member of the normal family that has zero mean and unit variance.

Standard normal:

$$\Phi(x) = \int_{-\infty}^{x} \phi(y)\, dy \qquad -\infty < x < \infty$$

$$= (2\pi)^{-1/2} \int_{-\infty}^{x} \exp\left(\frac{-y^2}{2}\right) dy \tag{5.3}$$

Student t:

$$T(x|n) = \int_{-\infty}^{x} f(y|n)\, dy \qquad -\infty < x < \infty$$

$$= \frac{1}{n^{1/2}B(1/2,\ n/2)} \int_{-\infty}^{x} \left(1 + \frac{y^2}{n}\right)^{-(n+1)/2} dy \tag{5.4}$$

where n is the degrees of freedom. B(a,b) is the *complete beta function* defined as

$$B(a,b) = \int_0^1 y^{a-1}(1 - y)^{b-1}\, dy \qquad a > 0,\ b > 0 \tag{5.5}$$

$$= \frac{\Gamma(a)\,\Gamma(b)}{\Gamma(a + b)}$$

and $\Gamma(c)$ is the *complete gama function* defined as

$$\Gamma(c) = \int_0^\infty y^{c-1} \exp(-y)\, dy \qquad (c > 0) \tag{5.6}$$

For integer c, $\Gamma(c) = (c - 1)!$

Beta distribution:

$$I_x(a,b) = \frac{1}{B(a,b)} \int_0^x y^{a-1}(1 - y)^{b-1}\, dy \tag{5.7}$$

$0 \leq x \leq 1$, $a > 0$, $b > 0$. The function $I_x(a,b)$ is called the *incomplete beta function* (or incomplete beta function ratio).

Gamma distribution:

$$G(x|a,b) = \frac{1}{b^a \Gamma(a)} \int_0^x y^{a-1} \exp\left(\frac{-y}{b}\right) dy \tag{5.8}$$

$a > 0$, $b > 0$, $x \geq 0$. The function $G(x|a,1)$ is called the *incomplete gamma function*. A special case for $a = n/2$ and $b = 2$ is the following.

Chi-square distribution:

$$H(x|n) = \left[2^{n/2}\Gamma\left(\frac{n}{2}\right)\right]^{-1} \int_0^x y^{(n-2)/2} \exp\left(\frac{-y}{2}\right) dy \tag{5.9}$$

where n is the degrees of freedom.

F distribution:

$$F(x|n_1,n_2) = \frac{(n_1)^{n_1/2}(n_2)^{n_2/2}}{B(n_1/2,\, n_2/2)} \int_0^x y^{(n_1/2-1)}(n_2 + n_1 y)^{-(n_1+n_2)/2}\, dy \tag{5.10}$$

$x > 0$, $n_1 > 0$, $n_2 > 0$. The parameters n_1 and n_2 are the numerator and denominator degrees of freedom, respectively.

5.1.2 Accuracy Considerations

The statement is often made that approximations to probabilities and percentage points need only be accurate to about three decimal places because to achieve greater accuracy is simply a waste of time since no one uses the additional trailing digits. This reasoning has led to the proliferation of "rough" approximations which do provide two-or three-figure accuracy over certain ranges of values of the parameters and variable in the probability function. These approximations normally perform, at best, poorly over other ranges of the parameter and variable space.

A similar statement might be made (admittedly with less justification) about the standard mathematical functions such as the square root, sine, cosine, and exponential. In cases where we simply desire the function value, we do not often use more than a few of the leading significant digits. However, when these function values are in turn used to evaluate other expressions, then we do want high accuracy in the function values, else the end result may be totally in error. Fortunately, approximations have been developed which evaluate these standard mathematical functions essentially to machine precision, and do so at relatively small cost to the computer user. FORTRAN programmers routinely make use of library functions, SQRT, EXP, ALOG, and so on and expect that the approximate value returned by the subprogram will be essentially accurate to working precision.

In statistical research there is a need for accuracy in computation of probabilities and percentage points because these values are often used in evaluating other expressions for which accurate results are required. An example is the expression (see David, Kennedy, and Knight 1977)

$$E(y) = \sum_{j=1}^{k} \left\{ \frac{(n-1)!}{(r-2)!(n-r)!} F^{r-2}(x_j)G(x_j)[1 - F(x_j)]^{n-r} f(x_j) + \frac{(n-1)!}{(r-1)!(n-r)!} F^{r-1}(x_j)[1 - F(x_j)]^{n-r} g(x_j) + \frac{(n-1)!}{(r-1)!(n-r-1)!} F^{r-1}(x_j)[1 - F(x_j)]^{n-r-1}[1 - G(x_j)]f(x_j) \right\} w_j$$

Here F(x) and G(x) are cdf's; f(x) and g(x) are the corresponding probability density functions; n, r, and k are parameters; w_j is a given weighting factor corresponding to a given point x_j. The value of E(y) is needed to (say) six-figure accuracy for a selected set of parameter and y values. For most values of k, n, and r there is no hope of obtaining the required level of accuracy in computation of E(y) unless F(x), G(x), f(x), and g(x) are evaluated to a much higher level of accuracy. *This is not an isolated example*; it is typical of computational problems that arise in research applications. There is indeed often a need for highly accurate solutions to the problems (5.1) and (5.2). Another example is the recurrence derived by Bozivich, Bancroft, Hartley, and Huntsberger (1956) for power computations. The function $P_2(a,b)$ has the form

$$P_2(a,b) = (1 - \tau)P_2(a - 1, b) + \tau P_2(a, b - 1) - \frac{\tau(1 - Q)^b Q^{n/2}}{bB(b, n/2)} I_x \left(b + \frac{n}{2}, a + 1\right)$$

Accuracy must be maintained in the computation of the incomplete beta function, else the value obtained for $P_2(a,b)$, even for moderate-size a and b, will be very inaccurate.

The position which we adopt in this chapter is that while "rough" methods are useful in many instances, they cannot satisfy computing requirements in the majority of cases; so we will concentrate on the more accurate methods. When considering each specific distribution, emphasis in the presentation will always be on achieving accuracy within moderate cost constraints.

5.2 GENERAL METHODS

There has been a large amount of research conducted into methods for approximating probability functions, and there is a substantial amount of literature on this subject contained in many different periodicals. Reviewing this area of statistical computing is difficult because a multitude of different numerical methods are employed, often in differing forms, to construct algorithms. It is, however, possible when viewing the overall situation to identify several general numerical methods or techniques which have proven to be especially useful in more than one case. With these in mind, gaining an understanding of individual algorithms and their relationship with other algorithms for similar applications is facilitated. The purpose of this section is to describe, in fairly general terms, some numerical methods and techniques which seem to be basic tools for use in this area of application. In subsequent sections applications to problems related to specific distributions will be given.

5.2.1 Approximate Transformation of Random Variables

One useful way to obtain an approximation for a given distribution is to utilize an approximate distributional relationship between the given random variable and another whose cdf is more easily approximated. For example, Fisher used the fact that for a chi-square random variable χ_n^2, the quantity $(2\chi_n^2)^{1/2}$ is approximately distributed normally with mean $(2n - 1)^{1/2}$ and unit variance. This leads to the approximate relationship

$$\chi_n^2(p) \cong \frac{1}{2} \cdot \left[x_p + (2n - 1)^{1/2}\right]^2$$

between the p percentage points of the standard normal (x_p) and the chi-square $[\chi_n^2(p)]$ distributions. Assuming that x_p is relatively easy to obtain, for given p, this approximation will obviously be desirable if it provides sufficient accuracy across a wide range of n and p values. The term *normalizing transformation* is often used to describe this situation where the normal distribution is used as the basis for the expression of values relative to some other distribution. Generally, such transformations provide satisfactory results at reasonable cost only over some intervals of the range of the random variable and for certain values of the parameters of the target distribution. The degree of accuracy obtained by these transformations is usually determined primarily by the mathematical relationship underlying the transformation rather than inaccuracies that are the result of rounding error and cancellation of significant digits in floating-point computations.

To indicate how the approximation given above actually performs, the following values were computed and compared with known percentage points of χ_{10}^2. Rounding was to three decimal places in each case.

p	Approximate	Exact
0.10	4.865	4.735
0.25	6.737	6.787
0.75	12.590	12.667
0.90	15.987	15.907
0.95	18.307	18.022

Although this "rough" approximation appears to be inadequate, its performance will improve for larger n and it might be useful in cases where an initial approximation is required for an iterative improvement procedure. If we were contemplating the use of this approximation, a much more extensive and thorough analysis of its properties would be needed.

5.2.2 Closed Form Approximations

Given a probability function F(x) it is possible to fit an approximating function that is simpler in form and which to some extent models the behavior of the given function. Typically, a polynomial or rational fraction approximating function is used. There are several methods which can be used to derive such approximating functions using various criteria for goodness of fit. Many examples of approximations of this kind are given by Abramowitz and Stegun (1964). A typical general closed form expression is

$$F(x) = \frac{\sum_{j=0}^{k} a_j x^j}{\sum_{\ell=0}^{m} b_\ell x^\ell} + e(x) \qquad |e(x)| \leq 10^{-2}$$

where the a_j and b_ℓ are given and e(x) represents the error in approximation.

One recurring problem encountered when implementing these and other approximations is that of evaluating a polynomial

$$h(x) = \sum_{i=0}^{n} a_i x^i$$

for given n, x, and coefficients a_j. The recommended method for performing this task is to use what is called *Horner's rule*. This rule calls for evaluating the polynomial using its nested form as a basis for proceeding. This form is

$$h(x) = \left\{ \cdots \left\{ \left[(a_n x + a_{n-1})x + a_{n-2} \right] x + a_{n-3} \right\} x + \cdots \right\} x + a_0 \tag{5.11}$$

The initial computation is $a_n x + a_{n-1}$, and the process proceeds in the obvious fashion from this beginning. It is easily shown that this approach is more efficient than straightforwardly taking a linear combination of powers of x, and it is one of the most stable methods available (see Newberg 1974).

An example of a closed form approximation is one given by Smillie and Anstey (1964) for computing probabilities for an F distribution. The expression is

$$F(x|\, n_1, n_2) \doteq \frac{1}{2}(1 + a_1 y + a_2 y^2 + a_3 y^3 + a_4 y^4)^{-4}$$

where

$$y = \frac{(1 - 2/(9n_2))x^{1/3} - (1 - 2/(9n_1))}{\left[2x^{2/3}/(9n_2) - 2/(9n_1) \right]^{1/2}}$$

$$a_1 = 0.278393 \qquad a_3 = 0.000972$$
$$a_2 = 0.230339 \qquad a_4 = 0.078108$$

This is a "rough" approximate method whose accuracy increases as n_2 becomes large and x remains at a moderate level.

Closed form approximations have been derived, for certain interesting functions, which are exceptionally accurate. We shall see an example of one such approximation when we study methods for dealing with the normal distribution.

5.2.3 General Series Expansion

Series expansion of various types can be derived and used as a basis for developing computing algorithms. In some cases the first few terms for an expansion are used to provide an approximation of the type described in Section 5.2.1.

Some especially useful results in this area were derived by Hill and Davis (1968), and we will now consider some of this work. To begin, suppose that $\{F_n(y)\}$ is a sequence of cdf's converging to a limiting distribution $F(y)$ as n increases. If x and u are corresponding percentage points of F_n and F, respectively, then

$$F_n(x) = F(u) \tag{5.12}$$

Let $f_n(y)$ and $f(y)$ be corresponding probability density functions (pdf's), and further assume that they have derivatives of all required orders. Defining the function

$$Z_n(x) = F_n(x) - F(x)$$

it follows from (5.12) that

$$Z_n(x) = \int_x^u f(t)\, dt \tag{5.13}$$

Since the equation

$$\xi = \int_x^u f(t)\, dt$$

defines a relation $u(\xi)$, for which $u(0) = x$, we can consider expanding $u(\xi)$ in a Taylor series about $\xi = 0$.

Derivatives needed for the series expansion are found through the following considerations. First, note that

$$\frac{du}{d\xi} = [f(u)]^{-1}$$

$$\frac{d^2u}{d\xi^2} = -\frac{(df/du)(du/d\xi)}{f^2(u)}$$

$$= -\frac{f'(u)}{f^3(u)}$$

$$= \frac{\psi(u)}{f^2(u)}$$

where $\psi(u) = -f'(u)/f(u)$. Continuing in this fashion, we find that

$$\frac{d^ku}{d\xi^k} = c_k(u)[f(u)]^{-k}$$

where $c_1(u) = 1$ and for $k = 1, 2, \ldots,$

$$c_{k+1}(u) = kc_k(u)\psi(u) + \frac{dc_k(u)}{du} \tag{5.14}$$

Since $u(0) = x$, the Taylor series can be written as

$$u = x + \sum_{k=1}^{\infty} \frac{c_k(x)\xi^k}{f^k(x)k!}$$

$$= x + \sum_{k=1}^{\infty} \frac{c_k(x)[Z_n(x)/f(x)]^k}{k!} \tag{5.15}$$

Hill and Davis (1968) also show that when the limiting distribution $F(y)$ is standard normal, then $\psi(x) = x$ and the $c_k(x)$ are polynomials

$$c_1(x) = 1, \quad c_2(x) = x, \quad c_3(x) = 2x^2 + 1, \quad c_4(x) = 6x^3 + 7x, \quad \ldots$$

This is a form of the Cornish-Fisher (1937) normalizing expansion.

One way in which Equation (5.15) can be used is to imagine a sequence $\{x_n\}$ of approximate p percentage points of F such that $\{x_n\}$ converges to the x for which $F(x) = p$. Then define $F_n(x)$ to be $F(x_n)$ so that $Z_n(x)$ is a measure of error incurred by using x_n instead of x.

Series expansions in the form of (5.15) will be considered in specific cases in subsequent discussion. Some examples of normalizing series for percentage points, due to Fisher and Cornish (1960), are given below. In each series, x is such that for given p

$$\Phi(x) = p$$

In terms of x, the corresponding p percentile of the t distribution having n degrees of freedom is expressible as

$$t_n(p) = \frac{x + (x^3 + x)}{4n} + \frac{5x^5 + 16x^3 + 3x}{96n^2}$$

$$+ \frac{3x^7 + 19x^5 + 17x^3 - 15x}{384n^3}$$

Similarly for the chi-square distribution, we have

$$\chi_n^2(p) = n + x(2n)^{1/2} + \frac{2(x^2 - 1)}{3} + \frac{(x^3 - 7x)}{9(2n)^{1/2}} - \frac{(6x^4 + 14x^2 - 32)}{(405n)} + \cdots$$

An example expansion of a cdf is given below for the chi-square having even degrees of freedom:

$$H(x|n) = 1 - \exp(-x) - \exp(-x) \sum_{k=1}^{(n/2)-1} \frac{x^k}{k!}$$

5.2.4 Exact Relationships Between Distributions

In many cases exact relationships exist between distributions and these can be exploited when deriving algorithms. For example, a relationship which holds between the F and beta distributions is expressed by

$$F(y|n_1,n_2) = 1 - I_x\left(\frac{n_1}{2},\frac{n_2}{2}\right) \tag{5.16}$$

where $x = n_2/(n_2 + n_1y)$. Thus a good approximating method for one of these cdf's can suffice for both functions.

5.2.5 Numerical Root Finding

Suppose that F(x) is a given cdf and we have a good algorithm for evaluating F(x) given x. One direct approach to computation of the percentage point x_p, for given p, is to use a standard numerical root finding method to compute the root x_p of the equation.

$$F(x) - p = 0 \tag{5.17}$$

There are several standard numerical methods available for solving this problem. In this subsection we shall describe three methods which are sometimes used in this situation. All are iterative and compute the next approximation x_{i+1} to x_p using previous values $(x_i,f(x_i))$ and possibly $(x_{i-1},f(x_{i-1}))$, where $f(x) = F(x) - p$.

Probably the most famous root finding method is *Newton's method.* Using the notation

$$f(x) = F(x) - p$$

and expanding f in a Taylor series about x_i, we have

$$f(x) = f(x_i) + (x - x_i)f'(x_i) + \frac{1}{2}(x - x_i)^2 f''(x_i) + \ldots$$

Truncating this series after two terms we have

$$f(x) \cong f(x_i) + (x - x_i)f'(x_i)$$

Newton's method consists of solving, at each iteration, the equation

$$f(x_i) + (x - x_i)f'(x_i) = 0$$

for the next iterate x_{i+1}. The solution takes the form

$$x_{i+1} = x_i - \frac{f(x_i)}{f'(x_i)} \tag{5.18}$$

Note that this amounts to selecting x_{i+1} as the point where the tangent line to f(x) at x_i intersects the x axis. For the problem defined by Eq. (5.17), use of Newton's method will require that we evaluate both the cdf and the pdf of the given distribution. This is not usually desirable, so Newton's method is not used as often as other methods which do not require evaluation of f'(x).

A second well-known root finding method which does not require use of derivatives is called the *regula falsi* or *secant method*. Given two values x_i and x_{i-1}, the secant of f(x) is the line intersecting the curve f(x) at $f(x_i)$ and $f(x_{i-1})$. The equation of this line is

$$h(x) = f(x_i) + \frac{(x - x_i)[f(x_i) - f(x_{i-1})]}{x_i - x_{i-1}}$$

The next approximation x_{i+1} to the zero of the function f(x) is obtained by solving for x in h(x) = 0, and the solution has the form

$$x_{i+1} = x_i - \frac{f(x_i)(x_i - x_{i-1})}{[f(x_i) - f(x_{i-1})]} \tag{5.19}$$

Note that the regula falsi method chooses x_{i+1} as the point where the secant to f(x) intersects the x axis.

A modification of the secant method, which is designed to speed convergence, is described by Dowell and Jarrett (1971). This method is called the *Illinois method*. It is defined by the following rules:

i. If $f(x_{i+1})f(x_i) < 0$, then $(x_{i-1}, f(x_{i-1}))$ is replaced by $(x_i, f(x_i))$, as usual.

ii. If $f(x_{i+1})f(x_i) > 0$, then $(x_{i-1}, f(x_{i-1}))$ is replaced by $(x_{i-1}, f(x_{i-1})/2)$. (5.20)

After these two rules are applied, $(x_{i+1}, f(x_{i+1}))$ replaces $(x_i, f(x_i))$, as in regula falsi. The modification prevents retention of an end point and, in many cases, this action reduces the number of iterations needed for convergence.

The final method to be described is called *bisection*. The idea here is to begin with two points x_i and x_{i+1} such that the sign of $f(x_i)$ is opposite that of $f(x_{i+1})$. The midpoint of the interval defined by these two points is found as

$$x_{i+2} = \frac{x_{i+1} + x_i}{2}$$

and $f(x_{i+2})$ is computed. If $f(x_{i+1}) \neq 0$, then either $f(x_{i+2})f(x_i) < 0$ or $f(x_{i+2})f(x_{i+1}) < 0$. That one of the two subintervals which contains x_p is thereby specified, and the next iteration is made using this subinterval. Bisection usually converges more slowly than the previous two methods, but it is desirable because it does not use derivatives, it avoids potential numeric problems which may occur using (5.19), and it maintains a bracket of x_p throughout all iterations.

Convergence criteria for these methods can be based on $|x_{i+1} - x_i|$ and/or $|f(x_{i+1})|$ or functions of these functions. Some specific criteria will be given when we consider application of these methods in later sections.

If we were to use bisection for finding the root of the equation (involving the cdf of the standard normal distribution)

$$\Phi(x) - 0.95 = 0$$

and the initial interval specified was $(x_1,x_2) = (1.5,2)$, the following results would be obtained:

i	x_i	x_{i+1}	$\Phi(x_i)$	$\Phi(x_{i+1})$
1	1.50000	2.00000	0.93319	0.97725
2	1.50000	1.75000	0.93319	0.95994
...	...	...	...	...
5	1.62500	1.65625	0.94792	0.95116
...	...	...	...	...
10	1.64453	1.64551	0.94997	0.95007
...	...	...	...	...
15	1.64484	1.64487	0.94999	0.95000

5.2.6 Continued Fractions

In some cases continued fraction expansions of desired functions are utilized by statistical computing algorithms. These fractions provide an accurate and efficient means for approximating the function which gives rise to the continued fraction. An example of a continued fraction representation is

$$\int_0^x e^{-t^2/2}\,dt = \frac{1}{2}(\pi)^{1/2} - \cfrac{\exp(-x^2/2)/2}{x + \cfrac{1}{2x + \cfrac{2}{x + \cfrac{3}{2x + \cfrac{4}{x + (\text{etc.})}}}}}$$

In general, a typical form of continued fractions is written as

$$\cfrac{a_1}{b_1 + \cfrac{a_2}{b_2 + \cfrac{a_3}{b_3 + a_4 \ddots}}}$$

or more compactly as

$$\frac{a_1}{b_1 +}\ \frac{a_2}{b_2 +}\ \frac{a_3}{b_3 +}\ \frac{a_4}{b_4 +}\ \cdots$$

The numbers a_k, b_k are called *elements* of the fraction.

One method for evaluating a continued fraction is defined through the equations

$$\begin{aligned} A_k &= b_k A_{k-1} + a_k A_{k-2} \\ B_k &= b_k B_{k-1} + a_k B_{k-2} \end{aligned} \qquad (k = 1, 2, \ldots) \tag{5.21}$$

with

$$A_{-1} = 1, \quad A_0 = 0, \quad B_{-1} = 0, \quad B_0 = 1$$

and the *nth convergent* (also called an *approximant*) is given by

$$v_n = \frac{A_n}{B_n}$$

This method for computing convergents is called the *forward recurrence* method.

A second method, called *backward recurrence*, proceeds according to the following, where n must be specified prior to initiation of the procedure:

$$U_{n+1} = 0$$

$$U_k = \frac{a_k}{b_k + U_{k+1}} \qquad k = n, n - 1, \ldots, 1 \tag{5.22}$$

$$v_n = U_1$$

Using either method, v_n, for given n, approximates the given function value. Continued fractions can be terminating or infinite. An infinite continued fraction is said to converge when the sequence $\{v_n\}$ converges as n approaches infinity.

To illustrate the use of Eq. (5.21), consider the continued fraction

$$\tan(x) = \frac{x}{1-}\,\frac{x^2}{3-}\,\frac{x^2}{5-}\,\frac{x^2}{7-}\cdots$$

The steps in the computation of the first three convergents are summarized as follows for x = 0.5:

i	a_i	b_i	A_i	B_i	v_i
1	0.50	1.0	0.500	1.000	0.50000
2	-0.25	3.0	1.500	2.750	0.54545
3	-0.25	5.0	7.375	13.500	0.54630

The true value of tan(0.5) agrees with the computed value v_3 to the five decimal places shown above.

To show one way in which continued fractions can be derived, we consider *Viskovatov's method* for converting a rational expression to a continued fraction. The basis for this method is the relation

$$\frac{c_0 + c_1x + c_2x^2 + c_3x^3 + \cdots}{d_0 + d_1x + d_2x^2 + d_3x^3 + \cdots} = \frac{c_0}{d_0} + \frac{c_0 + c_1x + c_2x^2 + \cdots}{d_0 + d_1x + d_2x^2 + \cdots} - \frac{c_0}{d_0}$$

$$= \frac{c_0}{d_0} \quad \cfrac{x}{\cfrac{d_0 + d_1x + d_2x^2 + \cdots}{c_0^1 + c_1^1x + c_2^1x^2 + \cdots}} \tag{5.23}$$

where the c_j^1 denote transformed values. Application of (5.23) to the special case of the power series

$$\exp(x) = \sum_{k=0}^{\infty} x^k/k!$$

proceeds as follows:

$$\exp(x) = \cfrac{1}{1 + \cfrac{1}{\sum\limits_{k=0}^{\infty} x^k/k!} - 1}$$

$$= \cfrac{1}{1 - \cfrac{x \sum_{k=0}^{\infty} x^k/(k+1)!}{\sum_{k=0}^{\infty} x^k/k!}}$$

$$= \cfrac{1}{1 - \cfrac{x}{1 + \cfrac{\sum_{k=0}^{\infty} x^k/k!}{\sum_{k=0}^{\infty} x^k/(k+1)!} - 1}}$$

$$= \cfrac{1}{1 - \cfrac{x}{1 + \cfrac{x}{2 + \cfrac{\sum_{k=0}^{\infty} x^k/(k+1)!}{\frac{1}{2} + \frac{x}{3} + \frac{x^2}{8} + \cdots} - 2}}}$$

$$\vdots$$

$$\cfrac{1}{1 - \cfrac{x}{1 + \cfrac{x}{2 - \cfrac{x}{3 + \cfrac{x}{2 - \cfrac{x}{5 + \ddots}}}}}}$$

A general description of this conversion is as follows. Given the rational expression

$$g(x) = \frac{a_{10} + a_{11}x + a_{12}x^2 + \cdots + a_{1n}x^n + \cdots}{a_{00} + a_{01}x + a_{02}x^2 + \cdots + a_{0n}x^n + \cdots}$$

the first step in reduction gives

$$g(x) = \cfrac{a_{10}}{a_{00} + x\left\{\cfrac{a_{20} + a_{21}x + a_{22}x^2 + \cdots}{a_{10} + a_{11}x + a_{12}x^2 + \cdots}\right\}}$$

where

$$a_{2k} = a_{10}a_{0,k+1} - a_{00}a_{1,k+1}$$

Continuing the reduction, we find

$$g(x) = \cfrac{a_{10}}{a_{00} + \cfrac{a_{20}x}{a_{10} + \cfrac{a_{30}x}{a_{20} + \ddots}}}$$

where

$$a_{m,n} = a_{m-1,0}a_{m-2,n+1} - a_{m-2,0}a_{m-1,n+1}$$

The basic expression of a power series as a continued fraction has the simple form

$$\sum_{i=0}^{\infty} c_i x^i = \cfrac{a_0}{1 - \cfrac{a_1 x}{(1 + a_1 x) - \cfrac{a_2 x}{(1 + a_2 x) - \ddots}}}$$

where

$$a_0 = c_0, \quad a_i = \frac{c_i}{c_{i-1}} \qquad i \geq 1$$

Convergence of continued fractions is a subject which we will not address except to say that the region of convergence of a given continued fraction is not necessarily the same as that of the series from which it is derived. Several continued fractions will be considered for various applications later in this chapter. All of these fractions have a known region of convergence which includes the range of values considered in the application. An extensive discussion of convergence of continued fractions is given by Wall (1948).

Normally, for a given continued fraction, the backward recurrence will be used if n is prespecified. The reason for this is that the backward recurrence is much more resistant to rounding error than is the forward recurrence (see Jones and Thron 1974). However, forward recurrence together with a convergence criterion are used when it is not known how large n must be in order to obtain an adequate approximation.

Many continued fractions have been derived and are given in the literature. In addition to the literature already cited, other material on this subject is given by Khovanskii (1963), Snyder (1966), Fike (1968), and Knuth (1969).

5.2.7 Gaussian Quadrature

Numerical integration (quadrature) methods are sometimes used to obtain approximate values for given cdf's of continuous distributions. In this subsection we consider one general class of methods for numerical integration. This class of methods is called *Gaussian quadrature.*

The general problem which we will consider is evaluation of the integral

$$\int_a^b w(x)f(x)\,dx$$

where the function w(x) in the integrand is called the *weight function.* The significance of the weight function will become apparent when we describe the various kinds of Gaussian quadrature.

Gaussian quadrature calls for approximating the given integral using the form

$$\begin{aligned}\int_a^b w(x)f(x)\,dx &= \sum_{j=1}^{n} m_j f(a_j) + E \\ &= \hat{I} + E\end{aligned}$$

where E is the error involved when $\hat{I}$ is taken as the integral value. The m_j and a_j are defined by the specific type of Gaussian quadrature formula used. Regardless of type of formula, the integer n is called the *order* of the quadrature formula.

In order to define the quantities m_j and a_j given above, consider a sequence of polynomials $\{p_k(x),\ k = 0, 1, \ldots\}$. We require that $p_k(x)$ have degree k and that the polynomials be orthogonal with respect to the weight function w(x) over the interval [a,b]; in other words,

$$\int_a^b w(x)p_j(x)p_i(x)\,dx = 0 \qquad i \neq j$$

Assuming that we have a sequence $\{p_k(x)\}$ of orthogonal polynomials, let c_k be the coefficient of x^k in $p_k(x)$ and define

$$\delta_k = \int_a^b w(x)p_k^2(x)\,dx$$

Under this definition it can be shown (see Ralston 1965 and Davis and Rabinowitz 1967) that the m_j are given as

$$m_j = \frac{-c_{n+1}\delta_n}{c_n p_{n+1}(a_j)p_n'(a_j)} \qquad j = 1, 2, \ldots, n$$

where the a_j are the n zeros of $p_n(x)$, and $p_n'(a_j)$ denotes the first derivative of $p_n(x)$ evaluated at a_j. The a_j are called *abscissas* and the m_j are called *weights* in Gaussian quadrature. The error E has the form

$$E = \frac{\delta_n}{c_n^2(2n)!}\, f^{(2n)}(\xi)$$

where $\xi \in (a,b)$ and $f^{(2n)}(x)$ denotes the 2nth derivative of f(x).

Thus the particular quadrature formula used depends upon the choice of a set of orthogonal polynomials and the order n of the formula. Given these two items, we must know the abscissas a_j in order to compute weights m_j and function values $f(a_j)$. Fortunately, for the most often used formulas, the m_j and a_j have been computed and are available in tabular form in books such as Abramowitz and Stegun (1964). Before we describe some of the different sets of orthogonal polynomials available, let us briefly consider intervals of integration. This has a bearing on the choice of quadrature formula.

When the interval of integration has infinite length, we have two choices: either truncate to a finite length interval or use a quadrature formula designed specifically for infinite intervals. In many cases we will be able to determine that truncation will introduce a negligible amount of error, so the problem of integration over an infinite length interval can be changed to one of integrating over a finite length interval. With regard to finite length intervals, it is often the case that the interval of integration is not the same as the interval required by the quadrature formula selected for use. This necessitates a change of variables which is often required in these applications. For example, the change of variable $y = [1/(b - a)](2x - a - b)$ transforms from the interval $[a,b]$ in x to the interval $[-1,1]$ in y.

We next describe some of the most popular and useful quadrature formulas. A small sample set of the tabulated abscissas and weights available in the literature cited is given in Table 5.1 (also see, e.g., Galant 1969 and Steen, Byrne, and Gelbard 1969).

Table 5.1 Abscissas and Weights for Selected Quadrature Formulas

	Legendre		Laguerre		Hermite	
Order	m_j	a_j	m_j	a_j	m_j	a_j
2	1.000000	0.577350	0.853553	0.585786	0.886227	0.707107
	1.000000	-0.577350	0.146447	3.414214	0.886227	-0.707107
3	8/9	0.0	0.711093	0.415775	1.181636	0.0
	5/9	0.774597	0.278518	2.294280	0.295409	1.224745
	5/9	-0.774597	0.010389	6.289945	0.295409	-1.224745

Formula: Legendre-Gauss

Interval: [-1,1]

Weight function: $w(x) \equiv 1$

Polynomials $p_j(x)$: Legendre polynomials

$$m_j = \frac{-2}{(n+1)p_{n+1}(a_j)p_n'(a_j)}$$

$$E = \frac{2^{2n+1}(n!)^4}{(2n+1)[(2n)!]^3} f^{(2n)}(\xi)$$

Formula: Jacobi-Gauss

Interval: [-1,1]

Weight function: $w(x) = (1-x)^{\alpha}(1+x)^{\beta}$ $\alpha,\beta > -1$

Polynomials $p_j(x)$: Jacobi polynomials

$$m_j = \frac{-(2n+\alpha+\beta+2)\Gamma(n+\alpha+1)\Gamma(n+\beta+1)2^{\alpha+\beta}}{(n+\alpha+\beta+1)\Gamma(n+\alpha+\beta+1)(n+1)!p_n'(a_j)p_{n+1}(a_j)}$$

$$E = \frac{\Gamma(n+\alpha+1)\Gamma(n+\beta+1)\Gamma(n+\alpha+\beta+1)n!2^{2n+\alpha+\beta+1}}{(2n+\alpha+\beta+1)[\Gamma(2n+\alpha+\beta+1)]^2(2n)!} f^{(2n)}(\xi)$$

Formula: Laguerre-Gauss

Interval: $(0,\infty)$

Weight function: $w(x) = \exp(-x)$

Polynomials $p_j(x)$: Laguerre polynomials

$$m_j = \frac{(n!)^2}{p_n'(a_j)p_{n+1}(a_j)}$$

$$E = \frac{(n!)^2}{(2n)!} f^{(2n)}(\xi)$$

Formula: Hermite-Gauss

Interval: $(-\infty,\infty)$

Weight function: $w(x) = \exp(-x^2)$

Polynomials $p_j(x)$: Hermite polynomials

$$m_j = \frac{2^{n+1}n!\pi^{1/2}}{p_n'(a_j)p_{n+1}(a_j)}$$

$$E = \frac{n!\pi}{2^n(2n)!} f^{(2n)}(\xi)$$

Formula: Chebyshev-Gauss

Interval: (-1,1)

Weight function: $w(x) = (1 - x^2)^{-1/2}$

Polynomials $p_j(x)$: Chebyshev

$p_j(x) = \cos[j \cos^{-1}(x)]$

$$\int_{-1}^{1} (1 - x^2)^{-1/2} f(x)\, dx = \frac{\pi}{n} \sum_{j=1}^{n} f(a_j) + E$$

$$a_j = \cos\frac{(2j - 1)\pi}{2n}$$

$$E = \frac{2\pi}{2^{2n}(2n)!} f^{(2n)}(\xi)$$

Example 5.1

Consider the integral

$$(2\pi)^{-1/2} \int_{-\infty}^{\infty} x \exp\left(\frac{-x^2}{2}\right) dx = \left(\frac{2}{\pi}\right)^{1/2} \int_{-\infty}^{\infty} y \exp(-y^2)\, dy$$

The Hermite-Gauss formula is appropriate in this case because the interval of integration $(-\infty,\infty)$ is as required by the formula. Also, the integrand contains the required $w(x)$ as a factor. If this were not the case, the factor $1 \equiv \exp(-y^2)/\exp(-y^2)$ could be introduced. Using the second-order Hermite Gauss formula gives

$$\left(\frac{2}{\pi}\right)^{1/2} \int_{-\infty}^{\infty} y \exp(-y^2)\, dy \simeq \left(\frac{2}{\pi}\right)^{1/2} \{(0.886227)(0.707107) - (0.886227)(0.707107)\} = 0$$

In this case the second-order formula gives the exact value of the integral because the function $f(y) = y$ in the integrand has degree less than $2n = 4$; hence the error term [which involves $f^{(2n)}(x)$ as a factor] is zero.

In each quadrature formula the error term involves the derivative of $f(x)$ evaluated at an unknown point ξ in the interval of integration. In some cases an upper bound for E can be determined as a function of ξ over the interval. In other cases $f^{(2n)}(x)$ is such a complicated function that we cannot obtain a bound for it, and sometimes E is unbounded over the interval. In these latter cases

E is not useful when we attempt to determine the accuracy of the approximate value obtained for the integral. Various devices can be employed to help in assessing error in approximation and to reduce such error. One of these is composite quadrature, which will now be discussed.

Suppose that we divide the given interval of integration [a,b] into $m > 1$ subintervals:

$$a = L_0 < L_1 < L_2 < \cdots < L_m = b$$

The integral in question can be expressed as the sum of integrals

$$\int_a^b w(x)f(x)\,dx = \int_a^{L_1} w(x)f(x)\,dx + \int_{L_1}^{L_2} w(x)f(x)\,dx + \cdots + \int_{L_{m-1}}^{b} w(x)f(x)\,dx$$

If, for example, we apply a Legendre-Gauss nth-order formula to approximate each integral, then, under fairly general regularity conditions, the error in this approximation has the form

$$E = \left(\frac{1}{m}\right)^{2n}\left(\frac{b-a}{2}\right)^{2n+1} \cdot \frac{2^{2n+1}(n!)^4}{(2n+1)[(2n)!]^3} f^{(2n)}(\eta) \qquad \eta \in (a,b)$$

Notice the factor $(1/m)^{2n}$, which will force E toward zero as m increases. This exhibits the utility of *composite quadrature* wherein we divide the original interval into subintervals and apply a relatively-low-order quadrature formula over each subinterval. The following example illustrates the difference in results which are obtained using simple ($m = 1$) and composite ($m > 1$) quadrature.

Example 5.2

The standard normal integral over (-3,3) has approximate value

$$(2\pi)^{-1/2} \int_{-3}^{3} \exp\left(\frac{-x^2}{2}\right) dx \cong 0.9973002039367$$

Legendre-Gauss quadrature, simple and composite, was used to obtain the following values. These computations were carried out using double-precision arithmetic on an IBM 360 computer. Only as many decimal places as are needed for illustration are recorded.

n	m	$\hat{I}$	n	m	$\hat{I}$
2	1	0.8192	4	4	0.9973005
2	10	0.99731	4	8	0.9973002037
3	5	0.9973007	6	1	0.995
3	10	0.997300208	6	4	0.997300204
4	1	0.95	6	8	0.9973002039367

Successive applications of composite quadrature are often used to obtain an indication of the amount of error in approximation. This is done by using an increasing sequence of m values and observing the behavior of significant digits in the sequence of $\hat{I}$ values. When relatively large changes in m do not alter the first k digits, these digits are taken as being correct. Stabilization of leading digits is illustrated in the example given above for n = 6. Obviously, such stabilization is not proof of correctness. It is used in conjunction with any other checks that can be made.

Numerical quadrature is generally a relatively expensive way to obtain an approximate value for a cdf. We always search for more efficient methods in approximation and use either Gaussian quadrature or Newton-Cotes quadrature (described in the next subsection) only when necessary.

5.2.8 Newton-Cotes Quadrature

In contrast with the Gaussian quadrature formulas which use abscissas that are zeros of orthogonal polynomials, the Newton-Cotes formulas are derived by constraining the abscissas so that they are equally spaced in the (finite-length) interval of integration. The formulas which we consider in this section are called *closed Newton-Cotes formulas*. In the case $w(x) \equiv 1$, which is the only one we will consider, the general closed Newton-Cotes formula is

$$\int_a^b f(x)\,dx = \sum_{j=0}^{n} m_j f(a + hj) + E$$

where $h = (b - a)/n$, and n is the *order* of the formula. Using the notation $a_0 = a$, $a_n = b$, and $a_j = a + hj$, we have

$$\int_a^b f(x)\,dx = \sum_{j=0}^{n} m_j f(a_j) + E$$

The term *closed formula* is used to convey the fact that a_0 and a_n are abscissas.

The weights m_j, for the first three orders of formulas, are the following:

n	m_0	m_1	m_2	m_3
1	h/2	h/2		
2	h/3	4h/3	h/3	
3	3h/8	9h/8	9h/8	3h/8

Abramowitz and Stegun (1964, p. 886) list several higher-order formulas.

Consider the use of composite quadrature with the first-order Newton-Cotes formula on each of m subintervals having length $h = (b - a)/m$. An algebraic simplification yields

$$\int_a^b f(x)\,dx = h\left(\tfrac{1}{2}f_0 + f_1 + \cdots + f_{m-1} + \tfrac{1}{2}f_m\right) + E$$

where $f_j = f(a + hj)$. The error in this case can be shown to have the form

$$E = -\frac{(b - a)^3}{12m^2} f^{(2)}(\xi) \qquad \xi \in (a,b)$$

These two equations define what is called the *trapezoidal rule* for numerical integration over a finite length interval.

Use of the second-order Newton-Cotes formula in the composite mode with m subintervals and $h = a_{i+1} - a_i$ yields

$$\int_a^b f(x)\,dx = \frac{h}{3}(f_0 + 4f_1 + 2f_2 + 4f_3 + \cdots + 4f_{2m-3} + 2f_{2m-2} + 4f_{2m-1} + f_{2m}) + E$$

The error term in this case is

$$E = -\frac{(b-a)^5}{180m^4} f^{(4)}(\xi)$$

This composite form defines *Simpson's rule*, which is a frequently used method for numerical integration.

An interesting and useful method of extrapolation, called *Romberg integration*, is often applied in conjunction with the trapezoidal rule. Romberg integration is intended to accelerate convergence of a sequence of trapezoidal rule approximations obtained by a successive doubling of numbers of subintervals. In order to describe this method, let $m = 2^k$ $(k = 0, 1, 2, \ldots)$ and define

$$T_{0,k} = \frac{b-a}{m}\left(\frac{1}{2}f_0 + f_1 + \cdots + \frac{1}{2}f_m\right)$$

for given k. Thus $T_{0,k}$ is the trapezoidal rule based on 2^k subintervals. Now define

$$T_{s,k} = \frac{1}{4^s - 1}\left(4^s T_{s-1,k+1} - T_{s-1,k}\right)$$

for s = 1, 2, ... and k = 0, 1, 2, ... etc. It is enlightening to view these values in a triangular arrangement as

$$\begin{array}{lllll} T_{0,0} & & & & \\ T_{0,1} & T_{1,0} & & & \\ T_{0,2} & T_{1,1} & T_{2,0} & & \\ \vdots & \vdots & \vdots & \ddots & \\ T_{0,s} & T_{1,s-1} & T_{2,s-2} & \cdots & T_{s,0} \end{array}$$

The sequence of values on the diagonal in the triangle will converge to the true integral value at least as rapidly as the usual sequence of trapezoidal rule approximations which appear in the first column. This procedure has proven to be desirable in many practical applications (see, e.g., De Boor 1971).

Example 5.3

Consider again the integral of Example 5.2. The sequence of values $T_{0,k}$ and $T_{k,0}$ for $m = 2^k$ were computed to be the following:

m	$T_{0,k}$	$T_{k,0}$
2	1.12	1.59
4	0.9936	0.876
8	0.9961	1.0039
16	0.99699	0.99720
32	0.997226	0.9973005
64	0.9972807	0.997300204
128	0.9972953	0.9973002039

Thus in this example the $T_{k,0}$ converge much more rapidly than the $T_{0,k}$.

Programming the Romberg integration method is not a difficult task. The primary considerations are efficiency and stopping criteria. One device which avoids many unnecessary integrand function evaluations is to take advantage of the fact that the procedure is halving intervals at each stage; so essentially one-half the needed function values at end points have already been computed in the preceding stages. With only minor additional effort, the appropriate sum of such values can be made available at each stage, and this effects a substantial savings in machine run time. Finally, with regard to a stopping criterion, one which is often employed is to stop when

$$|T_{k,0} - T_{k+1,0}| < \varepsilon T_{k+1,0}$$

where ε is a user-supplied tolerance.

5.3 THE NORMAL DISTRIBUTION

Among the distributions which are most often used in computer applications, the normal distribution is easily the leader. Not only are normal probabilities and percentage points useful in themselves, but functions of these values are often used in obtaining like values in other distributions. Fortunately, there are many good approximations for this distribution. We shall consider only a few of the more popular ones in this section.

5.3.1 Normal Probabilities

Most mathematical subprogram libraries contain programs which provide values of the error function and its complement essentially to machine precision. These functions are defined as

$$\mathrm{erf}(x) = 2(\pi)^{-1/2} \int_0^x \exp(-t^2)\, dt \qquad (x > 0)$$

$$\mathrm{erfc}(x) = 2(\pi)^{-1/2} \int_x^\infty \exp(-t^2)\, dt \qquad (x > 0)$$

$$= 1 - \mathrm{erf}(x)$$

If there is a good routine for evaluating the error function, it provides an accurate, efficient, and convenient way to evaluate the standard normal cdf $\Phi(x)$ because there is a mathematically exact relationship between these functions given by

$$\Phi(x) = \begin{cases} \dfrac{1 + \mathrm{erf}(x/2^{1/2})}{2} & x \geq 0 \\[2ex] \dfrac{\mathrm{erfc}(-x/2^{1/2})}{2} & x < 0 \end{cases} \qquad (5.24)$$

Since this method depends on accurate approximation of the functions erf(x) and erfc(x), let us briefly consider approximating these functions. Most available algorithms use rational fraction approximations over various intervals of x values. One class of algorithms of this type (classified according to level of accuracy) is given by Cody (1969). The defining expressions are

$$\mathrm{erf}(x) \cong xR_1(x) \qquad 0 < x \leq 0.5$$

$$\mathrm{erfc}(x) \cong \exp(-x^2)R_2(x) \qquad 0.46875 \leq x \leq 4.0$$

$$\mathrm{erfc}(x) \cong (\exp(-x^2)/x)\{\pi^{-1/2} + x^{-2}R_3(x^{-2})\} \qquad x \geq 4 \quad (5.25)$$

The rational fractions R_1, R_2, and R_3 (which provide substantially more than single precision accuracy on the IBM 370 computer) are given in Table 5.2. To avoid unnecessary cancellation of floating-point digits it is desirable to evaluate erf(x) directly and erfc(x) as 1 - erf(x) when $x < 0.46875$. Similarly, when erf(x) is larger than erfc(x) ($x \geq 0.46875$), erfc(x) can be evaluated directly and erf(x) computed as 1 - erfc(x).

Table 5.2 Definition of Rational Fractions in Equation (5.25)

$$R_1 = \frac{\sum_{j=0}^{3} p_j x^{2j}}{\sum_{j=0}^{3} q_j x^{2j}} \qquad 0 < x \leq 0.5$$

j	p_j	q_j
0	$2.4266795523053175 \times 10^2$	$2.1505887586986120 \times 10^2$
1	$2.1979261618294152 \times 10^1$	$9.1164905404514901 \times 10^1$
2	6.9963834886191355	$1.5082797630407787 \times 10^1$
3	$-3.5609843701815385 \times 10^{-2}$	1.0000000000000000

$$R_2 = \frac{\sum_{j=0}^{7} p_j x^{j}}{\sum_{j=0}^{7} q_j x^{j}} \qquad 0.46875 \leq x \leq 4.0$$

j	p_j	q_j
0	$3.004592610201616005 \times 10^2$	$3.004592609569832933 \times 10^2$
1	$4.519189537118729422 \times 10^2$	$7.909509253278980272 \times 10^2$
2	$3.393208167343436870 \times 10^2$	$9.313540948506096211 \times 10^2$
3	$1.529892850469404039 \times 10^2$	$6.389802644656311665 \times 10^2$
4	$4.316222722205673530 \times 10^1$	$2.775854447439876434 \times 10^2$
5	7.211758250883093659	$7.700015293522947295 \times 10^1$
6	$5.641955174789739711 \times 10^{-1}$	$1.278272731962942351 \times 10^1$
7	$-1.368648573827167067 \times 10^{-7}$	1.000000000000000000

$$R_3 = \frac{\sum_{j=0}^{4} p_j x^{-2j}}{\sum_{j=0}^{4} q_j x^{-2j}}$$

Table 5.2 (cont.)

j	p_j	q_j
0	$-2.99610707703542174 \times 10^{-3}$	$1.06209230528467918 \times 10^{-2}$
1	$-4.94730910623250734 \times 10^{-2}$	$1.91308926107829841 \times 10^{-1}$
2	$-2.26956593539686930 \times 10^{-1}$	1.05167510706793207
3	$-2.78661308609647788 \times 10^{-1}$	1.98733201817135256
4	$-2.23192459734184686 \times 10^{-2}$	1.00000000000000000

A second excellent method for approximating $\Phi(x)$ is given by Hill (1972). This method consists of using a Taylor series expansion

$$\Phi(x) = \frac{1}{2} + \phi(x)\left(x + \frac{x^3}{3} + \frac{x^5}{3\cdot 5} + \frac{x^7}{3\cdot 5\cdot 7} \cdots\right) \tag{5.26}$$

for $|x|$ small, and an adaptation of a continued fraction suggested by Adams (1969) when $|x|$ is large. The above series can be specified using a nested form as

$$W_{2s+1} = x$$

$$W_{2r-1} = \frac{W_{2r+1}x^2}{2r + 1} + x \qquad r = s, s - 1, \ldots, 1$$

$$\Phi(x) \simeq \phi(x)W_1 + 1/2$$

and $\Phi(x)$ is obtained from the continued fraction form through

$$W_{4s+3} = x^2$$

$$W_{4r-1} = x^2 + 4r - 1 - \frac{2r(2r + 1)}{W_{4r+3}} \qquad r = s, s - 1, \ldots, 1$$

$$\phi(x) \simeq \begin{cases} \dfrac{\phi(x)(1/W_3 - 1)}{x} & \text{for } x < 0 \\ 1 + \dfrac{\phi(x)(1/W_3 - 1)}{x} & \text{for } x > 0 \end{cases} \tag{5.27}$$

The form of the continued fraction used in (5.27) is

$$\frac{1}{x}\left\{1 + \frac{-1}{x^2 + 3 +}\ \frac{-6}{x^2 + 7 +}\ \frac{-20}{x^2 + 11 +}\ \frac{-42}{x^2 + 15 +}\ \frac{-72}{x^2 + 19 +} \cdots\right\}$$

Adams (1969) reports that the convergents of this continued fraction are equal to alternate convergents of the continued fraction

$$\frac{1}{x+}\ \frac{1}{x+}\ \frac{2}{x+}\ \frac{3}{x+}\ \frac{4}{x+}\ \cdots$$

which is the essential part of the fraction (see Wall 1948, p. 358)

$$\int_x^\infty \exp\left(\frac{-t^2}{2}\right) dt = \exp\left(\frac{-x^2}{2}\right)\left(\frac{1}{x+}\ \frac{1}{x+}\ \frac{2}{x+}\ \frac{3}{x+}\ \cdots\right) \qquad x > 0$$

Over the range $(10^{-20} < |x| < 10)$ the following choice of s above gives very accurate results on the IBM 370 computer:

For $-10. < x \le -1.5$, use (5.27) with $s = \left|\left[4 - \frac{52}{x} + \frac{345}{x^2}\right]\right|$

For $-1.5 < x \le 3.0$, use (5.26) with $s = [10 + 10|x|]$

For $3.0 < x \le 10.$, use (5.27) with $s = \left|\left[-10 + \frac{85}{x} + \frac{255}{x^2}\right]\right|$

where [b] denotes the integer part of b. (The [·] does not denote integer part in this text except where specifically stated.) These limits were designed to provide at least 23-decimal-digit precision on computers that carry the equivalent of 25 decimal digits in floating-point. They can be adjusted to give excellent results on shorter-word-length machines.

Many other approximations to the normal cdf have been proposed and used. Included among these are the polynomial and rational approximations derived by Hastings (1955; see Abramowitz and Stegun 1964) and by Milton and Hotchkiss (1969), the normal approximations of Peizer and Pratt (1968), and the algorithm of Hill and Joyce (1967).

5.3.2 Normal Percentage Points

The inverse error function is more difficult to approximate to machine precision than is the error function itself, so not all mathematical function program libraries contain this function. When it is available, percentage points of the normal distributions are easily obtained through the inverse of the relation (5.24). One of the best of available methods for use in approximating the inverse error function is given by Strecok (1968).

A second method which can be used to obtain extended precision results employs the series (5.15) in the following manner. Suppose that x_0 is an initial approximation to the desired value x_p which satisfies

$$\Phi(x_p) = p$$

for given p. Then the function $Z_n(x_0)$ of Eq. (5.13) can be taken as the error term

$$Z_n(x_0) = p - \Phi(x_0)$$

An approximation to x_p (call it $\hat{x}_p$) based on the first s + 1 terms of (5.15) is expressible in the form

$$x_s = \frac{c_s Z}{s}$$

$$x_k = \frac{(x_{k+1} + c_k)Z}{k} \qquad k = s - 1, s - 2, \ldots, 1$$

$$\hat{x}_p = x_0 + x_1 \tag{5.28}$$

where $Z = Z_n(x_0)/\phi(x_0)$. The coefficients are $c_1 = 1$, $c_2 = x_0$, $c_3 = 2x_0^2 + 1$, and so on. Assuming that $\Phi(x_0)$ and $\phi(x_0)$ are computed essentially to machine precision, one or more applications of (5.28), using $x_0 = \hat{x}_p$ from the previous application, will yield accuracy limited only by the computer hardware capability. Hill (1972) uses s = 2 and reports that one application of (5.28) yields about s + 1 times as many correct digits in $\hat{x}_p$ as are present in x_0. This method requires an initial value x_0 which approximates x_p. Some methods for obtaining x_0 are given below.

Among the many methods available for obtaining relatively rough approximations to x_p is the following closed form approximation due to Hastings (1955). This approximation finds y_p such that $1 - \Phi(y_p) = p$ for $0 < p \leq 0.5$. Its form is

$$y_p = t - \frac{c_0 + c_1 t + c_2 t^2}{1 + d_1 t + d_2 t^2 + d_3 t^3} + \varepsilon(p) \qquad t = [-2\ln(p)]^{1/2}$$

where

$|\varepsilon(p)| < 4.5 \times 10^{-4}$

$c_0 = 2.515517 \qquad d_1 = 1.432788$

$c_1 = 0.802853 \qquad d_2 = 0.189269$

$c_2 = 0.010328 \qquad d_3 = 0.001308$

Another more accurate method is given by Odeh and Evans (1974). This is a rational fraction approximation which provides approximately seven-decimal-place accuracy in $|\hat{x}_p|$ for $10^{-20} < p < 0.5$. An algorithm, named GAUINV, which specifies this method is shown below.

ALGORITHM GAUINV(p, x_p, error)

Step	*Description*
1.	Set lim = 10^{-20}, $p_0 = -0.322232431088$, $p_1 = -1.0$, $p_2 = -0.342242088547$, $p_3 = -0.0204231210245$, $p_4 = -0.453642210148 \times 10^{-4}$, $q_0 = 0.0993484626060$, $q_1 = 0.588581570495$, $q_2 = 0.531103462366$, $q_3 = 0.103537752850$, $q_4 = 0.38560700634 \times 10^{-2}$.
2.	Set error = 1 and $x_p = 0$.
3.	If $p > 0.5$, set $p = 1 - p$.
4.	If $p <$ lim, exit with x_p.
5.	Set error = 0.
6.	If $p = 0.5$, exit with x_p.
7.	Set $y = [\ln(1/p^2)]^{1/2}$.
8.	Set $x_p = y + ((((yp_4 + p_3)y + p_2)y + p_1)y + p_0)/ ((((yq_4 + q_3)y + q_2)y + q_1)y + q_0)$.
9.	If $p < 0.5$, $x_p = -x_p$.
10.	Exit with x_p.

Either of these latter two methods can be used to provide an initial value x_0 for use in (5.28); or if less accuracy is required, they will directly produce the desired percentage point. Neither requires use of an approximation of the cdf; hence they are desirable in this respect.

5.4 STUDENT'S t DISTRIBUTION

The standard normal distribution is only one member of a two-parameter family of distributions, but the characteristics of that family are such that it is sufficient to study approximation of only one member of the family. Such is not the case for the other distributions which we shall consider. In particular, the degrees-of-freedom parameter n serves to distinguish between specific members of the family of t distributions. The computing problem is thus somewhat more complicated in that a general computational method for the t distribution should be capable of approximating quantities from any specified member of the family. This fact will be reflected in the methods considered in this section.

In most cases, computational methods are derived to approximate the function

$$P(t|n) = 2[1 - T(t|n)] \qquad (t > 0)$$

$$= \frac{2\Gamma\left(\frac{n+1}{2}\right)}{\sqrt{n\pi}\,\Gamma\left(\frac{n}{2}\right)} \int_t^\infty \left(1 + \frac{y^2}{n}\right)^{-(n+1)/2} dy \qquad (5.29)$$

or its complement

$$Q(t|n) = 1 - P(t|n)$$

We shall concentrate on these functions since they can be used to provide any desired cdf value.

5.4.1 t Probabilities

Consider the cdf $T(t|n)$, which we now write in the form

$$T(t|n) = \frac{1}{2} + \frac{\Gamma\left(\frac{n+1}{2}\right)}{\sqrt{n\pi}\,\Gamma(\frac{n}{2})} \int_0^t \left(1 + \frac{y^2}{n}\right)^{-(n+1)/2} dy$$

Under the change of variables $y = \sqrt{n}\tan(z)$ it follows that

$$\int_0^t \left(1 + \frac{y^2}{n}\right)^{-(n+1)/2} dy = \int_0^\theta \sqrt{n}\cos^{n-1}(z)\,dz$$

where $\theta = \tan^{-1}(t/\sqrt{n})$. Using the facts that

$$\int_0^\theta \cos^{n-1}(z)\,dz = \begin{cases} \theta & n = 1 \\ \sin(\theta) & n = 2 \\ \dfrac{\sin(\theta)\cos^{n-2}(\theta)}{n-1} + \dfrac{n-2}{n-1}\displaystyle\int_0^\theta \cos^{n-3}(z)\,dz \end{cases}$$

$$\Gamma\left(\frac{n}{2} + \frac{1}{2}\right) = \frac{1 \cdot 3 \cdot 5 \cdot 7 \cdots (n-1)}{2^{n/2}}\sqrt{\pi} \qquad n \text{ even}$$

$$\Gamma\left(\frac{n}{2} + \frac{1}{2}\right) = \frac{2 \cdot 4 \cdot 6 \cdot 8 \cdots (n-1)}{2^{(n-1)/2}} \qquad n \text{ odd}$$

$$\Gamma(\tfrac{1}{2}) = \sqrt{\pi}$$

it can be shown that (see Abramowitz and Stegun 1964, p. 948)

$$Q(t|n) = \begin{cases} \dfrac{2\theta}{\pi} & n = 1 \\ \dfrac{2}{\pi}\left\{\theta + \sin(\theta)\left[\cos(\theta) + \dfrac{2}{3}\cos^3(\theta) + \cdots + \dfrac{2 \cdot 4 \cdot 6 \cdots (n-3)}{1 \cdot 3 \cdot 5 \cdots (n-2)}\cos^{n-2}(\theta)\right]\right\} & n \text{ odd} \\ \sin(\theta)\left\{1 + \dfrac{1}{2}\cos^2(\theta) + \dfrac{1 \cdot 3}{2 \cdot 4}\cos^4(\theta) + \cdots + \dfrac{1 \cdot 3 \cdot 5 \cdots (n-3)}{2 \cdot 4 \cdot 6 \cdots (n-2)}\cos^{n-2}(\theta)\right\} & n \text{ even} \end{cases} \tag{5.30}$$

This expansion has been used in several different algorithms, and it provides good results for certain ranges of t and n values. We will now describe one of the most successful methods available for approximating probabilities in the t distribution. This method is given by Hill (1970, 1972) and it makes use of the expansion given above, along with other approximations, to obtain reasonably accurate results over a wide range of t and n values.

The method to be described approximates $P(t|n)$ by considering three cases. These are:

1. small and moderate n and moderate t values (e.g., $n \leq 20$, $t < 4$)
2. large n and moderate t values (e.g., $n > 20$, $t < 4$)
3. large t values (e.g., t values not covered above)

We shall consider each of these cases in turn and outline the suggested method for approximating $P(t|n)$.

Case 1: Small and moderate n and moderate t values. The cosine expansions (5.30) are recommended for use in this case. A useful form of these expressions is given in terms of $b = 1 + t^2/n$ and $y = t/\sqrt{n}$ as

$$c_n = 1$$

$$c_k = 1 + \frac{c_{k+2}(k - 1)}{kb} \qquad k = n - 2, n - 4, \ldots, 3 \text{ or } 2$$

$$Q(t|n) = \begin{cases} c_2 y/\sqrt{b}, & n \text{ even} \\ \dfrac{2}{\pi}\left[\tan^{-1}(y) + \dfrac{c_3 y}{b}\right] & n \text{ odd} \end{cases} \tag{5.31}$$

where $c_3 = 0$ if $n = 1$. The subtraction required to obtain $P(t|n)$ should not result in cancellation of significant digits in this case.

Case 2: Large n and moderate t values. The amount of arithmetic required to use expressions (5.30) becomes undesirably large as n increases; hence a more economical method is desired for large n. One such method is given by the following normalizing expression of the Cornish-Fisher type:

$$\begin{aligned} x &= z + \frac{z^3 + 3z}{d} - \frac{4z^7 + 33z^5 + 240z^3 + 855z}{10d^2} \\ &\quad + \frac{64z^{11} + 788z^9 + 9801z^7 + 89775z^5 + 543375z^3 + 1788885z}{210d^3} \\ &\quad + \cdots \qquad\qquad (5.32) \\ &= z + \sum_r \frac{p_r(z)}{d^r} \end{aligned}$$

where $z = [a\ln(1 + t^2/n)]^{1/2}$, $a = n - 1/2$, $d = 48a^2$, and x is the normal percentage point at the same level as that required for t. Here it is assumed that a subprogram which computes normal probabilities is available. Hill (1970) uses the first three grouped terms in this series and replaces the third grouped term divisor ($10d^2$) with

$$10d(d + 0.8z^4 + 100)$$

Coefficients of higher-order terms in this expansion are given by Hill (1972). An efficient form for evaluation in this case, given coefficients c_j, in polynomials $p_r(z)$ is

$$y_s = \frac{p_s(z)}{d}$$

$$y_k = \frac{p_k(z) + y_{k+1}}{d} \qquad k = s - 1, s - 2, \ldots, 1$$

$$x = z + y_1$$

$$P(t|n) = 2\Phi(-x)$$

Case 3: Large t values. This case might more appropriately be called "small P(t|n) values," since this case is defined because of the loss of significant figures which will result when P(t|n) is computed as $1 - Q(t|n)$. Thus we wish to avoid the method used in case 1, and it has been found that the method described for case 2 also does not maintain good accuracy for large t.

An alternative method which has been adopted is a direct series expansion of P(t|n) in terms of $y = (1 + t^2/n)^{-1/2}$. This takes the form

$$P(t|n) = 2f(t|n)\left(\frac{\sqrt{n}}{y}\right)\left[\frac{1}{n} + \frac{y^2}{2(n+2)} + \frac{1 \cdot 3y^4}{2 \cdot 4(n+4)} + \frac{5y^6}{16(n+6)} + \cdots\right] \tag{5.33}$$

where f(t|n) is the pdf of t. A desirable method for evaluating f(t|n) is given by Hill (1973). Letting $u = 1 + t^2/n$, this expansion is evaluated according to

$$q_2 = 2f(t|n)(nu)^{1/2} \qquad p_2 = \frac{q_2}{n}$$

$$q_{k+2} = \frac{q_k(k - 1)}{ku} \qquad p_{k+2} = p_k + \frac{q_{k+2}}{n + k}$$

$$k = 2, 4, \ldots, s$$

The expansion is carried out until $p_s = p_{s-2}$ to working precision. Then the required probability is taken as

$$P(t|n) = p_s$$

Studies by Hill (1970) and others indicate that the composite method described above is easily capable of producing at least eight-decimal-place accuracy over a very wide range of t and n values and it can be made to give extended precision accuracy with some additional effort.

There are other possible approaches to obtaining t probabilities. One is to use the relationship between the t and beta distributions, which is

$$Q(t|n) = 1 - I_x\left(\frac{n}{2}, \frac{1}{2}\right) \qquad (0 \leq t < \infty) \tag{5.34}$$

where $x = n/(n + t^2)$. If $I_x(n/2, 1/2)$ can be evaluated more accurately than $P(t|n)$, then this will be useful. However, this is not the case if the best of available methods for the t distribution are used.

Other algorithms for this application are given by Cooper (1968b), Morris (1968), Gentleman and Jenkins (1968), Levine (1969), Hill and Loughhead (1970), Taylor (1970), and Dudewicz and Dalal (1972), to name only a few. None of these appears to be generally superior to an algorithm based on the methods given previously in this subsection.

5.4.2 t-Percentage Points

Given a method for obtaining a relatively rough approximate single-tail percentage point, at level 0.5p, one application of the method of Section 5.2.3 can be used to obtain a substantially improved value assuming that $P(t|n)$ is adequately approximated. To see how this is accomplished, let t_1 be the approximate value and let

$$Z(t_1) = \frac{1}{2}[P(t_1|n) - p]$$

Then for

$$f(t_1|n) = \frac{\Gamma\frac{(n+1)}{2}}{\sqrt{n\pi}\,\Gamma(\frac{n}{2})}\left(1 + \frac{t_1^2}{n}\right)^{-(n+1)/2}$$

and

$$W = \frac{Z(t_1)}{f(t_1|n)}$$

we have

$$\psi(t_1) = \frac{-f'(t_1|n)}{f(t_1|n)} = \frac{t_1(n+1)}{(n+t_1^2)}$$

$$\psi'(t_1) = \frac{(n+1)\,(n-t_1^2)}{(n+t_1^2)^2}$$

The first three coefficients c_j in the series expansion, as given by Eq. (5.14), are

$$c_1(t_1) = 1, \qquad c_2(t_1) = \psi(t_1), \qquad c_3(t_1) = 2\psi^2(t_1) + \psi'(t_1) \tag{5.35}$$

Therefore, Eq. (5.15) takes the form

$$t = t_1 + W + \frac{\psi(t_1)W^2}{2!} + \frac{[2\psi^2(t_1) + \psi'(t_1)]W^3}{3!} + \cdots \tag{5.36}$$

The above derivation presumed a method for obtaining t_1. One of the best of available methods is given by Hill (1970). This method is described as follows.

Two different expansions are utilized in this approximate method. These are an inverse asymptotic expansion of $y = [a\ln(1 + t^2/n)]^{1/2}$ in the form

$$y = x - \frac{(x^3 + 3x)}{b} + \frac{4x^7 + 63x^5 + 360x^3 + 945x}{10b^2} - \cdots \tag{5.37}$$

with x the normal percentage point at level p/2, $a = (n - 1/2)^{-1}$, and $b = 48/a^2$; and the expansion

$$t^2/n = \frac{1}{Z} + \frac{n + 1}{n + 2}\left\{-1 + \frac{Z}{2(n + 4)} + \frac{nZ^2}{3(n + 2)(n + 6)} + \frac{n(n + 3)(2n^2 + 9n - 2)Z^3}{8(n + 2)^2(n + 4)^2(n + 8)} + \cdots\right\} \tag{5.38}$$

where $Z = (pd)^{2/n}$ and $d = (n/2)\pi^{1/2}\Gamma(n/2)/\Gamma(n/2 + 1/2)$. Equation (5.37) is utilized when n is large and t moderate. Equation (5.38) is used when t is large. At least six figures of accuracy for $p > 10^{-24}$ and $n > 3$ can be achieved by using the following specializations of (5.37) and (5.38).

First, the value of d is approximated as

$$d = n\left(\frac{a\pi}{2}\right)^{1/2}\left[1 - \frac{3}{b} + \frac{94.5}{b(b + c)}\right]$$

where

$$c = 96.36 - 16a - 98a^2 + \frac{20700a^3}{b}$$

This approximation is designed to provide eight-figure accuracy in d for $n \geq 3$. Second, when $Z > a + 0.5$, a three-term modification of the series (5.37) is used in the form

$$y = x - \frac{x^3 + 3x}{b} + \frac{4x^7 + 63x^5 + 360x^3 + 945x}{u} \tag{5.39}$$

where

$$u = 10b(b + c - 2x - 7x^2 - 5x^3 + 0.05dx^4) \tag{5.40}$$

is utilized for $n \geq 5$. For $3 \leq n < 5$ the value of c in (5.40) is increased by $0.3(n - 4.5)(x + 0.6)$. Third, when $Z \leq a + 0.5$, a carefully fitted three-term approximation from (5.38) is used in the form

$$\frac{t^2}{n} = \frac{1}{Z} + \frac{n+1}{n+2}\left[-1 + \frac{Z}{2(n+4)} + \frac{Z}{3(n+2)\left(\frac{n+6}{nz} - 0.822 - 0.089d\right)}\right] \tag{5.41}$$

Finally, for $n = 1$ an exact result is obtained as $t = \cot(p\pi/2)$; and for $n = 2$, $t^2 = \frac{2}{p(2 - p)} - 2$ is easily shown to be the exact value.

Once an approximate value is obtained as indicated by Eqs. (5.37) through (5.41), Eq. (5.36) can be used to improve the value. Ignoring all terms after the fourth can provide at least 20 significant figures of accuracy for a wide range (but not all) of t and n values. For example, when $p < 10^{-20}$ and $n > 50$, at least five terms in (5.36) will be needed to have any hope of achieving more than 20-figure accuracy. Of course, such accuracy is possible in the IBM 370 computer only when extended (more than double) precision arithmetic is used and extremely close attention is given to all details when programming. Hill (1972) used forms of the methods of this subsection to prepare tables of percentage points correct to 20 decimal places. These include n as large as n = 120,000 and p as small as $p = 10^{-30}$.

The methods of Section 5.2.6 might also be used to improve an approximate percentage point. More accuracy can also be obtained by using more terms in the series (5.37) and (5.38), but this approach is generally less efficient than use of the expansion (5.36).

5.5 THE BETA DISTRIBUTION

Many research projects have been conducted to find good methods for approximating incomplete beta function values and beta percentage points. The beta distribution is not only interesting in itself, it is also related to several other important distributions. Some of these relationships are the following:

Binomial distribution:

$$I_p(a, k - a + 1) = \sum_{j=a}^{k} \binom{k}{j} p^j (1 - p)^{k-j} \tag{5.42}$$

Negative binomial distribution:

$$I_q(a,k) = \sum_{j=a}^{k} \binom{k + j - 1}{j} p^k q^j \tag{5.43}$$

Student's t distribution:

$$I_x\left(\frac{n}{2}, \frac{1}{2}\right) = [1 - T(t|n)] \qquad \text{where } x = \frac{n}{n + t^2} \tag{5.44}$$

F distribution:

$$I_x\left(\frac{n_2}{2}, \frac{n_1}{2}\right) = 1 - F(y|n_1,n_2) \qquad \text{where } x = \frac{n_2}{n_2 + n_1 y} \tag{5.45}$$

5.5.1 Evaluating the Incomplete Beta Function

It is apparent from the above that an algorithm for the incomplete beta function can be a multipurpose algorithm capable of computing probabilities in several important distributions. Unfortunately, this function has proven to be difficult to approximate accurately over a sufficiently wide range of the parameters, and no single algorithm has yet been developed which is entirely satisfactory. In the remainder of this subsection we will consider several different methods which are currently being used to approximate this function.

The first method which we consider forms the basis for an algorithm that was developed by Ludwig (1963) with subsequent modifications by Pike and Hill (1967), Bosten and Battiste (1974), and Pike and SooHoo (1976). Defining equations for this method are

$$I_x(a,b) = \frac{Ax^a \Gamma(s + a)}{\Gamma(s)\Gamma(a + 1)} + \frac{x^a(1 - x)^b \Gamma(a + b)B}{\Gamma(a)\Gamma(b + 1)} \tag{5.46}$$

where

$$A = \sum_{i=0}^{\infty} \frac{a(1 - s)_i}{a + i} \cdot \frac{x^i}{i!}$$

$$s = \begin{cases} 1 & \text{if } b \text{ is an integer} \\ b - [b] & \text{otherwise } ([\cdot] \text{ denotes greatest integer function}) \end{cases}$$

$$(1 - s)_i = \begin{cases} 1 & \text{if } i = 0 \\ \dfrac{\Gamma(1 + i - s)}{\Gamma(1 - s)} & \text{if } i > 0 \end{cases}$$

$$B = \sum_{i=1}^{[b]} \frac{b(b - 1) \cdots (b - i + 1)}{(a + b - 1)(a + b - 2) \cdots (a + b - i)} \frac{1}{(1 - x)^i}$$

and B = 0 whenever [b] = 0. The two series A and B are positive term series, which helps to give the method numerical stability. In application the series are expanded, scaling each term as shown in (5.46), until the contribution of the next term is less than a specified tolerance. Some care must be taken to rearrange terms appropriately. Suggestions for such rearrangement are provided by Bosten and Battiste (1974). Reports on the performance of this method (see Bosten and Battiste 1974 and Boardman and Kopitzke 1975) indicate that at least three significant digits of accuracy can be expected for a and b not exceeding 2000 under a FORTRAN single-precision implementation on an IBM 370 computer. The subprogram MDBETA, shown in Fig. 5.1, is the IMSL implementation of this method. This program, and the underlying mathematical method, is among the best available at this time.

A second method, which is very competitive, uses a continued fraction derived by Aroian (1941, 1959). The form of this fraction (see also Abramowitz and Stegun 1964) is

$$I_x(a,b) = \frac{\Gamma(a + b)x^a(1 - x)^b}{\Gamma(a + 1)\Gamma(b)} \left[\frac{1}{1 +} \; \frac{c_1}{1 +} \; \frac{c_2}{1 +} \; \cdots \right] \tag{5.47}$$

where

```
C     SUBROUTINE MDBETA (X,A,B,P,IER)
C
C-MDBETA--------S-------LIBRARY 1----------------------------------------
C
C   FUNCTION            - INCOMPLETE BETA PROBABILITY DISTRIBUTION
C                           FUNCTION
C   USAGE               - CALL MDBETA (X,A,B,P,IER)
C   PARAMETERS    X     - VALUE TO WHICH FUNCTION IS TO BE INTEGRATED.
C                           X MUST BE IN RANGE (0,1) INCLUSIVE
C                 A     - INPUT (1ST) PARAMETER (MUST BE GREATER THAN 0)
C                 B     - INPUT (2ND) PARAMETER (MUST BE GREATER THAN 0)
C                 P     - OUTPUT PROBABILITY THAT A RANDOM VARIABLE
C                           FROM A BETA DISTRIBUTION HAVING PARAMETERS
C                           A AND B WILL BE LESS THAN OR EQUAL TO X.
C                 IER   - ERROR PARAMETER.
C                         TERMINAL ERROR = 128+N
C                           N=1 INDICATES X IS NOT IN RANGE (0,1)
C                             INCLUSIVE
C                           N=2 INDICATES A AND/OR B LESS THAN OR
C                             EQUAL TO 0
C   PRECISION           - SINGLE
C   REQD. IMSL ROUTINES - UERTST
C   LANGUAGE            - FORTRAN
C-----------------------------------------------------------------------
C   LATEST REVISION     - FEBRUARY 18, 1977
C
      SUBROUTINE MDBETA  (X,A,B,P,IER)
C
      DOUBLE PRECISION   PS,PX,Y,P1,DA,XINT,CNT,WH,XB,DB,C,EPS,EPS1,
     *                   ALEPS,TOT,PQ,D4
C                                  MACHINE PRECISION
      DATA               EPS/1.D-6/
C                                  SMALLEST POSITIVE NUMBER
C                                  REPRESENTABLE
      DATA               EPS1/1.D-78/
C                                  NATURAL LOG OF EPS1
      DATA               ALEPS/-179.6016D0/
C                                  CHECK RANGES OF THE ARGUMENTS
      Y = X
      IF ((X .LE. 1.0) .AND. (X .GE. 0.0)) GO TO 5
      IER = 129
      GO TO 9000
    5 IF ((A .GT. 0.0) .AND. (B .GT. 0.0)) GO TO 10
      IER = 130
      GO TO 9000
   10 IER = 0
      AA = A
      BB = B
      IF (X .GT. 0.5) GO TO 15
      INT = 0
      GO TO 20
C                                  SWITCH ARGUMENTS FOR MORE EFFICIENT
C                                  USE OF THE POWER SERIES
   15 INT = 1
      TEMP = AA
      AA = BB
      BB = TEMP
      Y = 1.D0-Y
```

```
   20 IF (X .NE. 0. .AND. X .NE. 1.) GO TO 25
C                                      SPECIAL CASE - X IS 0. OR 1.
      P = 0.
      GO TO 60
   25 IB = BB
      TEMP = IB
      PS = BB-FLOAT(IB)
      IF (BB .EQ. TEMP) PS = 1.D0
      DA = AA
      DB = BB
      PX = DA*DLOG(Y)
      PQ = DLGAMA(DA+DB)
      P1 = DLGAMA(DA)
      C = DLGAMA(DB)
      D4 = DLOG(DA)
      XB = PX+DLGAMA(PS+DA)-DLGAMA(PS)-D4-P1
C                                      SCALING
      IB = XB/ALEPS
      XINT = 0.D0
C                                      FIRST TERM OF A DECREASING SERIES
C                                      WILL UNDERFLOW
      IF (IB .NE. 0) GO TO 35
      XINT = DEXP(XB)
      CNT = XINT*DA
C                                      CNT WILL EQUAL DEXP(TEMP)*(1.D0-PS)I*
C                                      P*Y**I/FACTORIAL(I)
      WH = 0.0D0
   30 WH = WH+1.D0
      CNT = CNT*(WH-PS)*Y/WH
      XB = CNT/(DA+WH)
      XINT = XINT+XB
      IF (XB/EPS .GT. XINT) GO TO 30
   35 TOT = 0.D0
      IF (DB .LE. 1.D0) GO TO 55
      XB = PX+DB*DLOG(1.D0-Y)+PQ-P1-DLOG(DB)-C
C                                      SCALING
      IB = XB/ALEPS
      IF (IB .LT. 0) IB = 0
      C = 1.D0/(1.D0-Y)
      CNT = DEXP(XB-DFLOAT(IB)*ALEPS)
      PS = DB
      WH = DB
   40 WH = WH-1.D0
      IF (WH .LE. 0.0D0) GO TO 55
      PX = (PS*C)/(DA+WH)
      IF (PX .GT. 1.D0) GO TO 45
      IF (CNT/EPS .LE. TOT .OR. CNT .LE. EPS1/PX) GO TO 55
   45 CNT = CNT*PX
      IF (CNT .LE. 1.D0) GO TO 50
C                                      RESCALE
      IB = IB-1
      CNT = CNT*EPS1
   50 PS = WH
      IF (IB .EQ. 0) TOT = TOT+CNT
      GO TO 40
   55 P = TOT+XINT
   60 IF (INT .NE. 0) P = 1.-P
      GO TO 9005
 9000 CONTINUE
      CALL UERTST(IER,6HMDBETA)
 9005 RETURN
      END
```

Figure 5.1 Incomplete beta function subprogram. (Reprinted with permission from IMSL Corporation.)

$$c_{2i} = \frac{i(b - i)x}{(a + 2i - 1)(a + 2i)}$$

$$c_{2i+1} = \frac{-(a + i)(a + b + i)x}{(a + 2i)(a + 2i + 1)}$$

Bouver and Bargmann (1978) report that unless $\min(a,b) \geq 70{,}000$ this continued fraction gives good results for $x < (a - 1)/(a + b - 2)$. When x is not in this range, it is suggested that the relationship

$$I_x(a,b) = 1 - I_{1-x}(b,a) \tag{5.48}$$

be used.

Another continued fraction due to Muller (1931) is used by Boardman and Kopitzke (1975). The form of this continued fraction (see also Aroian 1941, Abramowitz and Stegun 1964) is

$$I_x(a,b) = \frac{\Gamma(a + b)x^a(1 - x)^{b-1}}{\Gamma(a + 1)\Gamma(b)} \left[\frac{1}{1+}\ \frac{b_2}{1+}\ \frac{b_3}{1+}\ \frac{b_4}{1+}\ \cdots\right] \tag{5.49}$$

where

$$b_{2i} = \frac{-(a + i - 1)(b - i)}{(a + 2i - 2)(a + 2i - 1)} \cdot \left(\frac{x}{1 - x}\right)$$

$$b_{2i+1} = \frac{i(a + b + i - 1)}{(a + 2i - 1)(a + 2i)} \cdot \left(\frac{x}{1 - x}\right)$$

This continued fraction is only recommended for use when $x < (a-1)/(a+b-2)$; however, this is not a significant limitation in view of Eq. (5.48). Also, Aroian (1941) notes that the fraction (5.49) may not converge for small b ($b < 20$ say), and fraction (5.47) might best be used in such cases. However, Boardman and Kopitzke (1975) report no difficulty in using (5.49) except for a slow rate of convergence for x near one-half. They note that fewer than 30 approximants are normally required to obtain six-decimal-place accuracy.

A third method, due to Soper (1921) and programmed by Majumder and Bhattacharjee (1973a), utilizes the identities

$$I_x(a,b) = \frac{\Gamma(a + b)}{\Gamma(a + 1)\Gamma(b)} x^a(1 - x)^{b-1} + I_x(a + 1, b - 1) \tag{5.50}$$

$$I_x(a + i,\ b - i) = \frac{\Gamma(a + b)}{\Gamma(a + i + 1)\Gamma(b - i)}\, x^{a+i}(1 - x)^{b-i} + I_x(a + i + 1,\ b - i) \tag{5.51}$$

The first of these relations is termed "reduction by parts," and the second is called "reduction by raising a." This method proceeds as follows. If $a \geq (a + b)x$ and the greatest integer function value $i = [b + (i - x)(a + b)]$ is positive, then at least i "reductions by parts" are made followed by "reductions by raising a" until the absolute value of $I_x(a + i + 1,\ b - i)$ is negligible. If the integer part of $b + (1 - x)(a + b)$ is zero, only "reductions by raising a" are performed. When $a < (a + b)x$, $I_{1-x}(b,a)$ is computed using the procedure described above and Eq. (5.48) provides the final result. A summary equation describing the method (which shows both forms of reductions and the specific form of the resulting expression) is

$$\begin{aligned} I_x(a,b) = {} & \frac{\Gamma(a + b)}{\Gamma(a + 1)\Gamma(b)}\, x^a(1 - x)^{b-1} \left\{1 + \frac{b - 1}{a + 1}\,\frac{x}{1 - x}\right. \\ & \times\left\{1 + \frac{b - 2}{a + 2}\,\frac{x}{1 - x}\left\{1 + \cdots \left[1 + \frac{b - i + 1}{a + i - 1}\,\frac{x}{1 - x}\right.\right.\right. \\ & \times\left\{1 + \frac{b - i}{a + i}\, x \left\{1 + \frac{a + b}{a + i + 1}\, x \quad \left\{1 + \frac{a + b + 1}{a + i + 2}\, x\right.\right.\right. \\ & \times \left[1 + \cdots \quad \left\{1 + \frac{a + b + t - 2}{a + i + t - 1}\, x\right\} \cdots \right\} + I_x(a + i + t,\ b - i) \end{aligned} \tag{5.52}$$

For sufficiently large t, theoretical considerations indicate that high accuracy may be achieved. However, little is known at this time about how this method compares in accuracy with the previous methods, although it is closely related to the continued fractions (5.47) and (5.49) as noted by Aroian (1941).

Gautschi (1964b) provides another algorithm based on a power series expansion of $I_x(a,b)$ for small a and b, and use of formulas similar to (5.50) and (5.51) for moving up to given parameter values. This method has not been shown to be superior to either of the first two methods described in this subsection.

Other publications of interest relative to approximating the incomplete beta function include Di Donato and Jarnagin (1967), Tang (1963), Roe (1969), and Hill and Pike (1967). This last reference deals with approximating the logarithm of the gamma function.

Usually, ratios of gamma functions such as those that appear in Eq. (5.46) are computed as the exponential function value of the the appropriate linear combination of logarithms of gamma functions. The logarithm of the gamma function admits several different approximations such as Stirling's approximation (see, e.g. Erdelyi 1953). Fortunately, most standard mathematical subprogram libraries contain programs which will accurately compute the logarithm of the gamma function for a wide range of values.

5.5.2 Inverting the Incomplete Beta Function

Published algorithms for computing the inverse incomplete beta function utilize one of the numerical root finding methods given in Section 5.2.6. Majumder and Bhattacharjee (1973b) and Roe (1969) use Newton's method; Boardman and Kopitzke use the Illinois method; and the IMSL Library routine shown as Fig. 5.2 uses bisection.

All of the iterative root finding methods need a good starting value to use in defining an initial short interval if they are to converge rapidly. One way to obtain an approximate 100p% point x_0 of the beta distribution with cdf $I_x(a,b)$ is to utilize the approximate relationship

$$\frac{1 + x_0}{1 - x_0} = \frac{4a + 2b - 2}{\chi_p^2}$$

where χ_p^2 is the 100p% point of the chi-square distribution having 2b degrees of freedom. An approximation for χ_p^2 is given by Wilson and Hilferty (1931) as

$$\chi_p^2 = 2b\left[1 - \frac{2}{18b} + y_p\left(\frac{2}{18b}\right)^{1/2}\right]^3$$

```
C      SUBROUTINE MDBETI (P,A,B,X,IER)
C
C-MDBETI--------S-------LIBRARY 1----------------------------------------
C
C   FUNCTION            - INVERSE INCOMPLETE BETA PROBABILITY
C                       -   DISTRIBUTION FUNCTION
C   USAGE               - CALL MDBETI (P,A,B,X,IER)
C   PARAMETERS   P      - INPUT PROBABILITY IN THE EXCLUSIVE RANGE
C                           (0,1)
C                A      - INPUT--FIRST PARAMETER OF THE INCOMPLETE BETA
C                           PDF.
C                B      - INPUT--SECOND PARAMETER OF THE INCOMPLETE BETA
C                           PDF.
C                X      - OUTPUT--VALUE SUCH THAT THE PROBABILITY THAT
C                           A RANDOM VARIABLE DISTRIBUTED BETA(A,B) IS
C                           LESS THAN OR EQUAL TO P.
C                IER    - ERROR INDICATOR
C                         TERMINAL ERROR = 128+N
C                           N = 1 MEANS THAT AN ERROR OCCURRED IN MDBETA
C                           N = 2 MEANS THAT THE VALUE X COULD NOT BE
C                             FOUND WITHIN 30 (ITMAX) ITERATIONS, SUCH
C                             THAT THE ABSOLUTE VALUE OF P1-P WAS
C                             LESS THAN OR EQUAL TO EPS (P1= CALCULATED
C                             PROBABILITY AT X, EPS = .0001) AND THE
C                             ABSOLUTE VALUE OF TWO CONSECUTIVE APPROX.
C                             TO X WAS LESS THAN OR EQUAL TO SIG(1.E-5).
C                           N = 3 MEANS P WAS NOT IN THE RANGE (0,1).
C   PRECISION           - SINGLE
C   REQD. IMSL ROUTINES - MDBETA,UERTST
C   LANGUAGE            - FORTRAN
C-----------------------------------------------------------------------
C   LATEST REVISION     - APRIL 26, 1972
C
      SUBROUTINE MDBETI  (P,A,B,X,IER)
C
      DATA               EPS,SIG/.0001,1.E-5/
      DATA               ZERO,ITMAX/0.,30/
      IER = 0
      IC = 0
      AB = A/B
      XL=0.0
      XR=1.0
      FXL=-P
      FXR=1.0-P
      IF (FXL*FXR .GT. ZERO) GO TO 25
C                                  BISECTION METHOD
    5 X = (XL+XR) * .5
      CALL MDBETA(X,A,B,P1,IER)
      IF (IER .NE. 0) GO TO 20
      FCS = P1-P
      IF (FCS * FXL .GT. ZERO) GO TO 10
      XR = X
      FXR = FCS
      GO TO 15
   10 XL = X
      FXL = FCS
   15 XRMXL = XR - XL
      IF (XRMXL .LE. SIG .AND. ABS(FCS) .LE. EPS) GO TO 9005
      IC = IC+1
      IF (IC .LE. ITMAX) GO TO 5
      IER = 130
      GO TO 9000
C                                  ERROR RETURNED FROM MDBETA
   20 IER = 129
      GO TO 9000
   25 IER = 131
 9000 CONTINUE
      CALL UERTST (IER,6HMDBETI)
 9005 RETURN
      END
```

Figure 5.2 Inverse incomplete beta function subprogram. (Reprinted with permission of IMSL Corporation.)

where y_p is the corresponding percentage point of the standard normal distribution. If $\chi_p^2 < 0$, then choose

$$x_0 = 1 - \{(1 - p)bB(a,b)\}^{1/b}$$

and if $(4a + 2b - 2)/\chi_p^2 \leq 1$, choose

$$x_0 = \{paB(a,b)\}^{1/b}$$

Other possible starting approximations are given by Abramowitz and Stegun (1964).

Once x_0 is obtained and bisection is to be used, steps of preselected length are made away from x_0 in the direction of increase (or decrease) in cdf, depending upon the cdf value at x_0 compared with the desired p value.

5.6 F DISTRIBUTION

As was the case for the beta distribution, probabilities and percentage points for the F distribution are difficult to obtain to a high degree of accuracy using existing algorithms. The relationship between the F and beta distributions was given in the preceding section, and this relationship is often utilized to evaluate the cdf and inverse cdf of F. There are other available methods for obtaining F probabilities which are competitive in terms of speed and accuracy. One of these methods is given in the next subsection.

5.6.1 F Probabilities

Dorrer (1968) developed an algorithm, which was later improved by Davis and Khalil (1972), that is often used when the degrees of freedom n_1 and n_2 are integer. The basic integral form used is

$$F(x|n_1,n_2) = \frac{2\Gamma\left(\frac{n_1 + n_2}{2}\right)}{\Gamma\left(\frac{n_1}{2}\right)\Gamma\left(\frac{n_2}{2}\right)} \times \int_0^{\arcsin(\sqrt{a})} \sin^{n_1-1}(\theta)\cos^{n_2-1}(\theta)\,d\theta$$

where $a = n_1x/(n_2 + n_1x)$. A series expansion of this form, utilizing integration by parts, is

$$F(x|n_1,n_2) = a^{n_1/2} \sum_{i=R_2}^{\nu_2} \frac{\Gamma(n_1/2 + i)(1 - a)^i}{\Gamma(n_1/2)\Gamma(1 + i)}$$

$$- 2R_2(1 - a)^{1/2} \sum_{i=R_1}^{\nu_1} \frac{\Gamma(1/2 + i)a^i}{\Gamma(1/2)\Gamma(1 + i)}$$

$$+ 4R_2(R_2 - R_1) + \frac{8}{\pi} R_1 R_2 \arcsin(a^{1/2})$$

where $\nu_2 = (n_2 - 2)/2$, $\nu_1 = (n_1 - 2)/2$; R_1 and R_2 are the fractional parts of $n_1/2$ and $n_2/2$, respectively. Each of the summations can be expressed in nested polynomial form. The first is, for example,

$$\sum_{i=R_2}^{\nu_2} \frac{\Gamma(n_1/2 + i)(1 - a)^i}{\Gamma(n_1/2)\Gamma(1 + i)} =$$

$$\left\{\left\{\left[\frac{(1 - a)\left(\frac{n_1 + n_2 - 4}{2}\right)}{n_2/2 - 1} + 1\right] \frac{(1 - a)\left(\frac{n_1 + n_2 - 6}{2}\right)}{n_2/2 - 2}\right\}\right.$$

$$\left.\frac{(1 - a)\left(\frac{n_1 + n_2 - 8}{2}\right)}{n_2/2 - 3} + \cdots\right\} \frac{(n_1/2)(1 - a)}{1} + 1$$

for n_2 an even integer, and for n_2 odd the summation can be formulated as

$$\left\{\left\{\left[\frac{(1 - a)\left(\frac{n_1 + n_2 - 4}{2}\right)}{n_2/2 - 1} + 1\right] \frac{(1 - a)\left(\frac{n_1 + n_2 - 6}{2}\right)}{n_2/2 - 2} + \cdots\right\}\right.$$

$$\left.\frac{(1 - a)\left(\frac{n_1 + 1}{2}\right)}{3/2} + 1\right\} \cdot \frac{(1 - a)^{1/2}\,\Gamma\left(\frac{n_1 + 1}{2}\right)}{\Gamma(n_1/2)\Gamma(3/2)}$$

The subprogram listed in Fig. 5.3 is an example of an implementation of this method.

Another method which also is designed for use when n_1 and n_2 are positive integers is given by Morris (1969). This and the

```
C     SUBROUTINE PF(N1,N2,Z,POFF,IER)
C
C-PF--------D------EXAMPLE PROGRAM---------------------------------------
C
C   FUNCTION          - COMPUTE F PROBABILITY USING METHOD SUGGESTED BY
C                     - DAVIS AND KHALIL TO EVALUATE F-CDF.
C   USAGE             - CALL PF(N1,N2,Z,POFF,IER)
C   PARAMETERS N1     - NUMERATOR D.F.- INPUT
C              N2     - DENOMINATOR D.F.- INPUT
C               Z     - UPPER LIMIT OF INTEGRATION- INPUT
C              POFF-  COMPUTED PROBABILITY- OUTPUT
C              IER -  ERROR INDICATOR
C                   -  0 IF NORMAL TERMINATION
C                   -  1 IF Z<0 (POFF=0 RETURNED)
C                   -  2 IF N1 OR N2 IS NOT POSITIVE (POFF=0 RETURNED)
C                   -  3 IF COMPUTED VALUE POFF IS >1 OR <0.
C   PRECISION         - DOUBLE
C   LANGUAGE          - IBM FORTRAN
C-----------------------------------------------------------------------
C
      SUBROUTINE PF(N1,N2,Z,POFF,IER)
      IMPLICIT REAL*8 (A-H,O-Z)
      IER=0
      IF(Z .GT. 0.0D0) GC TO 5
      POFF=0.0D0
      IER=1
      RETURN
  5   IF(N1.GT.0 .AND. N2.GT.0) GO TO 10
      IER=2
      POFF=0.0
      RETURN
  10  CONTINUE
      AN1=N1
      AN2=N2
      A=AN1*Z/(AN1*Z+AN2)
      A1=1.D0-A
      IF(A1 .LT. 0.1D-36) A1=0.1D-36
      D1=AN1*0.5D0
      D2=AN2*0.5D0
      D3=D1+D2-1.D0
      R=0.D0
      S1=0.D0
      S2=0.D0
      DEL=1.D0
      XM=1.D0
      XK=1.D0
      C=0.25D0
      PI=3.141592653589793D0
      N=N2
C                            NOTE BEGINNING OF MAJOR LOOP.
   15 CONTINUE
C                            TO SEE IF DEGREES OF FREEDOM ARE ODD OR EVEN
        M=D2
        M=2*M
        IF(M .NE. N) GO TO 30
        N=D2-1
C                            IF DEGREES OF FREEDOM ARE EVEN
C                            N=D.F./2-1
C
        IF(N .EQ. 0) GO TO 25
```

```
      DO 20 I=1,N
        S1=DEL+S1*R
        D2=D2-1.D0
        D3=D3-1.D0
        TEM=A1/D2
        R=D3*TEM
        S2=(R+TEM)*S2
   20 CONTINUE
   25 S1=DEL+S1*R
      DEL=0.D0
      T=-1.D0
      D3=-1.D0
      S2=A*S2
      C=C+0.5D0
      GO TO 45
C                       IF DEGREES OF FREEDOM ARE ODD
C                       N=(D.F.-1)/2
   30 N=D2
C                       IF DEGREES OF FREEDOM EQUAL 1.
C                       DO NOT EXIT LOOP.
C
      IF(N .EQ. 0) GO TO 40
      DO 35 I=1,N
        S1=DEL+S1*R
        D2=D2-1.D0
        D3=D3-1.D0
        TEM=A1/D2
        R=D3*TEM
        S2=(R+TEM)*S2
   35 CONTINUE
   40 S1=XK*S1
      S2=XK*S2
      ART=DSQRT(A1)
      XM=XM*ART
      T=(XM-ART)/A1
      D3=-0.5D0
      XK=2.D0/PI
      C=C*2.0D0
   45 IF(C .GT. 0.875D0) GO TO 50
      D2=D1
      D3=D2+D3
      S2=S1
      S1=0.D0
      A1=A
      IF(A1 .LT. 0.1D-36) A1=0.1D-36
      N=N1
    GO TO 15
 50 IF(C .LT. 1.125D0) DEL=4.D0/PI*DATAN(T)
    POFF=XM*(S2-S1)-DEL
    IF(0.0D0 .LE. POFF .AND. 1.0D0 .GE. POFF) RETURN
    IF(POFF .LT. 0.0D0) POFF=0.0D0
    IF(POFF .GT. 1.0D0) POFF=1.0D0
    IER=3
    RETURN
    END
```

Figure 5.3 F-probability program.

preceding method are advertised as giving at least three-figure accuracy for $\max(n_1, n_2) \leq 500$; however, neither method has apparently been subjected to exhaustive testing (see Field 1969).

For noninteger degrees of freedom, F probabilities are usually computed by evaluating the incomplete beta function given in Eq. (5.45).

5.6.2 F Percentage Points

There are, at this time, no accurate closed form approximations to the percentage points specifically for this distribution. Many relatively inaccurate approximations have been derived, but we will not discuss these. The usual procedure for computing F percentage points currently is to use a numerical root finding method for this application. Normally, the inverse incomplete beta function is evaluated to produce F percentage points.

5.7 CHI-SQUARE DISTRIBUTION

As is the case for the Student's t distribution, the chi-square cdf can be approximated to a high degree of accuracy partially due to the fact that it is another of the single-parameter families of distributions which are generally easier to handle numerically. This fact then insures that percentage points can also be obtained to a high degree of accuracy.

5.7.1 Chi-Square Probabilities

There are two methods currently in use which give good results. The first of these applies for integer degrees of freedom and is described by Hill and Pike (1967). The method is based on the formula

$$1 - H(x|n) = 1 - H(x|n - 2) + \frac{(x/2)^{(n-2)/2} \exp(-x/2)}{\Gamma(n/2)} \quad (5.53)$$

which is also listed in Abramowitz and Stegun (1964). Starting values are obtained for even and odd degrees of freedom as

$$1 - H(x|2) = \exp\left(\frac{-x}{2}\right) \tag{5.54}$$

$$1 - H(x|1) = \frac{2}{(2\pi)^{1/2}} \int_{\sqrt{x}}^{\infty} \exp\left(\frac{-z^2}{2}\right) dz \tag{5.55}$$

Given the appropriate starting value, Eq. (5.53) is used to raise degrees of freedom to the desired level. Since each term in this expansion is positive, cancellation of significant digits will not occur.

For very large degrees of freedom (5.53) will be expensive to use. El Lozy (1976) recommends use of a normal approximation due to Peizer and Pratt (1968) in this case.

Another method, given by Battacharjee (1970), is applicable for real degrees of freedom. It evaluates the incomplete gamma function $G(x|a,1)$ in the following manner. If $a \leq x \leq 1$ or $x < a$, terms in the series

$$G(x|a,1) = \frac{\exp(-x)x^a}{\Gamma(a+1)} \left[1 + \sum_{i=1}^{\infty} \frac{x^i}{(a+1)(a+2)\cdots(a+i)}\right] \tag{5.56}$$

are accumulated until the contribution of a given term is less than a preset tolerance. For other cases (relationships between x and a), the continued fraction (see Abramowitz and Stegun 1964)

$$G(x|a,1) = 1 - \frac{\exp(-x)x^a}{\Gamma(a)} \left[\frac{1}{x+}\;\frac{1-a}{1+}\;\frac{1}{x+}\;\frac{2-a}{1+}\;\frac{2}{x+}\cdots\right]$$

is used. Successive approximations are computed until the difference between successive values is less than a given tolerance. The use of the continued fraction tends to be more efficient than the use of (5.56).

A mathematical relationship between the chi-square and Poisson distributions which is sometimes utilized is given by

$$1 - H(x|n) = \sum_{j=0}^{t-1} \frac{m^j}{j!} \exp(-m) \tag{5.57}$$

where $t = n/2$ and $m = x/2$ and n is an even integer.

5.7.2 Chi-Square Percentage Points

Best and Roberts (1975) take advantage of the accuracy available in computed chi-square probabilities by utilizing the Taylor series expansion of Hill and Davis (1968) which allows use of noninteger degrees of freedom. The series in this case is of the form

$$x = x_0 + \sum_i c_i(x_0)[Z/\gamma(x_0)]^i/i! \tag{5.58}$$

where

$$Z = p - \int_0^{x_0} \gamma(w)\, dw$$

$$\gamma(w) = 2^{-n/2}[\Gamma(n/2)]^{-1} \exp(-w/2)w^{(n-2)/2}$$

$$c_1(u) = 1 \qquad c_{r+1}(u) = r\psi(u)c_r(u) + \frac{dc_r(u)}{du}$$

with

$$\psi(u) = \frac{1}{2} - \frac{n/2 - 1}{u} \tag{5.59}$$

and x_0 is a starting approximation. The value of x_0 is obtained differently in various cases as is described by the following.

For the majority of x and n values the Wilson-Hilferty approximation (also used in Section 5.5.2)

$$x_0 = n\left[y_p\left(\frac{2}{9n}\right)^{1/2} + 1 - \frac{2}{9n}\right]^3 \tag{5.60}$$

is used, where y_p is the 100p% point of the standard normal distribution. If $n < -1.24\ln(p)$, then

$$x_0 = \left[pn\,\Gamma\left(\frac{n}{2}\right)2^{(n-2)/2}\right]^{2/n}$$

is used in place of (5.60). If the x_0 value computed in (5.60) exceeds $2.2n + 6$, then under the notation $x_0' = x_0$ in (5.60)

$$x_0 = -2\left\{\ln(1 - p) - \left(\frac{n}{2} - 1\right)\ln\left(\frac{x_0'}{2}\right) + \ln\left[\Gamma\left(\frac{n}{2}\right)\right]\right\}$$

is used to obtain a starting value. Other special cases can be defined for p near unity and n close to zero which provide better starting values than (5.60). These are also described by Best and Roberts.

Given a starting value, Eq. (5.58) is utilized one or more times to obtain the desired percentage point. Best and Roberts (1975) use as many as seven terms in the expansion (5.58).

Other methods are available for computing chi-square percentage points. One of the root finding methods could be applied with starting value x_0 obtained as given above. Goldstein (1973) also gives an alternative method. None of these other methods is considered to be as efficient and accurate as the Best and Roberts method. Narula and Li (1977) compare several "rough" methods and conclude that the most desirable with regard to accuracy and efficiency is that of Wilson and Hilferty.

EXERCISES

5.1 Derive the relation between $\Phi(x)$ and erf(x) as given in (5.25).

5.2 Expand $\Phi(x)$ in a Taylor series through four terms in such a way that the first four terms of Eq. (5.26) are obtained.

5.3 Verify that the recurrence which produces W_3 in (5.27) is evaluating the indicated continued fraction. Do this by computing (algebraically) the first three convergents in the continued fraction, and W_3, for s = 3, 2, 1. Also verify that the expression for $\Phi(x)$ is correct for each of the two cases based on sign(x).

5.4 Show that the computational method defined by (5.31) is evaluating the expressions given in (5.30).

5.5 Derive the series of Eq. (5.33) using change of variable in the integral $P(t|n)$ followed by expansion of the integrand in series and termwise integration. (You may wish to use the following expression.)

$$(1 + x)^{-1/2} = 1 - \frac{x}{2} + \frac{3x^2}{8} - \frac{5x^3}{16} + \cdots \qquad (-1 < x < 1)$$

5.6 Verify that the recurrence relations proposed for use with respect to the right member of Eq. (5.33) do correctly sum terms of this series.

5.7 Derive the relation (5.53) and show that (5.54) and (5.55) are correct starting values. Recommend a method for evaluating (5.55).

5.8 Assume that a subprogram is available which computes the inverse error function, call it inverf(p). Derive equations which give normal percentage points as a function of inverf(p).

5.9 Derive the basic integral form of the cdf of F used in Section 5.6.1. Expand this form by successive integration and show that the expansion can be placed in the finite summation form given in the subsection.

5.10 Use the Legendre-Gauss third-order formula to approximate the integral

$$\int_{-1}^{1} \frac{dx}{x + 2}$$

Obtain bounds for the error E and show that the error actually present in the computed value lies within the bounds.

5.11 Compute the first three convergents of the continued fraction

$$\ln[\Gamma(x)] = \left(x - \frac{1}{2}\right) \ln(x) - x + \frac{\ln(2\pi)}{2}$$
$$+ \frac{1}{12x +} \; \frac{2}{5x +} \; \frac{53}{42x +} \; \frac{1170}{53x +} \; \frac{22999}{429x +} \cdots$$

for x = 4 and compare your result with an accurate tabled value.

5.12 Consider the continued fraction

$$\int_{x}^{\infty} u^{a-1} \exp(-u)\, du = \cfrac{x^a \exp(-x)}{x + \cfrac{1-a}{1 + \cfrac{1}{x + \cfrac{2-a}{1 + \cfrac{2}{x + \cfrac{3-a}{1 + \cfrac{3}{x + \ddots}}}}}}}$$

Is this fraction related to any of the standard probability functions? (see Wall 1948.)

5.13 (a) Show that when b is a positive integer, repeated application of Eq. (5.50) will give a series that terminates after b terms. Also show that if the fractional part of b is not zero, application of (5.50) can continue but successive terms will begin to alternate in sign. (Soper 1921.)

(b) Show that repeated use of the relation (5.51) produces a series which converges for all $0 \le x \le 1$. (Soper 1921.)

5.14 The need to accurately evaluate $(1 + x)^{1/2}$ and $\ln(1 + x)$ is reflected in the algorithms given in this chapter. In order to see that these functions may need more than a small amount of attention, determine the error present when each function is evaluated using single-precision floating-point arithmetic for $x = 10^{-5}$. In general will small x values present an accuracy problem for these two functions? Give one way to achieve essentially single-precision accuracy when evaluating these functions for small x.

REFERENCES AND FURTHER READINGS

Abdel-Aty, S. H. (1954), Approximate Formula for the Percentage Points and the Probability Integral of the Non-Central χ^2 Distribution, *Biometrika 41*, 538-540.

Abramowitz, M., and I. Stegun (eds.) (1964), *Handbook of Mathematical Functions*, National Bureau of Standards Applied Mathematics Series 55, U.S. Government Printing Office, Washington, D.C.

Adams, A. G. (1969), Remark on Algorithm 304: Normal Curve Integral, *CACM 12*, 565-566.

Amos, D. E. (1964), Representations of the Central and Non-Central t Distributions, *Biometrika 51*, 451-458.

Amos, D. E. (1969), On Computation of the Bivariate Normal Integral, *Math. Comp. 23*, 655-660.

Amos, D. E. (1972), Evaluation of Some Cumulative Distribution Functions by Numerical Quadrature, *Proceedings of the Computer Science and Statistics: 6th Annual Symposium on the Interface*, edited by M. Tarter, University of California, Berkeley, 66-83.

Aroian, L. A. (1941), Continued Fractions for the Incomplete Beta Function, *AMS 12*, 218-223.

Aroian, L. A. (1950), On the Levels of Significance of the Incomplete Beta Function and F-Distributions, *Biometrika 37*, 219-223.

Aroian, L. A. (1959), Corrections to "Continued Fractions for the Incomplete Beta Functions," *AMS 30*, 1265.

Ashby, T. (1968), A Modification to Paulson's Approximation to the Variance Ratio Distribution, *Comp. J. 11*, 209-210.

Bancroft, T. A. (1949), Some Recurrence Formulae in the Incomplete Beta Function Ratio, *AMS 20*, 451-455.

Bargmann, R. E., and S. P. Ghosh (1964), *Noncentral Statistical Distribution Programs for a Computer Language*. IBM Research Report R.C.-1231, San Jose, Calif.

Beasley, J. D., and S. G. Springer (1977), Algorithm AS 111: The Percentage Points of the Normal Distribution, *Appl. Stat. 26*, 118-121.

Bergson, A. (1968), Certification of and Remarks on Algorithm 304: Normal Curve Integral, *CACM 11*, 271.

Best, D. J. and D. E. Roberts (1975), Algorithm AS 91: The Percentage Points of the χ^2 Distribution, *Appl. Stat. 24*, 385-388.

Bhattacharjee, G. P. (1970), Algorithm AS 32: The Incomplete Gamma Integral, *Appl. Stat. 19*, 285-287.

Boardman, T. J., and R. W. Kopitzke (1975), Probability and Table Values for Statistical Distributions, *Proceedings of the Statistical Computing Section, American Statistical Association*, Washington, D.C., 81-86.

Bosten, N. E., and T. J. Aird (1976), Remark on Algorithm 179: Incomplete Beta Ratio, *TOMS 2*, 207-208.

Bosten, N. E., and E. L. Battiste (1974), Remark on Algorithm 179, *CACM 17*, 156-157.

Bouver, H., and R. E. Bargmann (1978), Numerical Solution of the Beta Distribution, *Proceedings of the Computer Science and Statistics: Eleventh Annual Symposium on the Interface*, edited by A. R. Gallant and T. Gerig, North Carolina State University, Raleigh, 325-336.

Bozivich, H., T. A. Bancroft, H. O. Hartley and D. V. Huntsberger (1956), *Analysis of Variance: Preliminary Tests, Pooling, and Linear Models*, Wright Patterson Air Development Center, WADC ARL 55-244, Ohio.

Brownlee, J. (1923), Log ($\Gamma(x)$) from x = 1 to 50.9 by Interval .01, in *Tracts for Computers*, edited by K. Pearson, Cambridge University Press, London.

Bulgren, W. G. (1971), On Representations of the Doubly Non-Central F Distribution, *JASA 66*, 184-186.

Bulgren, W. G. (1974), Probability Integral of the Doubly Noncentral t Distribution with Degrees of Freedom n and Noncentrality Parameters δ and λ, in *Selected Tables in Mathematical Statistics, vol. 2*, edited by H. L. Harter and D. B. Owen, American Mathematical Society, Providence, R.I.

Bulgren, W. G., and D. E. Amos (1968), A Note on Representations of the Doubly Non-Central t Distribution, *JASA 63*, 1013-1019.

Cameron, M. A. (1975), An Extension of Ruben's Result on the Probability Integral of Non-Central Chi-Square, *Comm. Stat. (A) 4*, 783-786.

Carter, A. H. (1947), Approximations to Percentage Points of the z-Distribution, *Biometrika 34*, 352-358.

Chen, H. J. (1975), *On Approximations to the Inverse Student's t Distribution Function*, Technical Report no. 75-16, Memphis State University, Memphis, Tenn.

Chen, H. J., and A. B. Makowsky (1976), *On Approximations to the F-Distribution and Its Inverse*, Technical Report no. 76-3, Memphis State University, Memphis, Tenn.

Clenshaw, C. W. (1962), *Mathematical Tables, Chebyshev Series for Mathematical Functions*, National Physical Laboratory Mathematical Tables 5, London.

Cochran, W. G. (1940), Note on an Approximate Formula for the Significance Levels of Z, *AMS 11*, 93-95.

Cody, W. J. (1969), Rational Chebyshev Approximations for the Error Function, *Math. Comp. 23*, 631-638.

Cody, W. J. (1971), Software for Elementary Functions, in *Mathematical Software*, edited by J. R. Rice, Academic Press, New York, 171-186.

Cooper, B. E. (1968a), Algorithm AS 2: The Normal Integral, *Appl. Stat. 17*, 186-188.

Cooper, B. E. (1968b), Algorithm AS 3: The Integral of Student's t-Distribution, *Appl. Stat. 17*, 189-190.

Cooper, B. E. (1968c), Algorithm AS 4: An Auxiliary Function for Distribution Integrals, *Appl. Stat. 17*, 190-192.

Cooper, B. E. (1968d), Algorithm AS 5: The Integral of the Non-Central t-Distribution, *Appl. Stat. 17*, 193-194.

Cornish, E. A., and R. A. Fisher (1937), Moments and Cumulants in the Specification of Distributions, *Review of the International Statistical Institute 5*, 307-322.

Cran, G. W., K. J. Martin and G. E. Thomas (1977), Remark on Algorithm AS 63: The Incomplete Beta Integral and AS 64: Inverse of the Incomplete Beta Function Ratio, *Appl. Stat. 26*, 111-114.

Cromer, F. E. (1975), A Subroutine for Computing Exact Probabilities of F-Ratios, *Proceedings of the Statistical Computing Section, American Statistical Association*, Washington, D.C., 96-98.

Cunningham, S. W. (1969), Algorithm AS 24: From Normal Integral to Deviate, *Appl. Stat. 18*, 290-293.

David, H. A., W. J. Kennedy and R. D. Knight (1977), Means, Variances, and Covariances of Normal Order Statistics in the Presence of an Outlier, *Selected Tables in Mathematical Statistics, vol. 5*, edited by D. B. Owen and R. E. Odeh, American Mathematical Society, Providence, R.I.

David, J. L. (1978), *Approximations and Algorithms for Selected Probability Functions and Their Inverses*, M.S. Research Component, Iowa State University Library, Ames.

Davis, P. J., and P. Rabinowitz (1967), *Numerical Integration*, Blaisdell Publishing Co., Waltham, Mass.

Davis, W. E., and H. M. Khalil (1972), Algorithm and Series Expansion for the F-Distribution, *Signum Newsletter 7*, 21-23.

De Boor, C. (1971), CADRE: An Algorithm for Numerical Quadrature, in *Mathematical Software*, edited by J. R. Rice, Academic Press, New York, 417-450.

Dickey, J. M. (1967), Expansion of t Densities and Related Complete Integrals, *AMS 38*, 503-510.

Di Donato, A. R., and M. P. Jarnagin (1967), The Efficient Calculation of the Incomplete Beta Function Ratio for Half-Integer Values of the Parameters a,b, *Math. Comp. 21*, 652-662.

Dorrer, E. (1968), Algorithm 332: F-Distribution, *CACM 11*, 116-117.

Dowell, M., and P. Jarratt (1971), A Modified Regula Falsi Method for Computing the Root of an Equation, *BIT 11*, 168-174.

Drezner, Z. (1978), Computation of the Bivariate Normal Integral, *Math. Comp. 32*, 277-279.

Dudewicz, E. J. (1975), Chi-Square Percentage Points for Chi-Square Distribution Testing of Chi-Square Values, with Applications to Random Number Generators, *Proceedings of Computer Science and Statistics: 8th Annual Symposium on the Interface*, edited by J. Frane, Health Sciences Computing Facility, UCLA, 217-221.

Dudewicz, E. J., and S. R. Dalal (1972), On Approximations to the t-Distribution, *Journal of Quality Technology 4*, 196-198.

El Lozy, M. (1976), Remark on Algorithm 299: Chi-Square Integral, *TOMS 2*, 393-395.

Erdelyi, A. (ed.)(1953), *Higher Transcendental Functions I*, McGraw-Hill, New York.

Federighi, E. T. (1959), Extended Tables of the Percentage Points of Student's t-Distribution, *JASA 54*, 683-688.

Field, J. B. F. (1969), Certification of Algorithm 322: F-Distribution, *CACM 12*, 39.

Fike, C. T. (1968), *Computer Evaluation of Mathematical Functions*, Prentice-Hall, Englewood Cliffs, N.J.

Fisher, R. A., and E. A. Cornish (1960), The Percentile Points of Distributions Having Known Cumulants, *Techno. 2*, 209-225.

Galant, D. (1969), Gauss Quadrature Rules for the Evaluation of $2\pi^{-1/2} \int_0^\infty \exp(-x^2)f(x)\,dx$, *Math. Comp. 23*, 674.

Gardiner, D. A., and B. F. Bombay (1965), An Approximation to Student's t, *Techno. 7*, 71-72.

Gautschi, W. (1964a), Algorithm 221: Gamma Function, *CACM 7*, 143.

Gautschi, W. (1964b), Algorithm 222: Incomplete Beta Function Ratios, *CACM 7*, 143-144.

Gautschi, W. (1964c), Certification of Algorithm 222: Incomplete Beta Function Ratios, *CACM 7*, 244.

Gentleman, W. M., and M. A. Jenkins (1968), An Approximation for Student's t-Distribution, *Biometrika 55*, 571-572.

Goldberg, H., and H. Levine (1946), Approximate Formulas for the Percentage Points and Normalization of t and χ^2, *AMS 17*, 216-225.

Golden, R. R., D. J. Weiss and R. V. Davis (1968), An Evaluation of Jaspen's Approximation of the Distribution Functions of the F, t, and Chi-Square Statistics, *Educational and Psychological Measurement 28*, 163-165.

Goldstein, R. B. (1973), Algorithm 451: Chi-Square Quantiles, *CACM 16*, 483-485.

Gray, H. L., R. W. Thompson and G. V. McWilliams (1969), A New Approximation for the Chi-Square Integral, *Math. Comp. 23*, 85-89.

Greenwood, J. A, and H. O. Hartley (1962), *Guide to Tables in Mathematical Statistics*, Princeton University Press, Princeton, N.J.

Han, C. P. (1975), Some Relationships Between Noncentral Chi-Squared and Normal Distributions, *Biometrika 62*, 213-214.

Hart, J. F. (ed.)(1968), *Computer Approximations*, Wiley, New York.

Harter, H. L. (1964), *New Tables of the Incomplete Gamma Function Ratio and of Percentage Points of the Chi-Square and Beta Distributions*, U.S. Government Printing Office, Washington, D.C.

Hartley, H. O., and E. S. Pearson (1950), Tables of the Probability Integral of the t-Distribution, *Biometrika 37*, 169-172.

Hastings, C., Jr. (1955), *Approximations for Digital Computers*, Princeton University Press, Princeton, N.J.

Haynam, G. E., Z. Govindarajula, and F. C. Leone (1973), Tables of the Cumulative Non-Central Chi-Square Distribution, in *Selected Tables in Mathematical Statistics, vol. 1,* edited by H. L. Harter and D. B. Owen, American Mathematical Society, Providence, R.I.

Hendricks, W. A. (1937), An Approximation to Student's Distribution, *AMS 7*, 210-221.

Hill, G. W. (1970a), Algorithm 395: Student's t-Distribution, *CACM 13*, 617-619.

Hill, G. W. (1970b), Algorithm 396: Student's t-Quantiles, *CACM 13*, 619-620.

Hill, G. W. (1972), Reference Table: Student's t-Distribution Quantiles to 20D, Technical Paper no. 35, CSIRO, Melbourne, Australia.

Hill, G. W. (1973), Student's t Frequency, *CACM 16*, 690.

Hill, G. W., and A. W. Davis (1968), Generalized Asymptotic Expansions of Cornish-Fisher Type, *AMS 39*, 1264-1273.

Hill, G. W., and M. Loughhead (1970), Remarks on Algorithm 321: t-Test Probabilities, Algorithm 344: Student's t-Distribution, *CACM 13*, 124.

Hill, I. D. and S. A. Joyce (1967), Algorithm 304: Normal Curve Integral, *CACM 10*, 374.

Hill, I. D., and A. C. Pike (1967), Algorithm 299: Chi-Squared Integral, *CACM 10*, 243-244.

Hoaglin, D. C. (1977), Direct Approximations for Chi-Squared Percentage Points, *JASA 72*, 508-515.

Hogben, D., R. S. Pinkham, and M. B. Wilk (1964), An Approximation to the Distribution of Q (a variate related to the non-central t) *AMS 35*, 315-318.

Ibbetson, D. (1963), Algorithm 209: Gauss, *CACM 6*, 616.

IBM System/360 Operating System, *FORTRAN IV Library-Mathematical and Service Subprograms*, Order no. GC28-6818-1, San Jose, Calif.

IBM System 360/Scientific (1969), *Subroutine Package Version III Programmers Manual*, Order no. H20-0205-3, San Jose, Calif.

IMSL Library 1 Reference Manual (1977), International Mathematical and Statistical Libraries, Houston, Texas.

Jaspen, N. (1965), The Calculations of Probabilities Corresponding to Value of z, t, F, and Chi-Square, *Educational and Psychological Measurement 25*, 877-880.

Johnson, E. E. (1973), Empirical Equations for Approximating Tabular F Values, *Techno. 15*, 379-384.

Johnson, N. L., and S. Kotz (1970), *Continuous Univariate Distributions, vols. 1 and 2,* Houghton Mifflin, Boston.

Johnson, N. L., and E. S. Pearson (1969), Tables of Percentage Points of Non-Central χ^2, *Biometrika 56*, 255-272.

Jones, W. B., and W. J. Thron (1974), Numerical Stability in Evaluating Continued Fractions, *Math. Comp. 28*, 795-810.

Kahaner, D. K. (1971), Comparison of Numerical Quadrature Formulas, in *Mathematical Software*, edited by J. R. Rice, Academic Press, New York, 229-259.

Khovanskii, A. N. (1963), *The Application of Continued Fractions and Their Generalizations to Problems in Approximation Theory*, P. Noordhoff, Groningen, Netherlands.

Knight, W. (1975), Certification of Algorithm 451: Chi-Square Quantiles, *CACM 18*, 116.

Knuth, D. E. (1969), *The Art of Computer Programming, vol. 2: Seminumerical Algorithms*, Addison-Wesley, Reading, Mass.

Kogbetliantz, E. G. (1967), Generation of Elementary Functions. *Mathematical Methods for Digital Computers, vol. 1*, edited by A. Ralston and H. S. Wilf, Wiley, New York, 7-35.

Kramer, Clyde Y. (1966), Approximations to the Cumulative t-Distribution, *Techno. 8*, 358-359.

Krauth, J., and J. Steinebach (1976), Extended Tables of the Percentage Points of the Chi-Square Distribution for at Most Ten Degrees of Freedom, *Biometrische Zeitschrift 18*, 13-22.

Kuki, H. (1971), Mathematical Function Subprograms for Basic System Libraries: Objectives, Constraints, and Trade-Off, in *Mathematical Software*, edited by J. R. Rice, Academic Press, New York, 187-199.

Laubscher, N. F. (1960), Normalizing the Non-Central t and F Distributions, *AMS 31*, 1105-1112.

Levine, D. A. (1969), Algorithms 344: Student's t-Distribution, *CACM 12*, 37-38.

Ling, R. F. (1977), *A FORTRAN Subroutine for Normal, T, χ^2, and F Probabilities*, Technical Report no. 246, Clemson University. Clemson, S.C.

Ling, R. F. (1978), A Study of the Accuracy of Some Approximations for t, χ^2, and F Tail Probabilities, *JASA 73*, 274-283.

Locks, M. O., M. J. Alexander, and B. J. Byans (1963), *New Tables of the Non-Central t Distribution*, Aeronautical Research Laboratories Report ARL 63-19, Wright Patterson Air Force Base, Ohio.

Ludwig, O. G. (1963), Algorithm 179: Incomplete Beta Ratio, *CACM 6*, 314.

Luke, Y. L. (1975), *Mathematical Functions and Their Approximation*, Academic Press, New York.

Majumder, K. L., and G. P. Battacharjee (1973a), Algorithm AS 63: The Incomplete Beta Integral, *Appl. Stat. 22*, 409-411.

Majumder, K. L., and G. P. Battacharjee (1973b), Algorithm AS 64: Inverse of the Incomplete Beta Function Ratio, *Appl. Stat. 22*, 411-414.

Merrington, M. (1941), Numerical Approximations to the Percentage Points of the χ^2 Distribution, *Biometrika 2*, 200-202.

Merrington, M., and E. S. Pearson (1958), An Approximation to the Distribution of Non-Central t, *Biometrika 45*, 484-496.

Milton, R. C., and R. Hotchkiss (1969), Computer Evaluation of the Normal and Inverse Normal Distribution Functions, *Techno. 11*, 817-822.

Morris, J. (1968), Algorithm 321: t-Test Probabilities, *CACM 11*, 115-116.

Morris, J. (1969), Algorithm 346: F-Test Probabilities, *CACM 12*, 184-185.

Mudholkar, G. S., Y. P. Chaubrey, and C. Lin (1976a), Approximations for the Doubly Noncentral F-Distribution, *Comm. Stat. (A) 5*, 49-63.

Mudholkar, G. S., Y. P. Chaubrey, and C. Lin (1976b), Some Approximations for the Noncentral F-Distribution, *Techno. 18*, 351-358.

Muller, J. H. (1931), On the Application of Continued Fractions to the Evaluation of Certain Integrals, with Special Reference to the Incomplete Beta Function, *Biometrika 22*, 284-297.

Narula, S. C., and F. S. Li (1977), Approximations to the Chi-Square Distribution, *JSCS 5*, 267-277.

Newberg, A. C. R. (1974), Error Analysis for Polynomial Evaluation, *Math. Comp. 28*, 789-793.

O'Brien, W. and J. Wood (1968), Certification of Algorithm 299: Chi-Squared Integral, *CACM 11*, 271.

Odeh, R. E., and J. O. Evans (1974), Algorithm AS 70: Percentage Points of the Normal Distribution, *Appl. Stat. 23*, 96-97.

Owen, D. B. (1962), *Handbook of Statistical Tables*, Addison-Wesley, Reading, Mass.

Owen, D. B. (1965), A Special Case of a Bivariate Non-Central t-Distribution, *Biometrika 52*, 437-446.

Owen, D. B., and D. E. Amos (1963), *Programs for Computing Percentage Points of the Noncentral t-Distributions*, Sandia Corporation Monograph SCR-551, Albuquerque, N.M.

Patnaik, P. B. (1949), The Non-Central χ^2 and F-Distributions and Their Applications, *Biometrika 36*, 202-232.

Paulson, E. (1942), An Approximate Normalization of the Analysis of Variance Distribution, *AMS 13*, 233-235.

Pearson, E. S. (1922), *Table of the Logarithms of the Complete Γ-Function (for Arguments 2 to 1200, i.e., beyond Legendre's Range)*, in Tracts for Computers edited by K. Pearson, Cambridge University Press. London.

Pearson, E. S. (1959), Note on an Approximation to the Distribution of Non-Central χ^2, *Biometrika 46*, 364.

Pearson, E. S., and H. O. Hartley (1962), *Biometrika Tables for Statisticians, vol. 1*, Cambridge University Press, London.

Pearson, E. S., and M. L. Tiku (1970), Some Notes on the Relationship between the Distributions of Central and Non-Central F, *Biometrika 57*, 175-179.

Pearson, K. (1922), *Tables of the Incomplete Gamma Function*, Cambridge University Press, London.

Peiser, A. M. (1943), Asymptotic Formulas for Significance Levels of Certain Distributions, *AMS 14*, 56-62.

Peizer, D. B., and J. W. Pratt (1968), A Normal Approximation for Binomial, F, Beta, and Other Common, Related Tail Probabilities, I, *JASA 63*, 1416-1456.

Pike, M. C., and I. D. Hill (1966), Algorithm 291: Logarithms of Gamma Function, *CACM 9*, 684.

Pike, M. C., and I. D. Hill (1967), Remark on Algorithm 179: Incomplete Beta Ratio, *CACM 10*, 375-376.

Pike, M. C., and J. Soo Hoo (1976), Remark on Algorithm 179: Incomplete Beta Ratio, *TOMS 2*, 207-208.

Pinkham, R. S., and M. B. Wilk (1954), Tail Areas of the t-Distribution from a Mill's-Ratio-Like Expansion, *AMS 34*, 335-337.

Pratt, J. W. (1968), A Normal Approximation for Binomial, F, Beta, and Other Common, Related Tail Probabilities, II, *JASA 63*, 1457-1483.

Prescott, P. (1974), Normalizing Transformations of Student's t Distribution, *Biometrika 61*, 177-180.

Price, R. (1964), Some Non-Central F-Distributions Expressed in Closed Form, *Biometrika 51*, 107-122.

Ralston, A. (1965), *A First Course in Numerical Analysis*, McGraw-Hill, New York.

Resnikoff, G. J. and G. J. Liebermann (1957), *Tables of the Non-Central t-Distribution*, Stanford University Press, Palo Alto, Calif.

Robertson, G. H. (1969), Computation of the Noncentral Chi-Square Distribution, *Bell System Technical Journal 48*, 201-207.

Roe, G. M. (1969), *Programs for the Incomplete Beta and Gamma Functions and Their Inverses*, General Electric Research Report No. 69-C-045, Louisville, KY.

Ruben, H. (1974a), A New Result on the Probability Integral of Non-Central Chi-Square with Even Degrees of Freedom, *Comm. Stat. 3*, 473-476.

Ruben, H. (1974b), Non-Central Chi-Square and Gamma Revisited, *Comm. Stat. 3*, 607-633.

Rubin, H., and J. V. Zidek (1964), *Calculation of Upper Tail Percentiles of the Chi-Square Distribution*, Office of Naval Research Technical Report no. 102.

Sahai, H., and W. O. Thompson (1974), Comparison of Approximations to the Percentiles of the t, χ^2, and F Distributions, *JSCS 3*, 81-93.

Sankaran, M. (1959), On the Non-Central Chi-Square Distribution, *Biometrika 46*, 235-237.

Sankaran, M. (1963), Approximations to the Non-Central Chi-Square Distribution, *Biometrika 50*, 199-204.

Seber, G. A. F. (1963), The Non-Central Chi-Squared and Beta Distributions, *Biometrika 50*, 542-545.

Severo, N. C., and M. Zelen (1960), Normal Approximations to the Chi-Square and Non-Central F Probability Functions, *Biometrika 47*, 411-416.

Smillie, K. W., and T. H. Anstey (1964), A Note on the Calculation of Probabilties in an F-Distribution, *CACM 7*, 725.

Smirnov, N. V. (1961), *Tables for the Distribution and Density Functions of t-Distribution*, Permagon Press, New York.

Snyder, A. M. (1966), *Chebyshev Methods in Numerical Approximation*, Prentice-Hall, Englewood Cliffs, N.J.

Soms, A. P. (1976), An Asymptotic Expansion for the Tail Area of the t Distribution, *JASA 71*, 728-730.

Soper, H. E. (1921), *The Numerical Evaluation of the Incomplete B-Function, or of the Integral* $\int_0^x x^{p-1}(1 - x)^{q-1}\,dx$ *for Ranges of x Between 0 and 1*, in Tracts for Computers edited by K. Pearson, Cambridge University Press, London.

Steen, N. M., G. D. Byrne and E. M. Gelbard (1969), Gaussian Quadrature for the Integrals $\int_0^\infty \exp(-x^2)f(x)\,dx$ and $\int_0^b \exp(-x^2)f(x)\,dx$, *Math. Comp. 23*, 661-671.

Stecok, A. J. (1968), On Calculation of the Inverse of the Error Function, *Math Comp. 22*, 144-158.

Stroud, H. A., and D. Secrest (1966), *Gaussian Quadrature Formulas*, Prentice-Hall, Englewood Cliffs, N.J.

Tang, I. C. (1963), On the Computation of a Certain Type of Incomplete Beta Functions, *CACM 6*, 689.

Taylor, G. A. R. (1970), Algorithm AS 27: The Integral of Student's t-Distribution, *Appl. Stat. 19*, 113-114.

Thatcher, H. C. (1963), Algorithm 180: Error Function-Large X, *CACM 6*, 314-315.

Thatcher, H. C. (1964), Remark on Algorithm 60: Romberg Integration, *CACM 7*, 420.

Thompson, D. H. (1974), Approximate Formulae for the Percentage Points of the Incomplete Beta Function and the χ^2 Distribution, *Biometrika 34*, 368-372.

Tiku, M. L. (1963), Approximation to Student's t-Distribution in Terms of Hermite and Laguerre Polynomials, *J. Indian Math. Soc. 27*, 91-102.

Tiku, M. L. (1964), Approximating the General Non-Normal Variance Ratio Sampling Distributions, *Biometrika 51*, 83-96.

Tiku, M. L. (1965a), Laguerre Series Forms of the Non-Central χ^2 and F Distributions, *Biometrika 52*, 415-427.

Tiku, M. L. (1965b), Series Expansions of the Doubly Non-Central F-Distribution, *Australian Journal of Statistics 7*, 78-89.

Tiku, M. L. (1966), A Note on Approximating the Non-Central F Distribution, *Biometrika 53*, 606-610.

Tiku, M. L. (1967), Tables of the Power of the F-Test, *JASA 62*, 525-539.

Tiku, M. L. (1974), Doubly Noncentral F Distribution: Tables and Applications, in *Selected Tables in Mathematical Statistics, vol. 2*, edited by H. L. Harter and D. B. Owen, American Mathematical Society, Providence, R.I.

Tolman, H. (1971), Remark on Algorithm 322: F-Distribution, *CACM 14*, 117.

Vanderbeck, J. P., and J. R. Cooke (1961), *Extended Tables of Percentage Points of the Chi-Square Distribution*, U.S. Naval Ordinance Test Station, China Lake, Calif.

Van Eeeden, C. (1961), Some Approximations to the Percentage Points of the Non-Central t-Distribution, *Revue de L'Institut Internationale de Statistique 29*, 4-31.

Wall, H. S. (1948), *Analytic Theory of Continued Fractions*, Van Nostrand, New York.

White, J. S. (1970), Tables of Normal Percentile Points, *JASA 65*, 635-638.

Wilson, E. B., and M. M. Hilferty (1931), The Distribution of Chi-Square, *Proc. Nat. Acad. Sci. 17*, 684-688.

Wise, M. E. (1950), The Incomplete Beta Function as a Contour Integral and a Quickly Converging Series for Its Inverse, *Biometrika 37*, 208-218.

Wishart, J. (1957), An Approximate Formula for the Cumulative z-Distribution, *AMS 28*, 504-510.

Young, J. C., and C. E. Minder (1974), Algorithm AS 76: An Integral Useful in Calculating Non-Central t and Bivariate Normal Probabilities, *Appl. Stat. 23*, 455-457.

6 / RANDOM NUMBERS: GENERATION, TESTS AND APPLICATIONS

6.1 INTRODUCTION

Realizations of random processes are the raw materials of classical statistical inferences. Most applications of statistical methods in substantive research require a "random sample" or a "random assignment." Another, quite different way in which observations on random processes may be used is in the development of statistical methods and theory. The sampling experiment leading W. S. Gosset to the discovery of the distribution of the correlation coefficient is an early instance of this latter use of random processes (Student 1908a,b). Other early pioneers in statistics used "top-hat" methods of sampling, drawing numbered beads from a bag, to aid the development of statistical methods and theory. (See Tocher 1963 for a history.) More recently such sampling experiments using computer-generated random numbers have become commonplace in situations where the mathematics become intractable. (See, for example, Andrews et al. 1972, in which the performances of various estimators were compared using samples artificially generated from several different distributions.)

In the 1940s N. Metropolis, J. von Neumann, and S. M. Ulam popularized another use of sampling from random processes. It is the solution of deterministic problems by formulating an equivalent problem involving a stochastic process and then sampling from the stochastic problem to estimate the solution to the original deterministic problem. (See Metropolis and Ulam 1949 for one of the early discussions of the idea.) A simple example is the evaluation of the nonstochastic integral

$$\theta = \int_0^1 f(x)\, dx \tag{6.1}$$

If this integral exists, it is merely E[f(X)], where X has a uniform distribution over the interval (0,1); hence

$$\hat{\theta} = \frac{1}{n} \sum_{i=1}^{n} f(X_i) \tag{6.2}$$

where $X_1, X_2, \ldots, X_n$ is a random sample from that uniform distribution, is an unbiased estimate of the integral in Eq. (6.1). This idea can be applied in the solution of more general linear operator equations, including more complicated integrals, differential equations, matrix inversions, and eigen-systems. Von Neumann and Ulam introduced the term *Monte Carlo* to refer to the techniques employing this idea. In their use of this term, it would not apply to the sampling techniques such as that of Gossett referred to above. The appeal of the metaphor has, however, extended its use commonly to include not only the casting of deterministic problems such as (6.1) into transformations of stochastic processes, but more generally the solution of any problem by a simulated stochastic process. It should be remarked, however, that terminology in this area varies among the workers in the field; for example, Hammersley and Handscomb (1964) restrict the use of *Monte Carlo* to those artificial sampling schemes which employ a variance-reducing technique (see Section 6.6), and many authors, including Naylor et al. (1966) and Kleijnen (1974), distinguish between *Monte Carlo* and *simulation* based on whether the experimental model is changing over time. They use the latter term to refer to such situations in which time is an explicit component of the model. In the present book we will generally use the terms interchangeably to refer to solution of problems by simulating stochastic processes.

For all of the applications of techniques mentioned above, a mechanism for generating random processes is required. This mechanism is the topic of Section 6.2. The techniques which can be implemented practically on a digital computer are deterministic; hence, at best the process generated will be pseudostochastic. To emphasize this fact, many authors refer to the mechanism as a *pseudo random number generator*. Except occasionally as a reminder of the deterministic nature of the process, we will not use the *pseudo* qualifier. The question of randomness of a process generating a sequence of numbers has many ramifications, which will not be explored here. For our purposes we will adopt the attitude toward random sequences expressed in Lehmer's (1951) thoughts on such a sequence, "a vague notion embodying the idea of a sequence in which each term is unpredictable to the uninitiated and whose digits pass a certain number of tests traditional with statisticians and depending somewhat on the use to which the sequence is to be put." (See also Martin-Löf 1969).

Analysis of the techniques used in generating random numbers requires certain facts from number theory. Many of the results of Section 6.2, therefore, are stated without proof; but references to the relevant literature are given for the interested reader.

The basic random process which we consider simulating in Section 6.2 is one which yields independent variables from a uniform (0,1) distribution. In Section 6.3 we consider methods for evaluating the correspondence of the deterministic pseudostochastic process generated to the stochastic process being simulated. In Sections 6.4 and 6.5 we describe methods for transforming the basic random process to processes governed by distributions other than the uniform. In Section 6.6 we briefly consider some Monte Carlo applications. The References section contains references to most of the important papers on random number generation and testing, as well as to many papers on applications. Other extensive bibliographies on these topics are given by Jansson (1966), Nance and Overstreet (1972), and Sowey (1972, 1978).

6.2 GENERATION OF UNIFORM RANDOM NUMBERS

Given random samples from the standard uniform distribution denoted U(0,1), random samples from any other distribution may be obtained by transformations. For some distributions the transformations from the uniform are simple and can be made exactly; in other cases the more complicated transformations must be approximated. First, however, we must consider the generation of independent variates from U(0,1).

There are at present two basic techniques for generating numbers from U(0,1): the congruential methods and the feedback shift register methods. In both cases, random integers over some known range are first generated and then scaled to the interval (0,1). If the range of the integers is great enough, this discretization will be of little consequence in producing departures from the continuous distribution desired. It must be remembered in any case that only a relative few of the "real" numbers are available for representation in the computer.

6.2.1 Congruential Methods

A very useful source of pseudorandom integers is a linear congruential sequence. The congruential sequence can take many different forms, but the most commonly used method is defined by

$$x_i \equiv (\alpha x_{i-1} + c)(\text{mod } m) \qquad \text{for } i = 1, 2, 3, \ldots, \tag{6.3}$$

where x_i, α, c, and m are integers and $0 \leq x_i < m$. [The expression $y \equiv z$ (mod m), where y and z are integer-valued expressions, means that there is an integer k such that $mk = y - z$.] This method was apparently first proposed by Lehmer (1951), who suggested c could be taken as 0. Although Lehmer mentioned the possibility of a nonzero additive constant, the use of such a c was not considered in any detail until the papers by Thomson (1958), Franklin (1958), Rotenberg (1960), and Greenberger (1961a). When c = 0, the method is referred to as *multiplicative-congruential*; otherwise, to emphasize the presence of an additive constant, it is referred to as *mixed-congruential*.

The integers produced by (6.3) are in the interval [0,m). They are transformed by x_i/m into [0,1) over which they approximate a U(0,1) process. For some j and k, it must be the case that $x_j = x_k$; and it is obvious from the relationship in (6.3) that the subsequence produced beginning with x_j must be the same as that beginning with x_k. For a given generator and any given x_0, called a *seed*, the length of the shortest subsequence which by repetition forms the entire sequence from some point on is called the *period* of the seed for that generator. More generally, the *period of a sequence* is the minimal length of a subsequence which by repetition forms the entire sequence from some point on. The largest period of any seed for a given generator is called the *period of the generator*. It is intuitively desirable to choose α, c, m, and x_0 so as to maximize the period of sequences produced. The following theorem due to Hull and Dobell (1962) gives the conditions for achieving the maximum period (which is obviously m) for the generator in (6.3).

Theorem 6.1

The mixed-congruential generation (6.3) has a period of m if and only if

1. c is relatively prime to m.
2. $\alpha \equiv 1 \pmod{p}$ for every prime factor p of m.
3. $\alpha \equiv 1 \pmod{4}$ if 4 is a factor of m.

Obviously if a congruential generator has a period equal to the modulus m, any choice of x_0 would yield that period.

Although as seen by Theorem 6.1, a multiplicative-congruential generator cannot have a period of m, for m large enough, the simplicity gained by taking c = 0 may more than offset the loss in length of period. Hull and Dobell (1964) discussed some of the relative merits of the mixed and the multiplicative generators.

For a multiplicative-congruential generator with modulus m and *multiplier* α, the period is the smallest integer ν for which $\alpha^\nu \equiv 1 \pmod{m}$, since for any x_0, $x_\nu = \alpha^\nu x_0 \equiv x_0 \pmod{m}$ and if

x_0 and m are relatively prime, for $0 < i < \nu$, $x_i = \alpha^i x_0 \not\equiv x_0$ (mod m). The integer ν is referred to as the *order* of α modulo m. A *primitive element* modulo m is defined to be an α which has the maximum possible order; that is, a primitive element yields the maximum possible period in the multiplicative-congruential generator. The maximum possible order is denoted $\nu(m)$.

A large collection of facts concerning primitive elements has been developed in number theory. Knuth (1969) summarized some of the facts pertinent to multiplicative-congruential generators with moduli which are powers of a prime. We state these results in the following theorem.

Theorem 6.2

The integer α is a primitive element modulo p^e (p prime) if and only if

i. $p^e = 2$ and α is odd; or $p^e = 4$ and $\alpha \equiv 3 \pmod 4$; or $p^e = 8$ and $\alpha \equiv 3,5$, or $7 \pmod 8$, or $p = 2$, $e \geq 4$, and $\alpha \equiv 3$ or $5 \pmod 8$;

or ii. p is odd, $e = 1$, $\alpha \not\equiv 0 \pmod p$, and $\alpha^{(p-1)/q} \not\equiv 1 \pmod p$ for any prime divisor q of $p - 1$;

or iii. p is odd, $e > 1$, $\alpha \not\equiv 0 \pmod p$, $\alpha^{(p-1)} \not\equiv \pmod{p^2}$.

This theorem can be used to determine an α to yield maximum period for multiplicative-congruential generators having moduli of the given form. In practice the most commonly used moduli are 2^e for $e > 4$, and p, where p is a prime. For the former, α should be chosen so that $\alpha \equiv \pm 3 \pmod 8$; and for a prime modulus, α should be as in condition ii in Theorem 6.2. Further results of number theory state that the maximum period for a multiplicative generator with modulus 2^e is 2^{e-2}, and with modulus p(prime) the maximum period is $p - 1$.

Other commonly used moduli are of the form 10^e. The following theorem due to Knuth (1969) gives the requirements for α to yield a maximum period in this case.

Theorem 6.3

For modulus 10^e with $e \geq 5$, the multiplicative-congruential generator has maximum period if and only if α is congruent to

3, 11, 13, 19, 21, 27, 29, 37, 53, 59, 61, 67, 69, 77, 83, 91, 109, 117, 123, 131, 133, 139, 141, 147, 163, 171, 173, 179, 181, 187, 189, or 197 (mod 200)

Knuth also showed this maximal period to be $5 \times 10^{e-2}$.

Marsaglia (1972) gave a general formula for the period of a seed x_0 for a congruential generator (6.3) in which α and m are relatively prime. Define the *reduced modulus* to be m/d, where d is the greatest common divisor of m and $x_0(\alpha - 1) + c$. Let the reduced modulus be represented in canonical form, that is,

$$\frac{m}{d} = \prod_{i=1}^{k} p_i^{e_i} \qquad \text{where the } p_i \text{ are distinct primes}$$

Then the period is the least common multiple of $q_1, q_2, \ldots, q_k$, where q_i is the order of α modulo $p_i^{e_i}$ times the greatest common divisor of $\alpha - 1$ and $p_i^{e_i}$. (See also Fuller 1976).

The foregoing results may be used to guide the selection of the multiplier α so as to achieve the maximum period for a given modulus. There are other criteria, however, that may be employed in the selection of α. The relationships among various terms in a sequence of numbers produced by a generator are important considerations. The relationships of most interest are given by the *autocorrelation* of lag j, or jth-order autocorrelation, which for a stationary random process is defined as

$$\rho(x_t, x_{t+j}) = \frac{E\{(x_t - \mu)(x_{t+j} - \mu)\}}{\sigma^2}$$

where μ and σ^2 are the mean and variance, respectively. For a pseudorandom process which cycles over a finite set (such as the congruential generators do), expectations such as in the foregoing definition are taken by averaging over one complete cycle. Such an average over a finite set of numbers yields what is usually referred to as the *serial correlation* of lag j. Expectations are easier to compute when all values between 0 and m - 1 occur in the sequence, that is, when the period is m (see example below). Even in this case, however, the computations are not easy for expectations of products. Coveyou (1960) gave an approximation for the first-order autocorrelation of the sequence produced by (6.3) as

$$\rho(x_t, x_{t+1}) \approx \frac{1}{\alpha} - \frac{6c}{\alpha m}\left(1 - \frac{c}{m}\right)$$

Greenberger (1961b) improved on this approximation and Jansson (1964a) gave an exact expression. Dieter and Ahrens (1971), using some theory for generalized Dedekind sums, gave the first-order autocorrelation of the full-period mixed generator as

$$\rho_1 = \frac{S_{0,c}^{(m)}(\alpha, m) - \frac{1}{2}\left(\left(\frac{c}{m}\right)\right) + \frac{1}{2}\left(\left(\frac{c}{\alpha m}\right)\right) - \frac{1}{4m}}{\frac{m}{12}\left(1 - \frac{1}{m^2}\right)} \tag{6.4}$$

for the multiplicative-congruential generator with $m = 2^p$, $p \geq 3$, and $\alpha \equiv 5 \pmod 8$ as

$$\rho_1 = \frac{\frac{48}{2^p}\left[S_{1,0}^{(8)}\left(\alpha, 2^{p-3}\right) + S_{5,0}^{(8)}\left(\alpha, 2^{p-3}\right)\right] - \frac{12}{2^{2p}}}{1 - \frac{16}{2^{2p}}} \tag{6.5}$$

and for the multiplicative generator with $m = 2^p$, $p \geq 3$, and $\alpha \equiv 3 \pmod 8$ as

$$\rho_1 = \frac{\frac{48}{2^p}\left[S_{1,0}^{(8)}\left(\alpha, 2^{p-3}\right) + S_{3,0}^{(8)}\left(\alpha, 2^{p-3}\right) - \frac{1}{2^p}\right]}{1 - \frac{52}{2^{2p}}} \tag{6.6}$$

where the generalized Dedekind sum, $S_{g,h}^{(f)}(a,b)$ is given by

$$S_{g,h}^{(f)}(a,b) = \sum_{i=0}^{b-1} \left(\left(\frac{i}{b} + \frac{g}{bf}\right)\right)\left(\left(\frac{ai}{b} + \frac{ag + bh}{bf}\right)\right)$$

and

$$((x)) = \begin{cases} x - [x] - \frac{1}{2} & \text{for nonintegral } x \text{ ([}\cdot\text{] is the greatest integer function.)} \\ 0 & \text{for integral } x \end{cases}$$

These expressions for the first-order autocorrelations are sufficient for writing jth-order autocorrelations by merely replacing α in (6.4), (6.5), or (6.6) by α^j and replacing c by $c(\alpha^j - 1)/(\alpha - 1)$ [so long as $\alpha \not\equiv 1 \pmod{m}$]. The detailed derivations are given in Dieter and Ahrens (1971) and will not be considered here. Dieter (1971) considered the exact distribution of pairs of numbers in the sequence. Neuman and Merrick (1976) showed that autocorrelations of certain orders may be quite high, even 1, for lags of half the period. If the period is sufficiently great, however, this is of no concern.

Based on consideration of first-order autocorrelations, some authors have recommended α be chosen approximately equal to the square root of the modulus. Coveyou and MacPherson (1967) and Van Gelder (1967), however, have shown that such a choice leads to high correlations among triplets in the sequence.

For the multiplicative generator with $m = 2^p$, Ahrens, Dieter, and Grube (1970) suggested that α be chosen approximately equal to $2^{p-3}(\sqrt{5} - 1)$. The unusual factor in this expression comes from the golden section ratio, which, the theory of Diophantine approximation suggests, has certain optimal properties.

At present there do not appear to be any definite guidelines for the choice of an α that will guarantee good "random" sequences. One approach is to use the general guidelines suggested above to choose a tentative value for α and then to subject the generator to various tests as discussed in Section 6.3.

The autocorrelations of various orders are related to the concept of n-space uniformity, which refers basically to the uniformity with which a sequence of nk numbers would fill an n-dimensional hypercube, when the first n numbers are taken as one point, the next n numbers as a second point, and so on. The congruential generators have some serious shortcomings in this regard. Marsaglia's (1968) title succinctly describes the situation: "Random numbers fall mainly in the planes." We will see this property in Example 6.2 and will discuss it further in Section 6.3.

We now consider the calculation of moments of the variates produced by congruential generators. As indicated in the discussion on correlations, we understand "expectations" of the pseudorandom variables to be the values obtained by averaging over the finite set of values taken on by the variate. Thus, for a generator (6.3) yielding x over a full period,

$$E(x) = \frac{1}{m}\sum_{i=0}^{m-1} i = \frac{m-1}{2}$$

and

$$\mathrm{Var}(x) = \frac{1}{m}\sum_{i=0}^{m-1} i^2 - \frac{(m-1)^2}{4} = \frac{m^2-1}{12}$$

When these values are scaled into [0,1) by $U = x/m$, corresponding values are $E(U) = \frac{1}{2} - \frac{1}{2m}$ and $\mathrm{Var}(U) = \frac{1}{12} - \frac{1}{12m^2}$, which for large m are quite close to the mean and variance of a uniform (0,1) random variable.

To calculate the moments for generators which do not yield all m possible integers, we must determine which values are generated. For the multiplicative generator this requires consideration of *residue classes*. For a given modulus m, a residue class R_k^m is a set consisting of all integers which are congruent to each other modulo m. There are $\nu(m)$ residue classes in a generated sequence.

Let us consider the special case of the multiplicative generator with $m = 2^p$, $p \geq 3$, and α chosen so as to generate a full period, that is, a period of length 2^{p-2}. By Theorem 6.2, α (and odd x_0) will also generate a full period of length 2 using a modulus of 8. For any x and y in R_k^m, x and y are also in R_j^8 since $x \equiv y \pmod 8$; hence each of the $\nu(2^p) = 2^{p-2}$ residue classes modulo m which are represented in the generated sequence is contained in one of the two residue classes modulo 8. Each number in the full-period sequence can therefore be represented as either $8i + r_1$, or $8i + r_2$, for $i = 0, 1, 2, \ldots, 2^{p-3} - 1$, where r_1 is a member of one of the two residue classes generated by $x_n \equiv \alpha x_{n-1} \pmod 8$ and r_2 is a member of the other.

Now, with $w = 2^{p-3}$, we obtain the moments of the scaled $U = x/m$ as

$$E(U) = \frac{1}{2w} \sum_{i=0}^{w-1} \sum_{j=1}^{2} \left(\frac{8i + r_j}{m}\right)$$

$$= \frac{1}{2w} \sum_{i=0}^{w-1} \left(\frac{2i}{w} + \frac{\delta_1}{w}\right)$$

$$= \frac{1}{2} - \frac{1}{w}\left(\frac{1}{2} - \frac{\delta_1}{2}\right)$$

where $\delta_1 = \frac{1}{8} \sum_{j=1}^{2} r_j$, and

$$E(U)^2 = \frac{1}{2w} \sum_{i=0}^{w-1} \sum_{j=1}^{2} \left(\frac{8i + r_j}{m}\right)^2$$

$$= \frac{1}{2w} \sum_{i=0}^{w-1} \left(\frac{2i^2}{w^2} + \frac{2n\delta_1}{w^2} + \frac{\delta_2}{8w^2}\right)$$

$$= \frac{1}{3} - \frac{1}{2w} + \frac{1}{6w^2} + \frac{1}{2w} - \frac{\delta_1}{2w^2} + \frac{\delta_2}{16w^2}$$

where $\delta_2 = \frac{1}{8} \sum_{i=1}^{2} r_i^2$; hence

$$\mathrm{Var}(U) = \frac{1}{12}\left[1 - \frac{1}{w^2}\left(1 - \frac{3\delta_2}{4} - 3\delta_1^2\right)\right]$$

These moments depend on α through δ_1 and δ_2. For example, if $\alpha = 3$, possible values of r_1 and r_2 are 1 and 3 or 5 and 7, so $\delta_1 = 1/2$ or 15/8. For large m and hence large w, these moments tend to be close to 1/2 and 1/12 regardless of the value of α.

The coding for computer implementation of a congruential random number generator is usually done in such a way as to avoid an explicit operation for the modulo reduction by taking advantage of specific characteristics of the computer being used. On the IBM 370 computers moduli of 2^{32} and 2^{31} are particularly easy to use because of the carry-out feature of integer arithmetic described in Chapter 2. Since only the 32 low-order bits of the results of an integer operation are retained, a reduction modulo 2^{32} is done automatically following the integer operation. Because of the two's complement representation, this reduction is made to the interval $[-2^{31}, 2^{31}-1]$ instead of $[0, 2^{32}-1]$ as specified for Eq. (6.3). The results are equivalent, however. Hemmerle (1969) discussed the special features of IBM 370-type computers relevant to implementation of congruential generators.

Example 6.1

Consider the results of repeated execution of the FORTRAN statement

```
IX = IA * IX
```

on a machine with 5-bit words, but like the IBM 370 in other respects. Let IA = 5 and IX take an initial value of 9. Looping through the statement 9 times yields the following values, shown in binary and in ordinary decimal:

IA * IX	= IX Binary	Decimal	
00101 * 01001	= 01101	13	
00101 * 01101	= 00001	1	
00101 * 00001	= 00101	5	
00101 * 00101	= 11001	-7	$\equiv 25 \pmod{2^5}$
00101 * 11001	= 11101	-3	$\equiv 29 \pmod{2^5}$
00101 * 11101	= 10001	-15	$\equiv 17 \pmod{2^5}$
00101 * 10001	= 10101	-11	$\equiv 21 \pmod{2^5}$
00101 * 10101	= 01001	9	
00101 * 01001	= 01101	13	

In this example the period is seen to be 8. This is the maximum period obtainable with a multiplicative-congruential generator with a modulus of 2^5. The integers produced by this process could be scaled to the interval $\left[-\frac{1}{2},\frac{1}{2}\right)$ by division by 2^5. Seraphin (1969) gave an IBM 360/370 assembler language program for the method illustrated in Example 6.1. He used the multiplier 32781 and, of course, the modulus 2^{32}. His assembler language program is shown in Fig. 6.1.

```
RN      CSECT
        USING   *,15          SET BASE REGISTER
        L       1,SEED        LOAD U_k
        M       0,MULT        U_k+1 = αU_k
        ST      1,SEED        STORE U_k+1
        LD      0,FLOT        LOAD FLOAT CONSTANT AND SEED
        AD      0,ZERO        NORMALIZE
        BR      14            RETURN
ZERO    DC      D'0'
FLOT    DC      X'46000000'
SEED    DC      F'1'          INITIAL U_o
MULT    DC      F'32781'      MULTIPLIER α
        END     RN            TOTAL TIME

Fortran usage
     U = RN(0)
```

Figure 6.1 The Seraphin (1969) uniform number generator using modulus 2**32.

Inspection of the sequence in Example 6.1 may reveal certain "nonrandom" characteristics; for example, the four negative numbers in the sequence occur consecutively. This is a quirk that we would usually not expect to happen even for generators with relatively small moduli. We can, however, see a pattern in this example that is more characteristic of the technique. The sequence of first (highest-order) bits has a period of 8 (the period of the numbers themselves); the sequence of second bits has a period of 4; third bits, a period of 2; fourth, 1; and fifth, 1. Careful consideration of this generator allows us to generalize this for all multiplicative generators with modulus 2^p in the following way. The sequence formed by the lowest-order bit will always have a period of 1; the sequence formed by the two low-order bits will always have a period of 2 or less; the sequence formed by the three low-order bits will have a period no greater than 2 times the period of the sequence of two low-order bits; and so on; however, eventually the sequence formed by some k low-order bits will be less than 2^{k-1}, since the period of the sequence of all p bits cannot exceed 2^{p-2}. (This analysis also emphasizes the need to choose an odd-number seed x_0 whenever m is even, else a zero value may be generated which will cause all subsequent values to be zero.)

Patterns in sequences produced by generators with small moduli are readily discernible; such is generally not the case with larger moduli of the order of 2^{32}. Several methods to be discussed in Section 6.3 have been developed to detect departures from "randomness" when such may not be noticed by mere visual inspection.

As noted previously, 2^{31} is also frequently used as the modulus for a congruential generator on the IBM 370. When this modulus is used, the sign bit is tested at each iteration and changed whenever carry-out produces a negative sign. RANDU (IBM Corporation 1969), one of the most widely used subroutines for random number generation, uses this method. (As we will see later, RANDU is not a good generator.) This method can be illustrated using Example 6.1 and adding a statement:

```
IX = IA*IX
IF(IX.LT.0)IX = IX + 15 + 1
```

On the 5-bit word of the machine in Example 6.1, this has the effect of changing only the sign bit of negative numbers. Letting IA = 5 and IX take an initial value of 9 as before, we have the sequence 13, 1, 5, 9, 13, ... The first three terms are as before, but the -7 of Example 6.1 becomes a 9 by the addition of 16. (Note that 16 is represented as 15 + 1 in the coding.) The modulus here is 2^4.

Lehmer (1951) originally suggested the use of a prime modulus in the multiplicative-congruential generator. Hutchison (1966) recommended that the modulus be chosen as the largest prime representable in one machine word and that the multiplier be chosen as a primitive root of that prime (see Theorem 6.2 ii). Prime moduli have been discussed further by Downham and Roberts (1967) and Whittlesey (1968). Lewis, Goodman, and Miller (1969) and Payne, Rabung, and Bogyo (1969) have given algorithms for using the modulus $2^{31} - 1$ (which is a prime) on the IBM 360/370. Although use of this modulus is not as simple as use of 2^{32} or 2^{31}, the methods given by these authors are quite efficient by utilizing clever techniques that are best implemented in an assembler language. Lewis et al. showed that 7 is a primitive element modulo $2^{31} - 1$ and hence that a multiplier of 7^e will yield a period of $2^{31} - 2$. For the IBM 360/370 they recommended the multiplier 7^5, which is also a primitive element modulo $2^{31} - 1$.

Example 6.2 The Lattice of a Sequence from a Congruential Generator As mentioned previously, successive n-tuples of a sequence produced by a congruential generator will not be uniformly distributed over an n-dimensional hypercube. To illustrate this let us consider the generator $x_i \equiv \alpha x_{i-1}$ (mod 64). For $\alpha = 3$ and $\alpha = 11$ and beginning with $x_0 = 25$, we get the following two sequences:

α	sequence
3	25, 11, 33, 35, 41, 59, 49, 19, 57, 43, 01, 03, 09, 27, 17, 51, 25
11	25, 19, 17, 59, 09, 35, 01, 11, 57, 51, 49, 27, 41, 03, 33, 43, 25

The distributions of successive pairs from these sequences over the square are shown in Figure 6.2.

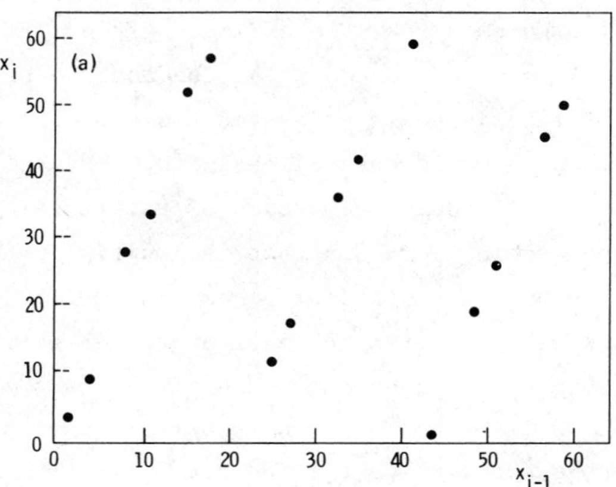

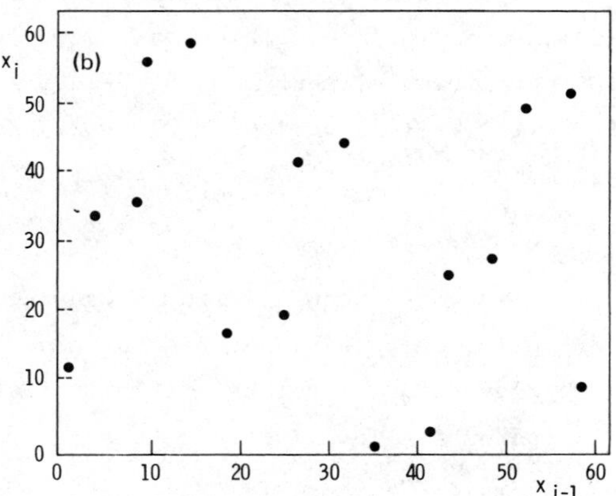

Figure 6.2 (a) Plot of successive pairs from $x_i = 3x_{i-1}$ (mod 64); (b) Plot of successive pairs from $x_i = 11x_{i-1}$ (mod 64)

In both cases the lack of uniformity is clearly visible. Figure 6.2a illustrates a more disturbing behavior as all pairs lie on essentially three lines and successive points tend to lie on the same line, while in Figure 6.2b there are essentially six lines and successive points tend to jump from one line to another. Pairs lie on lines determined by $x_i = \alpha x_{i-1}$, but in the case of the larger α, these lines more often "go over the edge" to reappear in a different position.

Extending these considerations to higher-dimension hypercubes reveals similar lattice structures for successive n-tuples.

The frequency of oscillation from one line to another is determined by the magnitude of the multiplier. Multipliers that may yield good pairwise properties may produce triplets with poor qualities. The IBM generator RANDU mentioned above gives an example of this fact. This particular generator has been discussed in detail by Blood (1976) and Coldwell (1974). The multiplier (65539) used by RANDU, which provides rapidly oscillating pairs, yields triplets with very poor patterns. This can be seen by considering the relationship between x_i, x_{i-1}, and x_{i-2}:

$$\begin{aligned} x_i &\equiv 65539^2 x_{i-2} \pmod{2^{31}} \\ &\equiv \left[(2^{16} + 3)^2 x_{i-2}\right] \pmod{2^{31}} \\ &\equiv (6x_{i-1} - 9x_{i-2}) \pmod{2^{31}} \end{aligned}$$

Successive triplets tend to remain on a plane given by

$$x_i = 6x_{i-1} - 9x_{i-2}$$

since the planes do not "go over the edge" rapidly.

This analysis indicates that a choice of multiplier based only on the rule $\alpha \approx m^{1/2}$ can cause problems. In Section 6.3 we will have occasion to discuss these ideas further. Niederreiter (1976, 1977) has discussed further considerations for the choice of the multiplier.

6.2.2 Feedback Shift Register Methods

Tausworthe (1965) suggested an interesting linear recurrence equation for generation of pseudorandom numbers. Some of the basic ideas for the method derive from feedback shift register (FSR) techniques developed in communications theory (Golomb 1967). FSR generators, like the methods discussed in the preceding section, use a modulo reduction to produce the pseudorandom sequence; but the FSR method takes successive positional values, usually bits, as residues congruent to a linear combination of the preceding positional values. An advantage of FSR generators over the congruential methods discussed in Section 6.2.1 is the higher-dimensional uniformity possible with the FSR methods.

The basic method of Tausworthe (1965) uses the sequence $\{a_k\}$ of 0's and 1's produced by the linear recurrence relation

$$a_k \equiv (c_p a_{k-p} + c_{p-1} a_{k-p+1} + \dots c_1 a_{k-1}) (\text{mod } 2) \tag{6.7}$$

for a fixed set of $c_i = 0$ or 1, with $c_p = 1$. Sequences of p-tuples $(a_{k-1}, a_{k-2}, \dots, a_{k-p})$ produced by (6.7) are equivalent to successive states of a shift register with feedback, that is, an ordered group of bits which is shifted left (or right) one position at a time with some combination of the bits being used to form the new rightmost (or leftmost) bit. (See Figure 6.3 for an example.)

The period of the sequence of p-tuples from (6.7) or of states of a shift register with p positions is the same as the period of the sequence $\{a_k\}$. Since the p-tuple (0, 0, ..., 0) is always followed by (0, 0, ..., 0), the maximum period is 2^{p-1}. Tausworthe (1965) showed that this maximal period is attained by the generator (6.7) if and only if the polynomial

$$f(x) = 1 + c_1 x + c_2 x^2 + \dots + c_p x^p$$

with $c_p = 1$ is irreducible over the Galois field consisting of {0,1} and the usual operations of addition and multiplication modulo 2. This field is generally denoted GF(2). Following the terminology of most authors who discuss the feedback shift register random

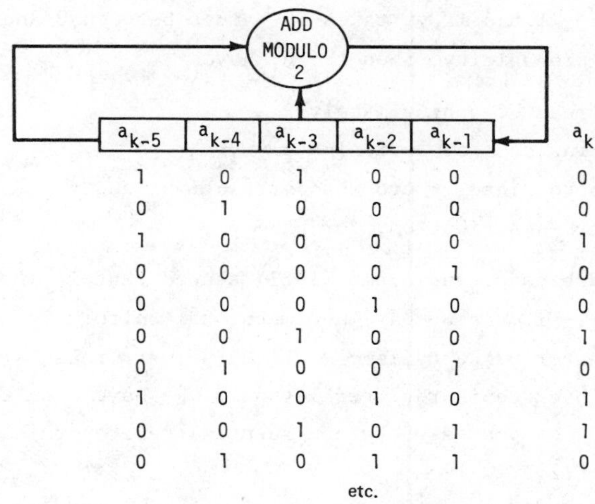

Figure 6.3 A feedback shift register and a generated Tausworthe sequence, with $p = 5$, $c_5 = c_3 = 1$ and $c_4 = c_2 = c_1 = 0$.

number generators, we use the term *primitive polynomial* to refer to an irreducible polynomial. Primitive polynomials in GF(2) up to degree 100 are given by Watson (1962), and primitive trinomials up to degree 1000 are given by Zierler and Brillhart (1968). (It should be noted that the term *primitive polynomial* is also used in a different sense, e.g., Knuth 1969).

The sequence of pseudorandom integers from the recurrence relation (6.7) are actually formed as L-bit binary integers by successive L-tuples of $\{a_k\}$, that is,

$$y_1 = a_1 a_2 \ldots a_L \qquad \text{(base 2)}$$

and $$y_i = a_{(i-1)L+1} a_{(i-1)L+2} \ldots a_{iL} \qquad \text{(base 2)}$$

This is referred to as an *L-wise decimation* of $\{a_k\}$. If L is relatively prime to $2^p - 1$, the decimation is *proper*. A generator implementing this method is referred to as a *Tausworthe generator*.

The integers y_i are then scaled into the interval [0,1) by taking $u_i = y_i/2^L$. Tausworthe (1965) showed that if L is relatively prime to $2^p - 1$ and if y_1 takes on values between 0 and $2^L - 1$ with equal probability, then the u_i have

1. a mean of approximately $\frac{1}{2}$
2. a variance of approximately $\frac{1}{12}$
3. approximately zero autocorrelation, and
4. an n-distribution, for $nL \leq p$

Toothill, Robinson, and Adams (1971) showed that if nL is relatively prime to $2^p - 1$, the u_i have n-dimensional uniformity.

The latter property is one of the primary reasons for interest in the feedback shift register methods. We have seen that this property is not possessed by congruential generators as discussed in Section 6.2.1.

Toothill, Robinson, and Adams (1971) gave an alternative description of the FSR generator based on a sequence of polynomials $\{u_k\}$, with coefficients in GF(2), generated by the recurrence relation

$$u_0 = 1$$
$$u_k = xu_{k-1} \pmod{x^p + x^q + 1} \tag{6.8}$$

For example, taking p = 5 and q = 2, we have

$$u_0 = 1$$
$$u_1 = x$$
$$u_2 = x^2$$
$$u_3 = x^3$$
$$u_4 = x^4$$
$$u_5 = x^5 = x^2 + 1$$
$$u_6 = x^3 + x$$

etc.

This alternative description depicts the polynomial generator as being vaguely similar to a multiplicative-congruential generator with multiplier x and modulus being a primitive polynomial.

If each p-tuple of coefficients in the polynomials $\{u_k\}$ is permuted by a circular shift of p - q to the left, the sequence of p-tuples $\{w_k\}$ formed in this manner is equivalent to the sequence of p-tuples formed from the Tausworthe sequence generated from (6.7) with $c_p = c_q = 1$ and $c_i = 0$ for $i \neq p$, $i \neq q$. (A circular shift is one in which the bits dropping off on one end are brought into the register from the other end.) In the example above, which uses Eq. (6.8), with p = 5 and q = 2, the sequence of coefficients are as shown in Fig. 6.4. Note that the p-tuples of Fig. 6.4 are the same as those of Fig. 6.3 beginning with the second row.

	x	1	x^4	x^3	x^2
w_0	0	1	0	0	0
w_1	1	0	0	0	0
w_2	0	0	0	0	1
w_3	0	0	0	1	0
w_4	0	0	1	0	0
w_5	0	1	0	0	1
w_6	1	0	0	1	0

etc.

Figure 6.4 Coefficients of polynomials generated by the relation (6.8) with p = 5 and q = 2.

The basic recurrence relation (6.8) gives

$$
\begin{aligned}
u_{n+p} &= x^n u_p \\
&= x^n(u_q + u_0) \pmod{x^p + x^q + 1} \\
&= (x^n u_q + x^n u_0) \pmod{x^p + x^q + 1} \\
&= (u_{n+q} + u_n) \pmod{x^p + x^q + 1}
\end{aligned}
$$

Hence, the elements of $\{w_k\}$ must satisfy the relation

$$w_{n+p} = w_{n+q} \oplus w_n \tag{6.9}$$

where $\oplus$ is the EXCLUSIVE-OR relation defined by

$$0 \oplus 0 = 0$$

$$0 \oplus 1 = 1 \oplus 0 = 1$$

$$1 \oplus 1 = 0$$

and $w_n \oplus w_m$ is understood to be an element-wise operation. Using the example in Fig. 6.4, we see

$$\begin{aligned} w_3 \oplus w_1 &= (00010) \oplus (10000) \\ &= 10010 \\ &= w_6 \end{aligned}$$

Use of (6.9) leads to an algorithm apparently first suggested by W. B. Kendall and described by Whittlesey (1968). In a slightly modified form the algorithm is as follows, wherein it is assumed that $q < p/2$ and n is an integer multiple of p:

1. Set w_n in register A in bit positions 1 through p (exclusive of sign bit).
2. Copy register A into register B and left shift register B q places, bringing 0's into the q rightmost places.
3. EXCLUSIVE-OR register A into register B and store the result back into register A.
4. Right shift register B $p - q$ places, bringing 0's in from the left.
5. EXCLUSIVE-OR B into A.

At the end of Step 5 register A contains the new pseudorandom integer. Often in implementation of this algorithm, the value of p is taken to be one less than the number of bits in a computer word. Thus, for the IBM 360/370 computers p is often chosen as 31. Figure 6.5 lists a FORTRAN subprogram, similar to one given by Payne (1970), which uses the algorithm above and then scales the random integer into the interval (0,1). This FORTRAN code assumes the use of the 360/370 G-level compiler. Other FORTRAN compilers may not perform the logical operations at the bit level and hence may not produce the object code needed to implement the procedure. The complexity of the coding in Fig. 6.5 is due to the fact that .OR. in FORTRAN is an INCLUSIVE-OR ($1 \oplus 1 = 1$); hence some logical manipulation is necessary to effect the EXCLUSIVE-OR. Note that the program uses the fact that the EXCLUSIVE-OR of values A and B is obtained through

$$(A \text{ a } \bar{B}) \oplus_I (B \text{ a } \bar{A})$$

where a is the AND relation defined by 0 a 1 = 0, 1 a 0 = 0, 1 a 1 = 1, 0 a 0 = 0, $\oplus_I$ denotes INCLUSIVE-OR, and $\bar{B}$ denotes the complement of B.

```
      FUNCTION TAUS(L)
      LOGICAL A,B,AC,BC
      EQUIVALENCE (I,A),(J,B),(AC,IC),(BC,JC)
      DATA K/2147483647/,N/262144/,M/8192/
      FK=K
C
C     I IS THE CURRENT INTEGER, M = 2**Q, N = 2**(P-Q)
C     WHERE  P = 31, K = 2**31 - 1.
C
      I = L
      J = I*M
      IC = K - I
      JC = K - J
      B = A.AND.BC.OR.AC.AND.B
      I = J
      J = J/N
      IC = K - I
      JC = K - J
      A = A.AND.BC.OR.AC.AND.B
      TAUS = I/FK
      L = I
      RETURN
      END
```

Figure 6.5 Implementation of a Tausworthe generator in IBM 360/370 FORTRAN G.

In order better to understand the details of this algorithm, consider the elements in the w's which we write as

$$w_n = a_n a_{n+1} a_{n+2} \cdots a_{n+p-1}$$

$$w_{n+q} = a_{n+q} a_{n+q+1} \cdots a_{n+p} \cdots a_{n+q+p-1}$$

$$w_{n+p} = a_{n+p} a_{n+p+1} \cdots a_{n+2p-1}$$

Using Eq. (6.9), we have that

$$a_{n+p} = (a_n \oplus a_{n+q})$$

$$a_{n+p+1} = (a_{n+1} \oplus a_{n+q+1})$$

$$\vdots \qquad \qquad \vdots$$

$$a_{n+p+q-1} = (a_{n+q-1} \oplus a_{n+2q-1})$$

Now we write w_{n+q} as

$$w_{n+q} = a_{n+q} a_{n+q+1} \cdots a_{n+p-1} (a_n \oplus a_{n+q}) \cdots (a_{n+q-1} \oplus a_{n+2q-1})$$

It follows that w_{n+p} is expressible in the form

$$w_{n+p} = (a_n \oplus a_{n+q}) \cdots (a_{n+p-q-1} \oplus a_{n+p-1})(a_n \oplus a_{n+q} \oplus a_{n+p-q})$$
$$\cdots (a_{n+q-1} \oplus a_{n+2q-1} \oplus a_{n+p-1})$$

This form of w_{n+p} is implemented by the algorithm.

The c_i's in the Tausworthe generator may be chosen so as to give a sequence with certain properties. An obviously desirable property is maximal period, which, as we have noted, can be achieved by selecting the c_i's as coefficients of primitive polynomials. For simplicity we would prefer polynomials containing few terms; and since the only primitive binomial over GF(2) is $x + 1$, the degree of which is too low to be of any interest (the maximal period for that polynomial would be 1), most studies of the Tausworthe generator have considered primitive trinomials, of which there are many. Tausworthe (1965) used $x^{35} + x^2 + 1$, Canavos (1967) used $x^{29} + x^2 + 1$, and Whittlesey (1968) used $x^{31} + x^3 + 1$, for example.

Fellen (1969) showed that sequences based on $x^{31} + x^3 + 1$ do not appear at all random. Toothill, Robinson, and Adams (1971) studied the runs up-and-down produced by several primitive trinomials of the form $x^p + x^q + 1$ for $p > q > 1$. In general, they concluded that if q is small, very close to p/2, or very close to p, the sequence is likely to have poor runs up-and-down properties. They also related the runs property to n-dimensional uniformity and showed that certain choices of p and q were likely to produce poor runs up-and-down of subsequences formed by every nth element of the entire sequence. Hence, they showed that some implementations of a Tausworthe-type generator may not produce sequences uniformly distributed in n-space. Uniformity in n-space will be achieved by a Tausworthe generator with L-wise decimation using the primitive trinomial $x^p + x^q + 1$ if $q = L$, $q < p/2$, $n = [p/q]$ ($[\cdot]$ denotes the greatest integer function), and nL is relatively prime to $2^p - 1$.

For applications in which the dimensionality is known in advance, such as an n-fold integration as discussed in Section 6.5, the rule of Toothill, Robinson, and Adams can be used to guide the selection of the primitive trinomial to be used. Toothill, Robinson, and Eagle (1973) considered the situation, such as may arise in inverse sampling, for example, in which the dimensionality is not known a priori. For this they defined two criteria for "asymptotic randomness," as follows.

Criterion 1: Maximal equidistribution. A Tausworthe s-bit number sequence of period $2^p - 1$ is maximally equidistributed if, for every k, $1 \le k \le p$, the sequence is k-distributed to the degree of resolution possible for t-bit numbers, where $t = \min(s,[p/k])$ and $[\cdot]$ denotes the greatest integer function.

Criterion 2: Maximal multidimensional equidistribution. A Tausworthe s-bit number sequence of period $2^p - 1$ is maximally multidimensionally equidistributed if, for every k, $1 \le k \le p$, and for every $n < 2^p - 1$, the sequence of every nth k-tuple has k-dimensional uniformity to the degree of resolution possible for t-bit numbers, where $t = \min(s,[p/k])$ and $[\cdot]$ denotes the greatest integer function.

In view of the condition for n-space uniformity noted above, Criterion 2 essentially imposes the condition that $2^p - 1$ be prime. [Such a prime is called a *Mersenne prime*, after Martin Mersenne (1588-1648) who studied primes of this type. It is interesting to note that for $c^n - 1$ to be a prime, it is necessary, but not sufficient, that $c = 2$ and n be a prime.] Toothill et al. (1973) described an algorithm for analyzing Tausworthe sequences with respect to the two criteria above. They showed a generator based on $x^{607} + x^{334} + 1$ and 512-wise decimation satisfies the two criteria. Their analysis, however, is in terms of the existence of purely empirical tests of a given sequence capable of distinguishing between the sequences and an ∞-distributed sequence tested in the same manner. In view of this interpretation of their analysis, they do not recommend that their techniques be used for evaluating sequences to be used in fixed-dimensional applications.

Lewis and Payne (1973) suggested a modification of the FSR generator which they called a *generalized feedback shift register* (GFSR) *generator*. The modification consists of introducing a suitably selected delay in the elements of the Tausworthe sequence in the successive columns. Figure 6.6 illustrates the delayed sequencing using a delay of 6. Since the bits in each column obey the recurrence relation

$$a_{n+p} = a_{n+q} \oplus a_n$$

(see Eq. 6.9), the p-tuples denoted as z's in the sequence generated by GFSR must obey this relation, that is,

$$z_{n+p} = z_{n+q} \oplus z_n$$

An arbitrarily chosen delay can be such, however, that the sequence $\{z_k\}$ is not of full period $2^p - 1$. Tausworthe's (1965) result guaranteeing full period applies if the first p of $\{z_k\}$ are linearly independent (assuming, of course, that $x^p + x^q + 1$ is a primitive trinomial modulo 2).

The number of different ways the first p of $\{z_k\}$ can be chosen to be linearly independent is the number of different pseudorandom sequences that can be generated by the GFSR using a given trinomial. For the trinomial $x^5 + x^2 + 1$, for example, there are

$$(2^5 - 1)(2^5 - 2)(2^5 - 4)(2^5 - 8)(2^5 - 16) = 9{,}999{,}360$$

different sequences possible, depending on the starting values. This fact could have important implications in generating permutations, for example, since the sample space of permutations of even a moderate number of elements may be extremely large. (See Salfi 1974 or Greenwood 1976a for further discussion of applications requiring very large periods.)

The algorithm for the GFSR, as described by Lewis and Payne (1973) is given below:

ALGORITHM GFSR

Step	*Description*
0.	If $k \neq 0$, go to 2 (k initially zero).
1.	Initialize $z_0, z_1, \ldots, z_{p-1}$ using a delayed basic sequence $\{a_i\}$ to obtain each column of $z_0, \ldots, z_{p-1}$.
2.	$k = k + 1$.
3.	If $k > p$, set $k = 1$.
4.	$j = k + q$.
5.	If $j > p$, set $j = j - p$.
6.	EXCLUSIVE-OR, $z_k \oplus z_j$.
7.	Store $z_k = z_k \oplus z_j$.

Note that a rotation table of p words is kept in this algorithm.

A FORTRAN implementation of the GFSR algorithm is given in Fig. 6.7. Initialization, and checking for required linear independence of starting vectors, are done in subroutine SETR which is shown in Fig. 6.8. Note that SETR used the starting p-tuple (1,1,...,1) and applied 5000p additional delays so that the leading (1,1,...,1) column is recurrenced also, thus giving a random pattern of leading bits rather than all 1's. In practice, SETR need be run only once and the table of constants entered in a DATA statement in RAND, or the table fed in initially and a new table saved at the end of each computer run, to be input in the next run.

$\underline{k}$	$\underline{z_k}$	$\underline{k}$	$\underline{z_k}$	$\underline{k}$	$\underline{z_k}$
0	11000	11	11001	21	01000
1	01011	12	00110	22	11101
2	01010	13	00010	23	11110
3	00111	14	11010	24	01001
4	01111	15	10001	25	10000
5	10010	16	11011	26	10110
6	01100	17	11100	27	10100
7	00101	18	10011	28	01110
8	10101	19	00001	29	11111
9	00011	20	01101	30	00100
10	10111				

Figure 6.6 GFSR sequence for polynomial $x^5 + x^2 + 1$ with delay of 6 between columns.

```
C...RAND...........................................................
C
C   USAGE:  X=RAND(M,P,Q,INTSIZ)
C
C   PARAMETERS:
C        M(P)=TABLE OF P PREVIOUS RANDOM INTEGERS
C        P,Q=POLYNOMIAL PARAMETERS X**P+X**Q+1
C        INTSIZ=NUMBER OF BITS IN HOST MACHINE INTEGER
C
C   RESULT:  THE VALUE OF X IS A PSEUDO RANDOM U(0,1) VARIATE.
C..................................................................
      FUNCTION RAND(M,P,Q,INTSIZ)
      LOGICAL AA,BB,LCOMPJ,LCOMPK
      INTEGER A,B,P,Q,INTSIZ,M(1)
      EQUIVALENCE(AA,A),(BB,B),(MCOMPJ,LCOMPJ),(MCOMPK,LCOMPK)
      DATA J/0/
      N=(2**(INTSIZ-1)-1)*2+1
      J=J+1
      IF(J.GT.P) J=1
      K=J+Q
      IF(K.GT.P) K=K-P
      MCOMPJ=N-M(J)
      MCOMPK=N-M(K)
      A=M(K)
      B=M(J)
      BB=LCOMPJ.AND.AA.OR.LCOMPK.AND.BB
      M(J)=B
      RAND=FLOAT(M(J))/FLOAT(N)
      RETURN
      END
```

Figure 6.7 GFSR algorithm in FORTRAN (from Lewis and Payne 1973).

```
C...SETR.........................................................
C
C   USAGE:  J=SETR(M,P,DELAY,Q,INTSIZ)
C
C   PARAMETERS:  M(P)=TABLE OF RANDOM NUMBERS TO BE FOUND
C                P,Q=POLYNOMIAL PARAMETERS AS IN RAND
C                DELAY=RELATIVE DELAY IN BITS
C                INTSIZ=BITS IN HOST MACHINE INTEGER WORD, E.G.
C                      IBM 360, 31; CDC 6000, 48; HP 2100, 15.
C
C   RESULT:  COLUMN NUMBER OF REPEATING ONE PATTERN.
C            IF J ≤ P, THEN AN IMPROPER DELAY HAS BEEN GIVEN
C
C................................................................
      FUNCTION SETR(M,P,DELAY,Q,INTSIZ)
      INTEGER DELAY,P,Q,ONE,INTSIZ,M(1)
      SETR=P+1
      ONE=2**(INTSIZ-1)
      DO 1 I=1,P
 1       M(I)=ONE
      DO 4 K=1, INTSIZ
         DO 2 J=1,DELAY
 2          X=RAND(M,P,Q,INTSIZ)
         KOUNT=0
         DO 3 I=1,P
            ITEMP=ONE/2**(K-1)
            ITEMP=(M(I)-M(I)/ONE*ONE)/ITEMP
            IF(ITEMP.EQ.1) KOUNT=KOUNT+1
               IF(K.EQ.INTSIZ) GO TO 3
            M(I)=M(I)/2+ONE
 3          CONTINUE
         IF(KOUNT.EQ.P) SETR=K
 4       CONTINUE
      DO 6 I=1,5000
         DO 5 J=1,P
 5          X=RAND(M,P,Q,INTSIZ)
 6       CONTINUE
      RETURN
      END
```

Figure 6.8 Initialization of GFSR algorithm (from Lewis and Payne 1973).

The parallel nature of columns in this algorithm means that it can automatically be used on L-bit machines regardless of the relationship between L and p. Whenever $L < p$, many repeated numbers will occur but the period will nevertheless be $2^p - 1$. Thus very long sequences can be generated on L-bit machines merely by selecting p large. Some examples of primitive trinomials with large p are

p	q	p	q	p	q
47	5	111	10	532	37
47	14	111	49		
47	20	124	37		
47	21	170	23		
95	11	250	103		
95	17	380	47		
98	11	476	15		
98	27	476	141		

The advantages of the GFSR are (1) speed in generation (after the SETR initialization), (2) generality, in that it is independent of computer word size, (3) good n-space uniformity, and (4) an extremely long period is readily achievable.

Arvillias and Maritsas (1978) described a procedure similar to the GFSR based on primitive trinomials with $q = 2^t$. For such trinomials their method could be implemented so as not to require the initialization procedure.

6.2.3 Coupled Generators

In order to extend the period of a random number generator or otherwise to improve the "randomness" of the generated sequences, a number of authors have suggested various combinations of two or more generators. The first useful method was suggested by MacLaren and Marsaglia (1965). The method assumes the existence of two different generators, yielding sequences $\{X_i\}$ and $\{Y_i\}$. An auxiliary table with entries T_j are taken as the first k values of $\{X_i\}$. Then, at each generation, (1) X and Y are obtained from their respective generators; (2) Y is used to select randomly an integer j, $1 \le j \le k$; and (3) the value T_j is taken as the new desired random number and the current value of X is set in T_j. The value of k is frequently taken as 128. Gebhardt (1967) has shown that this technique may yield sequences with much better randomness properties then either of the constituent sequences. Also, very long periods can be obtained by coupling generators with periods that are relatively prime. Greenwood (1976a) suggested a way to do this when mixed congruential generators are used. The moduli can be taken the same, conveniently 2^p; but whenever Y is generated as 0, Y is set to the next term in the Y-sequence, that is, the additive constant in the generator. If the generators meet the requirements of Theorem 6.1 the period of Greenwood's coupled generator would be $2^p(2^p - 1)$.

Nance and Overstreet (1978) reported on some empirical tests of the randomness properties of various coupled generators of the MacLaren-Marsaglia type. They found that a table size of two (i.e., k = 2) gave results comparable to coupled generators with larger tables and had the obvious advantage of less overhead.

Westlake (1967) gave a modification of the ideas of MacLaren and Marsaglia that did not require the storage of the table. In Westlake's method, first, p bits from the generated value of X are used to determine a left circular shift of the bits of Y. The shift will be between 0 and $2^p - 1$ bits. Then, the sign bit of the shifted Y is set to zero and the new desired random number is taken as the value obtained by an EXCLUSIVE-OR of X and the shifted Y. Westlake suggested p be taken as 6 and that the p bits from X used to determine the shift in Y be in the upper middle range of significance. For example, on the IBM 360/370, with the bits numbered 1 through 31 from most significant to least significant (the sign is in bit 0), bits 6 through 11 from X might be used to determine the amount of shift of Y.

Salfi (1974) described a method for combining the output of n generators so as to achieve an extremely long period. The method, for full-period mixed-congruential generators each with modulus 2^p and with different multipliers, is as follows. Set $a_1 = 0$, $a_2 = 1$, and determine a_i, i = 3, 4, ..., n, as the smallest integer for which

$$2^q - a_1,\ 2^q - a_2,\ \ldots,\ 2^q - a_n$$

are relatively prime, where q is some fixed value such that $1 \leq q < p$. The random numbers are then generated by the following scheme:

1. Choose i randomly between 1 and n.
2. Operate the ith generator, representing the output as $2^q b_i + d_i$ where $0 \le b_i < 2^{p-q}$ and $0 \le d_i < 2^q$.
3. If $d_i \ge a_i$, deliver b_i, else go to 1.

The period of this generator is

$$n \prod_{i=1}^{n} (2^q - a_i)$$

Note that the value delivered at each stage has only p - q bits. This method could be used effectively to replace the high-order p - q bits of the output of some other, simple generator to obtain the long period as well as a wide range of numbers.

Bays and Durham (1976) described a technique similar to the MacLaren-Marsaglia method except that only one basic generator is used. As before, initialize T_j, j = 1, 2, ...k, as the first k values output by the given generator. Then operate the generator once more to yield X. Subsequent generations are achieved by looping through the following steps:

1. Set j = [kX] + 1. ([·] denotes the greatest integer function.)
2. Deliver $X = T_j$.
3. Operate the given generator to yield Y. Set $T_j = Y$.

The properties of this generator are difficult to analyze, but the important thing is that it results in a longer-period sequence, and empirical studies (Bays and Durham 1976) indicate that the randomness properties of this method are at least as good as those of the basic generator used.

Marsaglia has developed a random number generator called *Super-Duper* which couples a multiplicative-congruential generator with a FSR generator. Some details of this generator, along with results of testing, of it and others, are given by Learmonth and Lewis (1973).

Lewis (1975, p. 21) briefly described a "masked" FSR generator to increase the period of a basic FSR generator, but the properties of the generator have not been analyzed.

Hurd (1974) described a generator consisting of several interconnected shift registers. His empirical studies of the performance of the generator indicated excellent randomness properties of the sequences produced.

The importance of techniques extending the period of random number generators as discussed above becomes greater as more and more minicomputers with smaller word lengths are being used.

Aside from considerations of period length, coupling generators or shuffling the output sequence can improve the randomness properties. Gebhardt (1967) showed that even a poor generator, the so-called Fibonacci generator $x_{i+1} \equiv (x_i + x_{i-1})(\text{mod } m)$, could be made generally acceptable by using it to shuffle its own output sequence.

Peskun (1978) considered the properties of a mixed-congruential generator in which the additive constant was randomly generated by a separate generator.

6.2.4 Portable Generators

In the preceding subsections most algorithms discussed made some essential use of the hardware and/or software characteristics, such as word size or how overflows are handled. While use of system peculiarities may increase computational efficiency, it militates against portability, that is, the ability of the software to run on a different system and yield the same answers.

Interest in portability of random number generators is growing with increasing use being made of results of sampling studies with artifically generated random numbers. (Hoaglin and Andrews 1975 report that during 1973 roughly 20 percent of the articles in two leading statistical journals contained simulation results; in 1978 almost 30 percent of articles in one of these journals (JASA) reported Monte Carlo studies.) As in any scientific endeavor, reproducibility of results is desirable; and while portable software is not required to allow this, it does make it more conveniently achieved.

Kruskal (1969), Greenwood (1977), and Schrage (1979) discuss some aspects of portability in random number generation and methods for achieving it without overdegradation of efficiency and while still maintaining desirable "randomness" aspects of the output. We will not pursue this matter other than to simply point out, as we have done, that portability is desirable and achievable but not without some cost to the user.

6.3 TESTS OF RANDOM NUMBER GENERATORS

As we have observed in the introductory section to this chapter, the process of generating random numbers on the computer involves somewhat contradictory activities and objectives. Since the basic processes are deterministic, it does not make sense to consider whether the numbers are really random. Depending on the use to which the numbers are to be put, however, it may be important that the numbers generated possess certain properties we may expect to be exhibited by a sequence of realizations from a true random process. Indeed, output from random number generators is sometimes used to provide confidence intervals around an estimate of some parameter. In doing this the practitioner is relying on his belief that the sequence generated by the deterministic process has certain variance characteristics of a random sequence.

There have been many methods suggested for evaluating the performance of random number generators and the "randomness" of sequences produced by them. Knuth (1969) gives an historical survey as well as discussions of various methods for testing random number generators. The methods are basically of two types, theoretical tests and empirical tests. We will briefly consider some examples of each type.

6.3.1 Theoretical Tests

Theoretical tests are tests that can be applied to the generator itself without obtaining samples of its output. A simple example would be a method for determining whether the period of the generator is sufficiently large. The results of Section 6.2 generally allow us to ascertain this quite readily. More complicated tests may evaluate the relationships among the terms of a sequence that would be generated by the random number generator. Such a test is discussed below.

Spectral Test

Coveyou and MacPherson (1967) devised a test based on the finite Fourier transform of the probability function of n-tuples generated. The details of the test depend on whether or not the generator is of full period. In the brief description below, which follows closely the discussion by Knuth (1969), we assume the generator is a mixed-congruential generator with modulus m, multiplier α, and additive constant c satisfying the hypotheses of Theorem 6.1. Knuth indicates the modifications necessary for a modulus of the form 2^e.

The basic idea of the test is to compare the finite Fourier transform of the probability function $f(X_i, X_{i+1}, \ldots, X_{i+n-1})$ of n-tuples generated to the probability function for random n-tuples the elements of which have identical independent discrete uniform distributions over the integers 0, 1, ..., m - 1. The probability function for such n-tuple would be the constant m^{-n}. The Fourier transform of f is, by definition,

$$T(s_1, s_2, \ldots, s_n) = \sum_{x_1=0}^{m-1} \sum_{x_2=0}^{m-1} \cdots \sum_{x_n=0}^{m-1} \exp\left\{-\frac{2\pi i}{m}(s_1x_1 + \cdots + s_nx_n)\right\} f(x_1, \ldots, x_n) \tag{6.10}$$

For the ideal random sequence, this is the transform of a constant, so T = 1 at the point (0, 0, ..., 0) and T = 0 elsewhere. For n-tuples from the mixed congruential generator, the probability function is

$$\begin{aligned} f(x_1, x_2, \ldots, x_n) &= \frac{1}{m} \qquad \text{if } x_{1+j} = \left(\alpha^j x_1 + \frac{\alpha^j - 1}{\alpha - 1}c\right) \pmod{m} \\ &\qquad\qquad \text{for } j = 0, 1, 2, \ldots, n - 1 \\ &= 0 \qquad \text{otherwise} \end{aligned}$$

Hence, the Fourier transform is (using k = j + 1)

$$T(s_1, s_2, \ldots, s_n) = \frac{1}{m} \sum_{x=0}^{m-1} \exp\left\{\frac{-2\pi i}{m} \sum_{j=0}^{n-1} s_k\left(\alpha^j x + \frac{\alpha^j - 1}{\alpha - 1}c\right) (\text{mod } m)\right\}$$

which is zero for all values of s_j, except when

$$\sum_{j=0}^{n-1} s_k \alpha^j \equiv 0 \ (\text{mod } m) \qquad (6.11)$$

and has a magnitude of 1 in this case. The nonzero values are of interest because they indicate the departures from random uniformity. The arguments of T yielding nonzero values correspond to the frequency of waves in the complex plane, and the quantity

$$\nu = \left(s_1^{\ 2} + s_2^{\ 2} + \ldots + s_n^{\ 2}\right)^{1/2} \qquad (6.12)$$

is referred to as the wave number. The spectral test is based on ν_n, the minimum ν in (6.12) for s_j not all zero such that (6.11) is satisfied.

Knuth (1969) proposed use of the quantity

$$C_n = \frac{\pi^{n/2} \nu_n^{\ n}}{(n/2)!m}$$

Larger values of C_n generally indicated more randomness. Knuth suggested that C_2, C_3, and C_4 all be ≥ 0.1 for minimal acceptability of the generator and that ideally each of the quantities be greater than 1.

The quantity ν_n is not simple to calculate. Knuth (1969) described an algorithm for obtaining this value, and Golder (1976) gave a program implementing Knuth's method. It should be noted, however, that Golder's program is applicable only to mixed-congruential generators of full period. Although Marsaglia (1972) has questioned the worth of the spectral test, it has proven useful in identifying many poor generators; in fact, Knuth (1969, p. 82) stated that all good generators pass it and "all linear congruential sequences now known to be bad actually fail it."

Lanczos and Gellai (1975) described the use of Fourier analysis for assessing the goodness of pseudorandom sequences of numbers supposedly simulating a normal process. (See Section 6.5.1.)

6.3.2 Empirical Tests

Empirical tests are those that are applied to samples of generated output. These are basically goodness-of-fit tests. Some of these tests are only for static properties of the sample, but others detect departures from the sequential nature to be expected in identically and independently distributed random variables. The familiar *chi-square test* and *Kolmogorov-Smirnov test* can be used on samples produced by the generator being tested to detect static departures from the hypothesized distribution. Various modifications of the chi-square test may be employed. The "digits" referred to by many of the tests described here may be in any base and can be any digitization of the basic sequence produced by the random number generator.

The *digit frequency test* compares the number of occurrences of each digit in a sequence with the expected number for that digit by means of a chi-square test. This test can also be used for testing the frequency of each pair of digits, or each triplet, and so on, thus attempting to test for serial correlations. In this case the test is usually referred to as the *serial test*. The groups of digits determine points in a k-dimensional hypercube. The groups may be allowed to overlap; however, when they do, the correction to the chi-square statistic given by Good (1957) must be used.

The *poker test* considers various patterns analogous to poker hands in groups of five consecutive digits in a sample. The number of groups in which all digits are distinct, the number containing exactly one pair, two pairs, three of a kind, a full house (three of one kind and two of another), four of a kind, and five of a kind are compared by a chi-square test with the expected number of each type in a sample of a given size. The poker test is a partition test, which was discussed by Butcher (1961a).

The *gap test* is a chi-square test described by Bofinger and Bofinger (1961a) comparing the number of intervening digits between successive occurrences of a given digit with the expected number. For example, considering the digit 5 in the sequence

5,3,2,4,9,5,8,5,5,7,6,...

The "gaps" are of length, 4, 1, 0, and so on. The distribution of the length of the gaps is geometric.

The *coupon collector's test* involves counting the lengths of the sequences of digits required to obtain a complete set of all digits and comparing these counts with the expected counts using the distribution worked out by Greenwood (1955), which for digits in base 10 is given by the probability function

$$P(n) = \frac{1}{10^{n-1}}\left[\sum_{j=0}^{8} (-1)^j \binom{9}{j} (9 - j)^{n-1}\right]$$

The d^2 *test* described by Gruenberger and Mark (1951) uses four consecutive numbers in the sample to form two pairs of coordinates in the two-dimensional plane. The distance between such pairs of points are then compared by means of a chi-square test with their expected values.

There are obviously any number of characteristics of samples that could be tested via chi-square for conformity to the expected characteristics. The ones mentioned so far have been useful for detecting at least some types of departures from "randomness."

The *serial correlation test* uses a form of the serial correlation coefficient of lag j. The quantity generally used is

$$r_j = \frac{\frac{1}{n}\sum_{i=1}^{n} (x_i - \bar{x})(x_k - \bar{x})}{\frac{1}{n}\sum_{i=1}^{n} (x_i - \bar{x})^2}$$

where $k = (i + j)(\text{mod } n)$, and $1 \leq k \leq n$. For a sequence of identically independently distributed variables, r_j has an expected value of 0. For performing tests, it is easier to use the first term in the expanded numerator,

$$c_j = \frac{1}{n} \sum_{i=1}^{n} x_i x_k$$

If the x's are from U(0,1) and are independent, $E(c_j) = 1/4$ for $j > 0$ and $E(c_0) = 1/3$, and the variance of c_j for any j is $1/[12(n - 1)]$. The central limit theorem can be used to yield

$$z = \frac{c_j - 0.25}{\sqrt{12(n - 1)}}$$

as a test statistic with an approximate N(0,1) distribution under the null hypothesis.

One important departure from randomness is a propensity for the occurrence of long monotonic subsequences. *Runs tests* are designed to detect nonrandom behavior of this type. A *run* is a monotonic subsequence. For assessing randomness "runs up" (i.e., increasing subsequences), "runs down," or "runs up and down" may be considered; and in any case, either just the total number of runs or the numbers of runs of each given length may be used. The simplest test statistic is the number of runs up and down, say T. For the sequence

1,5,4,1,3,1,3,4,7

the numbers of runs, T, is 5, as indicated. The use of tests based on this statistic was considered by Wallis and Moore (1941). Edgington (1961) gave a table of the null distribution for various sample sizes. The statistic is asymptotically normally distributed with mean $(2N - 1)/3$ and variance $(16N - 29)/90$, for sample size N. Gibbons (1976) suggested use of the approximate normality of the distribution of T (with continuity correction) for samples of size 25 or greater.

The numbers of runs of specific lengths can be used as a more powerful test of randomness based on the results of Levene and

Wolfowitz (1944). The *length* of a run is generally taken as one less than the number of terms in the monotonic subsequence. For example, in the sequence above the five runs are of length 1, 2, 1, 1, and 3. This definition of length is in accordance with that of Levene and Wolfowitz; other authors take the length to be equal to the number of terms in the subsequence (i.e., one more than the length as defined above). The latter definition is usually used when only runs up or runs down are being considered. For example, in the sequence above, the runs up may be indicated as

$$\underline{1,5},\underline{4},\underline{1,3},\underline{1,3,4,7}$$

that is, four runs of lengths 2, 1, 2, and 4.

Levene and Wolfowitz (1944) showed that in a sequence of length N the expected number of runs of length k, R_k, is

$$E(R_k) = \frac{(2k^2 + 6k + 2)N}{(k + 3)!} - \frac{2k^3 + 6k^2 - 2k - 8}{(k + 3)!} \qquad (6.13a)$$

A chi-square test cannot be based directly on this expectation, however, since the numbers of runs of various lengths are negatively correlated. Levene and Wolfowitz (1944) derived the variance-covariance matrix, C, of R_k, which like (6.13a) can be expressed as the sum of two parts, one involving N, that is, $C = NC_1 + C_2$.

The numbers of runs of lengths 6 or greater may conveniently be lumped into a single count. The expected value of the number of runs of length k or greater, R'_k, is

$$E(R'_k) = \frac{2}{(k + 2)!}\,[N(k + 1) - (k^2 + k - 1)] \qquad (6.13b)$$

The variance-covariance matrix C' for $R_1, \ldots, R_5$ and R'_6 also was derived by Levene and Wolfowitz.

A test statistic based on the results of Levene and Wolfowitz is

$$T = \frac{1}{N} \sum_{i=1}^{6} \sum_{j=1}^{6} (N_i - E_i)(N_j - E_j)a_{ij} \qquad (6.14)$$

where, for i = 1, 2, ..., 5, N_i is the number of runs of length i and $E_i = E(R_i)$; N_6 is the number of runs of length 6 or greater, and $E_6 = E(R_6')$; and a_{ij} is the (i,j)th element of the inverse of the variance-covariance matrix C'. Under the hypothesis of independence of the terms in the sequence, T has an approximate chi-square distribution with 6 d.f. for large N. Also for large N, the terms in $E(R_i)$, $E(R_i')$, and C' involving N dominate, so a good approximation may be obtained by using only the first terms in Eqs. (6.13a) and (6.13b), as well as the part of the variance-covariance matrix C' which involves N. The inverse of C_1', where $C' = NC_1' + C_2'$, is given in Table 6.1. These values were obtained from the formulas given for C_1' by Levene and Wolfowitz (1944).

Table 6.1 Values of a_{ij} for Use in Eq. (6.14)

18265.1	36527.2	54791.6	73056.0	91320.3	112193.9
	73061.2	109584.4	146113.1	182641.8	224388.8
		164396.3	219170.3	273963.3	336583.9
			292314.5	365285.0	448779.1
				457098.4	560974.4
					692077.5

Knuth (1969), who described a runs up test, recommended runs tests as very useful. Downham and Roberts (1967) and Prentice and Miller (1968), who used a runs test that did not take into account the covariances, found the version they used to be quite sensitive to departures from randomness. Learmonth and Lewis (1973), also using a version of the runs test ignoring the covariances, however, concluded that the test showed little power to detect poor random number generators. Downham (1969,1970) described the test ignoring the covariances and gave a program for calculating the test statistic. Levene and Wolfowitz (1944), Knuth (1969), and Wedderburn (1976) warned against use of the chi-square statistic which does not take account of the covariances. We recommend the use of the test statistic in Eq. (6.14) and Table 6.1.

An alternate runs test was proposed by Bofinger and Bofinger (1961b). The "runs" considered in their test were subsequences of a single value. A test for randomness of runs of this sort may be useful in situations in which the sequence itself contains only a small number of different values, as, for example, a sequence of single digits.

Ishiketa (1966, 1969) suggested an empirical test of random number generators based on one of the frequent applications of random numbers, a discrete two-dimensional random walk. The test is a chi-square test for the difference between the expected time for the walk to leave a given region and the observed time required. Rudolph and Hawkins (1976) extended this idea to assess the performance of a random number generator in simulating a more general Markov process.

Marsaglia (1972) described a *lattice test* for congruential generators, which, like the spectral test, assesses the uniformity with which an n-dimensional hypercube is filled by successive n-tuples generated. The derivation of the test will not be given here.

The lattice test is performed in the following steps:

1. Generate n + 1 n-tuples successively. Call these n-tuples $p_1, p_2, \ldots, p_{n+1}$.
2. Set D = the absolute value of the determinant of

$$\begin{matrix} p_2 - p_1 \\ p_3 - p_2 \\ \vdots \\ p_{n+1} - p_n \end{matrix}$$

   Unless this value is zero, in which case go to 1.
3. Repeat steps 1 and 2 as desired to form a sequence of D's, $D_1, D_2, \ldots$
4. Let $g_2 = \gcd(D_1, D_2)$

$$g_3 = \gcd(D_1, D_2, D_3)$$
$$g_4 = \gcd(D_1, D_2, D_3, D_4)$$
$$\vdots$$

5. If some $g_i = 1$, stop and declare the generator to have passed the test. If the sequence of g's becomes constant ($\neq 1$) for a number of successive iterations, stop and declare the generator to have failed the test.

Marsaglia recommended use of the test for n = 1,2,3,4,5.

Several other methods for assessing the goodness of generated sequences of pseudorandom numbers have been discussed. Examples may be found in Knuth (1969), Niederreiter (1972a), and Lanczos and Gellai (1975).

6.3.3 Selecting a Random Number Generator

The various tests described in the preceding subsections comprise a somewhat bewildering arsenal of techniques for evaluating random number generators. It seems likely that, given any generator, one or more tests could be formed that the generator would fail. Results of test batteries applied to some widely used generators are reported by Craddock and Farmer (1971), Marsaglia (1972), Learmonth and Lewis (1973), Rudolph and Hawkins (1976), and others. Gorenstein (1967) reported on extensive testing of the random number generator used in the GPSS simulation language system. McArdle (1976) conducted tests on multivariate generators. One general pattern that seems to emerge is that coupled generators, as described in Section 6.2.3, may be more likely to perform better than single generators whose output undergo no further randomization.

In selecting a random number generator, some thought may be given to the intended applications. For example, in some cases high autocorrelations may be of little consequence; in other cases (such as evaluation of multidimensional integrals), however, the autocorrelations may render the results useless. The statistician can frequently devise a chi-square test to assess some specific departure from randomness that would be of particular concern in a given application.

The type of computer on which the generator is to be implemented may affect the choice. The word length is generally the major consideration. The congruential methods may not perform well at all on computers with words of less than 20 bits in length. Kral (1972) has given an algorithm for random number generation that seems to work well on machines with short word lengths. Overstreet and Nance (1973) have tested Kral's generator and others and have concluded that Kral's generator produces sequences with good randomness properties. The theoretical properties of the generator, however, are not well understood.

Speed in generation and ease in programming are sometimes considerations for choice of a random number generator. We would suggest, however, that such considerations guide the choice only for preliminary work and that quality of the generator as determined by the various theoretical and empirical tests be the primary consideration.

6.4 GENERAL TECHNIQUES FOR GENERATION OF NONUNIFORM RANDOM DEVIATES

Frequently it is desirable to simulate random sampling from some distribution other than the uniform. This is most conveniently done by generating pseudorandom uniform (0,1) variates, as described in Section 6.2, and then performing some appropriate transformation to yield pseudorandom deviates from the distribution of interest. In this section we discuss some general techniques for doing this, and then in Section 6.5 we discuss efficient techniques designed for specific distributions.

6.4.1 Use of the Cumulative Distribution Function

If X is a continuous random variable with cumulative distribution function (cdf) F(x), it is easy to show that the random variable Y given by Y = F(X) has a uniform (0,1) distribution. This simple fact provides us with a general technique for generation of continuous random deviates from a distribution whose cdf can be inverted. The technique is generate Y from U(0,1) and take $X = F^{-1}(Y)$.

Example 6.3

It is desired to simulate sampling from the exponential distribution with probability density function (pdf)

$$f(x) = \frac{1}{\theta}e^{-x/\theta} \qquad x \geq 0; \quad \theta > 0$$

$$= 0 \qquad \text{otherwise}$$

Letting

$$Y = \int_0^X \frac{1}{\theta} e^{-t/\theta} dt$$

$$= 1 - e^{-X/\theta}$$

we have $Y \sim U(0,1)$ and

$$X = -\theta \ln (1 - Y)$$

or equivalently (in the sense of equivalent random variables),

$$X = -\theta \ln (Y)$$

There are, as we will see later, more efficient methods for generating variates from an exponential family, but this example serves to illustrate the general technique.

This technique can also be used to generate deviates from discrete distributions using U(0,1) deviates. Suppose the random variable X takes with nonzero probability only the values $x_1 < x_2 < x_3 \ldots$, and F(x) is the cdf of X. This distribution can be sampled by taking $X = x_i$, where

$$F(x_{i-1}) < Y < F(x_i)$$

with the arbitrary value $x_0 < x_1$, and $Y \sim U(0,1)$.

Example 6.4

It is desired to sample from the discrete uniform distribution defined by

$$\text{Prob}(X = k) = \frac{1}{n} \qquad k = 1, 2, 3, \ldots, n$$

This can be done by generating Y from U(0,1) and taking X as [nY] + 1. ([.] denotes greatest integer function.)

For general discrete distributions this technique may require a search, and the procedure will not be as simple as that in Example 6.4. Chen and Asau (1974) suggested forming k groups of values based on the first digit of the values of the cdf, and then searching first for the appropriate group and second for the particular value within the group. There are usually more efficient ways for sampling from discrete distributions, however.

The cdf method for generating variates from specified distributions is frequently implemented by means of a table. The cdf is tabulated to some suitably chosen precision and stored in primary memory. The uniform deviates generated are used to access values from the table. Depending on the accuracy desired, the value taken may be interpolated from the table, linearly or using higher-order splines. Shepherd and Hynes (1976) and Shepherd and Starner (1977), for example, described a table look-up method for normal deviates using quadratic interpolating splines. Mitchell and Stone (1977) discussed some general techniques.

The efficiency of a table look-up method becomes greater relative to other techniques as the number of variates to be generated increases, since there is a relatively constant overhead associated with construction of the table. The amount of primary memory available and its cost are other important considerations in the use of a table look-up procedure.

The use of the inverse cdf for generating variates is a special case of transformations of random variables to yield a variable with a desired distribution. If it is desired to generate X, and Y which is given by Y = g(X) can be generated, then it may be possible to obtain X by inverting the transformation. The use of transformations with multiple roots is discussed by Michael, Schucany, and Haas (1976). There is no problem for discrete random variables. Michael et al. give multinomial probabilities for selection of one of the multiple roots for generating from either a discrete or continuous distribution.

6.4.2 Use of Mixtures of Distributions

Marsaglia (1961a) suggested decomposing the cdf of the distribution to be sampled into a linear combination of other cdf's which may be easier to use in generating the random deviates. The technique has provided the basis for the development of several extremely efficient generators.

Suppose it is desired to sample from a distribution with cdf F(x) and that F(x) can be written as

$$F(x) = p_1F_1(x) + p_2F_2(x) + \dots + p_kF_k(x) \tag{6.15}$$

where $p_i > 0$, $\Sigma\, p_i = 1$, and each F_i is a cdf. Variates from F can be generated by generating from F_i with probability p_i. Efficiency is gained by finding a decomposition (6.15) such that p_i is large for F_i which corresponds to a distribution which is easy to simulate, such as a uniform distribution.

Example 6.5

Suppose it is desired to sample from the triangular distribution whose pdf is

$$\begin{aligned} f(x) &= 2 - 2x \qquad & 0 \le x \le 1 \\ &= 0 & \text{otherwise} \end{aligned}$$

The decomposition of a cdf in (6.15) is equivalent to a decomposition of the pdf. In the present case we can decompose f(x) into

$$f(x) = \frac{1}{2}f_1(x) + \frac{1}{4}f_2(x) + \frac{1}{4}f_3(x)$$

where

$$\begin{aligned} f_1(x) &= 2 \qquad & 0 \le x \le \tfrac{1}{2} \\ &= 0 & \text{otherwise} \\ f_2(x) &= 4 - 8x & 0 \le x \le \tfrac{1}{2} \\ &= 0 & \text{otherwise} \end{aligned}$$

and

$$\begin{aligned} f_3(x) &= 8 - 8x \qquad & \tfrac{1}{2} < x \le 1 \\ &= 0 & \text{otherwise} \end{aligned}$$

as illustrated in Fig. 6.9.

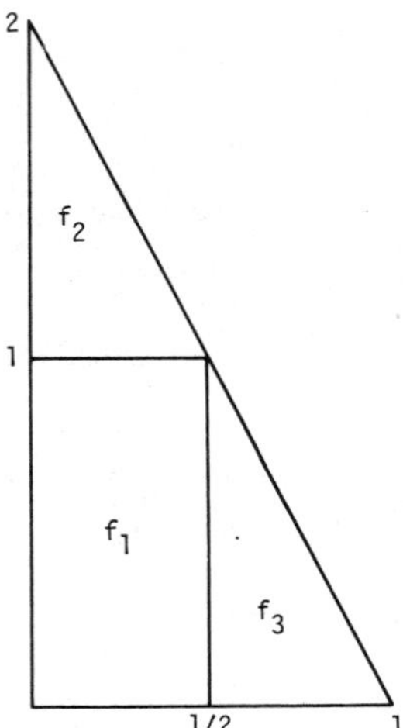

Figure 6.9 Decomposition of a triangular distribution.

Had we used the inverse cdf technique of Section 6.4.1 on the original distribution, we would have calculated a square root for each variate generated. The decomposition method illustrated here involved the generation of a uniform (0,1), u_1, to select a distribution and then the generation of a second uniform (0,1), u_2, to transform into a variate with pdf f_1, f_2, or f_3. On average, one-half of the time u_2 will merely be transformed into U(0,1/2), which does not require a square root. (See Exercise 6.10.)

This example does not illustrate any significant gain in efficiency; however, the technique should be clear from the example.

The efficiency of use of mixtures can be analyzed in the following way. Suppose it is desired to generate variates from distribution D and that an available direct method (such as inverting the cdf) takes t units of time (or t operations) for each variate produced. Now suppose the distribution D can be decomposed into other distributions D_1, D_2, ..., D_k by writing the cdf of D as in (6.15) in terms of the cdf's of the D_i. If it requires t_i units of time to produce a variate from distribution D_i, the time required for producing a variate from D using the mixture is approximately

$$t_M = \sum_{i=1}^{k} p_i t_i$$

There are also some overhead costs associated with use of the method of mixtures, such as the selection of a distribution D_i and the extra programming required. If $t_M \simeq t$, the direct method should probably be used. In many applications, however, t_M will be considerably smaller than t.

Example 6.6 describes an algorithm due to Butler (1970) which utilizes a combination of the inverse cdf technique and the method of mixtures as well as an alternate method of sampling from a triangular distribution (see also Exercise 6.4) or, more precisely, a trapezoidal distribution (such as $c[f_1(x) + f_2(x)]$ in Example 6.5).

Example 6.6 Butler's (1970) Method

Suppose it is desired to sample from the distribution with cdf F(x) and pdf f(x). A grid of k (where k may be of the order of 256 or so, depending on the smoothness of F) ordinate values of F(x) is formed and values x_j are obtained so that $F(x_{j+1}) - F(x_j) = 1/k$. [For a distribution of infinite length support, x_1 and x_k may be taken as finite values such that $F(x_1) \simeq 0$ and $F(x_k) \simeq 1$.] Next r_j, given by

$$r_j = \frac{f(x_j) - f(x_{j-1})}{f(x_j) + f(x_{j-1})} \qquad \text{for } j = 2, 3, \ldots, k$$

are calculated. If we envision the region under the graph of f(x) between x_j and x_{j-1} as being approximated by a trapezoid (consisting of a rectangular region as a base and a right triangle placed on top of the rectangle), then $|r_j|$ is the ratio of the triangular area to the total area of the trapezoid. Furthermore, the sign of r_j indicates whether the triangle slopes to the right or to the left. This completes the setup portion of the method. To generate variates from f(x), first an integer i is chosen between 2 and k with equal probability (as, e.g., in Example 6.4). Three uniform (0,1) variates u_1, u_2, and u_3 are then generated; and u_3 is used to choose

(1) $x = x_{i-1} + (x_i - x_{i-1})u_i$

(i.e., sample from the rectangle) with probability $1 - |r_i|$, or

$$(2) \quad x = x_{i-1} + (x_i - x_{i-1}) \cdot \begin{cases} \min(u_1,u_2) & \text{if } r_i < 0 \\ \max(u_1,u_2) & \text{if } r_i > 0 \end{cases}$$

(i.e., sample from the triangle) with probability $|r_i|$. The variate X delivered by this method will have a distribution approximately the same as that desired. (For the max used in this example, see Exercise 6.10). As Butler (1970) initially described the method, it was as intended for distributions whose pdf is known only in tabular form $[x_t, f(x_t)]$ for t = 1, 2, 3, ..., n. The percentage points (the x_j's above) were then obtained by approximation of the cdf at the points in the table followed by interpolation. Gates (1978) described a method for generating normal deviates based on Butler's method.

Ahrens and Dieter (1972) described an algorithm based on the method of mixtures for the special case when the density is a polynomial. Even when the density is not a polynomial, however, the method may have application, by a truncation of a power series expansion as an approximation. Alternatively, it may be the case that the density may be represented as different polynomials over different subsets of the support of the distribution.

Ahrens and Dieter considered the interval $(x_0, x_0 + d)$ over which the density is represented by the polynomial $f(x_0 + h) = c_0 + c_1h + c_2h^2 + \dots + c_rh^r$. Let a be the integral of the polynomial over $(x_0, x_0 + d)$. For $2 \leq i \leq r + 1$, if $c_{i-1} \geq 0$, define

$$p_i = \frac{c_{i-1}d^i}{(ai)}$$

and

$$s_i = 0$$

If $c_{i-1} < 0$, define

$$p_i = \frac{-c_{i-1}d^i(i-1)}{(ai)}$$

and

$$s_i = 1$$

Then define

$$p_1 = \frac{c_0 d}{a} + \frac{\Sigma\, s_i c_{i-1} d^i}{a}$$

If $p_1 < 0$, the method does not apply; however, it may be in this case that the interval $(x_0, x_0 + d)$ can be divided up and the p_1's for the subintervals will be nonnegative so that the method can be used over each of them. For the remainder of the discussion of the method here we assume $p_1 \geq 0$. Define

$$P_i = \sum_{k=1}^{i} p_k \qquad \text{for } i = 1, 2, \ldots, r + 1$$

Note that $P_{r+1} = 1$. In the following description of the algorithm PL due to Ahrens and Dieter, "generate u" refers to obtaining an observation from the U(0,1) distribution.

ALGORITHM PL(x_0,d,P_1,P_2,..,P_{r+1},s_2,s_3,...,s_{r+1},x)

Step	*Description*
1.	Generate u. If $u \leq P_1$, deliver $x = x_0 + du/P_1$.
2.	Set i = 2. Generate u_{prev} and u_{curr}. If $u_{curr} \geq u_{prev}$, set $u_{max} = u_{curr}$ and $\tilde{u} = u_{prev}$; else set $u_{max} = u_{prev}$ and $\tilde{u} = 0$.
3.	If $u \leq P_i$, then if $s_i = 0$, deliver $x = x_0 + du_{max}$; if $s_i = 1$ and $\tilde{u} > 0$, deliver $x = x_0 + d\tilde{u}$; if $s_i = 1$ and $\tilde{u} = 0$, deliver $x = x_0 + du_{curr}$.
4.	Set i = i + 1 and generate u_{curr}. If $u_{curr} \geq u_{max}$, set $u_{max} = u_{curr}$ and $\tilde{u} = u_{prev}$. Set $u_{prev} = u_{curr}$ and go to 3.

The validity of the algorithm can be shown by viewing the polynomial pdf as a composition of monomial densities and using the fact that the density of the maximum u_{max} of a set of k independent U(0,1) variables is ku_{max}^{k-1} and the density of any other variable u_i ($\neq u_{max}$) from the set is $k(1 - u_i^{k-i})/(k - 1)$. The details of the proof are left as an exercise.

If P_1 is very small, efficiency can be increased by substituting for steps 1 and 2 the following:

Step	*Description*
1!	Generate u and u_{prev}. If $u < P_1$, deliver $x = x_0 + du_{prev}$.
2!	Set i = 2. Generate u_{curr}. If $u_{curr} \geq u_{prev}$, set $u_{max} = u_{curr}$ and $\tilde{u} = u_{prev}$; else set $u_{max} = u_{prev}$ and $\tilde{u} = 0$.

6.4.3 Rejection Methods

One of the most widely used techniques for generating random numbers from a given distribution was described by von Neumann (1951). It is variously referred to as the *rejection method* or the *acceptance sampling method.*

The method can be described as follows. Let f(x) and g(y) be pdf's and let h(x) be a given function (not necessarily a pdf).

1. Generate X from the distribution having pdf f(x).
2. Generate independently Y from the distribution having pdf g(y).
3. If $Y \leq h(X)$, deliver Z = X; else go to 1.

The pdf of Z as generated by these steps is

$$\frac{f(z)G(h(z))}{\int_{-\infty}^{\infty} f(s)G(h(s))\, ds} \tag{6.16}$$

where G denotes the cdf corresponding to the pdf g. This pdf is easily derived by obtaining the cdf of Z and differentiating.

$$\mathrm{Prob}(Z \le r) = \mathrm{Prob}(X \le r | Y \le h(X))$$

$$= \frac{\int_{-\infty}^{r} \int_{-\infty}^{h(x)} f(x)g(y)\ dy\ dx}{\int_{-\infty}^{\infty} \int_{-\infty}^{h(x)} f(x)g(y)\ dy\ dx}$$

$$= \frac{\int_{-\infty}^{r} f(x)G(h(x))\ dx}{\int_{-\infty}^{\infty} f(x)G(h(x))\ dx}$$

Differentiation with respect to r of this cdf yields the pdf of Z as given.

It is of interest to know something about the efficiency of the rejection method. The time required to generate a variate by the method is a random variable. Let T be the number of executions of steps 1 through 3 before exit. T has the geometric distribution

$$p(T = t) = \tau(1 - \tau)^{t-1} \qquad t = 1, 2, \ldots,$$

where

$$\tau = \int_{-\infty}^{\infty} f(x)G(h(x))\ dx$$

Hence $E(T) = 1/\tau$ and $\mathrm{Var}(T) = (1 - \tau)/\tau^2$.

Sibuya (1962), in an attempt to accelerate the procedure, described a modification of the rejection method described above in which steps 1 and 2 are interchanged and the conditional branch of step 3 is to his step 2. Greenwood (1976c) considered the moments of the time to generate variates by this modified rejection method. Unfortunately, these moments may be infinite. (See Exercise 6.13.)

The foregoing description of the rejection method does not indicate the simplicity of its application, since the functions f, g, and h required may not be immediately apparent for a given distribution. We therefore will now give an alternate description that begins with the pdf of the desired distribution and indicates the way in which other needed functions can be obtained. Suppose it is desired to generate random numbers from the distribution having pdf p(x). Let M(x) be an integrable function such that $M(x) \ge p(x)$. The function M(x) is called a *majorizing function* or *upper envelope*. A pdf $q(x)$ is now formed from $M(x)$ by letting $q(x) = M(x)/c$, where $c = \int_{-\infty}^{\infty} M(x)\,dx$. The function analogous to $g(y)$ in step 2 of the previous description for simplicity is taken as the U(0,1) pdf, that is, the constant 1. The procedure is then as follows:

1. Generate X from the distribution having pdf $q(x)$.
2. Generate Y from U(0,1).
3. If $Y \leq p(X)/M(X)$, deliver $Z = X$; else go to 1.

Using (6.16), it is easy to see that the pdf of Z is $p(z)$, since

$$\frac{f(z)G(h(z))}{\int_{-\infty}^{\infty} f(s)G(h(s))\,ds} = \frac{q(z)p(z)/M(z)}{\int_{-\infty}^{\infty} \left[q(s)p(s)/M(s)\right]\,ds}$$
$$= \frac{cq(z)p(z)/q(z)}{\int_{-\infty}^{\infty} [cq(s)p(s)/q(s)]ds}$$
$$= p(z)$$

When using the rejection method, the efficiency of generating variates from $q(x)$ and the probability of acceptance in step 3 should be considered. The ideal would be to make the majorizing function $M(x)$ be very close to $p(x)$ (so as to have a high probability of acceptance), yet have $M(x)$ so that $q(x)$ is constant or linear. Schmeiser and Shalaby (1978) discuss the use of mixtures of distributions so as to arrive at piecewise linear majorizing functions for efficiency in the generation.

Efficiency may also be increased when evaluation of $p(x)$ is complicated, such as when $p(x)$ involves exponentials and so on, by making a simpler preliminary comparison with a *minorizing function* $m(x)$ in step 3. If $m(x) \leq p(x)$ and $m(x)$ is easy to evaluate, but $p(x)$ is complicated to evaluate, step 3 may be replaced by

3a. If $Y \leq m(x)/M(x)$, deliver $Z = x$.
3b. If $Y \leq p(x)/M(x)$, delivery $Z = x$; else go to 1.

(See Schmeiser and Shalaby 1978 for further discussion of this modification.)

Forsythe (1972) gave a modification of the rejection method for sampling from a continuous distribution whose density can be represented as

$$f(x) = ce^{-B(x)} \tag{6.17}$$

where $B(x)$ is an increasing function of x over $(0,\infty)$. The method is a modification of a technique used by von Neumann (1951) for exponential random deviates. For distributions not satisfying this form, suitable modifications can sometimes be made. For example, the normal can be generated from a half-normal, which does satisfy the conditions. Forsythe's method requires a grid on x, $q_0, q_1, \ldots, q_n$, such that

$$B(q_k) - B(q_{k-1}) \leq 1$$

and so that $P(X \leq q_n)$ is the largest representable number less than 1. Then for the kth interval, constants p_k and d_k are defined as $p_k = P(X \leq q_k)$ and $d_k = q_k - q_{k-1}$. Let $G_k(x) = B(q_{k-1} + x) - B(q_{k-1})$.

To generate a random number by this scheme, first select an interval randomly in such a way that the kth interval is chosen with probability $p_k - p_{k-1}$. Next sample within the selected interval by the following scheme:

1. Generate u from U(0,1), and set $w = ud_k$.
2. Set $t = G_k(w)$.
3. Generate independently from U(0,1), $u_1, u_2, \ldots, u_N$, where N is such that if $t \leq u_1$, $N = 1$; otherwise $u_1 > u_2 \ldots > u_{N-1} \leq u_N$.
4. If N is even, go to 1; otherwise deliver $x = w + q_{k-1}$.

Ahrens and Dieter (1973) gave a simple proof of the validity of Forsythe's rejection method. First they showed that the probability, $P(t)$, that N is odd is e^{-t}. This follows since $P(t) + \Pr(N \text{ is even} | t) = 1$ and $\Pr(N \text{ is even} | t)$ is $\int_0^t P(u_1)\, du_1$; hence

$$P(t) + \int_0^t P(u_1)\, du_1 = 1$$

Therefore, $P(0) = 1$ and $P'(t) = -P(t)$. The solution to this differential equation with the given boundary condition is $P(t) = e^{-t}$.

Once the interval is selected, the cdf of X generated by the above scheme is therefore

$$\begin{aligned} F_k(a) &= \Pr(X \le a) \\ &= \Pr(W \le a - q_{k-1} \mid t \text{ odd}) \\ &= \frac{\Pr(t \text{ odd} \mid W \le a - q_{k-1}) \cdot \Pr(W \le a - q_{k-1})}{\Pr(t \text{ odd})} \\ &= \frac{e^{-G_k(W)} \cdot \dfrac{a - q_{k-1}}{d_k}}{\int_0^{d_k} e^{-G_k(w)}\, dw} \\ &= \frac{e^{-B(X)} \cdot e^{B(q_{k-1})} \cdot \dfrac{a - q_{k-1}}{d_k}}{\int_0^{d_k} e^{-G_k(w)}\, dw} \end{aligned}$$

Differentiating with respect to a, we have the density

$$f_k(x) = c_k e^{-B(x)} \qquad q_{k-1} \le x \le q_k \; .$$

The division of the density (6.17) over the n intervals and the selection of the intervals with the same probability as that of the intervals under the given density, together with the fact just established, that the conditional density over the interval is proportional to the given density, ensure that the sampling is from this given density.

Ahrens and Dieter show that the average number of passes through the steps 1 to 4 before acceptance is

$$\frac{d_k + \int_0^{d_k} e^{G_k(w)}\, dw}{\int_0^{d_k} e^{-G_k(w)}\, dw}$$

Example 6.7

Rejection over a finite length interval.

Suppose p(x) is a pdf such that

$$p(x) = 0 \quad \text{if} \quad x \notin (a,b)$$

and

$$p(x) \leq M \quad \text{for all} \quad x \in (a,b)$$

where M is a finite real number. Using the rejection method as described above, we generate u_1 and u_2 independently from U(0,1), and if

$$u_2 \leq \frac{p[a + (b - a)u_1]}{M}$$

deliver $x = a + (b - a)u_1$; else repeat the process.

Example 6.8

It is desired to sample from the distribution with pdf p(z) which is nonzero over the interval (0,2). Suppose p(z) is of a form involving exponentials of z, nonintegral powers of z, and so on, so that its cdf is quite complicated and even the evaluation of p(z) at a given point cannot be performed easily. Suppose, however, that we know that $p(z) \leq \frac{1}{2}(z + 1)$ and that $p(z) \geq 0.3$ over (0,2). The situation is as shown below:

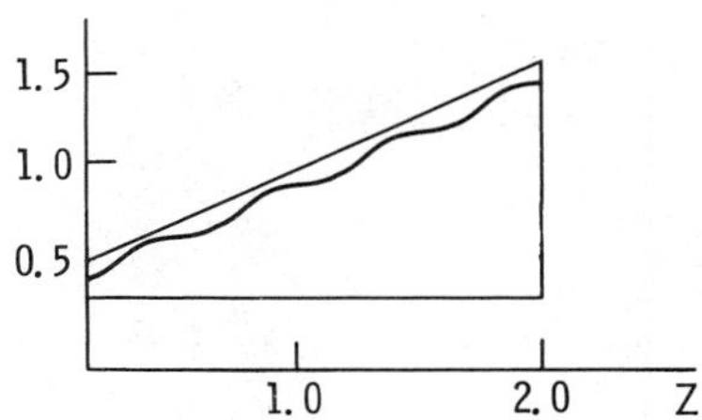

The rejection method may be implemented as follows, using a generator for U(0,1).

1. Generate u_1, u_2, and u_3 independently from U(0,1). If $u_3 \leq 0.5$ let $x = 2u_1$; else let $x = 2 \max(u_1,u_2)$.
2. Generate u_4 independently from U(0,1).
3. (a) If $u_4 \leq 0.6/(x + 1)$, deliver $z = x$.

   (b) If $u_4 \leq 2p(x)/(x + 1)$, deliver $z = x$; else go to 1.

Example 6.9 Tail of a Normal Distribution

There is often a need to generate samples from the tail of a distribution. Frequently when using the method of mixtures of distributions the tail of the desired distribution will be a residual term in the mixture. Marsaglia (1964) described the following procedure for generating random samples from the tail of the standard normal distribution, that is, generating x, where $x > a$ and x has density proportional to $e^{-x^2/2}$. Generate pairs of U(0,1) random variates u_1 and u_2 until

$$u_2 < a(a^2 - 2 \ln u_1)^{-1/2}$$

and then take $x = (a^2 - 2 \ln u_1)^{1/2}$.

Example 6.10 Nearly Linear Densities

Marsaglia (1962) discussed rejection procedures for intervals of a distribution over which the density is nearly linear. In this example we summarize Knuth's (1969) description of Marsaglia's method. Assume it is desired to generate random variates from the distribution with density g(x), which is nearly linear over the interval (s, s + h) and zero elsewhere. First, a band is formed by two parallel lines such that

$$a - \frac{b(x - s)}{h} \leq g(x) \leq b - \frac{b(x-s)}{h}$$

in the support of the distribution as shown in Fig. 6.10.

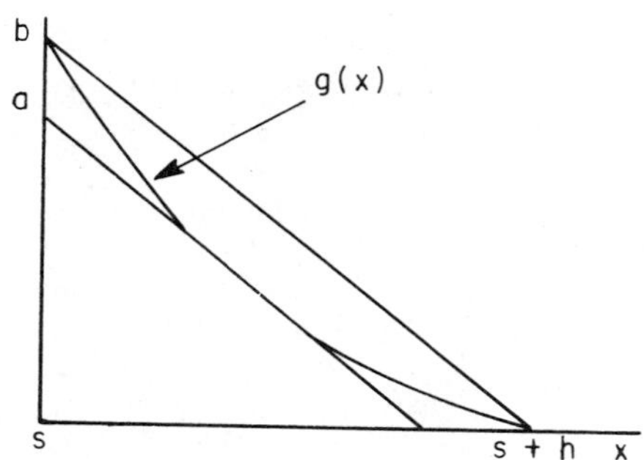

Figure 6.10 Nearly linear density.

The algorithm is as follows:

1. Generate independently two U(0,1) random variates, u_1 and u_2, and let $U = \min(u_1, u_2)$ and $V = \max(u_1, u_2)$.
2. If $V \leq a/b$, go to 4.
3. If $V > U + (1/b)g(s + hU)$, go to 1.
4. Deliver $x = s + hU$.

The validity of the algorithm can be seen by observing that

$$
\begin{aligned}
P(x \leq t) &= P(x \leq s + hr) \qquad (0 \leq r \leq 1) \\
&= P(U \leq r \mid V \leq U + \tfrac{1}{b}g(s + hU)) \\
&= \frac{\int_0^r \int_u^{u+(1/b)g(s+hu)} 2\, dv\, du}{\int_0^1 \int_u^{u+(1/b)g(s+hu)} 2\, dv\, du} \\
&= \frac{\int_0^r g(s + hu)\, du}{\int_0^1 g(s + hu)\, du} \\
&= \int_s^t g(x)\, dx
\end{aligned}
$$

The technique described in this example is frequently used in conjunction with the method of mixtures when a part of the distribution can be represented as a nearly linear function.

The rejection method also works for discrete random variables in a manner analogous to that for continuous random variables. The following example gives an application for Poisson variates. The method is not particularly efficient and is not recommended. (See Section 6.5 for better methods.)

Example 6.11. Rejection Method for Poisson Distribution (Slow)

The probability (density) function for the Poisson is

$$P(x) = \frac{\delta^x e^{-\delta}}{x!} \qquad x = 0, 1, 2, \ldots$$

0. Let $k = 0$ and $y_0 = u_0$, where u_0 is from U(0,1).
1. If $y_k \leq e^{-\delta}$, deliver $x = k$.
2. Generate u_{k+1} from U(0,1). Set $y_{k+1} = u_{k+1} y_k$. Set $k = k + 1$ and go to step 1.

The expected number of trials needed to obtain x under this procedure is $\delta + 1$.

6.4.4 Table Sampling Methods for Discrete Distribution

In Section 6.4.1 we mentioned a look-up procedure in a table of the cumulative distribution function for generation of random variates from a specified distribution. An alternative table look-up procedure for discrete distribution was described by Marsaglia (1963). We will refer to this technique as a *table sampling* procedure, since partitions of the table contain a number of each value of the desired random variable roughly proportional to the probability of the random variable's taking that value. A partition of the table is chosen with a specified probability and then that partition is sampled randomly. Norman and Cannon (1972) discussed the basic technique as well as various considerations for its implementation on a computer.

The idea underlying the method can be illustrated simply using the following small example. Suppose that the discrete random variable X assumes the values 1, 2, 3, and 4 with probabilities shown below:

x	P(X = x)
1	.6115
2	.2865
3	.0995
4	.0025

Let M_j denote the sum of the digits in the jth decimal position of all four probabilities; that is, $M_1 = 8$, $M_2 = 18$, $M_3 = 18$ and $M_4 = 20$. Next construct a table with four partitions, the first of which contains six 1's and two 2's, the second of which contains one 1, eight 2's, and nine 3's, and so on. Now choose partition j with probability $M_j \times 10^{-j}$, and select a number with equal probability from the set chosen. The partitioned table can be represented as in Table 6.2.

Table 6.2 Example of Partitions for Table Sampling

Location	Contents	Location	Contents	Location	Contents	Location	Contents
S_1 ⌈0	1	16	2	32	2	48	1
1	1	17	3	33	3	49	2
2	1	18	3	34	3	50	2
3	1	19	3	35	3	51	2
4	1	20	3	36	3	52	2
5	1	21	3	37	3	53	2
6	2	22	3	38	3	54	3
S_1 ⌊7	2	23	3	39	3	55	3
S_2 ⌈8	1	24	3	40	3	56	3
9	2	S_2 ⌊25	3	41	3	57	3
10	2	S_3 ⌈26	1	42	4	58	3
11	2	27	2	S_3 ⌊43	4	59	4
12	2	28	2	S_4 ⌈44	1	60	4
13	2	29	2	45	1	61	4
14	2	30	2	46	1	62	4
15	2	31	2	47	1	S_4 ⌊63	4

In general, we consider a discrete distribution for X taking values $\{x_1, x_2, \ldots, x_n\}$. Assume the probabilities can be represented in t positions in a numeral system with base β, that is, we have given

$$P(X = x_i) = 0.\delta_{i1}\delta_{i2} \cdots \delta_{it}$$

$$= \sum_{j=1}^{t} \delta_{ij}\beta^{-j} \qquad 0 \le \delta_{ij} < \beta$$

Define

$$P_0 = 0$$

$$P_j = \beta^{-j} \sum_{i=1}^{n} \delta_{ij} \qquad j = 1, 2, 3, \ldots, t$$

$$N_0 = 0$$

$$N_k = \sum_{j=1}^{k} \sum_{i=1}^{n} \delta_{ij} \qquad k = 1, 2, 3, \ldots, t$$

Form partitions S_j in which the values x_i are stored in δ_{ij} locations. Partition S_j is to be chosen with probability P_j. The number of storage locations required to hold the first k sets is N_k. Number the locations beginning with zero, so that the partition S_j will be composed of locations N_{j-1} to $N_j - 1$.

Now generate a random number, u', from U(0,1). Let $u = 0.d_1d_2 \ldots d_t$ be the value of u' to the nearest t places in base β. Let m be such that $1 \leq m \leq t$ and

$$\sum_{j=0}^{m-1} P_j \leq u \leq \sum_{j=0}^{m} P_j \tag{6.18}$$

and take X to be the value stored in location

$$\sum_{j=1}^{m} d_j\beta^{m-j} - \left(\beta^m \sum_{j=0}^{m-1} P_j - N_{m-1}\right) \tag{6.19}$$

From (6.18) we see that partition S_m is chosen with probability P_m. Next note that (6.18) is equivalent to

$$0 \leq \beta^m u - \beta^m \sum_{j=0}^{m-1} P_j < \beta^m \left(\sum_{j=0}^{m} P_j - \sum_{j=0}^{m-1} P_j\right)$$

or

$$0 \leq \sum_{j=1}^{m} d_j\beta^{m-j} - \beta^m \sum_{j=0}^{m-1} P_j < \beta^m P_m$$

The right-hand member of this inequality is $\sum_{i=1}^{n} \delta_{im}$, or, equivalently, $N_m - N_{m-1}$. Further, since u' is distributed as U(0,1) the first k digits (in base β) of u' have a discrete uniform distribution. Hence

$$\sum_{j=1}^{m} d_j\beta^{m-j} - \beta^m \sum_{j=0}^{m-1} P_j$$

has a discrete uniform distribution over the set {0, 1, 2, ..., $N_m - N_{m-1} - 1$}, and the location in (6.19) has a discrete uniform distribution over the integers from N_{m-1} to $N_m - 1$, that is, over the locations in partition S_m.

The validity of the basic technique is easily seen by observing that

$$\begin{aligned} P(X = x_i) &= \sum_{j=1}^{t} P(\text{partition } S_j \text{ is chosen}) \cdot P(x_i \text{ is chosen from partition } j) \\ &= \sum_{j=1}^{t} P_j \cdot \frac{\delta_{ij}}{\sum_{i=1}^{n} \delta_{ij}} \\ &= \sum_{j=1}^{t} \left(\beta^{-j} \sum_{i=1}^{n} \delta_{ij} \right) \frac{\delta_{ij}}{\sum_{i=1}^{n} \delta_{ij}} \\ &= \sum_{j=1}^{t} \delta_{ij} \beta^{-j} \end{aligned}$$

The construction of the table for sampling can be accomplished by a program as shown by Norman and Cannon (1972). Figure 6.11 is an adaptation from their description of the technique. In Fig. 6.11, P refers to the vector of probabilities expressed in the form $\delta_{i1}\delta_{i2}\delta_{i3}\delta_{i4}$ in base β, that is, the probabilities are rounded to four digits; Y refers to the vector of values taken by the random variable; and n is the number of these values. The coding may be done efficiently in assembler language using β as the base of the computer floating-point number system. For the IBM 360/370 computers, β would be taken to be 16.

Storage economy may be enhanced by manipulation of the table so as to reduce each partition to its smallest size containing the correct proportion and then to overlap the partitions. These operations generally must be done manually, and their worth would depend on how often the distribution is to be sampled. For example, Table 6.2 can be reduced by these techniques to the equivalent Table 6.3. The algorithm for sampling the table must be suitably modified.

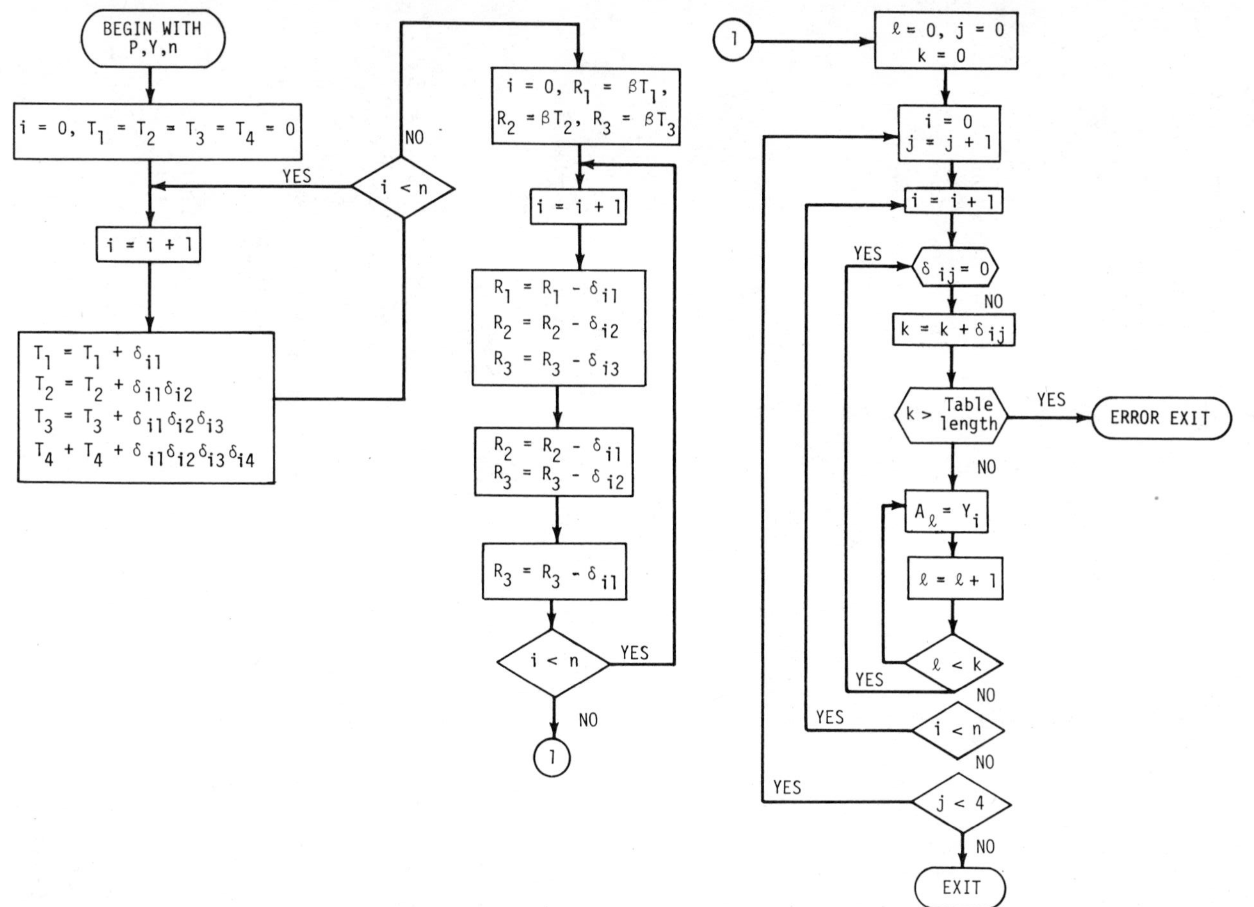

Figure 6.11 Setup portion of random number generator.

Table 6.3 Modification of Table 6.2 for Storage Efficiency

Location	Contents	Location	Contents	Location	Contents	Location	Contents
S_1 ⌈0	1	7	2	13	3	S_2 ⌊19	3
1	1	8	2	14	3	20	4
S_2 ⌈2	1	9	2	15	3	S_4 ⌈21	4
S_1 ⌊3	2	10	2	16	3	S_3 ⌊22	1
4	2	11	3	17	3	23	2
S_3 ⌈5	2	12	3	18	3	S_4 ⌊24	3
6	2						

In the example used to illustrate this method for generating random numbers, n = 4 and t = 4. It should be clear that the efficiency of the method relative to the use of the cdf improves when n is large relative to t.

6.4.5 The Alias Method for Discrete Distributions

Walker (1974a, 1977) introduced an efficient technique for generating variates from discrete distributions. The method uses a rejection procedure, but replaces rejected values by an alias. The probabilities of each value of the desired random variable are compared to the probability of a given value of the discrete uniform distribution having the same number of mass points, and tables of cutoff values and aliases are constructed during a setup stage of Walker's procedure. Then values are generated from U(0,1) and from the discrete uniform distribution and are used to select values from the desired distribution.

We now describe a modification of Walker's alias procedure. Consider the probability function $P(X = x_i) = p_i$, $i = 1, 2, \ldots, n$, and $\sum_{i=1}^{n} p_i = 1$. A list of aliases A_i and a list of cutoff values F_i are constructed by the following algorithm:

ALGORITHM SETUP

Step	*Description*
1.	For i = 1 to n, set $A_i = i$; $F_i = 0$; and $B_i = p_i - 1/n$.
2.	If $\Sigma \lvert B_i \rvert = 0$, exit.
3.	For i = 1 to n, find K such that $B_K \leq B_i$, and find L such that $B_L \geq B_i$.
4.	Set $C = B_k$, $D = B_L$.
5.	Set $B_K = 0$, $B_L = C + D$.
6.	Set $A_K = L$, $F_K = 1 + nC$.
7.	Go to 2.

To see that this procedure terminates, notice that at any stage $\Sigma B_i = 0$; hence there can never be exactly one $B_i \neq 0$, and, further, unless all $B_i = 0$ on the first pass, $C < 0$ and $D > 0$. Therefore, at each stage at least one more B_i is set to 0 and no zero B_i is changed, so at least by the (n - 1)th stage, there are exactly two $B_i \neq 0$.

These two nonzero B_i become C and D, C = -D, and at the beginning of the next pass, $\Sigma \lvert B_i \rvert = 0$. (Here we remind the reader that a test for equality of floating-point numbers must be implemented as a test for equality within a tolerance.)

Walker (1977) gave the following procedure to check that the setup procedure was successful:

ALGORITHM CHECK

Step	*Description*
1.	Set i = n.
2.	Set $q_i = F_i/n$.
3.	For j = 1 to n, if $A_j = i$, set $q_i = q_i + (1 - F_j)/n$.
4.	Set i = i - 1.
5.	If i > 0, go to 2.

In CHECK, A_i and F_i are the quantities produced by SETUP. The q_i's produced by CHECK should be the same as the original probabilities, the p_i's.

After setting up the A_i and F_i, the random number X is generated according to the following scheme:

ALGORITHM GENER

Step	*Description*
1.	Generate u from U(0,1).
2.	Generate i from {1, 2, ..., n} with equal probability and independently of u.
3.	If $u \le F_i$, deliver $X = x_i$; else deliver $X = x_{A_i}$.

To illustrate the alias method, consider the discrete random variable X with probability function given by

x	$P(X = x) = p_i$
0	.30
1	.05
2	.20
3	.40
4	.05

Three passes through steps 2 through 7 of the algorithm SETUP yields the following values:

	A_1	A_2	A_3	A_4	A_5	F_1	F_2	F_3	F_4	F_5
Initial	1	2	3	4	5	0	0	0	0	0
Pass 1	1	4	3	4	5	0	.25	0	0	0
Pass 2	1	4	3	4	1	0	.25	0	0	.25
Pass 3	4	4	3	4	1	.75	.25	0	0	.25

	B_1	B_2	B_3	B_4	B_5	K	L	C	D
Initial	.10	-.15	0	.20	-.15				
Pass 1	.10	0	0	.05	-.15	2	4	-.15	.20
Pass 2	-.05	0	0	.05	0	5	1	-.15	.10
Pass 3	0	0	0	0	0	1	4	-.05	.05

Algorithm CHECK yields each $q_i = p_i$.

To see that GENER will yield X = 0 with correct probability, for example, we write

$$\begin{aligned} P(X = 0) &= P(i = 1 \text{ and } u \leq F_1) + P(i = 5 \text{ and } u > F_5) \\ &= P(i = 1)P(u \leq F_1) + P(i = 5)P(u > F_5) \\ &= (.20)(.75) + (.20)(.75) \\ &= .30 \end{aligned}$$

Kronmal and Peterson (1978a) described an alternate but equivalent alias method and gave a general proof of the validity of the method.

The method can be programmed easily and it appears to be quite efficient for generating discrete random variables. Kronmal and Peterson (1978b) and Kronmal, Peterson and Lundberg (1978) extended the alias method to the generation of variates from continuous distributions. The reader is referred to these papers for a description of the technique.

6.5 GENERATION OF VARIATES FROM SPECIFIC DISTRIBUTIONS

While the previous section discussed several general techniques for transforming U(0,1) variates so as to form variates from other distributions, in the present section we consider some of the more common distributions in more detail so as to develop specific techniques with high efficiency. The general techniques of the previous section will frequently form the basis of the techniques discussed here, although occasionally some other, direct transformation may be of interest.

Of the various methods described for each of the distributions, the best choice may be made on considerations of specific parameter values and the proportion of time that may be required for setup. Some algorithms may require the construction of tables of constants; but once the tables are constructed, the algorithms may operate very efficiently. Such algorithms would be useful in applications requiring large numbers of random deviates.

Some of the methods described in the following subsections yield variates having only approximate distributions of the types desired. The approximation methods are sometimes used because of efficiency considerations. We would suggest that these approximate methods be used only very rarely, if ever; the cost of generating random numbers should not be allowed to vitiate the research being done. Usually the increased cost of using an exact method instead of an approximate method is not significant.

6.5.1 The Normal Distribution

Probably the most often used nonuniform distributions are those of the normal family with mean μ and variance σ^2, which we denote as $N(\mu,\sigma^2)$. A random variable from the standard normal distribution, $N(0,1)$, can easily be transformed so as to have the $N(\mu,\sigma^2)$ distribution; hence, we only consider generation of variates from the standard normal. Recent surveys and comparisons of methods for generation of normals are given by Ahrens and Dieter (1972), Atkinson and Pearce (1976), Kinderman and Ramage (1976), and Payne (1977).

Central Limit Theorem Approximation

Use of the central limit theorem on U(0,1) random variables provides a simple (and easily remembered) method for closely approximating normal random variates. If $U_1, U_2, \ldots, U_n$ are independently distributed as U(0,1),

$$X = \frac{(\Sigma\ U_i) - n/2}{\sqrt{n/12}}$$

has an approximate N(0,1) distribution. Because the U(0,1) is a "well-behaved" distribution, the approximation is fairly good even for small n. Choosing n = 12 leads to the simple form

$$X = (\Sigma\ U_i) - 6$$

This method would not generally be recommended, but it may have occasional application. Maritsas (1973) has recommended the use of a hardware peripheral implementing the method (on a larger number of uniforms) as a quick and convenient source of N(0,1) variates.

Box-Muller Transformation

Another method that is easy to remember was introduced by Box and Muller (1958). It is a direct transformation of two independent U(0,1) variates, U_1 and U_2, to two independent N(0,1) variates, X_1 and X_2:

$$X_1 = [-2 \ln(U_1)]^{1/2} \cos(2\pi U_2)$$

$$X_2 = [-2 \ln(U_1)]^{1/2} \sin(2\pi U_2)$$

Neave (1973) has warned against the use of the Box-Muller transformation on pseudorandom uniforms generated by a multiplicative-congruential generator with small multiplier. Although the difficulties Neave discussed did not occur when the multiplier was as large as is commonly used (over 1000, but considerably less than the modulus), Chay, Fardo, and Mazumdar (1975) and Golder and Settle (1976) have considered the problem further, recommending various modifications on the basic multiplicative-congruential generator. The most reasonable solution seems to be just to use a good generator of uniforms, such as a shuffled generator or a composite generator (see Section 6.2).

Marsaglia's Polar Method

Marsaglia (1962), using a method due to von Neumann (1951) for generating sines and cosines of random angles, gave the following modification of the Box-Muller transformation that avoids evaluation of the trigonometric function:

1. Generate u_1 and u_2 from U(0,1) and set $v_1 = 2u_1 - 1$ and $v_2 = 2u_2 - 1$.
2. If $v_1^2 + v_2^2 > 1$, go to 1.

3. Deliver $x_1 = v_1\left[-2 \ln\left(v_1^{\,2} + v_2^{\,2}\right)/\left(v_1^{\,2} + v_2^{\,2}\right)\right]^{1/2}$

and $x_2 = v_2\left[-2 \ln\left(v_1^{\,2} + v_2^{\,2}\right)/\left(v_1^{\,2} + v_2^{\,2}\right)\right]^{1/2}$

Sibuya (1961) showed how the logarithm computation could be avoided by generating from an exponential distribution using a rejection method.

Shafer (1962), Pike (1965), and Bell (1968) have given Algol programs for various modifications of the Box-Muller-polar method.

Marsaglia-Bray "Convenient" Method

Marsaglia and Bray (1964) described a fairly simple mixture and rejection method, such that with probability approximately .86 the normal is taken as the sum of three uniforms, with probability .11 it is taken as the sum of two uniforms, and with probability approximately .03 it is obtained by slower and more complicated methods, as shown below.

Probability for Choosing Method	Method
.8638	Deliver $X = 2(u_1 + u_2 + u_3 - 1.5)$.
.1107	Deliver $X = 1.5(u_1 + u_2 - 1)$.
.0228002039	Form $X = 6u_i - 3$ and $Y = .358u_{i+1}$ $(i = 4, 5, \ldots)$ until $Y < g(X)$ [$g(X)$ is given below], then deliver X.
.0026997961	Take $v_i = 2u_i - 1$ until $w = v_i^{\,2} + v_{i+1}^2 \leq 1$ and either $S = v_i[(9 - 2 \ln(w))/w]^{1/2}$ or $T = v_{i+1}[(9 - 2 \ln(w))/w]^{1/2}$ is > 3, then deliver X = S if S > 3 or X = T if $S \leq 3$ and T > 3.

The function g(x) in the method above is given by

$$17.49731196e^{-x^2/2} - 4.73570326(3 - x^2) - 2.15787544(1.5 - |x|) \quad \text{for } |x| < 1$$

$$17.49731196e^{-x^2/2} - 2.36785163(3 - |x|)^2 - 2.15787544(1.5 - |x|) \quad \text{for } 1 \leq |x| < 1.5$$

$$17.49731196e^{-x^2/2} - 2.36785163(3 - |x|)^2 \quad \text{for } 1.5 \leq |x| < 3$$

and

$$0 \quad \text{for } 3 \leq |x|$$

Ahrens and Dieter (1972) gave a modification of this method, called the "trapezoidal method," which we will not describe here.

Rectangle-Wedge-Tail Method

Marsaglia, MacLaren, and Bray (1964) described a very fast exact method of generating normal deviates by forming the normal as a mixture of several rectangular and wedge-shaped distributions and a tail related distribution as indicated in Fig. 6.12. Their procedure produces a deviate from the half-normal distribution, and a random sign is subsequently attached to obtain a normal random value.

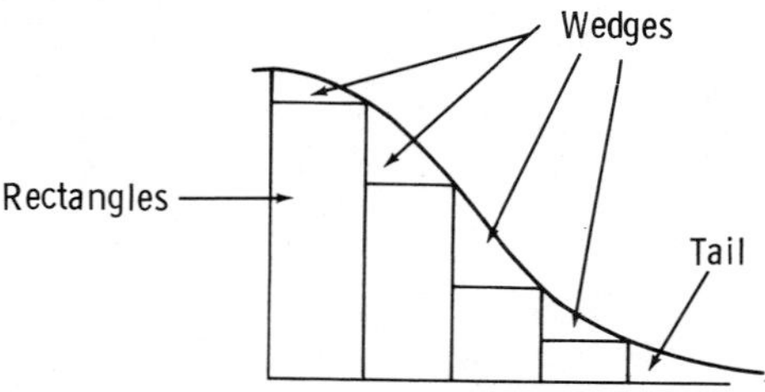

Figure 6.12 Rectangles, wedges, and tail of part of the normal density.

The technique can be implemented in different ways, but the basic ideas are to form the decomposition into a number of regions that allows easy selection of a region using only certain digits or bits of a random number from U(0,1) (in base 2, for example, $256 = 2^8$ may be a reasonable number of regions), and to sample from the easier rectangular densities a very large proportion of the time. For choosing the region, a table can be constructed and sampled using the techniques described in Section 6.4.4, since the indicator of the region is a discrete random variable. Only a few of the digits or bits of a U(0,1) number are used to select the region. The remaining bits [which form a U(0,1) number after the radix point is adjusted] are used to sample from the chosen region. If the region selected is rectangular, the sampling is done by a simple linear transformation of the U(0,1) number. If the region selected is a wedge, Marsaglia's technique for nearly linear densities (Example 6.10) or Ahrens and Dieter's polynomial sampling technique (Section 6.4.2) can be used to sample within the wedge. Either technique would require a table of the appropriate constants for each wedge. Either of these methods essentially adds another small rectangular density. (In the algorithm of Example 6.10, if step 3 is bypassed, or, in algorithm PL of Section 6.4.2, if $u \leq P_1$ in step 1, a rectangular density is sampled.) If the tail region is selected, the technique of Marsaglia (Example 6.9) can be used.

Ahrens and Dieter (1972) gave an implementation of this technique in which they divide the interval (0,3) into 12 equal parts over which the rectangles and wedges are formed. They supply tables M, for selecting a rectangle or a wedge in some cases; W, for selection of other wedges on the tail; and P, for polynomial sampling within the wedge if one is chosen.

Marsaglia, Ananthanarayanan, and Paul (1976) discussed methods for speeding up the procedure even more. Their techniques were described especially for IBM 370-type machines, but could be modified for any computer.

Kinderman-Ramage Procedure

Kinderman and Ramage (1976) also described a mixture of distributions that generate normal deviates very rapidly. The densities in the mixture are $f_1(x) = c_1(a - |x|)$, differences of the normal density $\phi(x)$ and $f_1(x)$, and the normal density over a tail region. Sampling from $f_1(x)$ over $(-a,a)$ is easy. The differences, $c_i[\phi(x) - f_1(x)]$ are nearly linear over subintervals, so the technique of Example 6.10 can be used. The tail region can be sampled as in Example 6.9. Kinderman and Ramage take the central interval to be approximately (-2.22,2.22). The differences over the positive half of this interval are nearly linear over the subintervals I_1, I_2 and I_3, which are respectively approximately (0,.48), (.48,1.59), and (1.59,2.22). The positive half of the normal density can then be represented as

$$\phi^+(x) \approx \underset{\text{over } I_1 \cup I_2 \cup I_3}{.884f_1(x)} + \underset{\text{over } I_1}{.027c_2[\phi(x) - f_1(x)]} + \underset{\text{over } I_2}{.047c_3[\phi(x) - f_1(x)]}$$

$$+ \underset{\text{over } I_3}{.015c_4[\phi(x) - f_1(x)]} + \underset{\text{over } (0,\infty) - I_1 \cup I_2 \cup I_3}{.027c_5\phi(x)}$$

where the c_i's are suitably chosen constants to make each part a pdf. The expression is approximate only because of the rounding of the constants involved. Kinderman and Ramage give the appropriate values to 14 decimal places.

With $a = 2.216035867166471$ and $g(x) = \phi(x) - .180025191068563 \times (a - |x|)$ for $|x| < a$, the algorithm of Kinderman and Ramage is as follows:

ALGORITHM KR(a,x)

Step	*Description*
1.	Generate u_1. If $u_1 < .884070402298758$, generate u_2 and deliver $x = a(1.13113163544180u_1 + u_2 - 1)$.
2.	If $u_1 < .973310954173898$, go to 4.
3.	[Tail Region] Generate u_2 and u_3 until $u_2^2 < a^2(a^2 - 2\ln(u_3))^{-1}$, then deliver $x = (a^2 - 2\ln(u_3))^{1/2}$ if $u_1 < .986655477086949$; else $x = -(a^2 - 2\ln(u_3))^{1/2}$.

4. If $u_1 < .958720824790463$, go to 6.
5. Generate u_2 and u_3. Set $t = a - [.630834801921960 \min(u_2,u_3)]$. If $\max(u_2,u_3) \leq .755591531667601$, go to 9. If $.034240503750111 (|u_2 - u_3|) \leq g(t)$, go to 9; otherwise repeat step 5.
6. If $u_1 < .911312780288703$, go to 8.
7. Generate u_2 and u_3. Set
$t = .479727404222441 + [1.105473661022070 \min(u_2,u_3)]$.
If $\max (u_2,u_3) \leq .872834976671790$, go to 9. If $.049264496373128(|u_2 - u_3|) \leq g(t)$, go to 9; otherwise repeat step 7.
8. Generate u_2 and u_3.
Set $t = .47972740422241 - [.595507138015940 \min(u_2,u_3)]$.
If $\max(u_2,u_3) \leq .805577924423817$, go to 9; otherwise repeat step 8.
9. If $u_2 < u_3$, deliver $x = t$; otherwise deliver $x = -t$.

Given the constants involved, the technique is easily recognizable as decomposition of the half-normal into a triangular density, three nearly linear densities, and a tail region. The standard methods are used on each part separately. The constants are determined so that the triangular density matches the normal at the peak of the scaled densities and is tangent to the normal at one point on each side as in Fig. 6.13.

Forsythe's Method and Modifications

Forsythe's (1972) method described in Section 6.4.3 can be used for generating normal deviates. In Forsythe's application of the technique approximately 4 uniforms are needed on the average to yield 1 normal. Ahrens and Dieter (1973) and Brent (1974) gave various modifications to the algorithm so as to reduce the expected number of uniforms required to generate 1 normal. In one of the

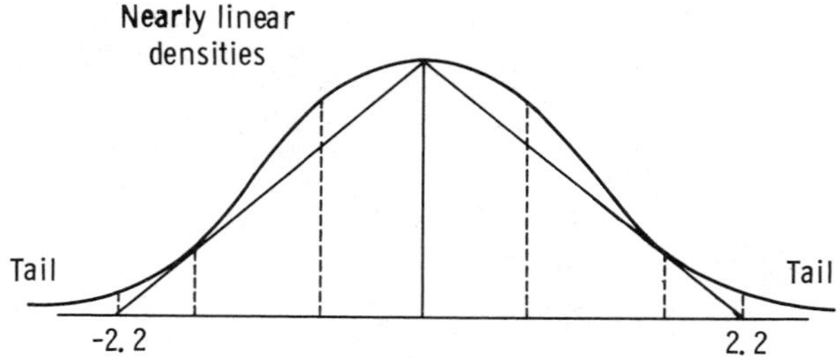

Figure 6.13 Decomposition of normal into triangular density, nearly linear densities, and tail areas.

modifications of Ahrens and Dieter and Brent, the expected number of uniforms was reduced to approximately 1.4 by the addition of a step 3 (in the notation used in Section 6.4.3): 3'. Set $u = (u_N - u_{N-1})/(1 - u_{N-1})$. The u of step 3' is U(0,1) and so, if necessary, a branch can be taken from step 4 to step 2, after forming $w = ud_k$, instead of branching back to step 1. The intervals are defined $P(Z > q_i) = 2^{-i-1}$, where $Z \sim N(0,1)$. (The q_i is the endpoint of the interval, as in Section 6.4.3.) Brent (1974) gave a FORTRAN function subprogram implementing this method. Timing comparisons by Kinderman and Ramage (1976) indicate Ahrens and Dieter's modification of the method is slightly faster than Brent's but that both are reasonably fast.

Several other methods for generating normal deviates have been suggested. Muller's (1959a) survey of the then extant procedures is of historical interest. Other methods that may be mentioned in the present study are a polynomial approximation to the inverse cdf by Muller (1958) and later by Gebhardt (1964); a rejection procedure of Butcher (1961b); a series approximation to the inverse cdf used by Wetherill (1965); a "center-tail" method of Dieter and Ahrens (1973), which employs a method similar to Forsythe's method on the central region and a method similar to Marsaglia's method for the tail region; a table look-up procedure using spline interpolations described by Shepherd and Hynes (1976) and Shepherd and Starner (1977); and a method using Butler's algorithm (see Example 6.6) by

Gates (1978). In comparative studies reported by Ahrens and Dieter (1972), Atkinson and Pearce (1976), Kinderman and Ramage (1976), and Payne (1977), the various algorithms have different execution efficiencies depending on the method of coding (particularly whether an assembler language or FORTRAN is used) and on the specific machine. In general, it appears that any of the last four methods discussed could be best, depending on the machine used and the coding, or on the criteria for evaluation such as execution efficiency or storage efficiency. For ease in programming, Marsaglia's polar method is easy to remember and can be written in just a few lines, but it is not competitive in execution efficiency.

6.5.2 The Gamma Distribution

The gamma family of distributions is probably the second most often generated nonuniform distribution. The pdf of this family can be written in the form

$$f(x) = \frac{1}{\Gamma(\alpha)\beta^{\alpha}}(x - \lambda)^{\alpha-1}e^{-(x-\lambda)/\beta} \qquad \text{if } x > \lambda$$

$$= 0 \qquad \text{otherwise}$$

with $\alpha, \beta > 0$. This family includes such widely used distributional models as the exponential ($\alpha = 1$), Erlang (α = positive integer), and chi-square with ν d.f. ($\alpha = \frac{\nu}{2}$, $\beta = 2$).

Since any gamma random variable can be obtained from the *standard gamma* (with $\lambda = 0$ and $\beta = 1$), we need consider only this special case, which has the single parameter α, called the *shape parameter*. We will denote the distribution of the standard gamma with parameter α by $G(\alpha)$.

We first consider the generation of exponential random variables, $G(1)$.

Logarithm Transformation; $\alpha = 1$

Use of the inverse cdf technique immediately yields the method: Generate u from $U(0,1)$ and take $X = -\ln(u)$. This method is not only simple, it is also fast. Studies by Ahrens and Dieter (1972) indicate it is as fast as any method coded in a high-level language, like FORTRAN.

Von Neumann (1951), Marsaglia (1961), and MacLaren, Marsaglia, and Bray (1964) gave various rejection algorithms for generating exponentials which avoided the explicit use of the logarithm function. When they are coded in an assembler language, these methods are faster then the logarithm transformation. Sibuya (1961) suggested some modifications to improve the efficiency of Marsaglia's (1961) method. Ahrens (in Ahrens and Dieter 1972) also gave a modification of Marsaglia's method.

Ahren's Method; $\alpha = 1$

The algorithm SA requires a table of constants whose entries $Q(i) = \sum_{j=1}^{i} [\ln(2)]^j/j!$ extend to the largest representable fraction less than 1.

ALGORITHM SA

Step	*Description*
1.	Initialize $a = 0$ and generate u from $U(0,1)$.
2.	If $u \geq .5$, go to 3; else increment $a = a + \ln(2)$, double $u = 2u$, and repeat this step.
3.	Set $u = 2u - 1$. If $u > \ln(2)$, go to 4; else deliver $x = a + u$.
4.	Initialize $i = 2$ and generate u^*. Set $u_{min} = u^*$.
5.	Generate a new u^*. If $u^* < u_{min}$, set $u_{min} = u^*$.
6.	If $u > Q(i)$, increase $i = i + 1$ and go to 5; else deliver $x = a + u_{min} \ln(2)$.

The proof of the validity of this method is given in Ahrens and Dieter (1972).

The operations involved in algorithm SA are generally very efficient to implement in an assembler language. The multiplications, for example, can be performed as simple shifts of binary integers.

In timing comparisons by Ahrens and Dieter (1972) this was the fastest algorithm (when coded in assembler language) of any for generating exponential random deviates.

For $G(\alpha)$ when α is an integer the methods above may be used, since if $Y_i \sim G(1)$, $i = 1, 2, 3, \ldots, \alpha$, and Y_i's are independent, then $X = \sum_{i=1}^{\alpha} Y_i$ has the $G(\alpha)$ distribution. This *summation method* is the recommended procedure (using a good algorithm for generating the exponentials) for generating a gamma random deviate with small integral shape parameter. For this case, when the logarithm transformation for the exponentials is used one may be tempted to replace the operations of taking logarithms of uniforms and adding to obtain X with the operations of multiplication of uniforms and a single logarithm evaluation, since $\Sigma \log(u_i) = \log(\Pi\, u_i)$. However, this equation may not hold in the floating-point number system and the product of the u_i may underflow, even for moderately small α.

Franklin and Sen (1975) gave an approximate method for the Erlang distribution (integral α) that involved truncation of the tails of the distribution and then employing a rejection technique. For larger values of α, their technique became more efficient than the summation method; and the approximation was fairly good.

The simplest algorithm for gammas with nonintegral shape parameter $\alpha > 1$ is the so-called *probability switch method* (see Naylor et al. 1966), which yields deviates with an *approximate* gamma distribution. The procedure is to generate a $G([\alpha])$ with probability $[\alpha] + 1 - \alpha$ and to generate a $G([\alpha] + 1)$ with probability $\alpha - [\alpha]$, where $[\cdot]$ is the greatest integer function.

A number of other approximate methods for generation of gamma variates have been given. Liittschwager (1965) suggested that the Weibull distribution, with pdf

$$f(x) = \frac{c}{b}\left(\frac{x-a}{b}\right)^{c-1} \exp\left\{-\left(\frac{x-a}{b}\right)^{c}\right\} \qquad x > a$$

could be used to approximate the gamma distribution. The cdf of the Weibull can be inverted easily as

$$x = a + b[-\ln(u)]^{1/c}$$

to generate a Weibull variate from a uniform (0,1), u; hence an approximate gamma variate can be generated by using the inverse cdf technique on the appropriate Weibull distribution. Phillips (1971) used a two-parameter (a = 0) Weibull to approximate the gamma. (The algorithm given in this reference contained an error. A correct version was given later by Phillips and Beightler 1972.) Ramberg and Tadikamalla (1974) used a three-parameter Weibull to obtain a better fit in all regions except the lower tail. [That this should be the case is clear by considering the fact that the range of the three-parameter Weibull is (a,∞) while the gamma of interest has range $(0,\infty)$.]

Another approximate method for generating gamma variates is by use of the Burr distribution (Burr 1942) with pdf

$$f(x) = kcx^{c-1}(1 + x^c)^{-(k+1)} \qquad x > 0$$

It is possible to invert the cdf to yield the Burr variate x from a uniform (0,1) variate u:

$$x = [(1 - u)^{-1/k} - 1]^{1/c} \tag{6.20}$$

Wheeler (1974, 1975) and Tadikamalla and Ramberg (1975) used this method and reported that better approximations were attained using the Burr distribution than by using the Weibull distribution. The parameters in (6.20) are determined by matching the first few moments of the Burr distribution as closely as possible. Obviously since there are only two degrees of freedom for the fit, only two moments can be matched exactly. Tadikamalla and Ramberg matched the third and fourth central moments of the gamma by those of the Burr. They then shifted and scaled the resulting variable so as to match the first and second moments of the gamma variate that was desired.

Ramberg and Schmeiser (1972) have suggested use of Tukey's lambda distribution as an approximate method for generating variates from various symmetric distributions. In Ramberg and Schmeiser

(1974) they generalized Tukey's lambda distribution to a four-parameter distribution, which allowed good approximations to asymmetric distributions, including the gamma. The inverse cdf of their generalized lambda distribution is

$$x = \lambda_1 + \frac{[u^{\lambda_3} - (1 - u)^{\lambda_4}]}{\lambda_2}$$

for u from uniform (0,1).

Schmeiser (1977) gave a survey of various approximate methods for generating variates from specified distributions. Bowman and Beauchamp (1975) considered some problems that may arise in use of approximate methods for gamma distributions; in particular, they found that Phillip's (1971) method using the Weibull approximation did not yield a distribution with proper skewness coefficient. Generally one of the exact methods discussed below would be preferred over any of the approximate methods.

The summation method for the Erlang distribution could be used for any gamma distribution if also a technique was available for generating gamma variates with $\alpha < 1$. Such a procedure was given by Ahrens in Ahrens and Dieter (1974). It is described as follows:

ALGORITHM GS; $0 < \alpha < 1$ (Ahrens)

Step	*Description*
1.	Generate u from U(0,1). Set $b = (e + \alpha)/e$. (e is the base of the natural log.) Set p = bu. If $p > 1$, go to 3.
2.	Set $x = p^{1/\alpha}$. Generate u_1 from U(0,1). If $u_1 > e^{-x}$, go to 1; otherwise deliver x.
3.	Set $x = -\ln[(b - p)/\alpha]$. Generate u_1 from U(0,1). If $u_1 > x^{\alpha-1}$, go to 1; otherwise deliver x.

Steps 2 and 3 in the algorithm GS correspond respectively to the cases $x \leq 1$ and $x > 1$. The reader may notice that the rejection method employed for $x \leq 1$ uses the majorizing function (see Section 6.4.3)

$$g(x) = \frac{x^{\alpha-1}}{\Gamma(\alpha)}$$

and the density

$$f(x) = \frac{x^{\alpha-1} e\alpha}{e + \alpha}$$

For $x > 1$ the majorizing function is

$$g(x) = \frac{e^{-x}}{\Gamma(\alpha)}$$

and the density is

$$f(x) = \frac{e^{-x} e\alpha}{e + \alpha}$$

The sampling procedure followed by the algorithm utilizes the mixture

$$q(x) = \left(\frac{e}{e+\alpha}\right) \alpha x^{\alpha-1} + \left(\frac{\alpha}{e+\alpha}\right) e^{1-x}$$

and the facts that $bu \sim U(0,1)$ conditional on $u \leq b$, and $[(e + \alpha)u - e]/\alpha \sim U(0,1)$ conditional on $u > b$. The expected number of trials per acceptance can be derived as $t = (e + \alpha)[e\alpha\Gamma(\alpha)]$. This quantity achieves its maximum of 1.39 (over $0 < \alpha \leq 1$) at $\alpha = .8$. The average value of t, if α is considered to be uniformly distributed between 0 and 1, is 1.27.

As noted above, the algorithm GS may be used in the summation method to yield gamma variates with arbitrary parameter.

Fishman (1976a) gave an efficient exact method for generating gamma deviates with $\alpha \geq 1$, described below.

ALGORITHM GF1; $\alpha > 1$ (Fishman)

Step	*Description*
1.	Generate y from G(1). (Algorithm SA described above could be used here.) Generate u from U(0,1).
2.	If $u \leq [y/(e^{y+1})]^{\alpha-1}$, deliver $x = \alpha y$; otherwise go to 1.

The straightforward derivation of this method was given in Fishman (1976a) and is omitted here.

The most efficient exact method for generating gamma deviates with larger values of α was given by Dieter and Ahrens. The algorithm, which requires $\alpha > 2.53278$, is given below without proof. The derivation was given in Ahrens and Dieter (1974).

ALGORITHM GO: $\alpha > 2.5327805161251$

Step	*Description*
1.	Set $\mu = \alpha - 1$, $V = \alpha^{1/2}$, $\sigma^2 = \alpha + 1.632993161855V$, $w = \sigma^2/\mu$, $d = 2.44948974278318\sigma$, and $b = \mu + d$.
2.	Generate u from U(0,1). If $u < 0.009572265238289$, go to 8.
3.	Generate s from N(0,1) (see Section 6.5.1). Set $x = \mu + \sigma s$. If $x < 0$ or $x > b$, go to 2.
4.	Generate u from U(0,1) and set $t = s^2/2$.
5.	If $u \leq 1 - t[(1 - 2s/V)w - 1]$, deliver x; else go to 7.
6.	If $u \leq 1 - t(w - 1)$, deliver x.
7.	If $\ln(u) \leq \mu[1 + \ln(x/\mu)] - x + t$, deliver x; else go to 2.
8.	Generate s from G(1) (see methods in first part of this subsection). Set $x = b(1 + s/d)$.
9.	Generate u from U(0,1). If $\ln(u) \leq \mu[2 + \ln(x/\mu) - x/b] + 3.7203284924588 - b - \ln(\sigma d/b)$, deliver x; else go to 2.

Ahrens and Dieter (1974) also discussed some other, similar techniques for using combinations of the normal and exponential distributions for generating gamma deviates.

In addition to the methods described above, useful algorithms for generating exact gamma deviates with nonintegral shape parameters (some requiring $\alpha \geq 1$) have been given by Jöhnk (1964) (who first generated a beta variate; see Section 6.5.3), Ahrens and Dieter (1974) (other than the ones presented above), Greenwood (1974), Wallace (1974), Wallace (1976), Atkinson and Pearce (1976) (an implementation of Forsythe's method is particularly noteworthy here), Atkinson (1977), Cheng (1977), Marsaglia (1977), Vaduva (1977), and Tadikamalla (1978). Greenwood and Marsaglia in the references above developed exact methods based on the Wilson-Hilferty approximation. (Wilson and Hilferty 1931.) Useful comparisons of methods were given by Ahrens and Dieter (1974), Atkinson and Pearce (1976), and Tadikamalla (1978). The methods presented in this subsection are all very competitive over the ranges of α indicated in the descriptions.

6.5.3 The Beta Distribution

The beta family of distributions with pdf (see also Eq. 5.7)

$$
\begin{aligned}
f(x) &= \frac{\Gamma(\alpha + \beta)}{\Gamma(\alpha)\Gamma(\beta)} x^{\alpha-1}(1 - x)^{\beta-1} \qquad 0 \le x \le 1 \\
&= 0 \quad \text{otherwise}
\end{aligned}
$$

provides very useful models for random processes with output restricted to a finite range. Various values of the parameters allow many different shapes of the density, and the variables can be linearly transformed into any finite range. Also of interest is the fact that ratios of beta random variables yield chi-square and F variables.

A simple way of obtaining a beta random variable with integral parameters is as an order statistic from a uniform distribution (Fox 1963). One of the first methods for arbitrary parameters was given by Jöhnk (1964). It is a rejection method, as follows:

ALGORITHM Jöhnk

Step	*Description*
1.	Generate u_1 and u_2 independently from U(0,1). Set $v_1 = u_1^{1/\alpha}$ and $v_2 = u_2^{1/\beta}$.
2.	Set $w = u_1 + u_2$. If $w \le 1$, set $x = v_1/w$ and deliver x; else go to 1.

Forsythe's method discussed in Section 6.4.3 also provides an efficient method for generating beta variates with arbitrary α and β. The $B(x)$ in (6.17) is taken as

$$B(x) = -(\alpha - 1)\ln(x) - (\beta - 1)\ln(1 - x)$$

and the $G_k(x)$ used in Forsythe's method is given by

$$G_k(x) = -(\alpha - 1)\ln[1 + (x/q_{k-1})] - (\beta - 1)\ln[1 - x/(1 - q_{k-1})]$$

Generation of beta variates is simpler if both α and β are greater than or equal to 1. One of the best methods for this situation was given by Ahrens and Dieter (1974). The basis for this method is given as follows.

First, consider the function

$$h_w(x) = \left(\frac{x}{A}\right)^A \left(\frac{1 - x}{B}\right)^B C^C \exp\left[\frac{wC(x - A/C)^2}{2}\right]$$

where $A = \alpha - 1$, $B = \beta - 1$, and $C = A + B$. Differentiating with respect to w yields

$$h_w'(x) = h_w(x)C\left(x - \frac{A}{C}\right)\left[w - \frac{1}{x(1-x)}\right]$$

Since $1/[x(1 - x)] \geq 4$ for all $x \in (0,1)$ and $h_w(A/C) = 1$ for all w, we see that whenever $w \leq 4$ the derivative is such that

$$h_w'(x) > 0 \quad \text{if } x < \frac{A}{C}$$

and

$$h_w'(x) < 0 \quad \text{if } x > \frac{A}{C}$$

Thus for $w \leq 4$, $h_w(x)$ is maximum at $x = A/C$. Using this fact we have, for $w = 4$, that

$$\left(\frac{x}{A}\right)^A \left(\frac{1 - x}{B}\right)^B C^C \leq \exp\left[-2C\left(x - \frac{A}{C}\right)^2\right]$$

for all $x \in (0,1)$. The left member of this inequality is a scalar multiple of the beta density function and the right member is proportional to a normal pdf. This suggests that a rejection method, using a multiple of some normal pdf as majoring function, can be applied to produce beta distributed random numbers. Algorithm BN, given below, provides details on the resultant rejection method.

ALGORITHM BN; $\alpha, \beta > 1$

Step	*Description*
1.	Set $A = \alpha - 1$, $B = \beta - 1$, $C = A + B$, $L = C \ln(C)$, $\mu = A/C$, and $\sigma = .5/C^{1/2}$.
2.	Generate y from N(0,1) and set $x = \sigma y + \mu$.
3.	If $x < 0$ or $x > 1$, go to 2.
4.	Generate u from U(0,1). If $\ln(u) \leq A \ln(x/A) + B \ln[(1 - x)/B] + L + .5y^2$, deliver x; else go to 2.

Algorithm BN is very efficient for large and nearly equal values of α and β. Most other algorithms tend to require more time as α and β increase, but the time required for BN in tests reported by Ahrens and Dieter decreased as α and β both increased.

Ahrens and Dieter (1974) gave a modification of Algorithm BN for the special case when $\alpha = \beta$. The modified algorithm, which they called BS, executes more rapidly than BN for the special case when it is appropriate. The modified algorithm is of interest in the algorithm BB for generating binomial random variates, which will be discussed in Section 6.5.5.

Schmeiser and Shalaby (1977) gave a modification of the Ahrens-Dieter algorithm BN that provides for earlier acceptance, resulting in a slight increase in efficiency.

Atkinson and Whittaker (1976) gave an algorithm that is very efficient relative to other methods when one parameter is less than 1 and the other parameter is greater than 1.

Timing comparisons as well as another rejection method for generating beta variates were given by Cheng (1978).

It should also be noted here that if X and Y are independently distributed as standard gammas with parameters a and b respectively, then $X/(X + Y)$ is distributed as beta with parameters a and b.

Hence, any of the techniques of Section 6.5.2 could be used to generate beta variates. The special methods discussed in this subsection are more efficient, however.

6.5.4 The F, t, and Chi-Square Distributions

We group the F, t, and chi-square distributions together in this subsection because they are frequently associated in statistical applications, not because the methods of generation are necessarily similar.

As mentioned in Section 6.5.2, the chi-square random variable is just a special case of a gamma variable. The most efficient method of generating a chi-square is one of the gamma routines described in Section 6.5.2, such as GS, GF1, or GO, depending on the value of the shape parameter of the gamma (which is twice the d.f. of the chi-square). A general subroutine for generating chi-square variates may include each of the algorithms mentioned above and branch into one or the other after testing the degrees of freedom. Forsythe's algorithm, which works for all values of the parameter may be a good general choice (Atkinson and Pearce 1976).

A Snedecor F random deviate may be obtained as a transformation of two chi-squares or as a transformation of one beta. The generation of two chi-squares takes longer than the generation of a single beta, using the good algorithms described in Sections 6.5.2 and 6.5.3; hence for generating an F with n_1 and n_2 d.f., the transformation

$$F = \frac{n_2}{n_1} \frac{x}{1 - x}$$

where x is a beta variate with parameters $n_1/2$ and $n_2/2$, is the recommended procedure.

The Student's t distribution may be obtained as the square root of an F with random sign attached or as the ratio of a normal and the square root of a chi-square. Special, faster algorithms have been given by Kinderman, Monahan, and Ramage (1977), using rejection and mixtures. Of the methods they describe, one of the best is given below. The constant c_ν, where ν is the d.f., is given by

$$c_\nu = \frac{(\nu + 1)/2}{(\pi\nu)^{1/2}(\nu/2)}$$

The function $f_\nu(x)$ is the unnormalized t density,

$$f_\nu(x) = \left[1 + \frac{x^2}{\nu}\right]^{-(\nu+1)/2}$$

The algorithms do not require integral d.f.

ALGORITHM TMX

Step	*Description*
1.	Generate u_1 from U(0,1). If $u_1 \le 2/\pi$, go to 2. If $u_1 \ge (2/\pi)^{1/2}$, go to 3. If $u_1 \ge 2c_\nu$, go to 3.
2.	Generate u_2 and u_3 from U(0,1). Deliver $x = 2(u_2 + u_3 - 1)$.
3.	Generate u_2 from U(0,1). If $u_2 \le .3622520694$, go to 5. Set $x = 1/(1.0680176321 - 1.5680176321u_2)$. Generate u_3 from U(0,1).
4.	If $u_3 \le x^2 f_\nu(x)$, deliver x; else go to 3.
5.	If $u_2 \le .0530096080$, go to 7. Set $x = 11.5909050257u_2 - 2.406629332$. Generate u_3 from U(0,1).
6.	If $.13528u_3 \le f_\nu(x) - 1 + \lvert x\rvert/2$, deliver x; else go to 3.
7.	Set $s = \text{sign}(7.840088159u_2 - .2078)$. Set $x = s(\lvert 7.840088159u_2 - .2078\rvert + 1.7922)$. Generate u_3 from U(0,1).
8.	If $.2u_3 \le f_\nu(x) - 1 + \lvert x\rvert/2$, deliver x; else go to 3.

The algorithm TMX uses a mixture of the triangular density $.5(1 - |x|/2)$ over the interval $|x| < 2$ and the difference in the t density and triangular density (see Kinderman-Ramage procedure in Section 6.5.1).

Kinderman, Monahan, and Ramage (1977) also gave a modification of TMX that requires some additional setup time, but that performs more efficiently for generating large numbers of t variates. The reader is referred to their paper for details as well as for the derivation of algorithm TMX, above.

6.5.5 The Binomial Distribution

The binomial distribution with parameters n and p is defined by the probability function

$$f(x) = \binom{n}{x} p^x (1 - p)^{n-x} \quad \text{for } x = 0, 1, 2, \ldots, n$$

The probability function leads immediately to the following direct algorithm:

ALGORITHM BU

Step	*Description*
1.	Set x = 0 and i = 0.
2.	Generate u from U(0,1). If $u \leq p$, then increment x = x + 1.
3.	Increment i = i + 1. If i < n, go to 2; else deliver x.

This is probably the most efficient method for generating a binomial variate with small parameter n. The computation time grows linearly in n, however, so a more efficient method would be desirable for larger values of n.

Ahrens and Dieter (1974) proved the following proposition, which leads to a more efficient algorithms for larger n.

Lemma. Let a and b be positive integers such that a + b - 1 = n. Let S be a random variable with a beta distribution with parameters a and b.

i. If $S \leq p$, let Y be a random variable with a binomial distribution with parameters b - 1 and (p - S)/(1 - S) and let X = a + Y.

ii. If $S > p$, let X be a random variable with a binomial distribution with parameters $a - 1$ and p/S.

Then X has a binomial distribution with parameters n and p.

The proof of this lemma is left to the exercises, or may be found in Ahrens and Dieter (1974).

Using this lemma, Ahrens and Dieter gave the following algorithm for generating binomials with parameters n and p:

ALGORITHM BB

Step	*Description*
1.	Set $m = n$, $q = p$, $k = 0$, $y = 0$, and $h = 1$.
2.	If $m \leq N$ (where N is a constant whose optimal value may depend on the machine and the language; Ahrens and Dieter report an optimal value of 38 for CDC-6400 assembler language), then generate j from the binomial distribution with parameters m and q, using the algorithm BU. Deliver $x = k + j$.
3.	If m is even, decrease $m = m - 1$ and generate u from $U(0,1)$. If $u \leq q$, increase $k = k + 1$.
4.	Set $a = (m + 1)/2$ and generate s from a beta distribution with parameters a and a (see Section 6.5.3). Set $g = hs$ and $z = y + g$.
5.	If $z \leq p$, update $y = z$, $h = h - g$, $q = (p - z)/h$ and $k = k + a$. Otherwise, update $h = g$ and $q = (p - y)/h$.
6.	Set $m = a - 1$ and go to 2.

Ahrens and Dieter compared the performance of implementations of BB and BU running on a CDC-6400 computer using various values of n. The following times in microseconds were reported by Ahrens and Dieter:

n =	1	2	5	10	20	50	100	200	500	1,000
BU	37	47	79	132	238	555	1,083	2,143	5,318	10,629
BB	48	60	91	144	250	501	693	889	1,155	1,352

These comparisons are for p = .5, but the value of p did not seem to affect the timings of either of the algorithms.

The computation times for BB grow proportional to log n, while, as previously noted, the times for BU increase proportional to n.

Relles (1972) gave a method similar to BB that used a normal approximation to the symmetric beta distribution. Use of the normal variates may be slightly faster than using exact betas; but in general the exact method would be preferred.

6.5.6 The Poisson Distribution

The Poisson distribution with parameter λ is defined by the probability function

$$f(x) = \frac{e^{-\lambda}\lambda^x}{x!} \quad \text{for } x = 0, 1, 2, \ldots$$

Snow (1968) gave an ALGOL program which generates variates from the Poisson distribution, using the inverse cdf technique. The program builds a table of the cdf as it executes so as to improve the efficiency on subsequent calls.

Knuth (1969) gave the following simple algorithm for generating Poisson variates with parameter λ:

ALGORITHM Q

Step	*Description*
1.	Set $L = e^{-\lambda}$, $x = 0$, and $p = 1$.
2.	Generate u from U(0,1) and set $p = pu$.
3.	If $p \geq L$, set $x = x + 1$ and go to 2; else deliver x.

This is a reasonably fast algorithm so long as λ is small. Computation time grows linearly with λ, however. Schaffer (1970) gave an ALGOL program implementing a modification of this algorithm.

The following lemma, proved by Ahrens and Dieter (1974), provides the basis for a more efficient algorithm for generating Poisson variates using the standard gamma distribution:

Lemma. Let Y have a (standard) gamma distribution with parameters n and 1, where n is a positive integer. Let $\lambda > 0$.

i. If $Y > \lambda$, let x have a binomial distribution with parameters $n - 1$ and λ/Y.

ii. If $Y \leq \lambda$, let z have a Poisson distribution with parameter $\lambda - Y$, and let $x = n + z$.

The x has a Poisson distribution with parameter λ.

The proof of this lemma is left to the exercises, or may be found in Ahrens and Dieter (1974). Ahrens and Dieter also gave the following algorithm for generating Poisson variates by the method suggested by the lemma.

ALGORITHM PG

Step	*Description*
1.	Initialize $x = 0$ and $w = \lambda$.
2.	If $w \geq c$ (where c is a machine/language-dependent constant; Ahrens and Dieter reported the optimal c was approximately 16 for FORTRAN coding and 24 for assembler coding), go to 6.
3.	Set $p = 1$ and $b = e^{-w}$.
4.	Generate u from U(0,1) and set $p = pu$. If $p < b$, deliver x.
5.	Increase $x = x + 1$ and go to 4.
6.	Set $n = [dw]$. (Ahrens and Dieter determined experimentally that $d = 7/8$ was a good choice for d.) Generate y from the standard gamma distribution with parameter n. If $y > w$, go to 8.
7.	Set $x = x + n$, $w = w - x$, and go to 2.
8.	Set $p = w/y$.
9.	Generate n from U(0,1). If $n \leq p$, increase $x = x + 1$.
10.	Set $n = n - 1$. If $n > 1$, go to 9; else deliver x.

Note that steps 3 through 5 are the algorithm Q, which is efficient for small values of w; and that steps 8 through 10 are essentially the simple algorithm BU for generating binomial variates as described in Section 6.5.5.

Ahrens and Dieter (1974) reported relative execution times in microseconds for assembler language implementations of algorithms Q and PG (with algorithm GO of Section 6.5.2 being used to generate the gamma variates) as follows:

λ	5	10	20	50	100	200	500
Q	192	265	415	858	1,599	3,081	7,527
PG	229	310	471	651	750	867	970

Other very efficient methods for generating Poisson variates were given by Ahrens and Dieter (1974) and by Fishman (1976).

6.5.7 Distributions of Order Statistics

Frequently in applications order statistics from a given distribution are needed. If the inverse cdf technique is appropriate for generating variates from the given distribution, the problem can be solved by first obtaining the order statistics from the uniform (0,1) distribution. If other techniques are used to transform the uniforms to variates from the given distribution, it may be necessary to obtain the full sample from the given distribution and then sort it to obtain the order statistics. We discuss below some methods for obtaining the order statistics from a uniform (0,1) distribution.

There are basically four methods for obtaining order statistics from U(0,1): sorting, use of the relationship to a beta variate, sequential generation, and generation using a random partitioning. The sorting method is quite simple conceptually. It consists of generating a sample of full size and ordering it as necessary to obtain the order statistics desired. Algorithms for sorting have been extensively analyzed in the literature of computer science (see Knuth 1973). The computational time for naive sorting methods is of order n^2, when n is the number of items to be sorted, while the better sorting algorithms have time asymptotically proportional to $n \cdot \log(n)$. Sorting algorithms are not considered in this book; for these the reader is referred to Knuth (1973). If only a subset of the order statistics is desired, such as the first k order statistics, there are partial sorting algorithms available to avoid the time required for a complete sort. As noted above, the sorting method may be the only choice for order statistics from a nonuniform distribution.

Since the i-th order statistic in a sample of size n from U(0,1) has a beta distribution with parameters i and $n - i + 1$, a single order statistic can be efficiently generated using this relationship and the techniques of Section 6.5.3.

When only a few of the extreme order statistics, say, the smallest k, or the largest k, are required, the sequential generation method is particularly useful. Lurie and Hartley (1972) described a method for generating the variates in ascending order, and Reeder (1972) and Schucany (1972) gave methods for descending order. Ramberg and Tadikamalla (1978) modified Schucany's method for generating subsets of the central order statistics. Their method was to generate first an appropriate beta variate and then generate the others desired in descending order as described below. Reeder (1972) and Mason and Lurie (1973) reported that the descending procedures were slightly faster when the full set of order statistics are to be generated. (In this case, the partitioning method described below is recommended, however.) A descending sequential method for generating all n order statistics from a random sample of size n from U(0,1) is as follows.

ALGORITHM OS

Step	*Description*
1.	Set $t = 1$ and $j = n$.
2.	Generate u from U(0,1) and preserve $x_j = tu^{1/j}$.
3.	Set $t = x_j$. Decrease $j = j - 1$. If $j > 0$, go to 2.
4.	Deliver x_j for $j = 1, \ldots, n$. [x_j is the $(n + 1 - j)$th order statistic.]

The random partitioning method, which apparently was first described in the literature by Lurie and Hartley (1972), is based on a well-known relationship between ratios of exponential variates and differences of order statistics from the uniform. If Y_1, Y_2, ..., Y_{n+1} are independently distributed as exponential (1), it is easy to show that, for i = 1, 2, ..., n, the ratio

$$R_i = \frac{Y_i}{Y_1 + Y_2 + \cdots + Y_{n+1}}$$

has the same distribution as

$$V_{(i)} - V_{(i-1)}$$

where $V_{(i)}$ is the ith order statistic in a sample of size n from U(0,1) and $V_{(0)} = 0$. This fact immediately leads to the following algorithm for n order statistics from U(0,1).

ALGORITHM OSP

Step	*Description*
1.	Set s = 0, j = 0.
2.	Generate y from G(1) [G(1) is the standard exponential distribution--see Section 6.5.2.]
3.	Increase s = s + y, j = j + 1, and preserve $x_j = s$.
4.	If $j \leq n$, go to 2.
5.	Set j = 1.
6.	Preserve $x_j = x_j/s$.
7.	Increase j = j + 1. If $j \leq n$, go to 6.
8.	Deliver x_j for j = 1, ..., n.

The algorithm OSP is more efficient than algorithm OS or sorting methods when the full set of order statistics is desired. Lurie and Mason (1973), Mason and Lurie (1973), and Rabinowitz and Berenson (1974) gave some empirical efficiency comparisons of the techniques described above.

6.5.8 Some Other Univariate Distributions

Most of the families of distributions commonly used have special efficient techniques for their generation, as seen in the previous subsections. There are special techniques also for some other, less common distributions such as the negative binomial family (Barrett 1965 and Ahrens and Dieter 1974), the stable family (Chambers, Mallows, and Stuck 1976 and Bartels 1978), the dipole distribution (Knop 1973), the inverse Gaussian (normal) distribution (Michael, Schucany, and Haas 1976), and the Cauchy distribution (Arnason 1974).

For other distributions, the general methods discussed in Section 6.4 may be employed. Sometimes, due to the intractability of the distribution which is to be simulated, it is necessary to approximate the desired distribution. This is generally done by matching some of the lower moments of the desired distribution with the moments of a member of a distributional family from which it is easy to generate variates. Ramberg and Schmeiser (1972, 1974) used the generalized lambda distribution, which has four parameters, to approximate other less tractible distributions. The Burr distribution, Eq. (6.20), (Burr 1942, 1973), the Pearson family (Cooper, Davis, and Dono 1965), and the Johnson system (Johnson 1965) also may be used to approximate the desired distribution. Johnson and Johnson (1978) described a useful distribution for approximation based on the cdf of the gamma distribution. The gamma itself and the beta also may be used to approximate other given distributions. Schmeiser (1977) gave a survey of the various methods for approximating distributions.

6.5.9 The Multivariate Normal Distribution

The p-variate normal distribution with mean $\underline{\mu}$ and variance-covariance matrix V (nonsingular) is defined by the pdf

$$f(\underline{x}) = (2\pi)^{-p/2}|V|^{-1/2} \exp\left\{-\frac{1}{2}(\underline{x} - \underline{\mu})'V^{-1}(\underline{x} - \underline{\mu})\right\} \qquad (6.21)$$

We denote this distribution $N_p(\underline{\mu},V)$. An obvious method to generate a p vector $\underline{x}$ with this distribution is to generate $z_1, z_2, \ldots, z_p$ as independent $N_1(0,1)$ variates, forming the p vector $\underline{z}$ having $N_p(\underline{0},I)$ then form $\underline{x}$ as

$$\underline{x} = A\underline{z} + \underline{\mu}$$

where A is such that AA' = V, for example, A could be from a Cholesky decomposition of V (see Section 7.4).

A second method builds up the p vector $\underline{x}$ from univariate conditional distributions, that is, x_3 is generated conditional on x_1 and x_2, and so on until finally x_n is generated. Without loss of generality we can take $\underline{\mu} = \underline{0}$ in (6.21). To establish the technique, we partition V as shown below so that the inverse of V is

$$V = \begin{pmatrix} v_{11} & \underline{v}'_{21} \\ \underline{v}_{21} & V_{22} \end{pmatrix} \qquad V^{-1} = \begin{pmatrix} w_{11} & \underline{w}'_{21} \\ \underline{w}_{21} & W_{22} \end{pmatrix}$$

The density (6.21) can thus be written as

$$c \cdot \exp\left\{-\frac{1}{2}(x_1,\underline{x}_2)' \begin{pmatrix} w_{11} & \underline{w}'_{21} \\ \underline{w}_{21} & W_{22} \end{pmatrix} (x_1,\underline{x}_2)\right\}$$

$$= c \cdot \exp\left\{-\frac{1}{2}(x_1 w_{11} x_1 + \underline{x}'_2 W_{22}\underline{x}_2 + x_1\underline{w}'_{21}\underline{x}_2 + \underline{x}'_2\underline{w}_{21}x_1)\right\}$$

$$= c \cdot \exp\left\{-\frac{1}{2}\left[x_1 w_{11} x_1 - x_1\underline{w}'_{21}W_{22}^{-1}\underline{w}_{21}x_1 + \underline{x}'_2 W_{22}\underline{x}_2 \right.\right.$$
$$\left.\left. + x_1\underline{w}'_{21}W_{22}^{-1}W_{22}\underline{x}_2 + \underline{x}'_2 W_{22}W_{22}^{-1}\underline{w}_{21}x_1 + x_1\underline{w}'_{21}W_{22}^{-1}\underline{w}_{21}x_1\right]\right\}$$

$$= c \cdot \exp\left\{-\frac{1}{2}x_1^{\,2}\left(w_{11} - \underline{w}'_{21}W_{22}^{-1}\underline{w}_{21}\right)\right.$$
$$\left. - \frac{1}{2}\left(\underline{x}_2 + x_1 W_{22}^{-1}\underline{w}_{21}\right)' W_{22}^{-1}(\underline{x}_2 + x_1 W_{22}\underline{w}_{21})\right\} \qquad (6.22)$$

Now, noting that

$$v_{11}^{-1} = w_{11} - \underline{w}'_{21}W_{22}^{-1}\underline{w}_{21}$$

$$W_{22} = \left\{V_{22} - \left(\frac{\underline{v}_{21}\underline{v}'_{21}}{v_{11}}\right)\right\}^{-1}$$

and

$$W_{22}\underline{w}_{21} = \frac{-\underline{v}_{21}}{v_{11}}$$

we can express (6.22) as the product of

$$c_1 \exp\left\{\frac{-\frac{1}{2}x_1^2}{v_{11}}\right\}$$

and

$$c_2 \exp\left\{-\frac{1}{2}\left(\underline{x}_2 - \frac{x_1\underline{v}_{21}}{v_{11}}\right)'\left(V_{22} - \frac{\underline{v}_{21}\underline{v}_{21}'}{v_{11}}\right)^{-1}\left(\underline{x}_2 - \frac{x_1\underline{v}_{21}}{v_{11}}\right)\right\}$$

Hence, the distribution $N_p(\underline{0},V)$ of $\underline{x}$ can be expressed as the marginal $N_1(0,v_{11})$ for x_1 and the conditional distribution of $\underline{x}_2|x_1$, $N_{p-1}(x_1\underline{v}_{21}/v_{11}, V_{22} - \underline{v}_{21}\underline{v}_{21}'/v_{11})$. This process is extended to the vector $\underline{x}_2$ and so on until the full vector $\underline{x}$ has been generated, one element at a time.

Schuer and Stoller (1962) described both of the techniques suggested above for generating random vectors from the multivariate normal distribuiton. Hurst and Knop (1972) also described the same two techniques and gave a FORTRAN program for the conditional distribution technique. Both methods require a certain amount of overhead in setting up and decomposing the various matrices involved. The computer timings reported by Hurst and Knop were for a decomposition of V that did not result in a triangular matrix A. Those timings indicated that the conditional distribution technique may be faster. Barr and Slezak (1972) reported on more extensive timing studies, including two different decompositions of V as well as the conditional distribution technique. Their empirical evidence indicated that the most efficient method of all was a decomposition method in which A in AA' = V was a triangular matrix. The Cholesky decomposition (Section 7.4) yields such a matrix. Special correlation patterns, however, can make the conditional distribution technique more efficient; hence, as so many times when there are alternate algorithms available, it is possible only to say that one algorithm is better sometimes, and the other is better other times.

6.5.10 Some Other Multivariate Distributions

Aside from the multivariate normal distribution, the most commonly used multivariate distribution is probably one derived from the normal, the Wishart distribution. The Wishart matrix is the p x p symmetric matrix of sums of squares and cross-products of a sample from the p-variate normal distribution, that is, it is essentially the sample covariance matrix. Hartley and Harris (1963) showed that for $p = 2$ the sums of squares and cross-products could be computed directly, without forming a sample from the normal distribution. Odell and Feiveson (1966) extended the method of Hartley and Harris to any size covariance matrix. The method of Odell and Feiveson for generating the p x p sample covariance matrix S from a sample of size n from a p-variate normal distribution with (nonsingular) variance-covariance matrix V is as follows:

1. Determine A such that $V = AA'$.
2. Generate a sequence of independent N(0,1) variates, z_{ij}, for $i = 1, 2, \ldots, j$ and $j = 2, 3, \ldots, p$.
3. Generate a sequence of independent chi-square variates, y_i, for $i = 1, \ldots, p$, such that y_i has $n - i$ d.f.
4. Compute $B = (b_{ij})$ as follows:

$$b_{11} = y_1$$

$$b_{jj} = y_j + \sum_{i=1}^{j-1} z_{ij}^2 \quad \text{for } j = 2, 3, \ldots, p$$

$$b_{1j} = z_{1j} y_1^{1/2}$$

$$b_{ij} = z_{ij} y_i^{1/2} + \sum_{k=1}^{i-1} z_{ki} z_{kj} \quad \text{for } i < j = 2, 3, \ldots, p$$

$$b_{ij} = b_{ji} \quad \text{for } j < i = 2, 3, \ldots, p$$

5. Form $S = ABA'/n$.

The matrix S is a Wishart matrix (aside from the factor 1/n) with n - p d.f.

A noncentral Wishart matrix (the matrix of uncorrected sums of squares and cross-products) with p d.f. can be generated from a central Wishart with n - p d.f. and p^2 standard univariate normals as shown by Chambers (1970).

Smith and Hocking (1972) gave a FORTRAN program for generating Wishart matrices.

Johnson and Hegemann (1974) and Gleser (1976) described efficient methods for generating noncentral Wishart matrices. Johnson and Hegemann also gave algorithms for generating the eigenvalues and determinants of random matrices.

Chalmers (1975) considered the generation of correlation matrices whose eigen-structure is specified.

Li and Hammond (1975) and Johnson and Ramberg (1977) extended some of the ideas of multivariate normal generation (Section 6.5.9) to the generation of other multivariate distributions with specified covariance structure. The early monograph of Fieller (1955) is also of interest in this connection.

Ronning (1977) pointed out that the techniques for generating multivariate normals apply immediately to the generation of multivariate lognormals.

Ronning (1977) considered a type of multivariate gamma distribution and gave an algorithm for generating variates from it.

Kemp and Loukas (1978a, 1978b) have studied the problems associated with generation of variates from general discrete bivariate distributions. They discussed in particular the generation of bivariate Poisson and bivariate Hermite variates.

Other than the work reported above, there has been very little research into methods for generation of variates from multivariate distributions.

6.6 APPLICATIONS

Random numbers play an important role in many areas of research. We will briefly indicate the nature of the applications in Monte Carlo and in sampling and randomization.

6.6.1 The Monte Carlo Method

Very good descriptions of the basics of the Monte Carlo method are given in the books by Hammersley and Handscomb (1964) and Shreider (1966). System simulation, which is an extension of the Monte Carlo method to more complex models, is discussed by several authors. The books by Fishman (1973), Kleijnen (1974), and Mihram (1972) are good introductions to this more general topic. Mihram considers some important differences in the techniques of simple Monte Carlo that may not be appropriate for simulation of more complex models.

Every quantitative result of a Monte Carlo experiment is essentially an estimate of the value of some integral. Moreover, the integral is generally over the hypercube with edges (0,1). This is the case because the random numbers being used are U(0,1), before transformation; the transformations form part of the integrand. Hence, the basic Monte Carlo procedure is to obtain an estimate of

$$\int_0^1 \cdots \int_0^1 T(u_1, u_2, \ldots, u_n)\, du_1\, du_2 \ldots du_n$$

The subsequent discussion may be simplified by considering instead the single integral (6.1),

$$\theta = \int_0^1 f(u)\, du$$

If $U_1, U_2, \ldots, U_n$ is a random sample from U(0,1), the estimator

$$\hat{\theta} = \frac{1}{n} \sum_{i=1}^{n} f(U_i) \tag{6.23}$$

is unbiased and has variance

$$\mathrm{Var}(\hat{\theta}) = \frac{1}{n} \int_0^1 [f(u) - \theta]^2\, du$$

The technique indicated by (6.23) is referred to as "crude" Monte Carlo.

Since Monte Carlo is the use of samples to obtain an estimate, the object would be to decrease the variance of the estimator, while preserving its unbiasedness and without increasing unduly the cost of obtaining the estimate. Such considerations constitute the core of the field of sampling and are discussed at length by Cochran (1977), for example. Specific types of variance reduction for Monte Carlo sampling are discussed by Hammersley and Handscomb (1964).

A very basic principle of variance reduction in Monte Carlo is: Replace random components by their expectations if the variance will be reduced thereby. For example, to integrate

$$\int_a^b [f(x) + g(x)]\, dx$$

if it is known that $\int_a^b g(x)\, dx = I$, it may be advantageous to use Monte Carlo on $\int_a^b f(x)\, dx$ and take the estimate of this latter integral plus I as an estimate of the integral of interest. This principle must be applied with care, and is, at best, only a general guide. In the example considered above, if the variation in $f(x) + g(x)$ is less than that of $f(x)$ alone, it may be better not to separate the integral into its component parts, even though one part is known.

Other techniques of variance reduction for which we refer the reader to Hammersley and Handscomb (1964) include stratified sampling, whose name indicates the nature of the technique; use of antithetic variates, in which, for example, for every u used, $1 - u$ is also used; use of control variates, and regression methods, in which combinations of variates are used to reduce the variance; and importance sampling.

The propensity for metaphor of Monte Carloists has led to the term *swindle* for any variance reducing method.

The framing of the general Monte Carlo problem as a problem of evaluating an integral suggests one of the straightforward applications of Monte Carlo, namely numerical quadrature. When Monte Carlo is to be used for quadrature various techniques of variance reduction mentioned above may be employed. Specific procedures are discussed by Haber (1966, 1967), Davis and Rabinowitz (1967), and Gallaher (1967, 1973), among others. Generally speaking, we would suggest that Monte Carlo be the technique of last resort for a straightforward quadrature problem. Techniques discussed in Chapter 5 should usually be preferred.

In the more common statistical applications of Monte Carlo (e.g., in Andrews et al. 1972 or Chase and Bulgran 1971), while it is true that the problem is essentially the evaluation of some integral, the integrand may not be immediately apparent. Nevertheless, the variance reduction techniques discussed above should be employed where possible. Discussions and examples of useful variance reductions may be seen in Relles (1970), Andrews et al. (1972), Gross (1973), and Simon (1976).

An important consideration in the conduct of a Monte Carlo experiment is the items to include in the report of the study. Frequently one sees in articles in the statistical journals reports of "Monte Carlo studies" that fail to supply any details of the study; only the "results" are given. Since the results are realizations of pseudorandom processes, however, the reader of the article may be left wondering about the magnitude of the pseudo-standard errors and, indeed, about the effect of the "pseudo" nature of the process itself. Hoaglin and Andrews (1975) have proposed standards for the reporting of Monte Carlo results. Among other things, they suggest that descriptions be given of the basic random number generators used as well as of the transformations to nonuniform variates and of any variance reducing technique used. Also descriptions of supporting software and the machine(s) used should be given. The basic suggestion is that the conduct and reporting of a Monte Carlo study be given as serious consideration as other types of scientific experimentation. The paper by Kleijnen (1977) is a good example of a well-designed Monte Carlo study of a statistical problem.

6.6.2 Sampling and Randomization

A process in which random numbers play an important role is random sampling. In many common sampling applications there is a finite population of size N from which it is desired to choose a sample of size n in such a way that any subset of size n of the original population has an equal chance of being the sample chosen. This is called *simple random sampling*. There are other useful types of sampling in which items are selected with probabilities chosen on the basis of some known characteristic of the item, or in which items are selected in clusters, or in some systematic fashion. Most of the general types of sampling allow for either *sampling with replacement*, in which items may be repeated in the sample, or *sampling without replacement*.

The population or subpopulation to be sampled can generally be assumed to be labeled by the integers 1, 2, ..., N, as subscripts in an array, indices on records, or just ordinal numbers of records; hence, the problem of taking a random sample becomes the problem of generating n random integers between 1 and N. For sampling without replacement these n integers must be unique. In the methodology of sampling an important distinction exists between sequential and nonsequential techniques. In sequential procedures the items of the population are encountered sequentially (and only once), and upon encounter must be included in the sample or rejected; there is no coming back to items once passed over. Nonsequential procedures allow for more flexibility of the selection process. Drawing numbers out of a hat is a nonsequential process, for example. In computer applications the sequential procedures are the most commonly employed due to the nature of storage and access of large files of data. Population data (such as U.S. Census data) generally are stored on magnetic tapes, since tapes are relatively inexpensive storage media. Tapes, however, are read sequentially, and so any practical method for sampling from a tape must be a sequential one.

There are several algorithms for random sampling, which are discussed by Fan, Muller, and Rezucha (1962). The criteria for a good algorithm are (1) efficiency in (a) storage requirements and (b) time, and (2) correct, or nearly correct, solution to the problem--that is, resulting in a sample that satisfies the definition of a random sample. The tradeoffs between storage requirements, time, relative sizes of n and N, and the desirability of achieving a sample that strictly meets all of the requirements of the definition of a random sample of size n are such that no algorithm can be said to be universally best. We now consider some of the possible methods.

1. A method which is very simple conceptually is to choose integers between 1 and N randomly (by generating a number U_k from Uniform (0,1) and taking $[U_k \cdot N] + 1$, for example) and each time to check the current sample to see if the integer is already included. Despite its simplicity, computer implementations of this method are quite inefficient unless n and n/N are small.

   A modification of this method omits the checking stage. This increases the efficiency, but the method is now sampling with replacement. If, however, N is very large relative to n, the probability of duplicating sample units is small and so this method is sometimes used, even for sampling that is intended to be without replacement.
2. A very efficient sequential method is to select each index with probability n/N unconditionally. In this procedure, each time a population unit is encountered a U(0,1) random number is generated. If this number is less than n/N, the unit is included in the sample; otherwise, it is rejected. Although this procedure would not usually result in a sample of size exactly n, its high efficiency sometimes recommends its use. In some cases achieving an exactly fixed-size sample is not extremely important.
3. An efficient sequential method that will result in a sample of size exactly n is the following. Select the ith population unit for inclusion in the sample with probability

$$\begin{cases} \dfrac{n - n_i}{N - i + 1} & \text{if } n_i < n \\ 0 & \text{otherwise} \end{cases}$$

where n_i is the number of units selected just prior to the encounter of the ith population unit; i = 1, 2, ..., N. The procedure is charted in Fig. 6.14.

4. If the population size N is not known in advance, it is still possible to select sequentially a random sample of size n. One way, of course, would be to read the file once just to determine N and then use method 3; and in fact, this should be done in some cases--the cost of reading the file would be the important consideration. An efficient sequential method requiring only

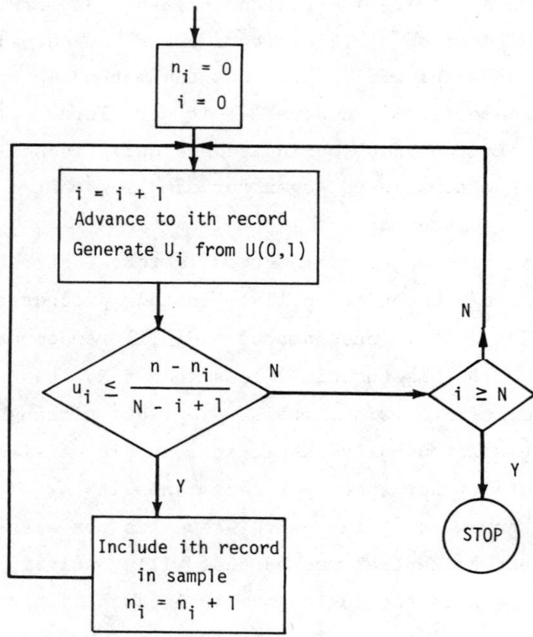

Figure 6.14 Sequential sampling without replacement to obtain a specified sample size when the population size is known.

one pass involves assigning random numbers independently to the population units and forming the same from a subset whose associated random numbers are n specified-order statistics, say the first n order statistics. It is easy to see that this yields a random sample of population units since the position in the

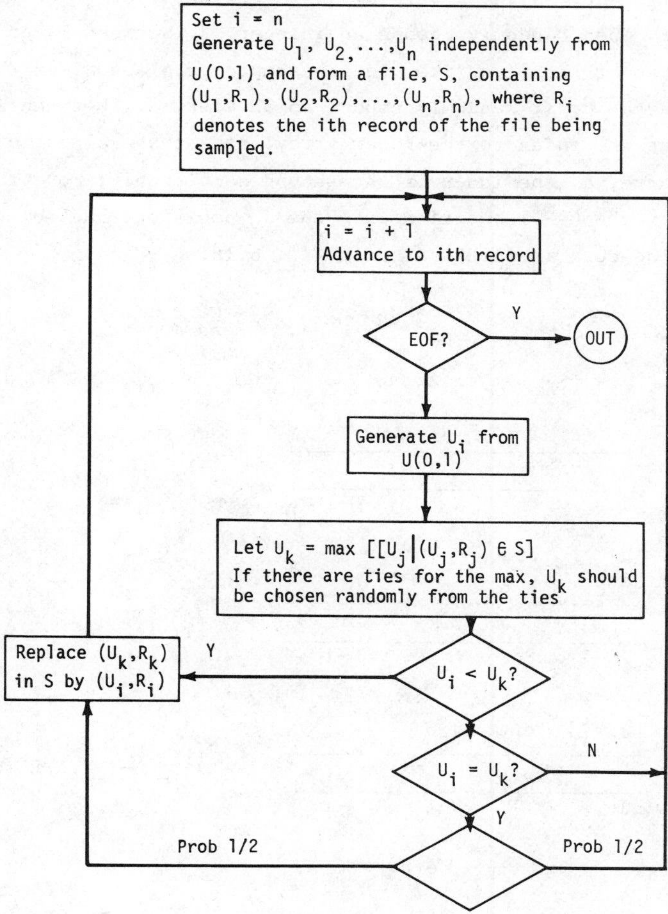

Figure 6.15 Sequential sampling without replacement to obtain a specified sample size when the population size is unknown.

sequence of any one first n order statistics of a random sequence of N random numbers occurs with equal probability and independently of the positions of any other order statistic in that sequence. It is also easy to see that this can be implemented in a sequential fashion. The flowchart of Fig. 6.15 describes the method.

5. A nonsequential method for selecting a random sample of size exactly n where N is known begins by filling an index array from 1 to N. Then randomly choose an integer, J_1, between 1 and N as the index of the first item in the sample; replace J_1 in the index array by N; randomly choose an integer, J_2, between 1 and N - 1 as the index to the index array; take that value from the index array as the index of the second sample unit; move the index in the N - 1 position into the J_2 position; and so on. This procedure is illustrated in Fig. 6.16.

	Index Array	*Random Number Generated*	*Sample*
Position	1 2 ... N-1 N		
Contents	1 2 ... N-1 N		
Position	1 2 ... J_1 ... N-1 N	$1 \leq J_1 \leq N$	INDEX(J_1)
Contents	1 2 ... N ... N-1 N		
Position	1 2 ... J_1 ... J_2 ... N-1 N	$1 \leq J_2 \leq N-1$	INDEX(J_2)
Contents	1 2 ... N ... N-1 ... N-1 N		

etc.

Figure 6.16 Randomization of the first N integers.

If N passes are made in the algorithm V, it has the effect of randomly rearranging the integers 1 through N. This makes the procedure useful in certain simulation problems. For example, the shuffling of a deck of playing cards can be simulated in this way. This method of random permutations was given by Page (1967). Rao (1961) discussed other methods of generating random permutations. Since the number of permutations N! can become quite large, random number generators with very long periods may be necessary to allow each of the possible permutations a chance to be realized. Salfi (1974) and Greenwood (1976a) discussed such problems (see also Section 6.2.2).

EXERCISES

6.1 In the usual computer implementation of a congruential generator, numbers in integer representation are produced and then changed into floating-point representation and scaled into the interval [0,1). In FORTRAN the statment to do the transformation may be

```
U = FLOAT(IX)/P
```

On the IBM 370 the IX in this statement may be between 0 and $2^{31} - 1$ and $P = f\ell(2^{31})$. Let I be the set of possible values of IX, R be the set of possible values of U, and S be the set of single-precision floating-point numbers in the interval [0,1). Assume the generator has period 2^{31}.

(a) What is the cardinality of S?

(b) What is the cardinality of I?

(c) How does the cardinality of R compare with those of S and I? Discuss.

(d) The sequence of IX's does not contain a repeat of any value in any contiguous subsequence of length 2^{31}. Does the sequence of U's? Support your assertion.

6.2 The Tausworthe method produces numbers in the range (0,1) which have the form

$$y_k = 0.a_{sk+r-1}a_{sk+r-2}\cdots a_{sk+r-L} \quad \text{(base 2)}$$

Define $\alpha_k = 1 - 2a_k$. Show that if w_k is defined as

$$w_k = \sum_{t=1}^{L} 2^{-t}\alpha_{sk+r-t}$$

then $w_k = 1 - 2^{-L} - 2y_k$.

6.3 Using w_k from Exercise 6.2, show that

$$E(w_k) = -2^{-p}\left(\frac{1 - 2^{-L}}{1 - 2^{-p}}\right)$$

and that $E(y_k)$ approaches 1/2 for large L and p.

6.4 Using the notation in Exercises 6.2 and 6.3, and the fact that $E(w_k) \cong 0$, show that the variance of w_k is approximately that of a standard uniform random variable by showing that

$$\frac{1}{n}\sum_{k=1}^{n} w_k^2 = \frac{1}{3} + 2^{-p}\left[\frac{1}{3}\frac{1 - 2^{-2L}}{1 - 2^{-p}} - \frac{(1 - 2^{-L})^2}{1 - 2^{-p}}\right]$$

6.5 Consider the sequence of integers $\{Z_j\}_{j=1}^{n}$ which are produced in binary form by the GFSR method. Show that on a L-bit machine (disregarding sign bit) the sequence $\{Z_j\}$ has the approximate mean and variance of a random sample of size $n = 2^p - 1$ from the standard uniform distribution. (*Hint:* Count the number of nonzero duplicates for each Z_j.)

6.6 In Section 6.5.7 it was mentioned that set of the central order statistics of a sample from U(0,1) could be generated by first generating an appropriate beta and then generating the others desired in descending order. Give the details of this technique for generating the ith- and (i + 1)th order statistics from a sample of size n.

6.7 Let $Y_1, Y_2, \ldots, Y_{n+1}$ be independently distributed as exponential with mean 1. Let

$$R_i = \frac{Y_i}{Y_1 + Y_2 + \cdots + Y_{n+1}} \qquad \text{for } i = 1, 2, \ldots, n$$

Prove that R_i has the same distribution as D_i given by

$$D_i = V_{(i)} - V_{(i-1)}$$

where $V_{(i)}$, $i = 1, 2, \ldots, n$, is the ith order statistic in a sample of size n from U(0,1) and $V_{(0)} = 0$.

6.8 Prove that Eq. (6.11) is the condition for the Fourier transform of the sequence from the mixed generator to be nonzero.

6.9 Derive an algorithm for generating variates from the Cauchy distribution with density

$$f(x) = \frac{1}{\pi(1 + x^2)} \qquad -\infty < x < \infty$$

Discuss the efficiency of your algorithm.

6.10 Let U_1 and U_2 be independent U(0,1) variates. Find the distribution of the variables $X = \sqrt{U_1}$ and $Y = \max(U_1, U_2)$. Comment on the applications of this exercise.

6.11 Modify algorithm PL of Section 6.4.2 for the simpler case in which all coefficients are ≥ 0.

6.12 Prove the validity of algorithm PL. (*Hint:* Consider the distribution of the "nonmaximum.")

6.13 Determine the moments of time to generate variates by Sibuya's modification of the rejection method described in Section 6.4.3. (See Greenwood 1976c.)

6.14 Consider the algorithm given in Example 6.11 and let $q = e^{-\delta}$. Note that on exit from the algorithm the conditions $u_1 > q$, $u_1u_2 > q, \ldots, u_1u_2\ldots u_n > q$, $u_1u_2\ldots u_{n+1} \leq q$ hold where the u_j are independent standard uniform random variables.

(a) Show that for n = 1 and n = 2 the probability that the set of inequalities given above hold is $q[-\ln(q)]^n/n!$ when $0 < q \leq 1$.

(b) Use induction on n to prove the validity of the algorithm in Example 6.11. You will, in the proof, probably evaluate the integral

$$\int_q^1 (q/x)\ [\ln (x/q)]^{n-1} \frac{dx}{(n - 1)!}$$

(c) Prove the expected number of trials to achieve a success in the given algorithm is $\delta + 1$. [You may find it easy to determine E(n) and note that n is 1 less than the number of trials.]

6.15 Prove the lemma in Section 6.5.5 that gives the algorithm BB. (Ahrens and Dieter 1974.)

6.16 Prove the lemma in Section 6.5.6 that gives the algorithm PG. (Ahrens and Dieter 1974.)

6.17 Write a FORTRAN subroutine to generate random numbers from the p-variate normal distribution with nonsingular variance-covariance matrix A using the conditional distribution technique (Section 6.5.9).

6.18 Devise an algorithm to generate variates from the noncentral chi-square distribution.

6.19 For estimating the value of the integral

$$\theta = \int_0^1 f(u)\ du$$

when $a \leq f(u) \leq b$ for $0 < u < 1$, consider the following method: Generate pairs of numbers (u_i, v_i), u_i from U(0,1) and v_i from U(a,b). Initialize n = 0. For i = 1 to N, if $v_i < f(u_i)$, increment n = n + 1. Take $\tilde{\theta} = n/N$. This is called "hit-or-miss" Monte Carlo.

(a) Prove $\tilde{\theta}$ is unbiased for θ.

(b) Prove that the variance of $\tilde{\theta}$ is at least as great as the variance of the crude estimator $\hat{\theta}$ in (6.23).

6.20 Prove that method 3 of Section 6.6.2 yields a simple random sample of size n from the population of size N.

6.21 Prove that method 5 of Section 6.6.2 produces a random permutation of the integers 1 to N.

6.22 In Example 6.6 derive the pdf of each of the following random variables.

(a) $x = q_j + (q_{j+1} - q_j) \max (u_1,u_2) \quad (r_j > 0)$

(b) $x = q_j + (q_{j+1} - q_j) \min (u_1,u_2) \quad (r_j < 0)$

(c) $x = q_j + (q_{j+1} - q_j)u_1$ where $u_1 \sim U(0,1)$

REFERENCES AND FURTHER READINGS

Abramowitz, A., and I. A. Stegun (eds.) (1964), *Handbook of Mathematical Functions*, National Bureau of Standards, Washington, D.C.

Ahrens, J. H., and U. Dieter (1972), Computer Methods for Sampling from the Exponential and Normal Distributions, *CACM 15*, 873-882.

Ahrens, J. H., and U. Dieter (1973), Extensions of Forsythe's Method for Random Sampling from the Normal Distribution, *Math. Comp. 27*, 927-937.

Ahrens, J. H., and U. Dieter (1974), Computer Methods for Sampling from Gamma, Beta, Poisson, and Binomial Distributions, *Computing 12*, 223-246.

Ahrens, J. H., U. Dieter and A. Grube (1970), Pseudo-random Numbers: A New Proposal for Choice of Multipliers, *Computing 6*, 121-138.

Allard, J. L., A. R. Dobell and T. E. Hull (1963), Mixed Congruential Random Number Generators for Decimal Machines, *JACM 10*, 131-141.

Andrews, D. F., P. J. Bickel, F. R. Hampel, P. J. Huber, W. H. Rogers and J. W. Tukey (1972), *Robust Estimates of Location: Survey and Advances*, Princeton University Press, Princeton, N.J.

Arnason, A. N. (1974), Computer Generation of Cauchy Variates, *Proceedings of the Fourth Manitoba Conference on Numerical Mathematics*, edited by H. C. Williams and B. L. Hartnell, Utilitas Mathematica Publishing, Winnipeg, 177-190.

Arvillias, A. C., and D. G. Maritsas (1978), Partitioning the Period of a Class m-Sequences and Applications to Pseudorandom Number Generation, *JACM 25*, 675-686.

Atkinson, A. C. (1977), An Easily Programmed Algorithm for Generating Gamma Random Variates, *JRSS (A) 140*, 232-234.

Atkinson, A. C., and M. C. Pearce (1976), The Computer Generation of Beta, Gamma, and Normal Random Variables (With Discussion), *JRSS (A) 139*, 431-461.

Atkinson, A. C., and J. Whittaker (1976), A Switching Algorithm for the Generation of Beta Random Variables with at Least One Parameter Less than 1, *JRSS (A) 139*, 462-467.

Barnett, V. D. (1962), The Behavior of Pseudo-random Sequences Generated on Computers by the Multiplicative Congruential Method, *Math. Comp. 16*, 63-69.

Barnett, V. D. (1965), *Random Negative Exponential Deviates*, Tracts for Computers 27, Cambridge University Press, London.

Barr, D. R., and N. L. Slezak (1972), A Comparison of Multivariate Normal Generators, *CACM 15*, 1048-1049.

Bartels, R. (1978), Generating Non-Normal Stable Variates Using Limit Theorem Properties, *JSCS 7*, 199-212.

Bauer, W. F. (1958), The Monte Carlo Method, *SIAM J. 6*, 438-451.

Bays, C., and S. D. Durham (1976), Improving a Poor Random Number Generator, *TOMS 2*, 59-64.

Bedall, F. K., and H. Zimmermann (1976), On the Generation of $N(\mu,\Sigma)$-distributed Random Vectors by $N(0,1)$-distributed Random Numbers, *Biometrische Zeitschrift 18*, 467-472.

Bell, J. R. (1968), Normal Random Deviates, *CACM 11*, 498.

Beyer, W. A., R. B. Roof, and D. Williamson (1971), The Lattice Structure of Multiplicative Congruential Pseudo-Random Vectors, *Math. Comp. 25*, 345-363.

Blood, F. A., Jr. (1976), Correlations in Power Residue Generated Random Numbers, *J. Computational Physics 20,* 372-379.

Bofinger, E., and V. J. Bofinger (1958), On a Periodic Property of Pseudo-random Sequences, *JACM 5*, 261-265.

Bofinger, E., and V. J. Bofinger (1961a), The Gap Test for Random Sequences, *AMS 32*, 524-534.

Bofinger, E., and V. J. Bofinger (1961b), A Runs Test for Sequences of Random Digits, *Australian J. of Statistics 3*, 37-41.

Bollman, D. A. (1965), Some Periodic Properties of Transformations on Vector Spaces Over Residue Class Rings, *SIAM J. Appl. Math. 13*, 902-912.

Bowman, K. O., and J. J. Beauchamp (1975), Pitfalls with Some Gamma Variate Simulation Routines, *JSCS 4*, 141-154.

Box, G. E. P., and M. E. Muller (1958), A Note on Generation of Normal Deviates, *AMS 28*, 610-611.

Brent, R. P. (1974), A Gaussian Pseudo-Random Number Generator, *CACM 17*, 704-706.

Brown, G. W. (1951), *History of RAND's Random Digits*, National Bureau of Standards Applied Mathematical Series 12, Washington, D.C., 31-32.

Burford, R. L. (1973), A Better Additive Congruential Random Number Generator, *Decision Sciences 4*, 190-193.

Burford, R. L. (1975), A Better Additive Congruential Random Number Generator?--A Reply, *Decision Sciences 6*, 199-201.

Burr, I. W. (1942), Cumulative Frequency Functions, *AMS 13*, 215-232.

Burr, I. W. (1967), A Useful Approximation to the Normal Distribution, with Application to Simulation, *Techno. 9*, 647-651.

Burr, I. W. (1973), Parameters for a General System of Distributions to Match a Grid of α_3 and α_4, *Comm. Stat. 2*, 1-21.

Butcher, J. C. (1961a), A Partition Test for Pseudo-Random Numbers, *Math. Comp. 15*, 198-199.

Butcher, J. C. (1961b), Random Sampling from the Normal Distribution, *Comp. J. 3*, 251-253.

Butler, E. L. (1970), Algorithm 370: General Random Number Generator, *CACM 13*, 49-52.

Canavos, G. C. (1967), A Comparative Analysis of Two Concepts in the Generation of Uniform Pseudorandom Numbers, *Proceedings of Twenty-second National Conference of the ACM*, 485-501.

Cenacchi, G., and A. Matteis (1970), Pseudo-Random Numbers for Comparative Monte Carlo Calculations, *Numer. Math. 16*, 11-15.

Certaine, J. (1958), On Sequences of Pseudo-Random Numbers of Maximal Length, *JACM 5*, 353-356.

Chalmers, C. P. (1975), Generation of Correlation Matrices with a Given Eigen-Structure, *JSCS 4*, 133-139.

Chamayou, J. M. F. (1976), On a Direct Algorithm for Generation of Log-Normal Pseudo-Random Numbers, *Computing 16*, 69-76.

Chambers, J. M. (1970), Computers in Statistical Research: Simulation and Computer-Aided Mathematics, *Techno. 12*, 1-15.

Chambers, J. M., C. L. Mallows, and B. W. Stuck (1976), A Method for Simulating Stable Random Variables, *JASA 71*, 340-344.

Chambers, R. P. (1967), Random Number Generation, *IEEE Spectrum 4*, part 1, no. 2, 48-56.

Chase, G. R., and W. G. Bulgran (1971), Monte Carlo Investigation of the Robustness of T^2, *JASA 66*, 499-502.

Chay, S. C., R. D. Fardo, and M. Mazumdar (1975), On Using the Box-Muller Transformation with Multiplicative Congruential Pseudo-Random Number Generators, *Appl. Stat. 24*, 132-135.

Chen, E. H. (1971), Random Normal Number Generator for 32-Bit-Word Computers, *JASA 66*, 400-403.

Chen, H. C., and Y. Asau (1974), On Generating Random Variates from an Empirical Distribution, *AIIE Transactions 6*, 163-166.

Cheng, R. C. H. (1977), The Generation of Gamma Variables with Non-Integral Shape Parameter, *Appl. Stat. 26*, 71-75.

Cheng, R. C. H. (1978), Generating Beta Variates with Nonintegral Shape Parameters, *CACM 21*, 317-322.

Claustriaux, J. J. (1976), Generation and Validity Control of Pseudo-Random Numbers on a Computer with 16 Bit Words, *Revue de Statistique Appliques 24*, 75-88.

Cochran, W. G. (1977), *Sampling Techniques*, 3rd ed., Wiley, New York.

Coldwell, R. L. (1974), Correlational Defects in the Standard IBM 360 Random Number Generator and the Classical Ideal Gas Correlational Function, *J. Computational Physics 14*, 223-226.

Cooper, J. D., S. A. Davis, and N. R. Dono (1965), Pearson Universal Random Distribution Generator (PURGE), *Amer. Soc. Quality Control Transactions*, 402-411.

Coveyou, R. R. (1960), Serial Correlation in the Generation of Pseudo-Random Numbers, *JACM 7*, 72-74.

Coveyou, R. R. (1970), Random Numbers Fall Mainly in the Planes (review), *Comp. Rev.*, 225.

Coveyou, R. R., and R. D. MacPherson (1967), Fourier Analysis of Uniform Random Number Generators, *JACM 14*, 100-119.

Craddock, J. M., and S. A. Farmer (1971), Two Robust Methods of Random Number Generation, *The Statistician 20*, September, 55-66.

Cranley, R., and T. N. L. Patterson (1970), A Regression Method for the Monte Carlo Evaluation of Multidimensional Integrals, *Numer. Math. 16*, 58-72.

Curtiss, J. H. (1953), Monte Carlo Methods for the Iteration of Linear Operations, *J. Math. Phys. 32*, 209-230.

Daley, R. P. (1975), Minimal-Program Complexity of Pseudo-Recursive and Pseudo-Random Sequences, *Math. System Theory 9*, 83-96.

Davis, P. J., and P. Rabinowitz (1967), *Numerical Integration*, Blaisdell, Waltham, Mass.

Dieter, U. (1969), Autokorrelation Multiplikativ-Erzeuter Pseudo-Zufallszahlen, *Operations Research-Verfahren 6*, 69-85.

Dieter, U. (1971), Pseudo-Random Numbers: The Exact Distribution of Pairs, *Math. Comp. 25*, 855-883.

Dieter, U. (1972), Statistical Interdependence of Pseudo-Random Numbers Generated by the Linear Congruential Method, in *Applications of Number Theory to Numerical Analysis*, edited by S. K. Zaremba, Academic Press, New York, 287-318.

Dieter, U. (1975), How to Calculate Shortest Vectors in a Lattice, *Math. Comp. 29*, 827-833.

Dieter, U., and J. Ahrens (1971), An Exact Determination of Serial Correlations of Pseudo-Random Numbers, *Numer. Math. 17*, 101-123.

Dieter, U., and J. H. Ahrens (1973), A Combinatorial Method for Generation of Normally Distributed Random Numbers, *Computing 11*, 137-146.

Dodd, E. L. (1942), Certain Tests for Randomness Applied to Data Grouped into Small Sets, *Econometrica 10*, 249-257.

Donnelly, T. (1969), Some Techniques for Using Pseudorandom Numbers in Computer Simulation, *CACM 12*, 392-394.

Downham, D. Y. (1969), The Runs Up and Down Test, *Comp. J. 12*, 373-376.

Downham, D. Y. (1970), The Runs Up and Down Test, *Appl. Stat. 19*, 190-192.

Downham, D. Y., and F. D. K. Roberts (1967), Multiplicative Congruential Pseudo-Random Number Generators, *Comp. J. 10*, 74-77.

Dudewicz, E. J. (1975a), Random Numbers: The Need, the History, the Generators, in *A Modern Course on Statistical Distributions in Scientific Work*, vol. 2, edited by G. P. Patil, S. Kotz, and J. K. Ord, D. Reidel, Boston, 25-36.

Dudewicz, E. J. (1975b), Chi-Square Percentage Points for Chi-Square Distribution Testing of Chi-Square Values, with an Application to Random Number Generators, *Proceedings of Computer Science and Statistics: Eighth Annual Symposium on the Interface*, edited by J. Frane, Health Science Computing Facility, UCLA, 217-221.

Dudewicz, E. J. (1976), Speed and Quality of Random Numbers for Simulation, *J. Quality Technology 8*, 171-178.

Edgington, E. S. (1961), Probability Table for Number of Runs of Signs of First Differences in Ordered Series, *JASA 56*, 156-159.

Edmons, A. R. (1960), The Generation of Pseudo-Random Numbers on Electronic Digital Computers, *Comp. J. 2*, 181-185.

Enison, R. L., and H. S. Bright (1976), Preliminary Testing of a 64-Bit Tausworthe-Lewis-Payne Generator of Mersenne Exponent Degree, *Proceedings of Ninth Interface Symposium on Computer Science and Statistics*, edited by D. Hoaglin and R. Welsch, Prindle, Weber and Schmidt, Boston, 208-211.

Fan, C. T., M. E. Muller, and I. Rezucha (1962), Development of Sampling Plans by Using Sequential (Item by Item) Selection Techniques and Digital Computers, *JASA 57*, 387-402.

Fellen, B. M. (1969), An Implementation of the Tausworthe Generator, *CACM 12*, 413.

Fieller, E. (1955), *Correlated Random Normal Deviates*. Tracts for Computers 26, Cambridge University Press, London.

Fishman, G. S. (1973a), *Concepts and Methods in Discrete Event Digital Simulation*, Wiley, New York.

Fishman, G. S. (1973b), Variance Reduction for Normal Variates in Monte Carlo Studies, *JSCS 2*, 365-374.

Fishman, G. S. (1976a), Sampling from the Gamma Distribution on a Computer, *CACM 19*, 407-409.

Fishman, G. S. (1976b), Sampling from the Poisson Distribution on a Computer, *Computing 17*, 147-156.

Fisser, H. (1961), Some Tests Applied to Pseudo-Random Numbers Generated by V. Hoerner's Rule, *Numer. Math. 3*, 247-249.

Forsythe, G. E. (1972), Von Neumann's Comparison Method for Random Sampling from the Normal and Other Distributions, *Math. Comp. 26*, 817-826.

Fox, B. L. (1963), Generation of Random Samples from the Beta and F Distributions, *Techno. 5*, 269-270.

Franklin, J. N. (1958), On the Equidistribution of Pseudo-Random Numbers, *Quart. Appl. Math. 16*, 183-188.

Franklin, J. N. (1965), Numerical Simulation of Stationary and Non-Stationary Gaussian Random Process, *SIAM Rev. 7*, 68-80.

Franklin, M. A., and A. Sen (1975), Comparison of Exact and Approximate Variate Generation Methods for the Erlang Distribution, *JSCS 4*, 1-18.

Friday, D. S. (1976), Computer Generation of Random Number from Bivariate Life Distributions, *Proceedings of Ninth Interface Symposium on Computer Science and Statistics*, edited by D. Hoaglin and R. Welsch, Prindle, Weber and Schmidt, Boston, 218-221.

Friday, D. S., G. P. Patil, and M. T. Boswell (1976), A Study of the Generation of Non-Uniform Random Numbers on a Computer, *Proceedings of Ninth Interface Symposium on Computer Science and Statistics*, edited by D. Hoaglin and R. Welsch, Prindle, Weber and Schmidt, Boston, 191-196.

Fuller, A. T. (1976), The Period of Pseudo-Random Numbers Generated by Lehmer's Congruential Method, *Comp. J. 19,* 173-177.

Gallaher, L. J. (1967), An Adaptive Quadrature Procedure with Random Panel Sizes, *CACM 10*, 373-374.

Gallaher, L. J. (1973), A Multidimensional Monte Carlo Quadrature with Adaptive Stratified Sampling, *CACM 16*, 49-50.

Galliher, H. P. (1959), *Simulation of Random Processes: Notes on Operations Research*. Technology Press, Massachusetts Institute of Technology, Cambridge, 231-254.

Gates, C. E. (1978), On Generating Random Normal Deviates Using the Butler Algorithm, *Proceedings of the Statistical Computing Section, American Statistical Association*, Washington, D.C., 111-114.

Gebhardt, F. (1964), Generating Normally Distributed Random Numbers by Inverting the Normal Distribution Function, *Math. Comp. 18*, 302-306.

Gebhardt, F. (1967), Generating Pseudo-Random Numbers by Shuffling a Fibonacci Sequence, *Math. Comp. 21*, 708-709.

George, R. (1963), Normal Random Variables, *CACM 6*, 444.

Gibbons, J. D. (1976), *Nonparametric Methods for Quantitative Analysis*, Holt, Rinehart and Winston, New York.

Gleser, L. J. (1976), A Canonical Representation for the Noncentral Wishart Distribution Useful for Simulation, *JASA 71*, 690-695.

Golder, E. R. (1976), The Spectral Test for the Evaluation of Congruential Pseudo-Random Generators, *Appl. Stat. 25*, 173-180.

Golder, E. R., and J. G. Settle (1976), The Box-Muller Method for Generating Pseudo-Random Normal Deviates, *Appl. Stat. 25*, 12-20.

Goldstein, N. (1963), Random Numbers from the Extreme Value Distributions, *Publications Inst. Statist. Univ. Paris 12*, 137-158.

Golomb, S. W. (1967), *Shift Register Sequences*, Holden-Day, San Francisco.

Good, I. J. (1953), The Serial Test for Sampling Numbers and Other Tests of Randomness, *Proc. Camb. Phil. Soc. 49*, 276-284.

Good, I. J. (1957), On the Serial Test for Random Sequences, *AMS 28*, 262-264.

Gorenstein, S. (1967), Testing a Random Number Generator, *CACM 10*, 111-118.

Greenberger, M. (1961a), Notes on a New Pseudo-Random Number Generator, *JACM 8*, 163-167.

Greenberger, M. (1961b), An A-priori Determination of Serial Correlation in Computer Generated Random Numbers, *Math. Comp. 15* 383-389.

Greenwood, J. A. (1974), A Fast Generator for Gamma-Distributed Random Variables, *Compstat 1974: Proceedings in Computational Statistics*, edited by G. Bruckmann, F. Ferschl, and L. Schmetterer, Physica Verlag, Vienna, 19-27.

Greenwood, J. A. (1976a), The Demands of Trivial Combinational Problems on Random Number Generators, *Proceedings of Ninth Interface Symposium on Computer Science and Statistics*, edited by D. Hoaglin and R. Welsch, Prindle, Weber and Schmidt, Boston, 222-227.

Greenwood, J. A. (1976b), A Fast Machine-Independent Long-Period Generator for 31-Bit Pseudo-Random Numbers, *Compstat 1976: Proceedings in Computational Statistics,* edited by J. Gordesch and P. Naeve, Physica Verlag, Vienna, 30-36.

Greenwood, J. A. (1976c), Moments of the Time to Generate Random Variables by Rejection, *Ann. Instit. Stat. Math. 28*, 399-401.

Greenwood, J. A. (1977), Portable Generators for the Random Variables Usual in Reliability Simulation, in *Recent Developments in Statistics*, edited by J. R. Barra et al., North-Holland Publishing Co., Amsterdam, 677-688.

Greenwood, R. E. (1955), Coupon Collector's Test for Random Digits, *MTAC 9*, 1-5, 224, 229.

Grosenbaugh, L. R. (1969), More on FORTRAN Random Number Generators, *CACM 11*, 639.

Gross, A. M. (1973), A Monte Carlo Swindle for Estimators of Location, *Appl. Stat. 22*, 347-353.

Gruenberger, F., and A. M. Mark (1951), The d^2 Test of Random Digits, *MTAC 5*, 109-110.

Guerra, V. O., R. A. Tapia, and J. R. Thompson (1976), A Random Number Generator for Continuous Random Variables Based on an Interpolation Procedure of Akima, *Proceedings of Ninth Interface Symposium on Computer Science and Statistics*, edited by D. Hoaglin and R. Welsch, Prindle, Weber and Schmidt, Boston, 228-230.

Gustavson, F. G., and W. Liniger (1970), A Fast Random Number Generator with Good Statistical Properties, *Computing 6*, 221-226.

Haber, S. A. (1966), A Modified Monte-Carlo Quadrature, *Math. Comp. 20*, 361-368.

Haber, S. A. (1967), A Modified Monte Carlo Quadrature II, *Math. Comp. 21*, 388-397.

Haber, S. A. (1970), Sequences of Numbers that Are Approximately Completely Equidistributed, *JACM 17*, 269-272.

Halton, J. H. (1970), A Retrospective and Prospective Survey of the Monte Carlo Method, *SIAM Rev. 12*, 1-63.

Hammersley, J. M., and D. C. Handscomb (1964), *Monte Carlo Methods*, Methuen, London.

Hartley, H. O., and D. L. Harris (1963), Monte Carlo Computations in Normal Correlation Problems, *JACM 10*, 302-306.

Hemmerle, W. J. (1967), *Statistical Computations on a Digital Computer*, Blaisdell, Waltham, Mass.

Hemmerle, W. J. (1969), Generating Pseudo-Random Numbers on a Two's Complement Machine such as the IBM 360, *CACM 12*, 382-383.

Hoaglin, D. C. (1975), Some Remarks on Implementing and Documenting Random Number Generators, *Proceedings of Computer Science and Statistics: Eighth Annual Symposium on the Interface*, edited by J. Frane, Health Sciences Computing Facility, UCLA, 213-216.

Hoaglin, D. C., and D. F. Andrews (1975), The Reporting of Computation-Based Results in Statistics, *Amer. Stat. 29*, 122-126.

Hull, T. E., and A. R. Dobell (1962), Random Number Generators, *SIAM Rev. 4*, 230-253.

Hull, T. E., and A. R. Dobell (1964), Mixed Congruential Random Number Generators for Binary Machines, *JACM 11*, 31-40.

Hunter, D. G. N. (1960), Note on a Test for Repeating Cycles in a Pseudo-Random Number Generator, *Comp. J. 3*, 9.

Hurd, W. J. (1974), Efficient Generation of Statistically Good Pseudonoise by Linearly Interconnected Shift Registers, *IEEE Transactions on Computers C-23*, 146-152.

Hurst, R. L., and R. E. Knop (1972), Generation of Random Normal Correlated Variables: Algorithm 425, *CACM 15*, 355-357.

Hutchison, D. W. (1966), A New Uniform Pseudorandom Number Generator, *CACM 9*, 432-433.

IBM Corporation (1968), IBM System/360 Scientific Subroutine Package, IBM Systems Reference Library, White Plains, N.Y.

Ishiketa, T. (1966), The Method for Testing Randomness and Transforming Distribution of Random Number Utilizing Random Walk Simulation, *Information Processing in Japan 6*, 59-71.

Ishiketa, T. (1969), Sensitivity of Method for Testing Randomness Utilizing Random Walk Process, *Information Processing in Japan 9*, 1-8.

Jagerman, D. L. (1965), Some Theorems Concerning Pseudo-Random Numbers, *Math. Comp. 19*, 418-426.

Jansson, B. (1964a), Autocorrelations Between Pseudorandom Numbers, *BIT 4*, 6-27.

Jansson, B. (1964b), Generation of Random Bivariate Normal Deviates and Computation of Related Integral, *BIT 4*, 205-212.

Jansson, B. (1966), *Random Number Generators*, Victor Pettersons, Stockholm.

Jöhnk, M. D. (1964), Erzeugung von Betaverteilter und Gammaverteilter Zufallszahlen, *Metrika 8*, 5-15.

Johnson, D. E., and V. Hegemann (1974), Procedure to Generate Random Matrices with Noncentral Distributions, *Comm. Stat. 3*, 691-699.

Johnson, M. E., and M. M. Johnson (1978), *A New Probability Distribution with Applications in Monte Carlo Studies*, Technical Report LA-7095-MS, Los Alamos Scientific Laboratory, Los Alamos, N.M.

Johnson, M. E., and J. S. Ramberg (1977), Elliptically Symmetric Distributions: Characterizations and Random Variate Generation, *Proceedings of the Statistical Computing Section, American Statistical Association*, Washington, D.C., 262-265.

Johnson, N. L. (1965), Tables to Facilitate Fitting S_u Frequency Curves, *Biometrika 52*, 547-558.

Kahn, H. (1956), Applications of Monte Carlo, Memorandum RM-1237-AEC, Rand Corporation, Santa Monica, Calif.

Kemp, C. D., and S. Loukas (1978a), Computer Generation of Bivariate Discrete Random Variables Using Ordered Probabilities, *Proceedings of Statistical Computing Section, American Statistical Association*, Washington, D.C., 115-116.

Kemp, C. D., and S. Loukas (1978b), The Computer Generation of Bivariate Discrete Random Variables, *JRSSA 141*, 513-519.

Kendall, M. G. (1941), A Theory of Randomness, *Biometrika 32*, 1-15.

Kendall, M. G., and B. Babington-Smith (1939), Randomness and Random Sampling Numbers, *JRSS 101*, 147-156.

Kinderman, A. J., and J. F. Monahan (1976), Generating Random Variables from the Ratio of Two Uniform Variates, *Proceedings of Ninth Interface Symposium on Computer Science and Statistics*, edited by D. Hoaglin and R. Welsch, Prindle, Weber and Schmidt, Boston, 197-200.

Kinderman, A. J., J. F. Monahan, and J. G. Ramage (1975), Computer Generation of Random Variables with Normal and Student's Distributions, *Proceedings of the Statistical Computing Section, American Statistical Association*, Washington, D.C., 128-131.

Kinderman, A. J., J. F. Monahan, and J. G. Ramage (1977), Computer Methods for Sampling from Student's t Distribution, *Math. Comp. 31*, 1009-1018.

Kinderman, A. J., and J. G. Ramage (1976), Computer Generation of Normal Random Variables, *JASA 71* 893-896.

Kleijnen, J. P. C. (1974), *Statistical Techniques in Simulation*, parts 1 and 2, Marcel Dekker, New York.

Kleijnen, J. P. C. (1975), Antithetic Variates, Common Random Numbers, and Optimal Computer Time Allocation in Simulation, *Management Science 21*, 1176-1185.

Knop, R. E. (1973), Random Deviates from the Dipole Distribution, *CACM 16*, 51.

Knuth, D. E. (1965), Construction of a Random Sequence, *BIT 5*, 246-250.

Knuth, D. E. (1969), *The Art of Computer Programming,* vol. 2: *Seminumerical Algorithms*, Addison-Wesley, Reading, Mass.

Knuth, D. E. (1973), *The Art of Computer Programming,* vol. 3: *Sorting and Searching,* Addison-Wesley, Reading, Mass.

Kral, J. (1972), A New Additive Pseudorandom Number Generator for Extremely Short Word-Length, *Information Processing Letters 1*, 164-167.

Kronmal, R. (1964), Evaluation of a Pseudo-Random Normal Number Generator, *JACM 11*, 357-363.

Kronmal, R. A., and A. V. Peterson, Jr. (1978a), *On the Alias Method for Generating Random Variables from a Discrete Distribution*, Technical Report no. 17, School of Public Health, University of Washington, Seattle.

Kronmal, R. A., and A. V. Peterson, Jr. (1978b), *A Modification of the Rejection Method for Generating Random Variables*, Technical Report, School of Public Health, University of Washington, Seattle.

Kronmal, R. A., A. V. Peterson, and E. D. Lundberg (1978), The Alias-Rejection-Mixture Method for Generating Random Variables from Continuous Distributions, *Proceedings of the Statistical Computing Section, American Statistical Association*, Washington, D.C., 106-110.

Kruskal, J. B. (1969), Extremely Portable Random Number Generators, *CACM 12*, 93-94.

Lanczos, C., and B. Gellai (1975), Fourier Analysis of Random Sequences, *Computers and Mathematics with Applications 1*, 269-276.

Learmonth, G. P. (1976), Empirical Tests of Multipliers for the Prime-Modulus Random Number Generator $X(I + 1) \equiv AX(I) \text{Mod } 2^{31} - 1$, *Proceedings of Ninth Interface Symposium on Computer Science and Statistics*, edited by D. Hoaglin and R. Welsch, Prindle, Weber and Schmidt, Boston, 178-183.

Learmonth, G. P., and P. A. W. Lewis (1973), Statistical Tests of Some Widely Used and Recently Proposed Uniform Random Number Generators, *Proceedings of Computer Science and Statistics: Seventh Annual Symposium on the Interface*, edited by W. J. Kennedy, Statistical Laboratory, Iowa State University, Ames, 163-171.

Lehmer, D. H. (1951), Mathematical Methods in Large-scale Computing Units, *Proceedings of the Second Symposium on Large Scale Digital Computing Machinery*, Harvard University Press, Cambridge, 141-146.

Levene, H., and J. Wolfowitz (1944), The Covariance Matrix of Runs Up and Down, *AMS 15*, 58-69.

Lewis, P. A. W., O. S. Goodman, and J. W. Miller (1969), A Pseudo-Random Number Generator for the System 360, *IBM Systems J. 8*, 136-146.

Lewis, T. G. (1975), *Distribution Sampling for Computer Simulation*, Lexington Books, Lexington, Mass.

Lewis, T. G., and W. H. Payne (1973), Generalized Feedback Shift Register Pseudorandom Number Algorithm, *JACM 20*, 456-468.

Li, S. T., and J. L. Hammond (1975), Generation of Pseudo-Random Numbers with Specified Univariate Distributions and Correlation Coefficients, *IEEE Transactions on Systems, Man, and Cybernetics 5*, 557-560.

Liittschwager, J. M. (1965), Results of a Gamma, Lognormal, and Weibull Sampling Experiment, *Industrial Quality Control 22*, 124-127.

Liniger, W. (1961), On a Method by D. H. Lehmer for the Generation of Pseudo-Random Numbers, *Numer. Math. 3*, 265-270.

Locks, M. O. (1976), Error Analysis of Various Methods for Generating Beta and Gamma Variates, *Proceedings of Ninth Interface Symposium on Computer Science and Statistics*, edited by D. Hoaglin and R. Welsch, Prindle, Weber and Schmidt, Boston, 184-188.

Lurie, D., and H. O. Hartley (1972), Machine Generation of Order Statistics for Monte Carlo Computations, *Amer. Stat. 26*, 26-27.

Lurie, D., and R. L. Mason (1973), Empirical Investigation of Several Techniques for Computer Generation of Order Statistics, *Comm. Stat. 2*, 363-371.

McArdle, J. J. (1976), Empirical Tests of Multivariate Generators, *Proceedings of the Ninth Annual Symposium on the Interface of Computer Science and Statistics*, edited by D. C. Hoaglin and R. Welsch, Prindle, Weber, and Schmidt, Boston, 263-267.

McArdle, J. J. (1978), The Statistical Analysis of Monte Carlo Simulation Data Using the Techniques of Discrete Multivariate Analysis. *Computer Science and Statistics: Tenth Annual Symposium on the Interface*, edited by D. Hogben and D. W. Fife, National Bureau of Standards, Washington, D. C., 241-246.

McGrath, E. J., S. L. Basin, R. W. Burton, D. C. Irving, S. C. Jaquette, W. R. Ketler, and C. A. Smith (1973), *Techniques for Efficient Monte Carlo Simulation*, vol. 1: *Selecting Probability Distributions*, National Technical Information Service, Springfield, Va.

McGrath, E. J., and D. C. Irving (1973), *Techniques for Efficient Monte Carlo Simulation*, vol. 2: *Random Number Generation for Selected Probability Distributions,* National Technical Information Service, Springfield, Va.

MacLaren, M. D., and G. Marsaglia (1965), Uniform Random Number Generators, *JACM 12*, 83-89.

MacLaren, M. D., G. Marsaglia, and T. A. Bray (1964), A Fast Procedure for Generating Exponential Random Variables, *CACM 7*, 298-300.

Maritsas, D. G. (1973), A High Speed and Accuracy Digital Gaussian Generator of Pseudorandom Numbers, *IEEE Trans. On Computers C-22*, 629-634.

Mars, P., and A. J. Miller (1977), Theory and Design of a Digital Stochastic Computer Random Number Generator, *Mathematics and Computers in Simulation 19,* 198-216.

Marsaglia, G. (1961a), Expressing a Random Variable in Terms of Uniform Random Variables, *AMS 32*, 894-898.

Marsaglia, G. (1961b), Generating Exponential Random Variables, *AMS 32*, 899-900.

Marsaglia, G. (1962), Random Variables and Computers, *Information Theory Statistical Decision Functions Random Processes: Transactions of the Third Prague Conf*, edited by J. Kozesnik, Czechoslovak Academy of Sciences, Prague, 499-510.

Marsaglia, G. (1963), Generating Discrete Random Variables in a Computer, *CACM 6*, 37-38.

Marsaglia, G. (1964), Generating a Variable from the Tail of the Normal Distribution, *Techno. 6*, 101-102.

Marsaglia, G. (1968), Random Numbers Fall Mainly in the Planes, *Proceedings of the National Academy of Sciences 61*, 25-28.

Marsaglia, G. (1970), Regularities in Congruential Random Number Generators, *Numer. Math. 16*, 8-10.

Marsaglia, G. (1972), The Structure of Linear Congruential Sequences, in *Applications of Number Theory to Numerical Analysis*, edited by S. K. Zaremba, Academic Press, New York, 249-286.

Marsaglia, G. (1977), The Squeeze Method for Generating Gamma Variates, *Computers and Mathematics with Applications 3*, 321-326.

Marsaglia, G., K. Ananthanarayanan, and N. J. Paul (1976), Improvements on Fast Methods for Generating Normal Random Variables, *Information Processing Letters 5*, 27-30.

Marsaglia, G., and T. A. Bray (1964), A Convenient Method for Generating Normal Variables, *SIAM Rev. 6*, 260-264.

Marsaglia, G., and T. A. Bray (1968), One-line Random Number Generators and Their Use in Combinations, *CACM 11*, 757-759.

Marsaglia, G., M. D. MacLaren, and T. A. Bray (1964), A Fast Procedure for Generating Normal Random Variables, *CACM 7*, 4-10.

Marshall, A. W. (1956), The Use of Multi-Stage Sampling Schemes in Monte Carlo Computations, in *Symposium on Monte Carlo Methods*, edited by H. A. Mayer, Wiley, New York, 123-140.

Martin-Löf, P. (1969), Algorithms and Randomness, *Review of the International Statistical Institute 37*, 265-272.

Mason, R. L., and D. Lurie (1973), Systematic Simulators of Joint Order Uniform Variates, *Proceedings of Computer Science and Statistics: Seventh Annual Symposium on the Interface*, edited by W. J. Kennedy, Statistical Laboratory, Iowa State University, Ames, 156-162.

Metropolis, N., and S. Ulam (1949), The Monte Carlo Method, *JASA 44*, 335-341.

Michael, J. R., W. R. Schucany, and R. W. Haas (1976), Generating Random Variates Using Transformations with Multiple Roots, *Amer. Stat. 30*, 88-90.

Mihram, G. A. (1972), *Simulation: Statistical Foundations and Methodology*, Academic Press, New York.

Miklich, D. R., and D. J. Austin (1976), A High-Speed Normal Random Number Generator Using Table Look-up, *Behavior Research Methods and Instrumentation 8*, 405.

Mitchell, R. L., and C. R. Stone (1977), Table Lookup Methods for Generating Arbitrary Random Numbers, *IEEE Trans. Computers C-26*, 1006-1008.

Muller, M. E. (1958), An Inverse Method for the Generation of Random Normal Deviates on Large-Scale Computers, *MTAC 12*, 167-174.

Muller, M. E. (1959a), A Comparison of Methods for Generating Normal Deviates on Digital Computers, *JACM 6*, 376-383.

Muller, M. E. (1959b), A Note on a Method for Generating Points Uniformly on N-dimensional Spheres, *CACM 2*, 19-20.

Murray, H. F. (1970), A General Approach for Generating Natural Random Variables, *IEEE Transactions on Computers 19*, 1210-1213.

Nance, R. E., and C. Overstreet (1972), A Bibliography on Random Number Generation, *Comp. Rev. 13*, 495-508.

Nance, R. E., and C. Overstreet (1975), Implementation of FORTRAN Random Number Generators on Computers with One's Complement Arithmetic, *JSCS 4*, 235-243.

Nance, R. E., and C. Overstreet (1978), Some Experimental Observations on the Behavior of Composite Random Number Generators, *Operations Research 26*, 915-935.

Naylor, T. H., J. L. Balintfy, D. S. Burdick, and K. Chu (1966), *Computer Simulation Techniques*, Wiley, New York.

Neave, H. R. (1973), On Using the Box-Muller Transformation with Multiplicative Congruential Pseudo-Random Number Generators, *Appl. Stat. 22*, 92-97.

Neuman, F., and C. F. Martin (1976), The Autocorrelation Structure of Tausworthe Pseudo-Random Number Generators, *IEEE Transactions on Computers 25*, 460-463.

Neuman, F., and R. B. Merrick (1976), Autocorrelation Peaks in Congruential Pseudo-Random Number Generators, *IEEE Transactions on Computers 25*, 457-459.

Niederreiter, H. (1972a), Methods for Estimating Discrepancy, in *Applications of Number Theory to Numerical Analysis*, edited by S. K. Zaremba, Academic Press, New York, 203-236.

Niederreiter, H. (1972b), On the Distribution of Pseudo-Random Numbers Generated by the Linear Congruential Method, *Math. Comp. 26*, 793-795.

Niederreiter, H. (1974), On the Distribution of Pseudo-Random Numbers Generated by the Linear Congruential Method, II, *Math. Comp. 28*, 1117-1132.

Niederreiter, H. (1976), On the Distribution of Pseudo-Random Numbers Generated by the Linear Congruential Method, III, *Math. Comp. 30*, 571-597.

Niederreiter, H. (1977), Pseudo-Random Numbers and Optimal Coefficients, *Advances in Math. 26*, 99-181.

Niederreiter, H. (1978), The Serial Test for Linear Congruential Pseudo-Random Numbers, *Bull. Amer. Math. Soc. 84*, 273-274.

Norman, J. E., and L. E. Cannon (1972), A Computer Program for the Generation of Random Variables from Any Discrete Distribution, *JSCS 1*, 331-348.

Odell, P. L., and A. H. Feiveson (1966), A Numerical Procedure to Generate a Sample Covariance Matrix, *JASA 61*, 199-203.

Overstreet, C., and R. E. Nance (1973), A Random Number Generator for Small Word-Length Computers, *Proceedings of ACM Annual Conference*, 219-223.

Page, E. S. (1954), The Monte Carlo Solution of Some Integral Equations, *Proc. Camb. Phil. Soc. 50*, 414-425.

Page, E. S. (1967), A Note on Generating Random Permutations, Appl. Stat. 16*, 273-274*.

Patil, G. P., M. Boswell, and D. Friday (1975), Chance Mechanisms in the Computer Generation of Random Variables, *A Modern Course on Statistical Distributions in Scientific Work*, vol. 2, edited by G. P. Patil, S. Kotz, and J. K. Ord, D. Reidel, Boston, 37-50.

Payne, W. H. (1970), Fortran Tausworthe Pseudo-random Number Generator, *CACM 13*, 57-58.

Payne, W. H. (1977), Normal Random Numbers: Using Machine Analysis to Choose the Best Algorithm, *TOMS 3*, 346-358.

Payne, W. H., and T. G. Lewis (1971), Continuous Distribution Sampling: Accuracy and Speed, in *Mathematical Software*, edited by J. R. Rice, Academic Press, New York, 331-345.

Payne, W., J. R. Rabung, and T. Bogyo (1969), Coding the Lehmer Pseudorandom Number Generator, *CACM 12*, 85-86.

Peach, P. (1961), Bias in Pseudo-Random Numbers, *JASA 56*, 610-618.

Perkins, W. C., and U. Menzefricke (1975), A Better Additive Congruential Random Number Generator? *Decision Sciences 6*, 194-198.

Peskun, P. (1978), Improving the Apparent Randomness of Pseudorandom Numbers Generated by the Mixed Congruential Method, *Proceedings of Computer Science and Statistics: Tenth Annual Symposium on the Interface*, edited by D. Hogben and D. W. Fife, National Bureau of Standards, Washington, D.C., 323-328.

Phillips, D. T. (1971), Generation of Random Gamma Variates from the Two-Parameter Gamma, *AIIE Transactions 3*, 191-198.

Phillips, D. T., and C. S. Beightler (1972), Procedure for Generating Gamma Variates with Non-integer Parameter Sets, *JSCS 1*, 197-208.

Pike, M. C. (1965), Random Normal Deviate, *CACM 8*, 606.

Prentice, M. J. and J. C. P. Miller (1968), Additive Congruential Pseudo-Random Number Generators, *Comp. J. 11*, 341-346.

Rabinowitz, M., and M. L. Berenson (1974), A Comparison of Various Methods of Obtaining Random Order Statistics for Monte Carlo Computations, *Amer. Stat. 28*, 27-29.

Ramberg, J. S., and B. W. Schmeiser (1972), An Approximate Method for Generating Symmetric Random Variables, *CACM 15*, 987-990.

Ramberg, J. S., and B. W. Schmeiser (1974), An Approximate Method for Generating Asymmetric Random Variables, *CACM 17*, 78-82.

Ramberg, J. S., and P. R. Tadikamalla (1974), An Algorithm for Generating Gamma Variates Based on the Weibull Distribution, *AIEE Transaction 6*, 257-260.

Ramberg, J. S., and P. R. Tadikamalla (1978), On the Generation of Subsets of Order Statistics, *JSCS 6*, 239-241.

Rao, C. R. (1961), Generation of Random Permutations of Given Number of Elements Using Random Sampling Numbers, *Sankhyā (A) 23*, 305-307.

Reeder, H. A. (1972), letter to the editor, *Amer. Stat. 26*, 56-57.

Relles, D. A. (1970), Variance Reduction Techniques for Monte Carlo Sampling from Student Distributions, *Techno. 12*, 499-515.

Relles, D. A. (1972), A Simple Algorithm for Generating Binomial Random Variables when N Is Large, *JASA 67*, 612-613.

Ronning, G. (1977), A Simple Scheme for Generating Multivariate Gamma Distributions with Non-Negative Covariance Matrix, *Techno. 19*, 179-183.

Rotenberg, A. (1960), A New Pseudo-Random Number Generator, *JACM 7*, 75-77.

Rudolph, E., and D. M. Hawkins (1976), Random Number Generators in Cyclic Queueing Applications, *JSCS 5*, 65-72.

Salfi, R. (1974), A Long-Period Random Number Generator with Application to Permutations, *Compstat 74: Proceedings in Computational Statistics*, edited by G. Bruckmann and F. Schmetterer, Physica Verlag, Vienna, 28-35.

Schaffer, H. E. (1970), Algorithm 369: Generator of Random Numbers Satisfying the Poisson Distribution, *CACM 13*, 49.

Schmeiser, B. W. (1977), Methods for Modelling and Generating Probabilistic Components in Digital Computer Simulation When the Standard Distributions Are Not Adequate: A Survey, *Proceedings of the Winter Simulation Conference*, Gaithersburg, Md., Dec. 5-7, 1977.

Schmeiser, B., and M. A. Shalaby (1977), *Rejection Methods for Beta Variate Generation*, Technical Report, Department of Industrial Engineering, Southern Methodist University, Dallas, Tx.

Schmeiser, B. W., and M. A. Shalaby (1978), Rejection Using Piecewise Linear Majorizing Functions in Random Variate Generation, *Proceedings of Computer Science and Statistics: Eleventh Annual Symposium on the Interface*, edited by A. R. Gallant and T. M. Gerig, North Carolina State University, Raleigh, 230-233.

Schrage, Linus (1979), A More Portable Fortran Random Number Generator, *TOMS 5*, 132-138.

Schucany, R. W. (1972), Order Statistics in Simulation, *JSCS 1*, 281-286.

Schuer, E. M., and D. S. Stoller (1962), On the Generation of Normal Random Vectors, *Techno. 4*, 278-281.

Seraphin, D. S. (1969), A Fast Random Number Generator for IBM 360, *CACM 12*, 695.

Shafer, D. (1962), NORMDEV, *CACM 5*, 485.

Shepherd, W. L., and J. N. Hynes (1976), Table Look-up and Interpolation for a Normal Random Number Generator, *Proceedings of the Twenty-Second Conference on the Design of Experiments in Army Research Development and Testing*, ARO Report 77-2, 153-164.

Shepherd, W. L., and J. W. Starner, Jr. (1977), Table Look-up and Interpolation for a Normal Random Number Generator, II, paper presented at SIAM annual meeting, Albuquerque, N. Mex., Oct. 31-Nov. 2.

Shreider, Y. A. (1966), *The Monte Carlo Method*, translated by G. J. Tee, Permagon Press, New York.

Sibuya, M. (1961), Exponential and Other Variable Generators, *Ann. Inst. Statist. Math. 13*, 231-237.

Sibuya, M. (1962), Further Consideration on Normal Random Variable Generator, *Ann. Inst. Statist. Math. 14*, 159-165.

Simon, G. (1976), Computer Simulation Swindles, with Applications to Estimates of Location and Dispersion, *Appl. Stat. 25*, 266-274.

Smith, C. S. (1971), Multiplicative Pseudo-Random Number Generators with Prime Modulus, *JACM 18*, 586-593.

Smith, W. B., and R. R. Hocking (1972), Wishart Variate Generator, *Appl. Stat. 21*, 341-345.

Snow, R. H. (1968), Generation of Random Numbers Satisfying the Poisson Distribution, *CACM 11*, 819-820.

Sowey, E. R. (1972), A Chronological and Classified Bibliography on Random Number Generation and Testing, *Int. Statist. Rev. 40*, 355-371.

Sowey, E. R. (1978), A Second Classified Bibliography on Random Number Generation and Testing, *Int. Statist. Rev. 46*, 89-102.

Spanier, J., and E. Gelbard (1969), *Monte Carlo Principles and Neutron Transport Problems*, Addison-Wesley, Reading, Mass.

Stoneham, R. G. (1970), On a New Class of Multiplicative Pseudorandom Number Generators, *BIT 10*, 481-500.

Student (1908a), On the Probable Error of a Mean, *Biometrika 6*, 1-25.

Student (1908b), Probable Error of a Correlation Coefficient, *Biometrika 6*, 302-310.

Tadikamalla, P. R. (1978), Computer Generation of Gamma Random Variables, *CACM 21*, 419-422.

Tadikamalla, P. R., and J. S. Ramberg (1975), An Approximate Method for Generating Gamma and Other Variates, *JSCS 3*, 275-282.

Tausworthe, R. C. (1965), Random Numbers Generated by Linear Recurrence Modulo Two, *Math. Comp. 19*, 201-209.

Thomson, W. E. (1958), A Modified Congruence Method of Generating Pseudo-Random Numbers, *Comp. J. 1*, 83, 86.

Tocher, K. D. (1963), *The Art of Simulation*, English University Press, London.

Toothill, J. P. R., W. D. Robinson, and A. G. Adams (1971), The Runs Up-and Down Performance of Tausworthe Pseudo-Random Number Generators, *JACM 18*, 381-399.

Toothill, J. P. R., W. D. Robinson, and D. J. Eagle (1973), An Asymptotically Random Tausworthe Sequence, *JACM 20,* 469-481.

Vaduva, I. (1977), On Computer Generation of Gamma Random Variables by Rejection and Composition Procedures, *Proceedings of the Fifth Conference on Probability Theory*, Editura Academiei, Bucharest, 131-142.

Van Gelder, A. (1967), Some New Results in Pseudo-Random Number Generation, *JACM 14*, 785-792.

Von Neumann, J. (1951), *Various Techniques Used in Connection with Random Digits*, National Bureau of Standards symposium, NBS Applied Mathematics Series 12, National Bureau of Standards, Washington, D.C.

Wakimoto, K. (1976), Algorithms for Generating a Random Vector with Restricted Integer Components and Their Extension to Matrix, in *Essays in Probability and Statistics*, Ogawa volume, edited by S. Ikeda et al., Shinko Tsusho, Tokyo, 179-188.

Walker, A. J. (1974a), New Fast Method for Generating Discrete Random Numbers with Arbitrary Frequency Distributions, *Electronics Letters 10*, 127-128.

Walker, A. J. (1974b), Fast Generation of Uniformly Distributed Pseudorandom Numbers with Floating-Point Representation, *Electronics Letters 10*, 533-534.

Walker, A. J. (1977), An Efficient Method for Generating Discrete Random Variables with General Distributions, *TOMS 3*, 253-256.

Wallace, C. S. (1976), Transformed Rejection Generators for Gamma and Normal Pseudo-Random Generators, *Australian Computer Journal 8*, 103-105.

Wallace, N. D. (1974), Computer Generation of Gamma Random Variables with Non-Integral Shape Parameters, *CACM 17*, 691-695.

Wallis, W. A., and G. H. Moore (1941), A Significance Test for Time Series, *JASA 36*, 401-409.

Walsh, J. E. (1957), An Experimental Method for Obtaining Random Digits and Permutations, *Sankhyā 17*, 355-360.

Watson, E. J. (1962), Primitive Polynomials (Mod 2), *Math. Comp. 16*, 368-369.

Wedderburn, R. W. M. (1976), A Remark on Algorithm AS29: "The Runs Up and Down Test," *Appl. Stat. 25*, 193.

Westlake, W. J. (1967), A Uniform Random Number Generator Based on the Combination of Two Congruential Generators, *JACM 14*, 337-340.

Wetherill, G. B. (1965), An Approximation to the Inverse Normal Function for the Generation of Random Normal Deviates on Electronic Computers, *Appl. Stat. 16*, 201-205.

Wheeler, D. J. (1974), Simulation of Arbitrary Gamma Distributions, *AIEE Transactions 6*, 167-169.

Wheeler, D. J. (1975), An Approximation for Simulation of Gamma Distributions, *JSCS 3*, 225-232.

Whittaker, J. (1974), Generating Gamma and Beta Random Variables with Non-Integral Shape Parameters, *Appl. Stat. 23*, 210-214.

Whittlesey, J. R. B. (1968), A Comparison of the Correlation Behavior of Random Number Generators for the IBM 360, *CACM 11*, 641-644.

Whittlesey, J. R. B. (1969), On the Multidimensional Uniformity of Pseudo-Random Generators, *CACM 12*, 247.

Wilson, E. B., and M. M. Hilferty (1931), The Distribution of Chi-square, *Proceedings of the National Academy of Sciences 17*, 684.

Wolfowitz, J. (1944), Asymptotic Distribution of Runs Up and Down, *AMS 15*, 163-172.

Yakowitz, S. J. (1977), *Computational Probability and Simulation*, Addison-Wesley, Reading, Mass.

Young, L. C. (1941), On Randomness in Order Sequences, *AMS 12*, 292-300.

Zaremba, K. S. (1968), The Mathematical Basis of Monte Carlo and Quasi-Monte Carlo Methods, *SIAM Rev. 10*, 303-314.

Zierler, N. (1959), Linear Recurring Sequences, *SIAM J. 7*, 31-48.

Zierler, N., and J. Brillhart (1968), On Primitive Trinomials (Mod 2), *Information and Control 13*, 541-554.

7 / SELECTED COMPUTATIONAL METHODS IN LINEAR ALGEBRA

7.1 INTRODUCTION

Matrix notation and large portions of the theory of linear algebra are used in many areas of the theory of statistics. One notable example is linear model theory. This being the case, it is certainly natural to expect that linear algebraic computational methods will be employed in statistical computing. As we shall see in subsequent chapters, several major areas in statistical computing do indeed make extensive use of such methods.

The purpose of this chapter is to describe and discuss some computational methods which involve matrices and have utility in statistical computing. To be more specific, various numerical methods for transforming a given matrix into some desirable form will be presented. These methods form the basis for several widely used numerical algorithms in statistical computing, and they rank high among the most satisfactory numerical methods in all areas of computing when numerical stability and efficiency are criteria used to make comparisons.

Transformations will be made on an $n \times p$ matrix X having real number elements x_{ij}. Unless otherwise specified it is assumed that $p \leq n$ and that X has rank $r \leq p$. In each case the method will consist of transforming by premultiplication, and sometimes postmultiplication, of X by a chosen set of nonsingular matrices to produce a matrix of desired form. In some cases the transformations will be termed *orthogonal* because orthogonal matrix multipliers are used and an orthogonal basis for the column space of X is either explicitly or implicitly involved in the method.

Statistical applications of the matrix methods given in this chapter are contained in subsequent chapters. This organization of the subject matter is made to insure that all aspects of given statistical computing methods are recognized and understood. It is, however, desirable to give some application for the methods at the time they are presented in this chapter. For this reason the related problems of matrix inversion and solving linear systems of equations will sometimes be considered.

7.2 METHODS BASED ON ORTHOGONAL TRANSFORMATIONS

In this section four different computational methods will be given. In each case the objective will be to transform X, using orthogonal transformations, to a form in which all elements below the main diagonal [the (i,i) positions for a nonsquare matrix] are zero. The amount of information available about the diagonal and superdiagonal elements will not be the same for each method. Methods based on orthogonal transformations do, in most cases, have good numerical stability, and the methods given in this section have proven to be the best methods of this type.

7.2.1 Householder Transformations

Consider an n x n symmetric matrix H having the form

$$H = I_n - \frac{\underline{u}\underline{u}'}{\beta} \tag{7.1}$$

where I_n is the n-square identity matrix, $\underline{u}$ is an n-component column vector such that $|\underline{u}'\underline{u}| \neq 0$, and β is a positive scalar. Householder (1958) proposed this form of transformation matrix under a choice of β and $\underline{u}$ which make H orthogonal. The way in which matrices in the form of H can be utilized for the problem at hand is best described by first considering the transformation of a column vector $\underline{w}$, that is, the product $H\underline{w}$. Having done this, it will be easy to see how H operates on the columns of X when the product HX is formed.

Let $\underline{w}$ be a given (n x 1) vector. Suppose that we want to define the matrix H in Eq. (7.1) such that the vector $H\underline{w}$ has zeros in positions t + 1, t + 2, ..., n and contains $\underline{w}$ vector elements w_1, w_2, ..., w_{t-1} in positions 1, 2, ..., t - 1, respectively, and such that H is orthogonal. To do this, consider the following choice of $\underline{u}$ vector components u_j and the scalar β:

$$\begin{aligned} u_j &= 0 \quad j = 1, 2, \ldots, t - 1 \\ u_t &= w_t + \text{sign}(w_t)S \\ u_{t+i} &= w_{t+i} \quad i = 1, 2, \ldots, n - t \\ \beta &= S^2 + |w_t|S \\ &= \text{sign}(w_t)u_tS \end{aligned} \tag{7.2}$$

where $S^2 = \sum_{i=t}^{n} w_i^2$

Note that the choice of sign in the second and fourth equations is determined by the sign which produces addition of like signed numbers in the formation of u_t, that is, the sign of w_t is selected. It can be shown that H is orthogonal under either choice of sign, but the choice indicated above turns out to be critical to the preservation of numerical stability. A subtraction in the formation of u_t often produces catastrophic cancellation.

It is readily verified that $H\underline{w}$ has zero as its last n - t components and contains the first t - 1 components of $\underline{w}$ as its first t - 1 components. The t-th component of $H\underline{w}$ turns out to be $\pm$ S, where the sign is the opposite of the sign of w_t. Thus we have

$$(H\underline{w})' = (w_1, w_2, \ldots, w_{t-1}, \pm S, 0, \ldots, 0) \tag{7.3}$$

as the transformed vector under this transformation.

Suppose, for the moment, that X has rank p and n > p. We can transform X to the required form using Householder transformations. One transformation will correspond to each of the p columns of X. Defining X_0 = X, the kth transformation is

$$X_k = H_kX_{k-1} \quad (k = 1, 2, \ldots, p) \tag{7.4}$$

where H_k is defined with respect to the kth column of X_{k-1} (column k of X_{k-1} is used as the $\underline{w}$ vector in defining the kth one of the p number of H matrices), using t = k. The first k - 1 rows and columns of X_k and X_{k-1} will be the same, and subdiagonal elements in the kth column of X_k will be zero. Upon completion of the p transformations we have an orthogonal transformation matrix H that is itself a product of p orthogonal matrices, which operates on X as

$$HX = H_p H_{p-1} \cdots H_1 X = \begin{pmatrix} T \\ \hline 0 \end{pmatrix} \tag{7.4}$$

In (7.4) T is a p x p upper-triangular matrix and 0 is an (n - p) x p zero matrix. Thus for full rank X we have accomplished the stated objective, which was to transform X to upper-triangular form.

Now assume that X has rank r < p. Let P be a p x p permutation matrix such that the first r columns of XP are linearly independent. Consider a partitioning of the columns of XP as

$$XP = (Z_0 | Z_1) \tag{7.5}$$

where Z_0 is n x r and has rank r so that each column of Z_1 is expressible as a linear combination of the columns of Z_0. Thus there exists a matrix K such that $Z_1 = Z_0K$ and (7.5) takes the form

$$XP = (Z_0 | Z_0K) \tag{7.6}$$

Suppose that we determine an orthogonal H which triangularizes Z_0 as in Eq. (7.4) with p = r. Then we have

$$\begin{aligned} HXP &= (HZ_0 | HZ_0K) \\ &= \begin{pmatrix} T_1 & R_1 \\ 0_1 & 0_2 \end{pmatrix} \end{aligned} \tag{7.7}$$

where T_1 is r x r upper-triangular and nonsingular, 0_1 is an (n - r) x r zero matrix, and 0_2 is an (n - r) x (p - r) zero matrix. Thus, if X is rank-deficient, the p x p upper-triangular matrix of Eq. (7.4) is singular and has (p - r) zero rows, and the triangularization of X requires both pre- and postmultiplication of X by an orthogonal matrix.

In practice, we seldom know which columns of X are independent before computations begin. Thus the rank of X, and the column interchanges which define P, must be found in the course of the computations. Since floating-point arithmetic is usually used when this method is implemented, we cannot distinguish between elements which are theoretically zero and elements which are very small in absolute value. The best we can do is to use a tolerance value of $\varepsilon > 0$ and at each stage decide whether a zero row has been encountered by comparing some function of elements in that row with ε. Lawson and Hanson (1974) use the term *pseudorank* to refer to the computed rank. They also utilize a column interchange at each stage of the computations so that the largest possible diagonal element is produced each time a transformation is made. This is done by computing the S^2 of Eq. (7.2) for each candidate column. Obviously the succession of diagonal elements of the submatrix T_1 shown in Eq. (7.7), under this column interchange strategy, will be nonincreasing.

To illustrate an algorithm, which might well form a part of a complete Householder triangularization algorithm, we now give algorithm BHOUSE. This algorithm accepts as inputs a vector $\underline{w}$, position pointer t, length n, and tolerance ε. With reference to Eq. (7.2), BHOUSE returns β in b, the t-th component of $\underline{u}$ in h, the remaining nonzero components of $\underline{u}$ in the corresponding positions of $\underline{w}$, and the first t components of $H\underline{w}$ in the first t positions of $\underline{w}$. If the S value is less than ε, a nonzero error switch E is returned.

ALGORITHM BHOUSE(w,n,t,b,h,ε,E)

Step	*Description*
1.	Set $S = (\sum_{i=t}^{n} w_i^2)^{1/2}$. If $S < \varepsilon$, set E = 1 and exit; otherwise set E = 0.
2.	If $w_t > 0$, set S = -S.
3.	Set $h = w_t - S$, and $w_t = S$. Comment: Now the nonzero elements of $H\underline{w}$ are in place in $\underline{w}$, and u_t is stored as h.
4.	Set $b = -w_t h$ and exit.

Computation of the square root of the sum of squares in step 1 sometimes gives problems due to underflow in floating-point computations. One recommended procedure to help in these cases is to find m = $\max\{|w_i|, i = t, \ldots, n\}$ and compute S as

$$S = m\left[\left(\frac{w_t}{m}\right)^2 + \cdots + \left(\frac{w_n}{m}\right)^2\right]^{1/2}$$

Blue (1978) gives a complete analysis of this computational problem.

One application of Householder transformation is to the problem of solving a nonsingular system of linear equations. If the system of linear equations is

$$X\underline{b} = \underline{Y}$$

where X is p x p nonsingular, $\underline{b}$ is a p x 1 vector of unknowns, and $\underline{Y}$ is a p x 1 vector of given constants, then under appropriate transformations HX = T and $H\underline{Y} = \underline{U}$; an equivalent system is

$$T\underline{b} = \underline{U}$$

This equivalent system has an upper-triangular coefficient matrix T; hence the components of $\underline{b}$ are easily determined through a simple back-substitution process working from the last equation up to the first equation.

Householder matrices are sometimes called *reflectors* in the geometrical sense, and are stated in the form

$$H = I - 2\underline{v}\underline{v}'$$

where $\underline{v}$ is n x 1 and $\underline{v}'\underline{v} = 1$. This form is equivalent to that given in Eq. (7.1) and (7.2) with $\underline{u} = (2\beta)^{1/2}\underline{v}$. The term *reflector* is applied because H operates on a given vector $\underline{w}$, in the plane defined by $\underline{v}$ and $\underline{w}$, to produce a reflection of $\underline{w}$ about a vector perpendicular to $\underline{v}$. To see this note that

$$\begin{aligned} H &= [I - \underline{v}(\underline{v}'\underline{v})^{-1}\underline{v}'] - \underline{v}(\underline{v}'\underline{v})^{-1}\underline{v}' \\ &= (I - R) - R, \quad \text{say} \end{aligned}$$

and

$$H\underline{w} = (I - R)\underline{w} - R\underline{w}$$

Thus since $(I - R)\underline{w}$ is the projection of $\underline{w}$ onto the space orthogonal to $\underline{v}$, and $R\underline{w}$ is the projection of $\underline{w}$ onto the space of $\underline{v}$, the result follows.

Again considering H in the above form

$$H = I - 2\underline{v}\underline{v}'$$

suppose that for a given i and j $(i \neq j)$ we have

$$v_i = \frac{1}{\sqrt{2}}$$

$$v_j = \frac{-1}{\sqrt{2}}$$

$$v_k = 0 \quad \text{for all } k \neq i, j$$

Then H is readily seen to differ from the identity matrix only in that rows (columns) i and j have been interchanged In other words, H is an *elementary permutation matrix* under this choice of $\underline{v}$.

There are some additional details relative to the programming of Householder triangularization which need to be mentioned before we leave the subject. Equation (7.6) specifies a permutation of columns of the X matrix, and the discussion of the possibly non-full-rank case referred throughout to such permutations in the course of computation. In practice, computer programs do not necessarily physically interchange columns (see Lawson and Hanson 1974). Rather, at each stage a transformation can be made on the matrix with respect to a selected column and an index vector used to keep a record of the order and column number of selected columns. This provides all information needed to specify the matrix of Eq. (7.7). Henceforth we will use the term *pivot element* to refer to the element w_t of Eq. (7.2), which serves to define the elements in $\underline{w}$ that are to be annihilated under the transformation. When operating on a matrix, the phrase "pivot on the fourth element in column 2" will mean that the Householder transformation is defined with respect to the second column of the matrix and the transformation will annihilate all elements in that column below the fourth element. The notation $H_i^{(j)}$ is sometimes used to specify that the Householder matrix H is defined with respect to column i and the jth element in that column is the pivot element. To illustrate these pivot operations, consider the form of the following transformations, which are the rounded output of a computer analysis:

$$X = \begin{pmatrix} 2 & 0 & 1 \\ 1 & 1 & 2 \\ 2 & 1 & 1 \\ 1 & 1 & 2 \\ 2 & 1 & 0 \\ 1 & 2 & 1 \end{pmatrix}$$

$$W_1 = H_3^{(3)} H_2^{(2)} H_1^{(1)} X = \begin{bmatrix} -3.87 & -2.07 & -2.32 \\ 0 & -1.93 & -1.13 \\ 0 & 0 & 2.07 \\ 0 & 0 & 0 \\ 0 & 0 & 0 \\ 0 & 0 & 0 \end{bmatrix}$$

$$W_2 = H_1^{(3)} H_3^{(2)} H_2^{(1)} W_1 = \begin{bmatrix} 2.83 & 3.38 & 2.47 \\ -0.91 & 0 & -2.21 \\ 2.48 & 0 & 0 \\ 0 & 0 & 0 \\ 0 & 0 & 0 \\ 0 & 0 & 0 \end{bmatrix}$$

Finally, when the various rows of X contain elements which vary widely in magnitude, it has been observed that severe loss of significant figures sometimes occurs if the pivot element, at some stage, is substantially smaller in magnitude than an element located below the pivot element and in the same column. This is so because perturbations made during the solving process are relatively more significant for smaller elements, and when these are used as pivot elements the error is magnified. To help avoid this problem it has been suggested (see, e.g., Lawson and Hanson 1974) that a row interchange strategy be adopted. Specifically, once a column k has been selected as the target for the next Householder transformation, find the largest element in column k on or below the kth row. If (say) row $\ell > k$ contains the largest element, then rows ℓ and k are interchanged, and the Householder transformation is made as usual. Obviously, such row interchanges are equivalent to premultiplication by elementary permutation matrices, hence performing these interchanges preserves the orthogonality of the transformation.

Row interchanges are not routinely done by most programs which implement the methods of this subsection. Rather, they are done optionally or are never considered in most programs. We will view row interchanges as special operations which are performed only in certain cases.

For datasets having n very large compared to p, storage of an n x p array in computer memory may not be desirable. It is possible to perform orthogonal triangularization by working only with successive subsets of the rows of X. A procedure for doing this is given by Hanson and Lawson (1969).

7.2.2 Givens Transformations

Another type of orthogonal matrix, popularized by Givens (1954), is sometimes used to triangularize X. In general, as in Section 7.2.1, the problem is to find an orthogonal matrix G and a permutation matrix P so that

$$GXP = \begin{bmatrix} T_1 & T_2 \\ 0_1 & 0_2 \end{bmatrix}$$

where T_1 is r x r, nonsingular, and upper-triangular. We now describe a way in which G is formed as a product of Givens transformation matrices. To do this we begin by defining a Givens transformation matrix G_{ij} which operates on an n x 1 vector $\underline{w}$ to annihilate the ith element of $\underline{w}$.

Assume $i > j$ for given i and j, and that the component w_j of $\underline{w}$ is nonzero. Then the elements of $G_{ij} = (g_{k\ell})$ are taken to be

$$
\begin{aligned}
g_{ii} &= g_{jj} = c \\
g_{kk} &= 1 \quad k \neq i, \quad k \neq j, \quad k = 1, 2, \ldots, n \\
g_{ij} &= -s \\
g_{ji} &= s \\
g_{k\ell} &= 0 \quad \text{for all } k \neq \ell \text{ except as indicated above}
\end{aligned} \tag{7.8}
$$

where

$$
\begin{aligned}
c &= \frac{w_j}{S} \\
s &= \frac{w_i}{S} \\
S &= \left(w_i^2 + w_j^2\right)^{1/2}
\end{aligned}
$$

Thus G_{ij} differs from the identity matrix in only four elements, and G_{ij} is easily seen to be orthogonal. What remains is to verify that G_{ij} transforms $\underline{w}$ as required. This also is easy since only components i and j in $\underline{w}$ are changed under the transformation. In particular, the ith component of $G_{ij}\underline{w}$ is

$$
\begin{aligned}
cw_i - sw_j &= \left(\frac{w_i w_j - w_i w_j}{S}\right) \\
&= 0
\end{aligned}
$$

The way in which the G_{ij} are used in operations on X is to associate the second subscript (j) with the column number in X and the first subscript (i) with the element in that column which is to be annihilated. Then $G_{ij}X$ has zero in the (i,j) position as required. Triangularization of X then proceeds, a column at a time, as

$$
GXP = G_{n,r} \cdots G_{n,2} \cdots G_{3,2} G_{n,1} \cdots G_{3,1} G_{2,1} XP = \begin{pmatrix} T_1 & T_2 \\ 0 & 0 \end{pmatrix}
$$

Each transformation annihilates only one element, whereas each Householder transformation annihilates all subdiagonal elements in the desired columns.

Example 7.1 illustrates the Givens transformation in a numerical problem. For general use, Householder transformations are preferred to Givens transformations because less computations are required and slightly better results are usually obtained. Givens transformations are useful in special situations such as cases where the matrix X is known to be very sparse. Geometrically, Givens transformations can be characterized as rotations in a plane defined by the subscript values (i,j). Verification of this fact will be left as an exercise for the reader.

Example 7.1

Consider the 3 x 3 matrix X given below and assume that the (2,1) element is to be annihilated:

$$X = \begin{bmatrix} 4 & 1 & 1 \\ 1 & 3 & 0 \\ 1 & 0 & 2 \end{bmatrix}$$

With respect to the (2,1) element we compute

$$S = (1^2 + 4^2)^{1/2} = (17)^{1/2} = 17a \quad [\text{letting } a = (17)^{-1/2}]$$

$$g_{11} = g_{22} = 4a$$
$$g_{21} = -a$$
$$g_{12} = -g_{21}$$

Thus

$$G_{21} = a \begin{bmatrix} 4 & 1 & 0 \\ -1 & 4 & 0 \\ 0 & 0 & a^{-1} \end{bmatrix}$$

$$G_{21}X = a \begin{bmatrix} 17 & 7 & 4 \\ 0 & 11 & -1 \\ a^{-1} & 0 & 2a^{-1} \end{bmatrix}$$

This transformation alters only two rows in X and produces the zero in position (2,1), as desired.

Many variations of the basic form and use of Givens rotations are available. One variation given by Lawson and Hanson (1974) avoids unnecessary overflow in the calculation of S in Eq. (7.8) by using

$$S = \begin{cases} a[1 + (b/a)^2]^{1/2} & \text{if } a \neq 0 \\ 0 & \text{if } a = 0 \end{cases}$$

where

$$a = \max(|w_i|, |w_j|)$$
$$b = \min(|w_i|, |w_j|)$$

7.2.3 The Modified Gram-Schmidt Method

A modification of the classical Gram-Schmidt vector orthogonalization process is sometimes used instead of Householder or Givens transformations for certain matrix applications. Longley (1977), for example, makes effective use of this method for least-squares computations. The modification of the classical method, which we abbreviate as M.G.S., was made by Bjorck (1967a), and it is now widely accepted that the M.G.S. does indeed provide significantly more stability than is present in the classical method for most applications.

Before providing details of this method, let us consider the general form for full-rank X. The method can be viewed as producing a factorization of X in the form

$$X = VT$$

where V is n x p, T is p x p nonsingular and upper-triangular, and $V'V = I_p$. Expressed as a transformation on X, we have

$$V'X = T$$

The columns of V obviously form an orthonormal set in the column space of X.

Now let us consider the details of the method. First, the procedure is carried out through a succession of modifications of the X matrix, taking each column in turn, through p stages. At each stage one column of V and one row of T are obtained. We will use the notation $\underline{x}_i^{(j)}$ to denote column i of the modified X matrix at stage j, $\underline{v}_i$ to denote the ith column of V, and the elements of T are written t_{ij}. With this notation, and beginning with $X^{(0)} = X$, we proceed as follows. For each k, $1 \leq k \leq p$, compute

$$d_k = \left[\underline{x}_k^{(k-1)'}\underline{x}_k^{(k-1)}\right]^{1/2}$$

$$\underline{v}_k = \underline{x}_k^{(k-1)} d_k^{-1}$$

$$t_{kk} = d_k$$

$$\left.\begin{aligned} t_{kj} &= \underline{v}_k'\underline{x}_j^{(k-1)} \\ \underline{x}_j^{(k)} &= \underline{x}_j^{(k-1)} - t_{kj}\underline{v}_k \end{aligned}\right\} \quad \text{for all } k + 1 \le j \le p \qquad (7.9)$$

In general applications it is usual to make a choice among candidates for $\underline{x}_k^{(k-1)}$ at stage k, so that d_k is maximized. This also allows entry of a linearly independent vector at each stage if one remains, and determination of the rank r of X at the beginning of stage r + 1. Example 7.2 illustrates these computations for a 3 x 3 matrix X. When $r < p$, the matrix V obtained in direct application has (p - r) zero columns and r orthonormal columns. These last (p - r) columns will be deleted for purpose of decomposition of X. Modifications can be made to produce orthogonal V if desired.

Under good implementations, the Householder and M.G.S. methods seem to produce results of comparable accuracy. One problem with the M.G.S. method is that the columns of V are often not sufficiently orthogonal when X is ill-conditioned. Reorthogonalization of V is often necessary if V is to be used subsequently.

Example 7.2

$$X = \begin{bmatrix} 4 & 1 & 2 \\ 2 & 3 & 6 \\ 4 & 1 & 2 \end{bmatrix}$$

(Stage 1) $d_1 = (4^2 + 2^2 + 4^2)^{1/2} = 6$

$$\underline{v}_1' = \left(\frac{2}{3}, \frac{1}{3}, \frac{2}{3}\right)$$

$$t_{11} = 6$$

$$t_{12} = \underline{v}_1'\underline{x}_2^{(0)} = \frac{7}{3}$$

$$t_{13} = \underline{v}_1'\underline{x}_3^{(0)} = \frac{14}{3}$$

$$\underline{x}_2^{(1)} = \underline{x}_2^{(0)} - t_{12}\underline{v}_1 = \left(-\frac{1}{9}\right)[5, -20, 5]'$$

$$\underline{x}_3^{(1)} = \underline{x}_3^{(0)} - t_{13}\underline{v}_1 = \left(-\frac{2}{9}\right)[5, -20, 5]'$$

(Stage 2) $$d_2 = \left[\underline{x}_2^{(1)'}\underline{x}_2^{(1)}\right]^{1/2} = \frac{\delta}{3} \quad \text{where } \delta = (50)^{1/2}$$

$$\underline{v}_2' = \left(\frac{9}{50}\right)^{1/2}\underline{x}_2^{(1)} = -\left[\frac{1}{(3\delta)}\right] \cdot [5, -20, 5]$$

$$t_{22} = d_2 = \frac{\delta}{3}$$

$$t_{23} = \underline{v}_2'\underline{x}_3^{(1)} = \frac{2\delta}{3}$$

$$\underline{x}_3^{(2)} = \underline{x}_3^{(1)} - t_{23}\,\underline{v}_2 = [0, 0, 0]'$$

(Stage 3) $d_3 = 0$

7.2.4 Singular-value Decomposition

One of the most numerically stable and generally satisfactory methods in the area of matrix computations is the singular-value decomposition (S.V.D.) method. The version in general use today is described in the work of Golub and Kahan (1965), Businger and Golub (1969), and Golub and Reinsch (1970). Before we describe the computational aspects of the S.V.D. method, we will discuss the decomposition itself.

Given the n x p real matrix X, there exists an n x n orthogonal matrix U', a p x p orthogonal matrix V, and an n x p diagonal matrix S having nonnegative elements such that

$$U'XV = S$$

or equivalently

$$X = USV' \tag{7.10}$$

The diagonal entries [(i,i) positions] of S are called the *singular values* of X, and all off-diagonal entries in S are zero. It is possible to select U and V so that the diagonal entries in S are nonincreasing. A concise proof of the existence of U, V, and S for any X is given by Lawson and Hanson (1974). Early work on this application can be found in Eckart and Young (1936). The expression of X as USV' is called the S.V.D. of X.

The first consequence of (7.10) which we note is that

$$S'S = V'X'XV \tag{7.11}$$

Since the left member of Eq. (7.11) is diagonal and V is orthogonal, it follows that the singular values of X are the square roots of the eigenvalues of the symmetric matrix X'X, and the columns of V form an orthonormal set of eigenvectors. Furthermore, forming

$$SS' = U'XX'U$$

it is easy to prove that the columns of U are eigenvectors of the symmetric matrix XX'. It follows that U, V, and S contain only real number elements, that the singular values of X are unique, and that the number of nonzero singular values is equal to the rank of X.

The ratio of the largest to the smallest nonzero singular value of X is often used to indicate the condition of X, and is called a *condition number* of X. Large condition numbers are usually taken as a warning that an inaccurate solution for the problem at hand may be forthcoming. Some suggested reading on the subject of matrix condition is Householder (1964), Wilkinson (1965), Stewart (1973), and Longley (1976).

Let us now turn to the details of a numerical method for computing the S.V.D. The first step in the method is to transform X to *upper-bidiagonal* form using pre- and postmultiplication by Householder transformation matrices. The result of this computation is

$$PXQ = \left[\begin{array}{ccccc} q_1 & e_2 & 0 & \cdots & 0 \\ & q_2 & e_3 & & \vdots \\ & & \ddots & \ddots & 0 \\ & 0 & & \ddots & e_p \\ & & & & q_p \\ \hline & & 0 & & \end{array}\right] = \left(\frac{A}{0}\right) = S_1 \tag{7.12}$$

The orthogonal matrix P can be obtained as the product of p orthogonal matrices as in Eq. (7.4). Q can be obtained as the product of p - 1 Householder transformations designed to zero parts of rows. The singular values of A are the same as those of X. To verify this, note that if the S.V.D. of A is implied by

$$U_1AQ_1 = \Sigma$$

then

$$\begin{pmatrix} U_1 & 0 \\ 0 & I_{n-p} \end{pmatrix} \begin{pmatrix} A \\ \hline 0 \end{pmatrix} Q_1 = \begin{pmatrix} \Sigma \\ \hline 0 \end{pmatrix} \tag{7.13}$$

Writing the left member of (7.13) as $P_1S_1Q_1$ and substituting into (7.12), we have

$$P_1PXQQ_1 = \begin{pmatrix} \Sigma \\ \hline 0 \end{pmatrix} \tag{7.14}$$

Since P_1P and QQ_1 are both orthogonal, it follows that the singular values of A are indeed those of X. Thus the problem of computing the S.V.D. of X reduces to one of computing the singular values of the upper-bidiagonal matrix A and forming the orthogonal matrices U_1 and Q_1. With these we can easily complete the solving process as indicated in Eq. (7.14). We now turn to the S.V.D. problem involving A.

Note that A'A is symmetric and tridiagonal. It would be possible for us to use the "QR method" of Francis (1961) to iteratively compute eigenvalues and eigenvectors of this matrix, and this would give us Σ^2 and Q_1. A like computation using AA' would provide U_1. However, it is possible to be more efficient by employing what is basically a factored form of the Francis method to obtain all needed quantities at one time. This is the essence of the Golub and Reinsch (1970) presentation which we are describing. The reader who is familiar with the QR method for the eigenvalue problem will note many similarities with the current method, and will be in a position to understand the reason why certain operations are performed. Other readers will need to accept, without proof, that certain operations are indeed correct and required.

The method which we employ is iterative and produces a sequence of upper-bidiagonal matrices A, $A^{(1)}$, $A^{(2)}$, ... which converges to the diagonal matrix Σ. At each iteration a transformation of the form

$$A^{(j+1)} = R^{(j)'}A^{(j)}T^{(j)} \tag{7.15}$$

is made, where $R^{(j)}$ and $T^{(j)}$ are orthogonal. In order to provide details of this transformation, let us drop the iteration designator and use

$$\bar{B} = A^{(j+1)}, \qquad R = R^{(j)}, \qquad B = A^{(j)}, \qquad \text{and } T = T^{(j)}$$

The transition from B to $\bar{B}$ is made by employing Givens transformations R_j and T_j (as described in Section 7.2.2) in the following fashion:

$$\bar{B} = \underbrace{R'_p\left\{R'_{p-1}\cdots[R'_2}_{R'}(\underbrace{BT_2)]T_3\cdots T_p\right\}}_{T} \tag{7.16}$$

The objective of each transformation serves to define the transformation matrix. The following objectives are stated for each of the transformations, and the additional result it produces is given. The main objective of T_2 will be discussed later.

Matrix	Objective	Additional result
T_2	Annihilates nothing	Generates an entry in (2,1) position
R'_2	Annihilates (2,1) entry	Generates an entry in (1,3) position
T_3	Annihilates (1,3) entry	Generates an entry in (3,2) position
⋮	⋮	
R'_p	Annihilates (p, p - 1) entry	Generates nothing

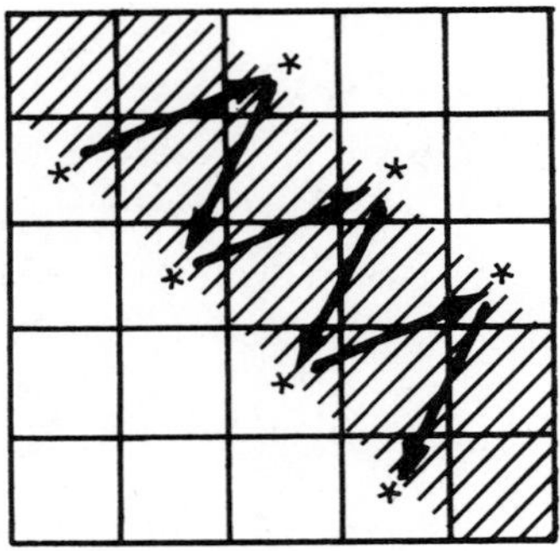

Figure 7.1 The process of chasing.

This process is called *chasing* because the nonzero element outside the bidiagonal positions is chased down the matrix, back and forth across the diagonal, until it finally disappears from the matrix. This is depicted in Fig. 7.1. The way in which Givens transformations are used to carry out this process should be apparent to the reader, given that T_2 is yet to be defined. Now we must deal with the definition of the Givens transformation T_2. Suppose that the elements of B are

$$B = \begin{bmatrix} q_1 & e_2 & 0 & \cdots & 0 \\ 0 & q_2 & e_3 & 0 & \\ & & \ddots & \ddots & 0 \\ & & & \ddots & e_p \\ & 0 & & & q_p \end{bmatrix}$$

and none of the e_j or q_j is zero. We will handle the case where $e_j = 0$ or $q_j = 0$ later. The Givens transformation T_2' is defined as that left transformation which annihilates the second element in the vector

$$\begin{bmatrix} q_1^2 - \delta \\ q_1 e_2 \\ 0 \\ \vdots \\ 0 \end{bmatrix} \qquad (7.17)$$

where δ is the eigenvalue of the lower-right-hand 2 x 2 submatrix of B'B that is closest in value to the (p,p) element of B'B. An algebraic expression for this eigenvalue is easily derived in terms of q_p, q_{p-1}, e_p, and e_{p-1}. This derivation is left as an exercise for the reader. Given that T_2' annihilates $q_1 e_2$ in (7.17), the transition from B to $\bar{B}$ in (7.16) can be completed using T_2 under the assumption that none of the q_j or e_j elements of B is zero. We now consider two cases which are violations of this assumption. We will see that additional transformations on B must sometimes be made before proceeding to compute $\bar{B}$ as in (7.16).

Case 1: $e_i = 0$ *for some* $2 \leq i \leq p$. If some one of the e_i is zero, then the matrix B separates into two disjoint bidiagonal blocks. Each block can be treated independently as a separate B matrix in the iterative procedure with the understanding that the overall situation is one of a block diagonal B matrix in the form

$$B = \begin{bmatrix} B_1 & 0 \\ \\ 0 & B_2 \end{bmatrix}$$

In practice, some tolerance γ is used and $|e_i| \leq \gamma$ replaces the theoretical case $e_i = 0$.

Case 2: $q_k = 0$. When one of the diagonal entries is zero, then at least one singular value of X is zero. In this case, an extra set of Givens transformations is applied to B from the left. The pairs of rows involved in each of the transformations are (k, k + 1), (k, k + 2), ..., (k,p); and each transformation is designed to zero the value in the B matrix position defined by the row-column pair. A summary of these operations is given below:

Matrix	Objective	Additional result
T_{k+1}	Annihilates (k, k + 1) entry	Generates (k, k + 2) entry
T_{k+2}	Annihilates (k, k + 2) entry	Generates (k, k + 3) entry
⋮	⋮	⋮
T_p	Annihilates (k,p) entry	Generates nothing

After these transformations are made, the new B has the form

$$\hat{B} = T_p T_{p-1} \cdots T_{k+1} B = \begin{bmatrix} q_1 & e_2 & & & & & \\ & \ddots & \ddots & & & & \\ & & q_{k-1} & e_k & & & \\ & & & 0 & 0 & & \\ & & & & q'_{k+1} & e'_{k+2} & \\ & & & & & \ddots & \ddots & \\ & & & & & & e'_p \\ & & & & & & q'_p \end{bmatrix}$$

Partitioning the matrix $\hat{B}$ into

$$\hat{B} = \begin{bmatrix} \hat{B}_1 & 0 \\ 0 & \hat{B}_2 \end{bmatrix}$$

we have that $\hat{B}_1$ has a zero row and $\hat{B}_2$ is bidiagonal with all q'_j and e'_j nonzero. $\hat{B}_1$ would now be the subject of right transformations using Givens rotations to zero e_k. This is described mathematically as

$$\hat{\hat{B}}_1 = \hat{B}_1 R_{k-1} R_{k-2} \cdots R_1$$

where R_i annihilates the element in the (i,k) location and for i > 1 a nonzero entry will be introduced in the (i - 1, k) location during the process. Thus, in a sense, we are chasing e_k up and out of the matrix. Completion of this process gives (say) a three-block diagonal matrix with the middle block being a zero scalar. This completes consideration of the second case.

Iteration according to (7.15) continues until all superdiagonal elements in the A matrix are less than some tolerance γ in absolute value. In the course of iteration the *product of all left multipliers is accumulated to give* U_1 *in (7.13), and the product of all right multipliers is accumulated to give the* Q_1 *in (7.13).*

There is one final detail to be considered. The limiting value of the matrix sequence $A^{(j)}$, call it $\hat{\Sigma}$, has diagonal elements which are not necessarily nonnegative and nonincreasing. To remedy this we postmultiply $\hat{\Sigma}$ by a diagonal matrix D having 1 or -1 in each diagonal position (note that D is orthogonal), and use a permutation matrix P to exchange rows and columns to achieve nonincreasing diagonal elements. This is described in matrix notation as

$$\Sigma = P'\hat{\Sigma}DP$$

Experience in computer applications using tolerance on the order of $\gamma = 10^{-15}\|A\|$ ($\|A\|$ denotes the Euclidean norm; see Exercise 7.14 for the definition) indicates that convergence is very rapid, often less than 2p iterations, so long as the matrix A is not extremely ill-conditioned. If, however, there is a wide spread between largest and smallest nonzero singular value, the process converges more slowly but still at a very satisfactory rate. The computer cost of using an efficient implementation of an algorithm based on this method is surprisingly small in view of the mathematical complexity of the method. One excellent FORTRAN subprogram for S.V.D., which uses this method, is contained in the IMSL Library. Another program is listed in the book by Lawson and Hanson (1974).

The S.V.D. is used in solving many problem types, including solving linear systems and computing the Moore-Penrose generalized inverse A^+ of a given matrix A. With regard to this latter application, recall that A^+ is the *unique* matrix which satisfies the Moore-Penrose conditions (Moore 1920 and Penrose 1955)

$$AA^+A = A, \quad A^+AA^+ = A^+, \quad (A^+A)' = A^+A, \quad (AA^+)' = AA^+$$

(Books by Pringle and Rayner 1971 and Rao and Mitra 1971 deal specifically with generalized inverses.) If the S.V.D. of A is

$$A = USV' \tag{7.18}$$

where

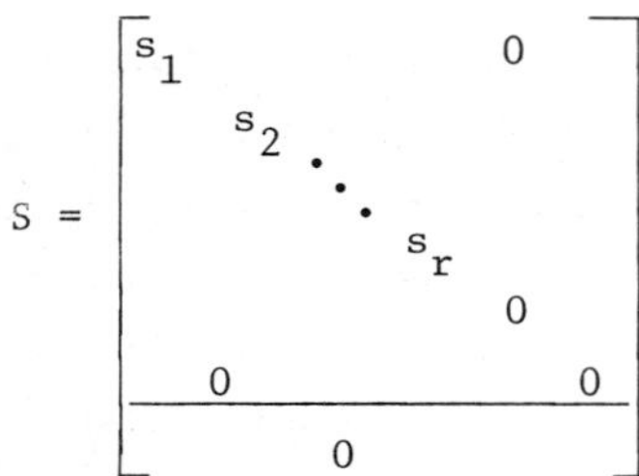

then

$$A^+ = VS^+U' \tag{7.19}$$

where S^+ is diagonal and is formed from S by taking the reciprocal of each nonzero diagonal element in S as the corresponding nonzero diagonal element in S^+.

7.3 GAUSSIAN ELIMINATION AND THE SWEEP OPERATOR

This section deals with transformations made using what are usually called *elementary row operations*. We will discuss these matrix operations with the purpose of showing how they are used in matrix inversion operations. We will deal mostly with symmetric matrices of the form X'X because most statistical computing uses of these matrix operations feature this symmetric form. Elementary row operations will be seen to be equivalent to matrix transformations using nonsingular, but not orthogonal, matrices. Methods which use elementary row operations to zero selected elements in the matrix are called *Gaussian elimination* methods.

As in the preceding section, let X be an n x p matrix of rank r so that X'X is p x p and has rank r. The elementary row operations which we may utilize are:

1. Interchange any two rows.
2. Multiply any row by a constant (the constant must be nonzero to achieve nonsingular transformations).
3. Add one row to another.

The effect of these row operations can be obtained by first performing the operation on the identity matrix and then using this altered identity as a premultiplier.

To illustrate the use of these operations, consider the *Gauss-Jordan* method applied to the following 3 x 3 matrix augmented on the right with an identity matrix. Elementary row operations will be performed which create a third-order identity matrix in the left-hand 3 x 3 matrix positions

$$[X'X|I] = \begin{bmatrix} 2 & 0 & 1 & 1 & 0 & 0 \\ 0 & 1 & 0 & 0 & 1 & 0 \\ 1 & 0 & 2 & 0 & 0 & 1 \end{bmatrix}$$

Reduction 1: Produces unit vector in column 1.

(a) Multiply the first row by -1/2.

(b) Add the first row to the third row.

(c) Multiply the first row by -1.

The resulting matrix is

$$\begin{bmatrix} 1 & 0 & 1/2 & 1/2 & 0 & 0 \\ 0 & 1 & 0 & 0 & 1 & 0 \\ 0 & 0 & 3/2 & -1/2 & 0 & 1 \end{bmatrix}$$

Reduction 2: No operations, since column 2 is in the required form.

Reduction 3: Produces unit vector in column 3.

(a) Multiply the third row by -1/3.

(b) Add the third row to the first row.

(c) Multiply the third row by -2.

The result is

$$\begin{bmatrix} 1 & 0 & 0 & 2/3 & 0 & -1/3 \\ 0 & 1 & 0 & 0 & 1 & 0 \\ 0 & 0 & 1 & -1/3 & 0 & 2/3 \end{bmatrix} = [I|(X'X)^{-1}]$$

We have, of course, produced $(X'X)^{-1}$ through use of elementary row operations. Symbolically we have

$$[X'X|I] \underset{E}{\rightarrow\ \rightarrow} [I|(X'X)^{-1}]$$

One feature of these types of operations is that after each reduction all columns of the identity matrix are still contained in the transformed matrix. They simply change location with each nonnull reduction. This is illustrated in the computations given above. Also, for each reduction k, the kth column of the current matrix is the center of interest but the kth row is the focal point for operations. For this reason the kth row is called the *pivot row* and the kth element in that row (it is on the diagonal of the matrix) is called the *pivot element*. Choice of other than diagonal elements as pivot elements is possible and this is sometimes done in applications. We will continue to "pivot on diagonal elements."

Since each reduction introduces one unit column vector in the area occupied originally by X'X and it modifies one of the unit column vectors in the area originally occupied by the identity, we can take advantage of this fact which obviates the need for augmenting X'X with I when $(X'X)^{-1}$ is desired. Using this space-conserving modification, the results of reduction in the preceding example are

$$\begin{bmatrix} 2 & 0 & 1 \\ 0 & 1 & 0 \\ 1 & 0 & 2 \end{bmatrix} \xrightarrow{\text{reduction 1}} \begin{bmatrix} 1/2 & 0 & 1/2 \\ 0 & 1 & 0 \\ -1/2 & 0 & 3/2 \end{bmatrix} \xrightarrow{\text{reduction 3}} \begin{bmatrix} 2/3 & 0 & -1/3 \\ 0 & 1 & 0 \\ -1/3 & 0 & 2/3 \end{bmatrix}$$

The term *sweep operator* was coined by Beaton (1964) to describe these modified elementary operators. A "sweep of row k" amounts to the kth reduction coupled with use of the storage conservation idea to overwrite the unit vector created by the reduction. Actually, Beaton's sweep differed slightly from this, but we will ignore the difference.

Applications of the sweep operator usually occur in connection with partitioned matrices. Since results for nonsingular matrices are easier to describe, we will first consider the full-rank case.

Suppose that X is n x p, has rank p, and is partitioned as $(X_1|X_2)$ where X_1 is n x m and X_2 is n x (p - m). Forming the sum-of-cross-products matrix in partitioned form gives

$$X'X = \begin{bmatrix} X_1'X_1 & X_1'X_2 \\ X_2'X_1 & X_2'X_2 \end{bmatrix} = \begin{bmatrix} A & B \\ B' & C \end{bmatrix} \tag{7.20}$$

If the first m rows of X'X are swept, pivoting on the diagonal positions in $X_1'X_1$, it can be shown that the resulting matrix is of the form

$$(X'X) \underset{S(1,\ \ldots,\ m)}{\rightarrow} \begin{bmatrix} A^{-1} & A^{-1}B \\ -B'A^{-1} & C-B'A^{-1}B \end{bmatrix} \tag{7.21}$$

If we once again sweep the first m rows of the matrix (7.21), then (7.20) will be the result. This demonstrates the *reversibility of the sweep operator* for full-rank matrices. Similarly, if the last (p - m) rows of (7.20) are swept, pivoting on the diagonal entries in C, the resulting matrix is

$$X'X \underset{S(m+1,\ \ldots,\ p)}{\rightarrow} \begin{bmatrix} A-BC^{-1}B' & -BC^{-1} \\ C^{-1}B' & C^{-1} \end{bmatrix}$$

These results are readily extended to higher-order partitions.

When X has rank $r < p$, an attempt to sweep all rows of X'X will produce, at some stage, a zero pivot element. The utility of sweep in this case is in the production of matrices which have some of the properties of the Moore-Penrose generalized inverse given in Section 7.2.4. Specifically, we can consider, for a given matrix A, any matrix A^- which is such that

$$AA^-A = A \tag{7.22}$$

Any matrix A^- which satisfies (7.22) will be called a g_1 inverse of A. Similarly, any matrix A* which satisfies the first two Moore-Penrose conditions

$$\begin{aligned} AA^*A &= A \\ A^*AA^* &= A^* \end{aligned} \tag{7.23}$$

will be called a g_2 inverse of A. In general, neither A^- nor A^* is unique, but they each coincide with A^{-1} when A is nonsingular.

Consider a partitioning of X wherein the r linearly independent columns are collected to form an n x r submatrix X_1 and the remaining (p - r) columns form X_2. Then $X = (X_1|X_2)$ and

$$X'X = \begin{bmatrix} X_1'X_1 & X_1'X_2 \\ X_2'X_1 & X_2'X_2 \end{bmatrix} = \begin{bmatrix} A & B \\ B' & C \end{bmatrix}$$

where A is nonsingular, C = M'AM for matrix M such that $X_2 = X_1M$, and B = AM. Sweeping the first r rows, and using (7.21), gives

$$X'X \xrightarrow[S(1,\ \ldots,\ r)]{} \begin{bmatrix} A^{-1} & A^{-1}B \\ -B'A^{-1} & C-B'A^{-1}B \end{bmatrix} = \begin{bmatrix} A^{-1} & A^{-1}B \\ -B'A^{-1} & 0 \end{bmatrix} \quad (7.24)$$

The submatrix $C-B'A^{-1}B$ is the zero matrix. It is easy to verify that the matrix given in (7.24) is $(X'X)^-$. Furthermore, if we zero the (2,1) and (1,2) submatrices in (7.24), the resulting matrix is

$$(X'X)^* = \begin{bmatrix} A^{-1} & 0 \\ 0 & 0 \end{bmatrix}$$

Whenever the independent rows of X'X appear as the first r rows, it is obvious that we produce $(X'X)^-$ if we simply proceed to sweep all rows of the matrix, but skip over any row that has a zero pivot element. Similarly, $(X'X)^*$ is produced if we proceed to sweep all rows of X'X, and whenever a zero pivot element is encountered, set all elements of that row and the like-numbered column to zero before moving on to the next row. These latter operations involve a zeroing of rows and columns as well as a sweep on each row which has a nonzero pivot. The term *generalized sweep* will be used to describe this extension of the sweeping operation.

The foregoing description assumed that the linearly independent columns were placed first in the matrix. In practical applications this is not necessary. The obvious procedure to follow in computing

$(X'X)^-$ or $(X'X)^*$ is to sweep each row of the given matrix so long as the row has a nonzero pivot element, and when a zero pivot is encountered either skip the row or do the zeroing of row and column which is needed for generalized sweeps.

When generalized sweeps are used, a result analogous to (7.21) can be obtained for partitioned matrices having singular submatrices along the diagonal. Consider the partitioned matrix X'X with X in the form

$$\begin{aligned} X &= (X_1 | X_2) \\ &= (Z_1 | Z_2 | Z_3 | Z_4) \end{aligned}$$

where

$$\begin{aligned} X_1 &= (Z_1 | Z_2) \\ X_2 &= (Z_3 | Z_4) \end{aligned}$$

and $Z_1'Z_1$ is nonsingular, has rank r, and Z_1 has rank r:

$$X'X = \begin{bmatrix} Z_1'Z_1 & Z_1'Z_2 & Z_1'Z_3 & Z_1'Z_4 \\ Z_2'Z_1 & Z_2'Z_2 & Z_2'Z_3 & Z_2'Z_4 \\ Z_3'Z_1 & Z_3'Z_2 & Z_3'Z_3 & Z_3'Z_4 \\ Z_4'Z_1 & Z_4'Z_2 & Z_4'Z_3 & Z_4'Z_4 \end{bmatrix} = \begin{bmatrix} X_1'X_1 & X_1'X_2 \\ X_2'X_1 & X_2'X_2 \end{bmatrix} \tag{7.25}$$

Using generalized sweeps on the first r rows of this matrix gives

$$X'X \xrightarrow[GS(1,\ \ldots,\ r)]{} \begin{bmatrix} (Z_1'Z_1)^{-1} & 0 & (Z_1'Z_1)^{-1}Z_1'Z_3 & (Z_1'Z_1)^{-1}Z_1'Z_4 \\ 0 & 0 & 0 & 0 \\ -Z_3'Z_1(Z_1'Z_1)^{-1} & 0 & Z_3'MZ_3 & Z_3'MZ_4 \\ -Z_4'Z_1(Z_1'Z_1)^{-1} & 0 & Z_4'MZ_3 & Z_4'MZ_4 \end{bmatrix}$$

where $M = I - Z_1(Z_1'Z_1)^{-1}Z_1'$. This matrix is in the form

$$X'X \xrightarrow[GS(1,\ \ldots,\ r)]{} \begin{bmatrix} (X_1'X_1)^* & (X_1'X_1)^*X_1'X_2 \\ -X_2'X_1(X_1'X_1)^* & X_2'X_2 - X_2'X_1(X_1'X_1)^*X_1'X_2 \end{bmatrix} \tag{7.26}$$

Thus we see that (7.26) has the same form, in terms of a g_2 inverse, as the form displayed in (7.21) for a more restrictive situation.

A characteristic of the g_2 inverse which deserves mentioning at this time is that $(X'X)^*X'X$, under a suitable permutation of rows, is in what is sometimes called the Hermite canonical form of $X'X$. This form is defined by the following four conditions:

1. The first r rows, for some $r \geq 0$, are nonzero and all remaining rows are zero.
2. In the ith row $(i = 1, 2, \ldots, r)$, the first nonzero element is equal to unity, the column in which it occurs being numbered c_i.
3. $c_1 < c_2 < \ldots < c_r$.
4. In column c_i the only nonzero element is the 1 in row i.

To see this for one case, assume that $(X'X)^*$ is

$$(X'X)^* = \begin{bmatrix} (X_1'X_1)^{-1} & 0 \\ 0 & 0 \end{bmatrix}$$

and that

$$X'X = \begin{bmatrix} X_1'X_1 & X_1'X_2 \\ X_2'X_1 & X_2'X_2 \end{bmatrix}$$

Then the product $(X'X)^*(X'X)$ is

$$(X'X)^*X'X = \begin{bmatrix} I & (X_1'X_1)^{-1}X_1'X_2 \\ 0 & 0 \end{bmatrix} \tag{7.27}$$

The matrix (7.27) obviously satisfies the four conditions given above, but in general the rows of $(X'X)^*X'X$ will often need to be permuted to achieve this form. The computation of $(X'X)^*X'X$ can be accomplished by the following modification of the procedure for computing $(X'X)^*$. For each row of $X'X$, if a zero pivot is not encountered, do a normal sweep. When a zero pivot element is encountered, zero the entire row, but introduce zeros *only in positions* *below the diagonal* in that column. The result of this modified procedure, when the linearly independent rows appear first in the matrix, is

$$\begin{bmatrix} (X_1'X_1)^{-1} & (X_1'X_1)^{-1}X_1'X_2 \\ 0 & 0 \end{bmatrix} \tag{7.28}$$

Comparing this matrix with (7.27) we see that, in fact, in (7.28) we have all the information required to obtain $(X'X)^*$ *and* $(X'X)^*X'X$. The nonzero diagonal elements in (7.28) define rows and columns containing elements of $(X_1'X_1)^{-1}$ in $(X'X)^*$ and the identity matrix in (7.27).

Algorithms based on sweep operations are simple and easy to implement. Naturally, as in other numerical methods, in practice some tolerance ε must be used to decide whether a pivot element is sufficiently small in magnitude to be called zero. When sweeps or generalized sweeps are performed on a symmetric matrix, a significant savings in computer storage can be made by carrying only the triangular (upper or lower) portion of the matrix. This is possible because, although symmetry is not preserved at each stage, either the (i,j) element is equal to the (j,i) element or they differ only in sign. If the ith row has been swept and $j > i$, then the elements differ in sign.

Algorithm GSWP is designed to perform a generalized sweep of each row in a given p x p array $A = (a_{ij})$. The result is returned in the A array. Test for zero pivot is made against ε. The rank of A is returned as r.

ALGORITHM GSWP(A,p,r,ε)

Step	*Description*
1.	Set k = 1 and r = 0.
2.	Set $d = a_{kk}$.
3.	If $\lvert d \rvert < \varepsilon$, then zero row k and column k, and go to step 7.
4.	Divide row k by d, and set r = r + 1.

5. For each row $i \neq k$, set $c = a_{ik}$ and subtract c times row k from row i. Then set $a_{ik} = -c/d$.
6. Set $a_{kk} = 1/d$.
7. Set $k = k + 1$. If $k \leq p$, go to step 2; otherwise exit.

7.4 CHOLESKY DECOMPOSITION AND RANK-ONE UPDATE

In this section we consider a special form of triangular factorization which will be used in later chapters. The matrix A on which we operate is assumed to be n x n symmetric and positive definite.

Since A is positive definite there exists an n x n nonsingular matrix B such that $A = BB'$. Using orthogonal transformations, from the right, on B we have

$$BQ' = \hat{L}$$

where $\hat{L}$ is lower-triangular and Q' is orthogonal. To obtain a unit lower-triangular matrix, $\hat{L}$ can be factored as

$$\hat{L} = LD_1$$

where D_1 is diagonal and nonsingular and L has ones on the diagonal. Thus we have

$$B = LD_1Q$$

and

$$A = \hat{L}\hat{L}' \tag{7.29}$$

$$= LDL' \tag{7.30}$$

where $D = D_1^2$. This form of decomposition is called the *Cholesky decomposition* of A. The second form (7.30) is usually preferred because square roots are not needed in the decomposition.

This decomposition is useful when solving systems of linear equations. Consider the linear system

$$A\underline{\beta} = \underline{y}$$

Using the Cholesky decomposition of A, the solution $\hat{\underline{\beta}}$ can be obtained in two steps by first solving for $\underline{\gamma}$ in

$$LD\underline{\gamma} = \underline{y}$$

and then solving

$$L'\underline{\beta} = \underline{\gamma}$$

for $\hat{\underline{\beta}}$. Since each of these linear systems has a triangular coefficient matrix, a back-substitution is all that is needed in each case.

The Cholesky decomposition is especially easy to perform using the computer. One procedure is to use n steps and at step j compute

$$d_j = a_{jj} - \sum_{k=1}^{j-1} \ell_{jk}^2 d_k$$

$$\ell_{ij} = (a_{ij} - \sum_{k=1}^{j-1} \ell_{jk}\ell_{ik}d_k)d_j^{-1} \qquad i = j + 1, \ldots, n \qquad (7.31)$$

where the notation is $D = \text{diag}(d_i)$ and $L = (\ell_{ij})$ and the summation terms are taken as 0 when $j = 1$. Not only is this a simple procedure, it also is efficient and has good numerical stability. The formulas given above, are not difficult to derive, and their derivation is left as an exercise for the reader. Example 7.3 illustrates their use.

Example 7.3

Suppose that we wish to decompose the following matrix A using a Cholesky decomposition:

$$A = \begin{bmatrix} 2 & 0 & 1 \\ 0 & 1 & 0 \\ 1 & 0 & 1 \end{bmatrix}$$

Use of Eqs. (7.31) yields

$$d_1 = a_{11} \qquad \ell_{21} = \frac{a_{21}}{d_1} \qquad \ell_{31} = \frac{a_{31}}{d_1}$$

$$= 2 \qquad\qquad = 0 \qquad\qquad = \frac{1}{2}$$

$$d_2 = a_{22} - \ell_{21}^2 d_1 \qquad \ell_{32} = \frac{a_{32} - \ell_{21}\ell_{31}d_1}{d_2}$$

$$= 1 \qquad\qquad = 0$$

$$d_3 = a_{33} - \ell_{31}^2 d_1 - \ell_{32}^2 d_2$$

$$= \frac{1}{2}$$

Thus we have that

$$L = \begin{bmatrix} 1 & 0 & 0 \\ 0 & 1 & 0 \\ 1/2 & 0 & 1 \end{bmatrix} \qquad D = \begin{bmatrix} 2 & 0 & 0 \\ 0 & 1 & 0 \\ 0 & 0 & 1/2 \end{bmatrix}$$

and a numerical check will show that LDL' = A.

When A is not positive definite it cannot be expressed in this form. An attempt to decompose a semidefinite or indefinite symmetric matrix will result in a zero or negative d_j at some stage.

It sometimes happens that we wish to modify a given matrix A according to

$$\bar{A} = A + \sigma \underline{z}\underline{z}'$$

where σ is a scalar and both A and $\bar{A}$ are positive definite. This is usually called a *rank-one update* whenever $\sigma \neq 0$ and $\underline{z} \neq \underline{0}$. Situations where updating is required will be described in Chapter 10. If A has been previously decomposed and we want to update the decomposition, that is, we want $\bar{L}$ and $\bar{D}$ such that

$$\overline{LDL'} = LDL' + \text{sign}(\sigma)\underline{v}\underline{v}'$$

where $\underline{v} = |\sigma|^{1/2}\underline{z}$, then the problem is one which has been considered by Gill, Golub, Murray, and Saunders (1974) and Gill, Murray, and Saunders (1975). In the latter reference two algorithms are given which find $\bar{L}$ and $\bar{D}$ for each of the two possible signs of σ. These algorithms are very efficient, and attention has been given to questions of numerical stability, so we will describe them in the remainder of this section. Derivation of these algorithms can be found in the literature cited.

When sign(σ) is positive the problem can be simplified using the identity

$$\bar{A} = L(D + \underline{p}\underline{p}')L'$$

where $\underline{p}$ satisfies

$$L\underline{p} = \underline{v}$$

If the matrix $D + \underline{p}\underline{p}'$ is first decomposed as

$$D + \underline{p}\underline{p}' = \tilde{M}\tilde{D}\tilde{M}'$$

then $\bar{L} = L\tilde{M}$ and $\bar{D} = \tilde{D}$ are easily obtained. For this reason, we will next consider computing $\tilde{M}$ and $\tilde{D}$.

It can be shown that the matrix $\tilde{M} = (M_{ij})$ has a special structure of the form

$$M_{ij} = \begin{cases} 0 & i < j \\ 1 & i = j \\ p_i\beta_j & i > j \end{cases} \tag{7.32}$$

where β_j is a function of the elements in the matrix $D = \text{diag}(d_j)$ and the vector $\underline{p} = (p_1, p_2, \ldots, p_n)'$. The following algorithm produces the factors $\bar{D}$ and $\tilde{M}$.

ALGORITHM DP($\underline{d}$,$\underline{p}$,n,$\underline{\beta}$,$\underline{\bar{d}}$,$\underline{t}$)

Step	*Description*
1.	Set $t_0 = 1$
2.	Set $j = 1$
3.	Compute $t_j = t_{j-1} + \frac{p_j^2}{d_j}$, $\bar{d}_j = \frac{d_j t_j}{t_{j-1}}$, $\beta_j = \frac{p_j}{d_j t_j}$
4.	If $j < n$, set $j = j + 1$ and go to step 3; otherwise exit with $\underline{\beta}$, $\underline{p}$, and $\underline{\bar{d}}$ which define $\tilde{M}$ and $\bar{D} = \tilde{D}$.

The special structure of $\tilde{M}$ insures that the product $L\tilde{M} = \bar{L}$ can be computed efficiently. An efficient way to form $\bar{L}$ is as follows:

1. Let $\underline{V}^{(1)} = L\underline{p}$.
2. For each j = 1, 2, ..., n - 1, set

$$\left.\begin{aligned} V_k^{(j+1)} &= V_k^{(j)} - p_j \ell_{kj} \\ \bar{\ell}_{kj} &= \ell_{kj} + \beta_j V_k^{(j+1)} \end{aligned}\right\} \quad k = j + 1, \ldots, n \tag{7.33}$$

where $\underline{V}' = (V_1, V_2, \ldots, V_n)$. An alternative expression for $\bar{\ell}_{kj}$ is

$$\bar{\ell}_{kj} = \left(\frac{d_j}{\bar{d}_j}\right)\ell_{kj} + \beta_j V_k^{(j)} \tag{7.34}$$

This alternative provides better numerical stability because $\bar{d}_j > d_j$, as is easily seen by inspecting algorithm DP; thus $d_j/\bar{d}_j$ will tend to help keep numerical values near the same order of magnitude. However, the volume of computation increases by $n^2/2$ multiplications using this latter expression. Hence, we can increase numerical stability at the expense of increased computer costs. This, of course, is normally the case. The suggested procedure in the present situation is to use Eq. (7.33) whenever $\bar{d}_j/d_j \leq 4$ and use (7.34) otherwise. Example 7.4 illustrates the use of these equations.

Example 7.4

Consider the matrices A, L, and D of Example 7.3, and assume that we wish to perform the update

$$\begin{aligned} \bar{L}\bar{D}\bar{L}' &= LDL' + \sigma\underline{z}\underline{z}' \\ &= LDL' + 4\begin{pmatrix} 1 \\ -1 \\ 1 \end{pmatrix} (1 \quad -1 \quad 1) \end{aligned}$$

Thus we have $\sigma = 4$, $\underline{z}' = (1 \quad -1 \quad 1)$, and

$$\begin{aligned} \underline{v} &= \sigma^{1/2}\underline{z} \\ &= (2 \quad -2 \quad 2)' \end{aligned}$$

To find $\underline{p}$ we solve the system of linear equations $L\underline{p} = \underline{v}$, which in this example is the system

$$\begin{pmatrix} 1 & 0 & 0 \\ 0 & 1 & 0 \\ 1/2 & 0 & 1 \end{pmatrix} \begin{pmatrix} p_1 \\ p_2 \\ p_3 \end{pmatrix} = \begin{pmatrix} 2 \\ -2 \\ 2 \end{pmatrix}$$

The solution is

$$\underline{p}' = (2 \quad -2 \quad 1)$$

To find $\tilde{M}$ and $\tilde{D}$ we use algorithm DP to obtain the following results:

$$t_1 = t_0 + \frac{p_1^2}{d_1} \qquad \bar{d}_1 = \frac{d_1 t_1}{t_0} \qquad \beta_1 = \frac{p_1}{d_1 t_1}$$

$$= 3 \qquad\qquad = 6 \qquad\qquad = \frac{1}{3}$$

$$t_2 = t_1 + \frac{p_2^2}{d_2} \qquad \bar{d}_2 = \frac{d_2 t_2}{t_1} \qquad \beta_2 = \frac{p_2}{d_2 t_2}$$

$$= 7 \qquad\qquad = \frac{7}{3} \qquad\qquad = \frac{-2}{7}$$

$$t_3 = t_2 + \frac{p_3^2}{d_3} \qquad \bar{d}_3 = \frac{d_3 t_3}{t_2} \qquad \beta_3 = \frac{p_3}{d_3 t_3}$$

$$= 9 \qquad\qquad = \frac{9}{14} \qquad\qquad = \frac{2}{9}$$

Now $\tilde{M}$ and $\bar{D}$ are formed using Eqs. (7.32):

$$\tilde{M} = \begin{pmatrix} 1 & 0 & 0 \\ -2/3 & 1 & 0 \\ 1/3 & -2/7 & 1 \end{pmatrix} \qquad \bar{D} = \begin{pmatrix} 6 & 0 & 0 \\ 0 & 7/3 & 0 \\ 0 & 0 & 9/14 \end{pmatrix}$$

The updated decomposition can now be specified by computing $\bar{L}$ and using $\bar{D}$ from above. The matrix $\bar{L}$ can be found directly through the matrix multiplication

$$\bar{L} = L\tilde{M}$$

$$= \begin{pmatrix} 1 & 0 & 0 \\ -2/3 & 1 & 0 \\ 5/6 & -2/7 & 1 \end{pmatrix}$$

Using Eqs. (7.33), with $\bar{\ell}_{kj}$ given by (7.34), to compute the elements of $\bar{L}$ yields the following results:

$$\underline{v}^{(1)} = (2 \quad -2 \quad 2)'$$

$$v_2^{(2)} = v_2^{(1)} - p_1\ell_{21} \qquad \bar{\ell}_{21} = \left(\frac{d_1}{\bar{d}_1}\right)\ell_{21} + \beta_1 v_2^{(1)}$$

$$= -2 \qquad\qquad = -\frac{2}{3}$$

$$v_3^{(2)} = v_3^{(1)} - p_1\ell_{31} \qquad \bar{\ell}_{31} = \left(\frac{d_1}{\bar{d}_1}\right)\ell_{31} + \beta_1 v_3^{(1)}$$

$$= 1 \qquad\qquad = \frac{5}{6}$$

$$v_3^{(3)} = v_3^{(2)} - p_2\ell_{32} \qquad \bar{\ell}_{32} = \left(\frac{d_2}{\bar{d}_2}\right)\ell_{32} + \beta_2 v_3^{(2)}$$

$$= 1 \qquad\qquad = \frac{-2}{7}$$

Thus we see that Eqs. (7.33) and (7.34) yield the correct $\bar{L}$.

Turning to the case of sign(σ) negative, the analogous factorization of $D-\underline{p}\underline{p}'$ is

$$D - \underline{p}\underline{p}' = \tilde{M}\bar{D}\tilde{M}'$$

where $\tilde{M}$ is in the form (7.32). The elements $\bar{d}_j$ and β_j in this case are formed according to the following algorithm:

ALGORITHM DM($\underline{d}$,$\underline{p}$,n,$\underline{\beta}$,$\underline{\bar{d}}$,$\underline{t}$)

Step	*Description*
1.	Set $t_{n+1} = 1 - \underline{p}'D^{-1}\underline{p}$.
2.	Set $j = n$.

Algorithm cont.

Step	*Description*
3.	Compute $$t_j = t_{j+1} + \frac{p_j^2}{d_j}$$ $$\bar{d}_j = \frac{d_j t_{j+1}}{t_j}$$ $$\beta_j = \frac{-p_j}{d_j t_{j+1}}$$
4.	If j = 1, exit; otherwise set j = j - 1 and go to step 3.

Finally, the matrix product $L\tilde{M} = \bar{L}$ is formed in the following steps:

For each j = n, n - 1, ..., 1,

1. $V_j^{(j)} = p_j$

2. $$\left.\begin{aligned} \bar{\ell}_{kj} &= \ell_{kj} + \beta_j V_k^{(j+1)} \\ V_k^{(j)} &= V_k^{(j+1)} + p_j \ell_{kj} \end{aligned}\right\} \quad k = j + 1, \ldots, n.$$

7.5 SUMMARY

Several numerical methods for matrix applications have been described. Any one of these methods can be used for performing many tasks such as matrix inversion or solving linear systems of equations, for example, and these problems do frequently arise in statistical applications.

The question as to which one among competing methods to utilize in a given situation is difficult if not impossible to answer satisfactorily. Considerations such as amount of storage required and relative number of arithmetic operations are not as important in today's high-speed, large-capacity computers. The most important consideration is the stability of the method, and all of the methods described in this chapter are generally satisfactory in this respect.

On the IBM 370 computer, using double-precision floating-point arithmetic and accumulating vector inner products in higher precision, severe numerical inaccuracies are not often encountered in research applications. Orthogonal transformations usually provide more accurate results than do sweep operations, and some people recommend that sweeps never be used. Nevertheless, many large software systems in general use today do utilize sweep operators for relevant matrix computations. More will be said about these methods in connection with specific applications in subsequent chapters.

EXERCISES

7.1 Prove that the Householder transformation matrix H defined through Eqs. (7.1) and (7.2)

(a) is orthogonal under either choice of sign in (7.2);

(b) produces a vector $\underline{v} = H\underline{w}$ which has zero as components $t + 1, t + 2, \ldots, n$.

7.2 Explain why the argument b is actually not needed in algorithm BHOUSE.

7.3 Show, using Eqs. (7.2), that $\beta = \underline{u}'\underline{u}/2$.

7.4 Show how the basic definition of a Givens transformation can be redefined to produce a minor modification that makes the G_{ij} symmetric.

7.5 Show, by a simple example, that the Givens transformation does produce a rotation in a plane. Characterize the plane and the angle of rotation.

7.6 Use Householder reflections and Givens rotations in two separate applications to triangularize the following matrix:

$$\begin{pmatrix} 5 & 7 & 6 \\ 7 & 10 & 8 \\ 6 & 8 & 10 \end{pmatrix}$$

7.7 Givens rotations can be used to triangularize a matrix by zeroing elements one row at a time. Show that this process will not change any previously created zero elements. In which type of situation might one prefer to annihilate elements by row instead of by column?

7.8 Let X be a given n x p matrix of rank p. Suppose that it is desired to transform X to upper-triangular form using orthogonal transformations. Mathematically this is described as

$$QX = \left(\frac{T}{0}\right)$$

where T is p x p upper-triangular. The question of efficiency in terms of number of arithmetic operations is often posed when choice of method is under consideration. As a partial answer to this question, a comparison of competing methods is made based on the number of multiplications and divisions required by each method. With these facts in mind, do the following:

(a) Compare use of Householder reflections with Givens rotations to determine which method requires less arithmetic.

(b) Compare use of Householder reflections with Gram-Schmidt orthogonalization to determine which method requires less arithmetic.

You will find that the details of implementation of each method are important and that operation counts will vary depending on the details of implementation. This fact helps to explain why two algorithms based on the same mathematical computational method can display significantly different numerical stability. When solving (a) and (b) above, use the most efficient procedure that you can imagine. You will probably determine that the Householder method is most efficient.

7.9 Use the defining Eqs. (7.9) and show that the matrix V is indeed orthogonal when X has full-rank p and the Gram-Schmidt method is applied to X.

7.10 If X is an n x p matrix containing real number elements, what guarantees that the singular values of X will be real numbers?

7.11 Why don't we simply use a few additional Householder or Givens transformations on the right of the matrix A in Eq. (7.12) to diagonalize A and thus obtain S?

7.12 Consider the transformation of B, to obtain $\bar{B}$ in Eq. (7.16), and verify that $\bar{B}$ is indeed upper-bidiagonal and that the annihilation and generation process does operate as described after the initial transformation using T_2.

7.13 Verify that the matrix given in Eq. (7.19) does satisfy the four Moore-Penrose conditions.

7.14 The Euclidean norm of a matrix X is usually defined as the square root of the sum of squared elements in X. This is written as

$$||X|| = \left(\sum_{i=1}^{n} \sum_{j=1}^{p} x_{ij}^2 \right)^{1/2}$$

Prove that

$$||X|| = \left(\sum_{j=1}^{p} s_j^2 \right)^{1/2}$$

where the s_j are the singular values of X.

7.15 Show that the square of the largest singular value of X (i.e., s_1^2) is equal to the maximum value of the quadratic form $\underline{v}'X'X\underline{v}$ where $\underline{v}$ ranges over all p x 1 vectors such that $||\underline{v}|| = (\underline{v}'\underline{v})^{1/2} = 1$. Also show that if the first column of the V matrix in Eq. (7.10) is used as the $\underline{v}$ vector, then s_1^2 is the value of the quadratic form.

7.16 Suppose X is p x p and has rank p. Prove that there always exist orthogonal matrices U and V such that

$$UXV = S$$

where S is diagonal and has positive diagonal elements which are nonincreasing down the diagonal.

7.17 Use the results of Exercise 7.16 to show that a S.V.D. always exists for a given n x p matrix X.

7.18 We know that the singular values of a given matrix X are unique under the S.V.D.

$$X = USV'$$

Are the orthogonal matrices U and V always unique? (*Hint:* review the situation wherein a symmetric matrix has an eigenvalue of multiplicity greater than 1.)

7.19 Prove that the partitioned matrix (7.21) is indeed the result of sweeping the first m rows of (7.20). Also show that the matrix of (7.24) is $(X'X)^-$.

7.20 Use Eq. (7.20) and the matrix (7.21) and show the result of resweeping the first m rows of (7.21).

7.21 Is the generalized sweep operator, applied to matrices of less than full rank, reversible? Is the regular sweep, as used to produce (7.24), reversible?

7.22 The Moore-Penrose conditions as given in Section 7.2.4 for an n x p matrix X of rank r are

1. $XX^+X = X$
2. $X^+XX^+ = X^+$
3. $(XX^+)' = XX^+$
4. $(X^+X)' = X^+X$

Define a g_1 inverse to be any matrix B which satisfies only condition 1, a g_2 inverse to be any matrix B which satisfies conditions 1 and 2, a g_3 inverse to be any B which satisfies 1, 2, and 3, and the g_4 inverse to be the unique B which satisfies all four conditions. Under these definitions, prove the following:

(a) If B is a g_3 inverse of X, then BB' is a g_2 inverse of X'X. (Golub and Styan 1973.)

(b) Let $C = (X'X)^-$ be a g_1 inverse of X'X. Show that $X^* = CX'$ is a g_3 inverse of X.

(c) If X is decomposed as

$$X = H \begin{pmatrix} T_1 & T_2 \\ 0 & 0 \end{pmatrix} P$$

where H is n x n orthogonal, T_1 is r x r upper-triangular, T_2 is r x (p - r), and P is a p x p permutation matrix, then

$$B = P'\begin{pmatrix} T_1^{-1} & 0 \\ 0 & 0 \end{pmatrix}H'$$

is a g_3 inverse of X. (Golub and Styan 1973.)

(d) Additional Householder transformations, applied from the right to the matrix X of part (c), can be used to produce the decomposition

$$X = H\begin{pmatrix} R_1 & 0 \\ 0 & 0 \end{pmatrix}Q'$$

where H and Q are orthogonal and R_1 is r x r nonsingular and lower-triangular. (R_1 can also be made upper-triangular if we modify our procedure slightly.) Show that the g_4 inverse of X is given by

$$X^+ = Q\begin{pmatrix} R_1^{-1} & 0 \\ 0 & 0 \end{pmatrix}H'$$

Also show that for any orthogonal decomposition of X as

$$X = HRQ'$$

where H and Q are orthogonal, we have

$$X^+ = QR^+H'$$

7.23 Use the results of Exercise 7.22(d) to show that a consistent but overdetermined system of linear equations (X is n x p of rank r with n > p)

$$X\underline{\beta} = \underline{Y}$$

has, among its solutions $\hat{\underline{\beta}}$, a *unique minimum-length solution* (i.e., there is one $\hat{\underline{\beta}}$ having smallest euclidean vector norm) and that that solution is $\hat{\underline{\beta}}_m = \underline{b} = X^+\underline{Y}$. (See, e.g., Lawson and Hanson 1974.)

7.24 Show that if X^- is a g_1 inverse of X, and if the overdetermined system of linear equations

$$X\underline{\beta} = \underline{Y}$$

is consistent, then $\hat{\underline{\beta}} = X^-\underline{Y}$ is a solution. (See, e.g., Rao and Mitra 1971.)

7.25 Show that the set of all solutions to the system of linear equations in Exercise 7.23 is expressible in the form

$$\hat{\underline{\beta}} = X^+\underline{Y} + (I_p - X^+X)\underline{z}$$

where the vector $\underline{z}$ determines any particular solution.

7.26 Use sweep operations to invert the matrix

$$\begin{pmatrix} 5 & 7 & 6 \\ 7 & 10 & 8 \\ 6 & 8 & 10 \end{pmatrix}$$

7.27 Use Householder transformations to solve the linear system

$$\begin{pmatrix} 5 & 7 & 6 \\ 7 & 10 & 8 \\ 6 & 8 & 10 \end{pmatrix} \begin{pmatrix} X_1 \\ X_2 \\ X_3 \end{pmatrix} = \begin{pmatrix} 0 \\ 1 \\ 2 \end{pmatrix}$$

7.28 The famous mathematical method, called Cramer's rule, for solving the linear system

$$\underset{n \times n}{A}\underline{X} = \underset{n \times 1}{\underline{b}}$$

proceeds as follows.

If the determinant of A [denoted by det(A)] is nonzero, then for each j = 1, 2, ..., n the components of the solution vector $\underline{X}$ are obtained as

$$x_j = \frac{\det(A_j)}{\det(A)}$$

where A_j is obtained by replacing column j of A with the elements of the $\underline{b}$ vector.

The determinant of a matrix A can be written as

$$\det(A) = \sum_{(j)} e_{j_1,j_2,\cdots,j_n} a_{1j_1} a_{2j_2} \cdots a_{nj_n}$$

where $(j) = j_1, j_2, \cdots, j_n$ denotes a permutation of the integers 1 through n, $e_{j_1, j_2, \cdots, j_n}$ is +1 or -1 according as the permutation is even or odd, and the summation $\sum_{(j)}$ extends over all n! number of possible permutations.

(a) Obtain an estimate of the number of multiplications required to solve an nth-order linear system using Cramer's rule and the above expression to evaluate the determinants.

(b) For n = 25, how many human generations do you think will pass during the computer solution of this problem on the computer that you use?

(c) Estimate the number of seconds required to obtain a solution for size n = 25 using sweep operations.

REFERENCES AND FURTHER READINGS

Beaton, A. (1964), The Use of Special Matrix Operators in Statistical Calculus, Educational Testing Service, RB 64-51, Princeton, N.J.

Bjorck, A. (1967a), Solving Least Squares Problems by Gram-Schmidt Orthogonalization, *BIT 7*, 1-21.

Bjorck, A. (1967b), Iterative Refinement of Linear Least Squares Solutions, *BIT 7*, 257-278.

Bjorck, A., and G. H. Golub (1973), Numerical Methods for Computing Angles Between Linear Subspaces, *Math. Comp. 27*, 579-594.

Blue, J. L. (1978), A Portable FORTRAN Program to Find the Euclidean Norm of a Vector, *TOMS 4*, 15-23.

Businger, P. A. (1970), Updating a Singular Value Decomposition, *BIT 10*, 376-385.

Businger, P. A., and G. H. Golub (1969), Algorithm 358: Singular Value Decomposition of a Complex Matrix, *CACM 12*, 564-565.

Chambers, J. M. (1977), *Computational Methods for Data Analysis*, Wiley, New York.

Clayton, D. G. (1971), Algorithm AS46: Gram-Schmidt Orthogonalization, *Appl. Stat. 20*, 335-338.

Dahlquist, G., and A. Bjorck (1974), *Numerical Methods*, trans. N. Anderson, Prentice-Hall, Englewood Cliffs, N.J.

Daniel, J. W., W. B. Gregg, L. Kaufman, and G. W. Stewart (1976), Reorthogonalization and Stable Algorithms for Updating the Gram-Schmidt QR Factorization, *Math. Comp. 30*, 772-795.

Eckart, C., and G. Young (1936), The Approximation of One Matrix by Another of Lower Rank, *Psychometrika 1*, 211-218.

Forsythe, G. E., and C. B. Moler (1967), *Computer Solution of Linear Algebraic Systems*, Prentice-Hall, Englewood Cliffs, N.J.

Francis, J. (1961, 1962), The QR Transformation, parts I and II, *Comp. J. 4*, 265-271, 332-345.

Gentleman, W. M. (1973), Least Squares Computations by Givens Transformations Without Square Roots, *J. Inst. Maths. Applics. 12*, 329-336.

Gill, P. E., G. H. Golub, W. Murray, and M. A. Saunders (1974), Methods for Modifying Matrix Factorizations, *Math. Comp. 28*, 505-536.

Gill, P. E., W. Murray, and M. A. Saunders (1975), Methods for Computing and Modifying the LDV Factors of a Matrix, *Math. Comp. 29*, 1051-1077.

Givens, W. (1954), *Numerical Computation of the Characteristic Values of a Real Symmetric Matrix*, Oak Ridge National Laboratory report ORNL-1574, Oak Ridge, Tenn.

Givens, W. (1958), Computation of Plane Unitary Rotations Transforming a General Matrix to Triangular Form, *SIAM J. Appl. Math. 6*, 26.

Golub, G. H. (1968), Least Squares, Singular Values, and Matrix Approximations, *Aplikace Matematiky 13*, 44-51.

Golub, G. H. (1969), Matrix Decompositions and Statistical Calculations, in *Statistical Computation,* edited by R. C. Milton and J. A. Nelder, Academic Press, New York, 365-397.

Golub, G. H., and W. Kahan (1965), Calculating the Singular Values and Pseudoinverse of a Matrix, *SIAM J. Num. Anal. 2*, 205-224.

Golub, G. H., and C. Reinsch (1970), Singular Value Decompositions and Least Squares Solutions, *Numer. Math. 15*, 403-420.

Golub, G. H., and G. P. H. Styan (1973), Numerical Computations for Univariate Linear Models, *JSCS 2*, 253-274.

Goodnight, J. H. (1978), The Sweep Operator: Its Importance in Statistical Computing, *Proceedings of Computer Science and Statistics: Eleventh Annual Symposium on the Interface*, edited by T. Gerig and A. R. Gallant, Institute of Statistics, North Carolina State University, Raleigh, 218-229.

Greville, T. N. E. (1959), The Pseudoinverse of a Rectangular or Singular Matrix and Its Application to the Solution of Systems of Linear Equations, *SIAM Rev. 1*, 38-43.

Greville, T. N. E. (1960), Some Applications of the Pseudoinverse of a Matrix, *SIAM Rev. 2*, 15-22.

Hanson, R. J. (1973), Is the Fast Givens Transformation Really Fast? *Signum Newsletter, ACM*, 7.

Hanson, R. J., and C. L. Lawson (1969), Extensions and Applications of the Householder Algorithm for Solving Linear Least Squares Problems, *Math. Comp. 23*, 787-812.

Healy, M. J. R. (1968), Triangular Decomposition of a Symmetric Matrix, *Appl. Stat. 17*, 195-197.

Hestenes, M. R. (1958), Inversion of Matrices by Biorthogonalization and Related Results, *SIAM J. Appl. Math. 6*, 51-90.

Householder, A. S. (1958), Unitary Triangularization of a Nonsymmetric Matrix, *JACM 5*, 339-342.

Householder, A. S. (1964), *The Theory of Matrices in Numerical Analysis*, Blaisdell, New York.

IMSL Corporation (1977), *IMSL Reference Manual*, Houston.

Lawson, C. L., and R. J. Hanson (1974), *Solving Least Squares Problems*, Prentice-Hall, Englewood Cliffs, N.J.

Longley, J. W. (1975), Normalized Modified Gram-Schmidt Algorithm for the Solution of Linear Least Squares Equations, *Proceedings of the Computer Science and Statistics: Eighth Annual Symposium on the Interface*, edited by J. Frane, Health Sciences Computing Facility, UCLA, 178-187.

Longley, J. W. (1976), A Subroutine for Singular Value Decomposition by Normalized Modified Gram-Schmidt Algorithm for the Solution of Linear Least Squares Equations, *Proceedings of the Statistical Computing Section, American Statistical Association*, 207-212.

Longley, J. W. (1977), Least Squares Computations and the Condition of the Matrix, paper presented at the International Symposium on Extremal Methods and Systems Analysis, University of Texas, Austin.

Maindonald, J. H. (1977), Least Squares Computations Based on Cholesky Decomposition of the Correlation Matrix, *JSCS 5*, 247-258.

Martin, R. S., C. Reinsch, and J. H. Wilkinson (1968), Householder's Tridiagonalization of a Symmetric Matrix, *Numer. Math. 11*, 181-195.

Moore, E. H. (1920), On the Reciprocal of the General Algebraic Matrix, *Bulletin AMS. 26*, 394-395.

Parlett, B. N. (1967), The LU and QR Transformations, in *Mathematical Methods for Digital Computers II*, edited by A. Ralston and H. Wilf, Wiley, New York.

Penrose, R. (1955), A Generalized Inverse for Matrices, *Proc. Cambridge Phil. Soc. 51*, 406-413.

Pringle, R. M., and A. A. Rayner (1971), *Generalized Inverse Matrices with Applications to Statistics,* Griffin and Co., London.

Ralston, A., and H. S. Wilf (1967), *Mathematical Methods for Digital Computers*, vol. 2, Wiley, New York.

Rao, C. R., and S. K. Mitra (1971), *Generalized Inverse of Matrices and Its Applications*, Wiley, New York.

Rice, J. R. (1966), Experiments on Gram-Schmidt Orthogonalization, *Math. Comp. 20*, 325-328.

Stewart, G. W. (1970), Incorporating Origin Shifts into the QR Algorithm for Symmetric Tridiagonal Matrices, *CACM 13*, 365-371.

Stewart, G. W. (1973), *Introduction to Matrix Computations*, Academic Press, New York.

Stewart, G. W. (1974), Modifying Pivot Elements in Gaussian Elimination, *Math. Comp. 28*, 537-542.

Tsao, N. K. (1975), A Note on Implementing the Householder Transformation, *SIAM J. Num. Anal. 12*, 53-58.

Van Loan, C. F. (1976), Generalizing the Singular Value Decomposition, *SIAM J. Num. Anal. 13*, 76-83.

Wilkinson, J. H. (1962), Householder's Method for Symmetric Matrices, *Numer. Math. 4*, 354-361.

Wilkinson, J. H. (1965), *The Algebraic Eigenvalue Problem*, Clarendon Press, Oxford.

Wilkinson, J. H., and C. Reinsch (1971), *Handbook for Automatic Computation,* vol. 2: *Linear Algebra*, Springer-Verlag, Berlin.

8 / COMPUTATIONAL METHODS FOR MULTIPLE LINEAR REGRESSION ANALYSIS

8.1 BASIC COMPUTATIONAL METHODS

This chapter deals with computational methods relative to the usual linear model

$$\underline{Y} = X\underline{\beta} + \underline{e} \tag{8.1}$$

where X is a fixed known n x p matrix containing $n \geq p$ values for each of p independent variables, $\underline{Y}$ is an n x 1 vector of observed values of the dependent variable y, and $\underline{e}$ is a vector of random disturbances under the usual conditions of zero mean and covariance matrix $\sigma^2 I_n$. Discussion of computations will presume that the least-squares criterion is used for the underlying minimization problem, which is to find an estimate $\underline{b}$ of $\underline{\beta}$ that minimizes $||\underline{Y} - X\underline{b}||^2$, where $||\cdot||$ denotes the Euclidean norm.

A typical mathematical treatment of this subject finds $\underline{b}$ as a solution to the *normal equations*

$$X'X\underline{b} = X'\underline{Y} \tag{8.2}$$

and often utilizes the *vector of residuals*

$$\underline{Z} = \underline{Y} - X\underline{b} \tag{8.3}$$

in subsequent discussion. The inverse of X'X, either $(X'X)^{-1}$ or $(X'X)^{-}$, is another quantity of direct interest in these discussions.

Two different basic approaches are used for regression computations. One approach is to form the normal equations (Eq. 8.2) and proceed with these equations as a starting point. This approach is rejected by some people because it may be very difficult to form the matrices in (8.2) to a high degree of accuracy. The other approach is to avoid formation of the normal equations. Instead, computations proceed based on the given data $(X|\underline{Y})$ and utilize orthogonal triangularization in the process of performing regression analysis. This latter approach generally provides more accurate results, and we shall begin with a discussion of this approach in Section 8.1.1. Then methods based on use of normal equations will be given in Section 8.1.2.

8.1.1 Methods Using Orthogonal Triangularization of X

The basic problem in regression computations is to find $\underline{b}$ such that $\underline{Z}'\underline{Z}$ in (8.3) is minimized. Consider a decomposition of X as

$$X = QR \tag{8.4}$$

where Q is n x n orthogonal and R is n x p of the form

$$R = \begin{pmatrix} R_1 \\ \hline 0 \end{pmatrix}$$

where R_1 is p x p upper-triangular. Using this decomposition, the sum of squared residuals can be written as

$$\begin{aligned} ||\underline{Y} - X\underline{b}||^2 &= ||\underline{Y} - QR\underline{b}||^2 \\ &= ||Q'\underline{Y} - R\underline{b}||^2 \\ &= ||\underline{c}_1 - R_1\underline{b}||^2 + ||\underline{c}_2||^2 \end{aligned} \tag{8.5}$$

where $\underline{c}_1$ is a p x 1 vector and $\underline{c}_2$ is an (n - p) x 1 vector such that

$$Q'\underline{Y} = \begin{pmatrix} \underline{c}_1 \\ \hline \underline{c}_2 \end{pmatrix}$$

From (8.5) it is apparent that the sum of squared residuals is minimum when $\underline{b}$ is chosen to satisfy the system of linear equations

$$R_1\underline{b} = \underline{c}_1 \tag{8.6}$$

Thus we have in (8.6) an alternative to the normal equations for computation of $\underline{b}$. We now proceed, under this general approach, to show how various desired quantities in regression analyses can be computed using this approach. The first part of the discussion will assume full-rank X and give the mathematical basis for the general case of possibly rank-deficient X.

Suppose that we begin with the data matrix $(X|\underline{Y})$ and perform p Householder transformations Q_j' to obtain

$$Q'(X|\underline{Y}) = (R|\underline{c})$$

where

$$R = \left(\frac{R_1}{0}\right) \quad \text{and} \quad \underline{c} = \left(\frac{\underline{c}_1}{\underline{c}_2}\right)$$

with R_1 p x p upper-triangular. The vector of regression coefficients $\underline{b}$ is obtained as in (8.6) by solving the system of linear equations

$$R_1\underline{b} = \underline{c}_1 \tag{8.7}$$

The residual sum of squares from this regression is

$$(n - p)s^2 = \underline{c}_2'\underline{c}_2$$

as shown in Eq. (8.5). The residuals can be obtained as

$$\begin{aligned} \underline{Y} - X\underline{b} &= \underline{Y} - QR\underline{b} \\ &= Q(\underline{c} - R\underline{b}) \\ &= Q\left(\frac{0}{\underline{c}_2}\right) \end{aligned} \tag{8.8}$$

which means that Q must be available for use. (The vector and scalar which define each Householder reflection are usually stored as indicated in algorithm BHOUSE given in Chapter 7.) Predicted values $\hat{\underline{Y}} = X\underline{b}$ can be obtained as

$$\begin{aligned} X\underline{b} &= QR\underline{b} \\ &= Q\left(\frac{R_1}{0}\right)\underline{b} \\ &= Q\left(\frac{\underline{c}_1}{0}\right) \end{aligned} \tag{8.9}$$

The estimate of the covariance matrix of $\underline{b}$ is

$$s^2(X'X)^{-1} = s^2(R_1'R_1)^{-1}$$
$$= s^2R_1^{-1}(R_1')^{-1}$$

We need R_1^{-1} which is easily obtained using back-substitution in the system of linear equations

$$R_1W = I_p$$

to give $W = R_1^{-1}$.

Fitting models consisting of subsets of the overall set of independent variables is easily accomplished by selecting appropriate columns of X and pivoting on the desired element within each selected column. All columns of X are transformed, but only those of the selected subset are considered when Eqs. (8.7), (8.8), and (8.9) are used to compute needed statistics. Selection of pivots was discussed in Section 7.2.1.

Next suppose that we wish to test the hypothesis H_0: $L'\underline{\beta} = \underline{\gamma}$, where L' is q x p of rank q and $\underline{\gamma}$ is a given q vector. We assume throughout this chapter that the $L'\underline{\beta}$ are *estimable functions* of $\underline{\beta}$; in other words, $L' = KX$ for some matrix K. ($L'\underline{\beta}$ is estimable in the present case in which L is of rank q and X is of rank $p \geq q$. A method for checking to see whether $L'\underline{\beta}$ is estimable in general is included as an exercise.) The test statistic for this hypothesis is

$$F = \frac{(n - p)}{q} \cdot \frac{RSS_H - RSS}{RSS} \tag{8.10}$$

where RSS is the residual sum of squares from fitting the full model, and RSS_H is the like quantity for the model restricted by the hypothesis. It is well known that

$$RSS_H - RSS = (L'\underline{b} - \underline{\gamma})'[L'(X'X)^{-1}L]^{-1}(L'\underline{b} - \underline{\gamma}) \tag{8.11}$$

which is the quantity we must compute in addition to RSS in order to obtain F. Note first that

$$
\begin{aligned}
L'(X'X)^{-1}L &= L'(R'R)^{-1}L \\
&= [(R^{-1})'L]'[(R^{-1})'L] \\
&= A'A
\end{aligned}
$$

where $A = (R^{-1})'L$ is p x q and has rank q. Now decompose A, using Householder transformations to obtain

$$A = U\begin{pmatrix} T \\ \hline 0 \end{pmatrix} \tag{8.12}$$

where U is p x p orthogonal, and T is q x q upper-triangular. Thus we have

$$A'A = T'T$$

and (8.11) can be written as

$$
\begin{aligned}
RSS_H - RSS &= (L'\underline{b} - \underline{\gamma})'[T'T]^{-1}(L'\underline{b} - \underline{\gamma}) \\
&= [(T')^{-1}(L'\underline{b} - \underline{\gamma})]'[(T')^{-1}(L'\underline{b} - \underline{\gamma})] \\
&= \underline{v}'\underline{v} \qquad (8.13)
\end{aligned}
$$

where $\underline{v} = (T')^{-1}(L'\underline{b} - \underline{\gamma})$.

Using the above, the procedure to follow when computing (8.11) is

1. Form $\underline{w} = L'\underline{b} - \underline{\gamma}$
2. Find A as the solution of

   $$R'A = L$$

3. Perform the transformation (8.12) to obtain T.
4. Compute $\underline{v}$ by solving

   $$T'\underline{v} = \underline{w}$$

5. Form $\underline{v}'\underline{v}$, which is the value required in (8.11).

Thus far we have assumed X has rank p. In practice, the computed rank of X will be obtained based on some tolerance $\varepsilon > 0$ which is used to decide whether any given value will be declared zero. The computed rank may not, of course, be the "true" rank, but it will be the best we can determine from the data using the given algorithm. Thus the question of how to proceed when it is determined that X has rank $r < p$ must now be considered. This is really the key question in applied work because we are seldom willing to develop programs that cannot handle less than full-rank X.

The usual computational method, as described in Chapter 7 for Householder transformations, features a possible column permutation at each stage in the triangularization process to maximize the next diagonal element in the triangular matrix R. With column permutations the decomposition of X can be formed as

$$X = QTP \tag{8.14}$$

where Q and P are orthogonal and

$$T = \begin{pmatrix} T_1 & T_2 \\ 0 & 0 \end{pmatrix}$$

with T_1 being $r \times r$ upper-triangular and nonsingular. As before, let

$$Q'\underline{Y} = \begin{pmatrix} \underline{c}_1 \\ \hline \underline{c}_2 \end{pmatrix}$$

where $\underline{c}_1$ has r components. Now consider additional Householder transformations applied from the right and designed to zero elements in the upper-triangular portion of T (this zeros T_2). The form achieved is

$$X = QRU' \tag{8.15}$$

where

$$R = \begin{pmatrix} R_2 & 0 \\ 0 & 0 \end{pmatrix}$$

and R_2 is $r \times r$ lower-triangular. Introducing a change of variables

$$U'\underline{\beta} = \underline{a} = \begin{pmatrix} \underline{a}_1 \\ \hline \underline{a}_2 \end{pmatrix}$$

where $\underline{a}_1$ is r x 1, we can readily verify that all least-squares solutions can be found by solving

$$R_2\underline{a}_1 = \underline{c}_1$$

to obtain $\hat{\underline{a}}_1$ and computing

$$\underline{b} = U \begin{pmatrix} \hat{\underline{a}}_1 \\ \underline{a}_2 \end{pmatrix} \tag{8.16}$$

where $\underline{a}_2$ is arbitrary. The solution having minimum length is obviously obtained when $\underline{a}_2 = \underline{0}$. For this choice (8.16) is easily seen to be $\underline{b} = X^+\underline{Y}$. ($X^+$ is the Moore-Penrose inverse. See Section 7.2.4.)

As before, the residual sum of squares is $\underline{c}_2'\underline{c}_2$, and (8.9) holds in the same form. However, if we wish to test the same hypothesis considered in (8.10), we must compute

$$RSS_H - RSS = (L'\underline{b} - \underline{\gamma})'[L'(X'X)^*L]^{-1}(L'\underline{b} - \underline{\gamma}) \tag{8.17}$$

where $(X'X)^*$ is any g_2 inverse of X'X. With this computation in mind, let $B = (X'X)^*X'$. Then we have

$$\begin{aligned} L'(X'X)^*L &= L'BB'L \\ &= (B'L)'(B'L) \\ &= A'A \end{aligned} \tag{8.18}$$

where $A = X(X'X)^*L$. Now the analog of Eq. (8.13) is easily derived. The only thing we need at this point to have a satisfactory computing method for (8.17) is some efficient method of finding $X(X'X)^*$. To this end, consider (8.15) and form

$$X^+ = U \begin{pmatrix} R_2^{-1} & 0 \\ 0 & 0 \end{pmatrix} Q'$$

Now it is easily verified that

$$X^+(X^+)' = (X'X)^* = U \begin{pmatrix} (R_2'R_2)^{-1} & 0 \\ 0 & 0 \end{pmatrix} U'$$

and

$$X(X'X)^{*} = Q\begin{bmatrix} (R_2')^{-1} & 0 \\ 0 & 0 \end{bmatrix}U' \tag{8.19}$$

The matrix in (8.19) can be quickly computed using the available decomposition shown in (8.15); hence we can easily form the A matrix shown in Eq. (8.18). Thus (8.17) is obtainable under this procedure.

The results presented thus far in this subsection are similar to ones given in Golub and Styan (1973a) and Hanson and Lawson (1969). Much of the computer software in use today, which performs regression computations using orthogonal transformations, utilizes what are essentially the methods given in this subsection. There are, however, some excellent regression programs which use M.G.S. instead of Householder's method. The S.V.D. can be substituted for the decompositions given above, but it is significantly more expensive to use. S.V.D. often gives better results for very ill-conditioned problems, but for a majority of data sets S.V.D. will probably not provide significantly better results than M.G.S., Givens, or Householder transformation-based methods.

It must be emphasized that regression data *are* sometimes very ill-conditioned. For example, X is sometimes rank-deficient, or nearly so, due to inclusion of variables that are highly or perfectly correlated. Longley (1967) gave an example of economic data wherein the X matrix condition number was approximately 10^9, and Wampler (1970) generated data from polynomial models which had still larger X condition numbers. Error analyses, such as those given by Lawson and Hanson (1974), do indicate that the methods of this section have good numerical stability. Nevertheless, for a given floating-point level of precision, there will always be data sets that are sufficiently ill-conditioned so that they cannot be processed accurately. Hence, computed results should be inspected with accuracy in mind and, if it seems appropriate, some additional checks made to verify accuracy. We will discuss the checking of results in Section 8.1.3.

One simple device which is often used to enhance accuracy is to "center" the data X and $\underline{Y}$. A form of centered data is that wherein each variable is expressed as the deviation from its mean. Golub and Styan (1973b) discuss this subject and show that the condition number of a centered X will, in most cases, be strictly less than that of the original X. Computational details using centered data are considered in an exercise and in the next subsection.

The approach to regression computations given in this section is criticized by some on the grounds that it requires very large amounts of memory storage for even moderate-sized problems, and the amount of arithmetic involved is larger than that required when the normal equations are used. Advocates of orthogonalization methods cite their numerical stability and mention that large quantities of storage are available in modern computers at relatively small cost. The next subsection deals with use of the normal equations and sweep operations to perform regression analyses. Additional comments relative to comparison of the two general approaches will be made after describing this second approach.

8.1.2 Sweep Operations and Normal Equations

When normal equations form the starting point for regression computations, most algorithms in existence today use sweep operations to solve the system of equations, invert the coefficient matrix, and produce, during the process, several other desired quantities. In this subsection we will consider the details of this procedure.

Usually, the first column of the X matrix is a column of 1's which are coefficients of the intercept in the regression model. We will deal, in this subsection, with values expressed in deviation form (i.e., centered). The first column of the X matrix, in deviation form, is of course a column of zeros. This creates a small notation problem which we solve as follows. Henceforth in this subsection X_R will denote an $n \times (p + 1)$ matrix of raw values of the independent variables (including the intercept variable $x_0 \equiv 1$), and X will denote the $n \times p$ matrix of deviations of each variable (except the intercept) from its mean. Similarly, $\underline{Y}_R$ and $\underline{Y}$ will signify raw and corrected values, respectively. We will begin by describing the use of *reduced normal equations* which are formed from X and $\underline{Y}$. Then we describe the few additional computations needed to update computed values to make them the quantities which would have been obtained using X_R and $\underline{Y}_R$. The reason that X and $\underline{Y}$ are used instead of X_R and $\underline{Y}_R$ is that the elements of the former tend to be much less variable in magnitude, which usually means a better-conditioned problem, as was indicated in the preceding subsection.

The potential numerical inaccuracies involved in *formation* of $(X'X|X'\underline{Y})$ are apparent because vector inner products must be taken. An efficient single-pass method for computing corrected sums of squares and cross-products, which has good numerical stability, is described by Welford (1962). Neely (1966), Van Reeken (1968), Young and Cramer (1971), Ling (1974), and Hanson (1975) discuss various aspects of the algorithm. The method avoids use of the "machine formula," which is most unstable, and the costly two-pass method which forms means on the first pass and corrected sums of cross-products on the second pass. We will describe this method with formation of $(X|\underline{Y})'(X|\underline{Y}) = A = (a_{ij})$ as our objective using values from the last p columns of $(X_R|\underline{Y}_R) = (z_{ij})$. The method proceeds by using each row t of $(X_R|\underline{Y}_R)$, in turn, as follows. For each i, j = 2, 3, ..., p + 2 and $j \geq i$, set

$$\bar{z}_i^{(1)} = z_{1,i}$$

and

$$a_{i-1,j-1}^{(1)} = 0$$

then, for t = 2, 3, ..., n, compute

$$\bar{z}_i^{(t)} = \bar{z}_i^{(t-1)} + \frac{z_{t,i} - \bar{z}_i^{(t-1)}}{t}$$

and (8.20)

$$a_{i-1,j-1}^{(t)} = a_{i-1,j-1}^{(t-1)} + \frac{(t - 1)(z_{t,j} - \bar{z}_j^{(t-1)})(z_{t,i} - \bar{z}_i^{(t-1)})}{t}$$

The final values $\bar{z}_i^{(n)}$ and $a_{ij}^{(n)}$ with $j \geq i$ are the column means and corrected sums of cross-products, respectively.

Herraman (1968) described, and gave FORTRAN coding for, a similar algorithm for weighted data.

Once the matrices $X'X$ and $X'\underline{Y}$ are computed, the augmented matrix

$$A = \begin{bmatrix} X'X & X'\underline{Y} \\ \underline{Y}'X & \underline{Y}'\underline{Y} \end{bmatrix}$$

is formed. When X is of full rank, sweeping the first p rows of A gives

$$A \xrightarrow[S(1,\ \dots,\ p)]{} \begin{bmatrix} (X'X)^{-1} & (X'X)^{-1}X'\underline{Y} \\ -\underline{Y}'X(X'X)^{-1} & \underline{Y}'M\underline{Y} \end{bmatrix}$$

$$= \begin{bmatrix} (X'X)^{-1} & \hat{\underline{\alpha}} \\ -\hat{\underline{\alpha}}' & (n - p)s^2 \end{bmatrix} \qquad (8.21)$$

where $M = I - X(X'X)^{-1}X'$. Recovery to the situation in terms of X_R and $\underline{Y}_R$ is done using the means

$$\bar{\underline{z}}' = (\bar{z}_2, \bar{z}_3, \dots, \bar{z}_{p+1}) \text{ and } \bar{z}_{p+2} = \bar{y} = \frac{\sum_{i=1}^{n} y_i}{n}$$

computed from (8.20), and forming

$$(X_R'X_R)^{-1} = \begin{bmatrix} u & \underline{t}' \\ \underline{t} & (X'X)^{-1} \end{bmatrix}$$

and (8.22)

$$\underline{b} = \begin{bmatrix} \bar{y} - \bar{\underline{z}}'\hat{\underline{\alpha}} \\ \hline \hat{\underline{\alpha}} \end{bmatrix}$$

where

$$u = \frac{1}{n} + \bar{\underline{z}}'(X'X)^{-1}\bar{\underline{z}}$$

$$\underline{t} = -(X'X)^{-1}\bar{\underline{z}}$$

Note that we obtain the sum of squared residuals directly from (8.21). The residuals themselves must be computed using initial data and the $\underline{b}$ vector.

Verification of (8.22) is immediate using the results given in Chapter 7 relative to the sweep operator applied to partitioned matrices. Notice that sweeping the first row of $X_R'X_R$ yields X'X and $\bar{\underline{z}}$ as submatrices, and then sweeping the remaining p rows yields (8.22).

Suppose we wish to test an hypothesis of the form H_0: $\underline{\beta}_1 = \underline{0}$, where $\underline{\beta}_1$ is some subset of the parameters in $\underline{\beta}$. To simplify the notation, let $\underline{\beta}_1$ consist of the last q < p components in $\underline{\beta}$. Expressed in the general form of Eq. (8.11), the numerator sum of squares is

$$RSS_H - RSS = (L'\underline{b})'[L'(X'X)^{-1}L]^{-1}(L'\underline{b})$$

where L' is q x p in the form

$$L' = (0|I_q)$$

To obtain RSS_H we sweep the first (p - q) rows of the matrix A. At that point RSS_H appears in the (p + 1, p + 1) position of A. [We will describe this by saying that variables 1, 2, ..., (p - q) have been entered into the model.] Then the remaining q rows are swept, that is, the remaining variables are entered, and RSS is obtained. Since sweeps are reversible we might have swept all p rows to obtain RSS first, and then swept the last q rows to give RSS_H. It should be obvious to the reader that any subset of the parameters in $\underline{\beta}$ can be easily tested in this fashion. The subset will define the rows to be swept initially. Variables can also be moved in and out of the model at will. In the case of more general hypotheses, sweeps will not automatically give the required sums of squares. Some additional matrix manipulations will need to be performed.

When X is not of full rank, the g_1 inverse can be used and the sweep operations carried out in the same way except that pivot elements which fall below tolerance $\varepsilon > 0$ will, of course, be skipped. We will not discuss this case since its details follow easily from Section 7.3.

Corrected sums of squares and cross-products matrices X'X and X'$\underline{Y}$ were used in the previous discussion. In some statistical software packages an additional transformation is made to produce correlation matrices. The reason for doing this is that there is additional uniformity present since none of the matrix elements exceed 1 in magnitude. Whether the cost of doing this is worth the potential increased accuracy is an open question. Using data in deviation form (centered data) is, however, certainly recommended when the normal equations are used to obtain the regression statistics.

The compactness and convenience of sweep operations on normal equations for regression computations makes them popular with programmers. Many large software systems available today use the general procedures described in this section. There are, however, some aspects of this approach of which we must be aware. First, the condition number of X'X is the square of that of X' therefore, forming normal equations may produce a more difficult numerical problem than one which simply uses X. It is also difficult to compute X'X and X'$\underline{Y}$ accurately under any procedure. (See discussion by Youngs and Cramer 1971.) Hence the perturbations that we make in the basic problem, by forming the normal equations, may have disastrous consequences. Also, sweeps are not as stable as orthogonal operators; hence this may compound the numerical accuracy problems.

Many authors recommend that normal equations never be formed or used. Chambers (1977) expressed the thoughts of many people when he stated that the normal equations "have limited practical relevance." Advocates of the use of normal equations, on the other hand, argue that statistical software packages use such high precision in floating-point computations that the difference in accuracy of the two general computational approaches will seldom be worth considering.

Several points have been made for and against each of the general approaches to regression computations described in this and the preceding subsection. No clearcut choice seems possible at present.

8.1.3 Checking Programs, Computed Results, and Improving Solutions Iteratively

Knowing that regression computations may be in error due to program deficiencies or ill-conditioned data prompts a search for ways of checking computations in practical applications. Also, when poor results are obtained, we would like some means of improving them to an acceptable level of accuracy. These subjects are addressed in this subsection.

Programs are best checked by solving problems for which the answers are known and which have varying types of troublesome numerical characteristics. The most convenient way to obtain such problems is to generate them using the computer. Hastings (1972), Velleman and Allen (1976), and Kennedy, Gentle, and Sposito (1977) give methods for doing this. Longley (1967) and Wampler (1970) give individual problems which can also be used in testing. We now consider one method for forming test problems.

The method which we consider is flexible in that it allows choice of solution vector $\underline{b}$, the column means of the X matrix for all columns after the first, and the order of magnitude of the condition number of X. We will first consider the steps in the data generation procedure and then provide needed details. These steps are:

1. Generate at random an $n \times p$ matrix $Z (n > p)$ having rank p and the scalar 1 in all positions in its first column.
2. Decompose Z as

   $$Z = Q\begin{pmatrix} R \\ \hline 0 \end{pmatrix}$$

   where Q is orthogonal and R is $p \times p$ upper-triangular.
3. Form $U_0 = ZR^{-1}$. Note that $U_0'U_0 = I_p$.

4. Generate at random a (p - 1)-square matrix W and decompose as

$$W = UT$$

where U is orthogonal. Then form the p-square orthogonal matrix

$$V' = \begin{bmatrix} 1 & \underline{0}' \\ \underline{0} & U \end{bmatrix}$$

5. Select column means $m_2, m_3, \ldots, m_p$ and form the n x p matrix

$$E = \begin{bmatrix} 0 & m_2 & m_3 & \cdots & m_p \\ \cdot & m_2 & m_3 & & m_p \\ \cdot & \cdot & \cdot & & \cdot \\ \cdot & \cdot & \cdot & & \cdot \\ \cdot & \cdot & \cdot & & \cdot \\ 0 & m_2 & m_3 & \cdots & m_p \end{bmatrix}$$

6. Select a diagonal matrix $D = \mathrm{diag}(\alpha_1, \alpha_2, \ldots, \alpha_p)$ and form the n x p matrix as

$$\begin{aligned} X &= U_0 DV' + E \\ &= A + E \end{aligned}$$

7. Select a solution vector $\underline{b}$ and compute $\underline{Y} = X\underline{b}$.

In practice, the random matrices needed in steps 1 and 4 can be obtained using a random number generator. A single generation of elements for each matrix will almost always provide a nonsingular matrix in one pass. The decomposition in step 2 is easily accomplished using Householder transformations. Matrix inversion in step 3 is a simple back substitution process since R is triangular. Note that the first column of U_0 is constant as is the first column of X in step 6. Since the columns of U_0 are mutually orthogonal and its first column is constant, the mean of each column of the A matrix (except column 1) is zero. Therefore, X has the m_j (j = 2, ..., p) as its column means.

Step 6 requires further consideration. We desire an X matrix having specified column means *and* a condition number which falls within a given range. We are at liberty to choose the singular values of A (i.e., the matrix D), and this is the only freedom that we have in step 6. It is, however, possible to relate the singular values of X and A so as to allow a choice of D which will satisfy all requirements. The following theorem given by Lawson and Hanson (1974, p. 25) will help in this regard.

Theorem

Let X, A, and E be n x p matrices, (n > p), with X - A = E. Denote their singular values by $\gamma_1 \geq \gamma_2 \geq \cdots \geq \gamma_p$; $\alpha_1 \geq \alpha_2 \geq \cdots \geq \alpha_p$; and $\varepsilon_1 \geq \varepsilon_2 \geq \cdots \geq \varepsilon_p$, respectively. Then

$$|\gamma_i - \alpha_i| \leq \varepsilon_1 \qquad i = 1, 2, \ldots, p$$

If we select $\alpha_p > \varepsilon_1$, then bounds for the condition number of X are obtainable, using the theorem, in the form

$$\frac{\alpha_1 - \varepsilon_1}{\alpha_p + \varepsilon_1} \leq \frac{\gamma_1}{\gamma_p} \leq \frac{\alpha_1 + \varepsilon_1}{\alpha_p - \varepsilon_1}$$

The E matrix in our case has $\varepsilon_1 = (n \sum_{j=2}^{p} m_j^2)^{1/2}$. Thus, it is possible to choose singular values $\alpha_1, \alpha_2, \ldots, \alpha_p$ which will guarantee that the condition number of X does not exceed a given value greater than unity. This is indicated by the upper bound of the inequalities given above. In practice, some trials for varying α_1 and α_p will sometimes need to be made to raise the X condition number to a desired level without exceeding a given upper bound.

We have gone to a lot of trouble, in the method given above, to produce test data with allowance for varying several characteristics of the data. Obviously, data $(X|\underline{Y})$ which simply have known solution vector $\underline{b}$, for example, are easy to form. However, many times we want to "study an algorithm under stress," which means we want to use a variety of problems having numerical characteristics that vary in as many dimensions as possible. Therein lies the utility of the rather complicated method given above, as well as extensions of the method.

We have mentioned that the condition of X may not be a good predictor of the accuracy which we achieve in practice. We do, however, generally expect that regression problems will tend to be more difficult to solve accurately as the condition number of X increases; but we will not be surprised if sometimes we achieve very accurate results for a problem in which the X condition number is quite large.

Since we know that "correct" programs can give extremely bad answers for certain data sets, how do we determine when this has occurred? Some suggested ways to check for potentially bad answers are (see also Section 3.8):

1. Permute the columns of the X matrix and rerun.
2. Check to see whether the residuals sum to zero within rounding error. [Good only for models having an intercept term; in general, check whether $X'(\underline{Y} - X\underline{b}) = \underline{0}$.]
3. Inspect the correlation matrix to see if any off-diagonal element is very close to one in magnitude. (Feasible only for smaller problems.)
4. Form linear combinations of subsets of the independent variables (e.g., $w = x_1 + x_2$) and rerun the problem using the same $\underline{Y}$. The regression coefficients between the two runs should be such that $b_w = b_1 + b_2$.
5. Use the following technique, which is essentially the same one suggested by Mullet and Murray (1971):
   (a) Regress y on $x_1, x_2, \ldots, x_p$ and obtain the g_2 inverse $(X'X)^*$ shown in Eq. (7.27).
   (b) For chosen scalar $d \neq 0$ and j, where x_j is one of the linearly independent columns of X, regress $y + dx_j$ on $x_1, x_2, \ldots, x_p$ to obtain a solution based on the g_2 inverse.
   (c) Compare the regression coefficients obtained in (a) and (b) above. They should agree except for b_j, which is d units larger in the second regression. Also, the residuals should be the same in each case.

6. Rerun the problem using higher-precision floating-point arithmetic.

All of these checks will help to some extent. The Mullet and Murray procedure is probably the all-around best choice.

Suppose now that we are convinced that the regression coefficients $\underline{b}$ have probably not been obtained to a satisfactory degree of accuracy for some given data. In this situation we can sometimes use a method called *iterative refinement* to (hopefully) improve the first solution. This refinement procedure is described as follows. Let $\underline{b}^{(1)}$ be the initial solution. Suppose we determine $\underline{\delta}^{(1)}$ such that

$$||(\underline{Y} - X\underline{b}^{(1)}) - X\underline{\delta}^{(1)}|| \tag{8.23}$$

is minimized. Collecting terms differently in (8.23) shows that

$$||\underline{Y} - X(\underline{b}^{(1)} + \underline{\delta}^{(1)})||$$

is minimized; hence $\underline{b}^{(1)} + \underline{\delta}^{(1)}$ is a least-squares solution for regression coefficients. Calling the new solution

$$\underline{b}^{(2)} = \underline{b}^{(1)} + \underline{\delta}^{(1)}$$

the same procedure can be used iteratively to produce $\underline{b}^{(3)}$, $\underline{b}^{(4)}$, ..., and so on. A summary of the iteration is as follows. For each $s = 1, 2, \ldots,$

1. Compute $\underline{R}^{(s)} = \underline{Y} - X\underline{b}^{(s)}$.
2. Solve $X\underline{\delta}^{(s)} = \underline{R}^{(s)}$ for $\underline{\delta}^{(s)}$.
3. Compute $\underline{b}^{(s+1)} = \underline{b}^{(s)} + \underline{\delta}^{(s)}$.

Iteration continues until $||\underline{\delta}^{(s)}||/||\underline{b}^{(s)}|| < \varepsilon$, for some tolerance $\varepsilon > 0$, or until a specified number of iterations is made.

Iterative refinement is discussed by Golub and Wilkinson (1966), Forsythe and Moler (1967), Bjorck and Golub (1967), and Wilkinson (1967). This procedure will usually converge to the true solution (to machine working precision) if $\underline{b}^{(1)}$ is "sufficiently close" and iterative refinement steps 1 and 2 are carried out to *higher precision* than the working precision used to obtain $\underline{b}^{(1)}$. If X is not extremely ill-conditioned, only a few steps of iterative refinement are normally needed to obtain significant improvement in the computed solution. Some people advocate the use of iterative refinement as a standard part of regression algorithms. Most, however, view iterative refinement as a technique for use only in special cases. In lieu of iterative refinement, the problem can simply be solved in higher precision in an attempt to obtain a satisfactory degree of accuracy as was suggested for a check in step 6 above.

The primary motivation for consideration of the subjects dealt with in this subsection is the ever-present need to achieve accuracy in computed values and to detect the presence of inaccuracies. However, there are still many open questions in this area. For example, there is not complete agreement among researchers on the meaning of the term *accuracy* with regard to computed results in regressions. Suggested reading on this subject includes Berk (1976), Frane (1978), Freund (1963), Jennings and Osborne (1974), Stewart (1978), Swindel and Bower (1972), and Wampler (1978). Beaton, Rubin, and Barone (1976) and Beaton and Barone (1977) point out that the precision of the original data may reduce the importance of achieving exceptionally accurate results in regression computations.

8.2 REGRESSION MODEL BUILDING

It is often the case in research that an experimenter has data for some large superset of p independent variables (predictors) in a regression model and believes that a subset of these predictors may be entirely adequate for his purposes. In such cases the experimenter needs a method for selecting a "good" subset of predictors.

Selecting a "good" subset of independent variates implies the existence of some measure of goodness. The criterion which is normally used is the magnitude of the residual sum of squares for each of the various subsets. One subset of p_1 variates is usually taken to be better than another containing p_2 variates if $p_1 \leq p_2$ and the difference between the residual sum of squares obtained in regression using the two subsets is "insignificant." Theoretical results in regression model building are not as plentiful as we would like, and the question of significance in tests which compare regressions on subsets is normally answered using some threshold value which is not strictly justifiable in a probabilistic sense. The reader is referred to Kennedy and Bancroft (1971) for a discussion of this subject. The methods which will be discussed in this section have proven useful in application despite this shortage of relevant distribution theory.

In practical applications, wherein more than one model-building procedure is applied to a given set of data, it is often the case that different methods will select different subsets of the overall set of independent variates for inclusion in the final model. When this happens the data are usually such that each of the selected models is "good" in the sense of small residual sum of squares. However, from the experimenter's point of view one of the selected models will probably be called "best" for one reason or another. For this reason, in data analysis situations, several model-building methods are sometimes applied, or all possible regressions are performed, in an effort to find the most desirable model as indicated by the observed data.

Our discussion in this section will be with respect to the full-rank model. However, the methods given can easily be used when X is rank-deficient. The minor modifications required, for each method, will be apparent in light of the methods presented thus far in this chapter.

8.2.1 All Possible Regressions

The most obvious way to choose an appropriate submodel using the available data is to do all of the possible $2^p - 1$ regressions on subsets. However, this approach is not feasible unless p is small. For example, $p = 15$ requires that $2^{15} - 1 = 65{,}535$ regressions be computed. For small p, say $p \leq 10$, the amount of computation is not excessive and all possible regressions are sometimes computed.

To compute all possible regressions we first need some method for designating the $2^p - 1$ nonempty subsets of the p predictors. Garside (1965) gives such an algorithm which designates $2^p - 1$ subsets in $2^p - 1$ steps, and does so in an order which requires the alteration of only one element at each step. This algorithm generates a sequence of integers $\{s_j\}$ using the following procedure:

1. Initially let $\underline{A}$ be a vector of length p having components

$$a_i = 2^{i-1} \quad \text{for } i = 1, 2, \ldots, p$$

2. For each $j = 1, 2, \ldots, 2^p - 1$, set s_j equal to the position number i of the minimum element of $\underline{A}$ and set $a_i = a_i + 2^i$.

The sequence of integers $\{s_j\}$ generated appears as

$$1213121412 \quad \text{etc.} \tag{8.24}$$

We interpret the successive integers in the set $\{s_j\}$ by saying for each i that predictor number s_i is to be entered into the current subset if it is not already in, and deleted from the subset if it is present. Thus we interpret the first few terms in the sequence (8.24) as shown in Fig. 8.1.

s_j	*Operation*	*Subset defined*
1	enter x_1	x_1
2	enter x_2	x_1, x_2
1	delete x_1	x_2
3	enter x_3	x_2, x_3
1	enter x_1	x_1, x_2, x_3.

Figure 8.1 Application of the Garside method.

The calculation of all possible regressions can readily be carried out using the Garside algorithm in conjunction with sweep operations. The matrix on which we operate is

$$B = \begin{bmatrix} X'X & X'\underline{Y} \\ \underline{Y}'X & \underline{Y}'\underline{Y} \end{bmatrix}$$

The operations of Fig. 8.1 become designations of sweeps. Thus to enter predictor x_j (not already in the model) we pivot on the jth diagonal element and to delete x_k we pivot on the kth diagonal element. At each stage the residual sum of squares is given in the (p + 1, p + 1) position of the transformed B matrix. If predictors numbered $i_1, i_2, \ldots, i_k$ are currently in the regression model, then the regression coefficient estimates for this current fit will be located in matrix positions $(i_1, p + 1), (i_2, p + 1), \ldots, (i_k, p + 1)$.

There is a definite possibility that a large number of pivot and repivot operations will cause excessive accumulation of error. Experience with this method, utilizing double-precision arithmetic, indicates that excessive loss of accuracy does not frequently occur. Nevertheless, if a large number of operations are to be performed, it is recommended that periodically the original matrix should be input and updated to reflect the current stage before continuing with the current sequence of pivot operations.

Schatzoff, Tsao, and Fienberg (1968) describe a similar procedure to reduce the number of computations. Furnival (1971) gave another method based on this general procedure that reduced the amount of computation still further. However, Furnival utilizes p + 1 matrices having sizes varying from one to p. His procedure requires $2^p - 1$ sweep operations to perform the $2^p - 1$ regressions. Despite their efficiency the cost of these methods, for even moderately large p, will still be great.

Computation of regressions need not, of course, be based on sweep operations on the sums of cross-products matrix. Householder transformations on $(X|\underline{Y})$ can be used to accomplish the same purpose. This approach will, however, prove to be more expensive in most cases.

Since it is not generally economically feasible to do all possible regressions for large problems, we need a method which somehow selects a "good" subset by performing only a small fraction of all the possible regressions. One such method is the stepwise regression method which will now be described.

8.2.2 Stepwise Regression

The stepwise regression algorithm due to Efroymson (1960) is an automated procedure which is particularly well suited for model building with *large* models. This algorithm, as described by Hemmerle (1967) and Jennrich (1977), utilizes sweeps on the corrected sums of cross-products matrix, but incorporates an entry and exit criterion that differs from the Garside sequence so as to avoid (hopefully) doing all possible regressions.

We will follow the development by Hemmerle (1967) in the following description of the stepwise regression procedure. Suppose that we begin with the matrices X and $\underline{Y}$ in deviation form and partition X as (X_1, X_2) where X_1 is $(n \times k)$ and X_2 is $n \times (p - k)$. Under these conditions the overall corrected sums of squares and cross-products matrix is

$$\begin{bmatrix} X_1'X_1 & X_1'X_2 & X_1'\underline{Y} \\ X_2'X_1 & X_2'X_2 & X_2'\underline{Y} \\ \underline{Y}'X_1 & \underline{Y}'X_2 & \underline{Y}'\underline{Y} \end{bmatrix}$$

If we perform k reductions on this matrix by sweeping the first k rows, the resultant matrix is

$$\begin{bmatrix} (X_1'X_1)^{-1} & (X_1'X_1)^{-1}X_1'X_2 & \underline{\hat{\beta}} \\ -X_2'X_1(X_1'X_1)^{-1} & X_2'X_2-X_2'X_1(X_1'X_1)^{-1}X_1'X_2 & X_2'\underline{Y}-X_2'X_1(X_1'X_1)^{-1}X_1'\underline{Y} \\ -\underline{\hat{\beta}}' & \underline{Y}'X_2-\underline{Y}'X_1(X_1'X_1)^{-1}X_1'X_2 & \underline{Y}'\underline{Y}-\underline{\hat{\beta}}'X_1'\underline{Y} \end{bmatrix}$$

$$= \left[\begin{array}{c|cc} (X_1'X_1)^{-1} & A & \underline{\hat{\beta}} \\ \hline -A' & & \\ \cline{1-1} -\underline{\hat{\beta}}' & & C \end{array}\right] \qquad (8.25)$$

where $\hat{\beta}$ is used to denote $\underline{b}$ in this section, to emphasize that the full model (8.1) is not being fitted in most cases. At this stage we say that the current model contains the k independent variates $x_1, x_2, \ldots, x_k$.

Let $q = p-k+1$ so that the submatrix C is q x q. If we consider, at this stage, entering one of the variables $x_{k+1}, x_{k+2}, \ldots, x_p$ into the current model, we will want to enter that predictor which will produce the smallest residual sum of squares among all possible choices. Since c_{qq} is the residual sum of squares for the current fit, we want the choice to produce the smallest c_{qq} element after the reduction is performed. We can look ahead and easily make the proper choice by noting that if we enter x_{k+j} $(j = 1, 2, \ldots, p - k)$, then the residual sum of squares will become

$$c_{qq} - \frac{c_{jq}^2}{c_{jj}} \tag{8.26}$$

Thus we say that x_{k+j} is a best candidate for entry if j is such that

$$\frac{c_{jq}^2}{c_{jj}} \geq \frac{c_{iq}^2}{c_{ii}} \quad \text{(for all } i \neq j = 1, 2, \ldots, p - k) \tag{8.27}$$

Suppose that we consider deletion of one of the predictors $x_1, x_2, \ldots, x_k$ which are currently in the model. As with entry, we can look ahead easily and make the most appropriate choice by noting that the residual sum of squares which results when x_i $(i = 1, 2, \ldots, k)$ is deleted is

$$c_{qq} + \frac{\hat{\beta}_i^2}{[(X_1'X_1)^{-1}]_{ii}} \tag{8.28}$$

Thus the most appropriate choice x_i is the one for which

$$\frac{\hat{\beta}_i^2}{[(X_1'X_1)^{-1}]_{ii}} \leq \frac{\hat{\beta}_j^2}{[(X_1'X_1)^{-1}]_{jj}} \quad \text{(for all } j \neq i = 1, 2, \ldots, k) \tag{8.29}$$

The appropriate choice of predictor for either entry or deletion is indicated by (8.27) and (8.29), respectively. We may not, however, want to delete or enter at some stage. If the residual sum of squares will increase too much for a deletion or not be decreased a sufficient amount for an entry, we will not want to effect the transfer. Some criteria which can be used to make these decisions are needed. The suggested criteria, assuming the user is not interacting with the computer system, are as follows:

1. Delete x_i if

$$\frac{\hat{\beta}_i^{\,2}}{s_k^{\,2}[(X_1'X_1)^{-1}]_{ii}} < F^{(\alpha)}_{1,n-k-1} \tag{8.30}$$

2. Enter x_i if

$$\frac{(n - k - 2)c_{iq}^2}{c_{ii}c_{qq} - c_{iq}^2} > F^{(\alpha)}_{1,n-k-2} \tag{8.31}$$

where $F^{(\alpha)}_{1,n-k-1}$ and $F^{(\alpha)}_{1,n-k-2}$ are either chosen tabular Snedecor F values or user-designated constants, and $s_k^{\,2}$ denotes the residual mean square from the current fit using k independent variables. Under certain assumptions about the joint distribution of predictors, these statistics will in fact be distributed as F, but the level α is questionable. In interactive applications the user may wish to decide whether to proceed based on other criteria that may be available to the user while computations are taking place.

The overall stepwise procedure proceeds, as the name implies, in a series of steps. The first step consists of entering the predictor x_i which satisfies (8.27) if (8.31) is also satisfied. If none can enter, the final model contains only the intercept. In general, all steps after the first are carried out as follows:

1. Choose the best predictor for deletion using the criterion (8.29) and delete if (8.30) is satisfied.

2. Choose the best predictor for entry using the criterion (8.27). Enter that predictor if (8.31) is satisfied; otherwise do not enter a variable into the current model.

We terminate the stepwise procedure when either of the following events occurs:

1. We cannot enter or delete according to the given criteria. This includes the case where we enter all predictors and cannot delete one.
2. The procedure dictates that the same predictor be entered and deleted in successive operations.

In practice some maximum number of allowable steps is usually prescribed to take care of a remote possibility that cycling will occur. Example (8.1) gives the results of an application of stepwise regression. The partial correlations shown are correlations between y and the indicated predictor. These are computed using the C matrix elements as if they were elements in a covariance matrix which, again, will be the case under certain distribution assumptions about the predictors. We are not, however, making such assumptions but are going ahead to calculate the quantities anyway.

Example 8.1

The following stepwise regression used 3.0 as the value of $F_s^{(\alpha)}$ both to enter and delete variates. (Some programs for stepwise regression do actually compute the F's at each step, but others use given fixed values.) The predictor labeled X0 is the intercept in the model. (The data in this example originally appeared in a 1932 article by Woods, Steinour, and Starke in *Industrial and Engineering Chemistry 24*, pp. 1207-1214. The data set has been discussed frequently by later authors.)

Step number 1
Variable entered X4
Multiple R square 0.6745

Analysis of Variance

Source	d.f.	Sum of Squares	Mean Square	F
Regression	1	1831.9860	1831.9860	22.7985
Residual	11	883.8669	80.3515	

Variables in equation

No.	Coefficient	F to remove
X4	-0.73816	22.799
X0	117.56793	

Variables not in equation

No.	Partial Corr.	F to enter
X1	0.95677	108.22
X2	0.13021	0.172
X3	-0.89508	40.295

Step number 2

Variable entered X1

Multiple R square 0.9725

Analysis of Variance

Source	d.f.	Sum of Squares	Mean Square	F
Regression	2	2641.0007	1320.5002	176.627
Residual	10	74.7621	7.4762	

Variables in equation

No.	Coefficient	F to remove
X1	1.4400	108.22
X4	-0.6140	159.30

Variables not in equation

No.	Partial Corr.	F to enter
X2	0.59861	5.0259
X3	-0.56571	4.2358

Step number 3

Variable entered X2

Multiple R square 0.9823

Analysis of Variance

Source	d.f.	Sum of Squares	Mean Square	F
Regression	3	2667.7900	889.2632	166.8316
Residual	9	47.9727	5.3303	

Variables in equation

No.	Coefficient	F to remove
X1	1.4519	154.01
X2	0.4161	5.03
X4	-0.2365	1.86
X0	71.6483	

Variables not in equation

No.	Partial Corr.	F to enter
X3	0.0477	0.0182

Step number 4
Variable removed X4
Multiple R square 0.9787

Analysis of Variance

Source	d.f.	Sum of Squares	Mean Square	F
Regression	2	2657.8584	1328.9292	229.50
Residual	10	54.9045	5.7904	

Variables in equation			*Variables not in equation*		
No.	Coefficient	F to remove	No.	Parital Corr.	F to enter
X1	1.4683	146.52	X3	0.41126	1.8321
X2	0.6622	208.58	X4	-0.41415	1.8633
X0	52.5774				

At this point, based on the F to enter and remove predictors, no variable could be entered or removed, so the program execution was terminated.

Stepwise regression is popular with many people because almost all needed decisions are made by the algorithm. The user receives one suggested submodel as output from the algorithm. This subset may be "best" only in the sense that it is the subset which terminated the algorithm. However, the way in which the algorithm arrives at a final subset is certainly intuitively appealing, and experience with the algorithm has demonstrated that it fairly consistently produces "desirable" results.

8.2.3 Other Methods

There are several good methods for subset selection in addition to the methods mentioned in preceding subsections. Some use the sum of squared residuals as a comparative measure between subsets. Others utilize different functions of the data in the decision-making process.

One efficient method was developed by LaMotte and Hocking (1970). This method utilizes a "branch and bound algorithm" which minimizes the number of subsets of size k that need to be considered in order to find a best subset having size k, based on the minimum sum of squared residuals criterion. For small- to medium-sized models ($p \leq 40$, say) this method identified best subsets of all sizes one through p at a reasonable cost. Solution of larger problems is very expensive using this method (though of course not nearly as expensive as computing all possible regressions). Furnival and Wilson (1974) also proposed a branch and bound procedure to identify the best subset of each given size. Descriptions of these methods were given by Hocking (1977).

Mallows (1965) proposed a statistic C_p as an alternative to the sum of squared residuals for use in deciding which subset of the p predictors is best. Basically, C_p accounts for the total mean squared error of prediction over the observed data. Thus it attempts to include bias in prediction into the criterion for comparison of subsets. Detailed discussions of the use of C_p are contained in Gorman and Toman (1966), Daniel and Wood (1971), and Mallows (1973). Again, these methods are very costly for larger problems.

Another proposed criterion is based on a measure of the total error in prediction. Allen (1971) described the model-building procedure called PRESS which is based on this criterion. This method is also relatively expensive for large problems.

All of the methods mentioned in this subsection provide more information about submodels but are considerably more expensive to use for large problems than is stepwise regression. Also, for small problems, most people prefer to compute all possible regressions. However, for medium-size problems such methods as stepwise selection are frequently used because they provide a large amount of information for reasonable cost.

We next turn to a special case in the area of regression models and model building, that of the polynomial model. Somewhat different methods are often employed for these models.

8.2.4 A Special Case--Polynomial Models

The general polynomial regression model is

$$y_i = \alpha_0 + \alpha_1 x_i + \alpha_2 x_i^2 + \cdots + \alpha_q x_i^q + e_i \quad (i = 1, 2, \ldots, n) \tag{8.32}$$

This model is the usual model for multiple linear regression; the independent variates in this case are powers of x. Thus we can, at least theoretically, perform all of the computations relative to this model using the techniques described in the preceding subsections. In practice, however, a multiple regression approach which uses X'X and X'$\underline{Y}$ often produces poor results in cases where higher-order ($q > 3$, say) polynomials are fit. Disastrous numerical difficulties almost always arise when $q > 9$ if standard multiple-regression methods are employed using single- or double-precision arithmetic on machines such as the IBM 370 (see Wampler 1970). For this reason some special method, or one which utilizes orthogonal transformations in higher precision on the matrices X and $\underline{Y}$, is preferred in this case.

In model-building situations the usual purpose is to find the degree of polynomial which provides a satisfactory fit to the data based on the RSS criterion. There are basically two schools of thought on methods of procedure. Some say to begin with the intercept model and include successively higher-degree terms, one by one, until a satisfactory fit results. Usually, two or three additional terms are tried after a "good" one is found, to insure that a substantially better model is not lurking nearby. The other approach calls for beginning with the highest-degree model that one would be willing to accept. Then reduce the degree in successive steps until, at some step, the amount of increase in the sum of squared residuals is judged to be intolerable.

The method which we will describe utilizes a reparametrization of (8.32) in terms of polynomials that are orthogonal over the n given x values. The reparametrized model is then such that X'X is diagonal; hence model-building operations are facilitated using either of the approaches defined above. Once regression coefficients are obtained for the reparametrized model, a transformation is made back to the original model. This transformation leaves the degree unchanged between the two models.

Use of orthogonal polynomials is closely related to use of M.G.S. or Householder transformations in this case. Normally these latter methods are not used for polynomial models because they require more arithmetic and do not always produce as good results. It is not true, however, that we totally overcome the inherent ill condition in polynomial regression data by employing orthogonal polynomials. They help to some extent, but these models are hard to handle numerically. Orthogonal polynomials are sometimes used in other statistical applications, so the method for generating these polynomials, which will be described in this subsection, can be used in other situations.

We begin the description of the use of orthogonal polynomials in regression by expressing each power of x in terms of a polynomial $p_j(x)$, where the $p_j(x)$ are such that

$$\sum_{i=1}^{n} p_j^2(x_i) = c_j \neq 0 \quad (j = 0, 1, \ldots, q)$$

$$\sum_{i=1}^{n} p_j(x_i)p_k(x_i) = 0 \quad j \neq k$$

and the c_j's are constants. In other words, the polynomials $\{p_j(x), j = 0, 1, \ldots, q\}$ are orthogonal with respect to the given values $x_i (i = 1, 2, \ldots, n)$. A set of monic polynomials which satisfy these requirements are given by the following recurrence (see Forsythe 1957 and Ralston 1965). [The reader should note that $p_j(x)$ has degree j; hence solution for x^j in terms of $p_j(x)$ and lower-order polynomials $p_k(x)$ is readily accomplished.] Let

$$p_{-1}(x) = 0$$

$$p_0(x) = 1$$

$$p_{j+1}(x) = (x - \rho_{j+1})p_j(x) - \gamma_j p_{j-1}(x) \quad (j = 0, 1, \ldots, q - 1)$$

where, for $j > 0$,

$$\rho_{j+1} = \frac{\sum_{i=1}^{n} x_i [p_j(x_i)]^2}{\sum_{i=1}^{n} [p_{j-1}(x_i)]^2} \quad (\text{note } \rho_1 = \bar{x}) \tag{8.33}$$

and

$$\gamma_j = \frac{\sum_{i=1}^{n} [p_j(x_i)]^2}{\sum_{i=1}^{n} [p_{j-1}(x_i)]^2} \quad (j = 1, 2, \ldots, q - 1)$$

with $\gamma_0 = 0$. We will leave the details of an algorithm for determining the coefficients in each $p_j(x)$ as an exercise. The recurrence relations (8.33) apply to sets of x data which are unequally spaced on the real line. In the special case of equal spacing there are tables of orthogonal polynomial coefficients which are available and might be utilized.

The reparametrization of the model (8.32) in terms of the $p_j(x)$ is accomplished by solving for x^j in $p_j(x)$ and substituting into (8.32). The resulting model is in the form of a linear combination of the $p_j(x)$ and is given as

$$y_i = \beta_0 p_0(x_i) + \beta_1 p_1(x_i) + \cdots + \beta_q p_q(x_i) + e_i \tag{8.34}$$

The relationship between the β and the α coefficients will be given below. Using this model, with the restriction $q < n$ to avoid the case wherein $p_j(x) = 0$, the regression coefficient estimates are readily derived as

$$\hat{\beta}_j = \frac{\sum_{i=1}^{n} y_i p_j(x_i)}{c_j} \quad (j = 0, 1, \ldots, q) \tag{8.35}$$

since X'X for the model (8.34) is diagonal.

Most often it is the case in polynomial regression that the α coefficient estimates are desired instead of the $\hat{\beta}$'s. Consider the previously computed polynomial coefficients in the polynomials (8.33). Calling these coefficients δ_{ij}, we have

$$p_j(x) = \sum_{k=0}^{j} \delta_{jk} x^k$$

Due to the reparametrization, we have

$$\sum_{j=0}^{q} \beta_j p_j(x) = \sum_{j=0}^{q} \beta_j \left(\sum_{k=0}^{j} \delta_{jk} x^k \right)$$

$$= \sum_{j=0}^{q} x^j \left(\sum_{k=j}^{q} \beta_k \delta_{kj} \right)$$

Thus

$$\alpha_j = \sum_{k=j}^{q} \beta_k \delta_{kj}$$

and we estimate $\hat{\alpha}_j$ as

$$\hat{\alpha}_j = \sum_{k=j}^{q} \hat{\beta}_k \delta_{kj}$$

If we define the upper-triangular matrix Δ as

$$\Delta = \begin{bmatrix} \delta_{00} & \delta_{10} & \delta_{20} & \cdots & \delta_{q0} \\ 0 & \delta_{11} & \delta_{21} & \cdots & \delta_{q1} \\ 0 & 0 & \delta_{22} & \cdots & \delta_{q2} \\ \vdots & \vdots & \vdots & & \vdots \\ 0 & 0 & 0 & \cdots & \delta_{qq} \end{bmatrix} \tag{8.36}$$

then

$$\underline{\hat{\alpha}} = \Delta \underline{\hat{\beta}}$$

and the variance-covariance matrix of $\underline{\hat{\alpha}}$ is

$$E[\underline{\hat{\alpha}} - E(\underline{\hat{\alpha}})][\underline{\hat{\alpha}} - E(\underline{\hat{\alpha}})]' = \sigma^2 \Delta D^{-1} \Delta' = \sigma^2 R^{-1}(R')^{-1}$$

where D is the $(q + 1) \times (q + 1)$ diagonal matrix

$$D = \text{diag}(c_i)$$

and

$$R^{-1} = \Delta D^{-1/2}$$

The generation and use of orthogonal polynomials is not difficult on a computer. The recurrence relations (8.33) automatically give the successive columns in the Δ array and these can be used, along with the original data, in subsequent computations. Forsythe (1957) recommends that the independent variable x be normalized so that all x_i lie in the interval (-2,2). This device tends to reduce the effect of cancellation in the computational process. All polynomial evaluations should utilize some form such as the "nested" polynomial form to minimize the volume of computation (see Eq. 5.11).

Shampine (1975) demonstrates that a desirable modification of the Forsythe method given above is to form residuals

$$r_i = y_i - p_{j-1}(x_i)$$

at each stage and compute $\hat{\beta}_j$ in (8.35) as

$$\hat{\beta}_j = \frac{\sum_{i=1}^{n} r_i p_j(x_i)}{c_j}$$

This does require more computation, but it helps to compensate for the fact that we cannot usually maintain as much orthogonality as we would like if more than a few polynomials are generated. Seber (1977) discusses this as well as other aspects of polynomial regression.

Example 8.2

The following data were used to generate the first five polynomials:

x	y
0.0	0.980
0.0	1.000
1.0	0.000
1.0	-0.015
2.0	29.000
2.0	28.600
3.0	138.101
3.0	141.000

The orthogonal polynomials in this case are:

$$p_0(x) = 1$$

$$p_1(x) = x - 1.5$$

$$p_2(x) = x^2 - 3x + 1$$

$$p_3(x) = x^3 - 4.5x^2 + 4.7x - 0.3$$

$$p_4(x) = x^4 - 6x^3 + 11x^2 - 6x$$

The prediction equation for the third-degree polynomial was found to be

$$\hat{y} = 42.3p_0(x) + 44.4p_1(x) + 27.9p_2(x) + 8.68p_3(x)$$

$$= 0.99 + 1.46x - 11.1x^2 + 8.68x^3$$

8.3 MULTIPLE REGRESSION UNDER LINEAR RESTRICTIONS

Many regression models incorporate additional information in the form of restrictions on the parameters in the model. In such cases we say that the model is a *restricted model*. For example, it may be known that the parameters sum to a fixed quantity, or that they are nonnegative. A least-squares solution vector $\underline{b}$ relative to a restricted model is called a *restricted least-squares solution*.

In practical applications most researchers require that the solution vector $\underline{b}$, obtained for a restricted regression problem, be such that it satisfies the form of the restrictions placed on the parameters in the model (e.g., if $\underline{\beta} \geq \underline{0}$ is a restriction, then an acceptable solution must satisfy $\underline{b} \geq \underline{0}$). In other words, constraints are placed on the solution. Henceforth, in this section, it will be assumed that a solution under the obvious constraints is required relative to the restricted model under consideration.

We will consider two general kinds of *linear* restrictions, which require different techniques for solution. The first of these, linear *equality* restrictions, is discussed in Section 8.3.1. For least-squares problems with linear equality restrictions, we will see that there is a closed-form solution. For the other type of problem with linear *inequality* restrictions, no closed form is available for the solution. Methods of computation for least-squares problems with linear inequality restrictions are discussed in Section 8.3.2. *Nonlinear* restrictions, such as that the sum of squares of the parameters be a fixed quantity, are much more difficult to handle. Such restrictions occur less frequently in practical applications and these will not be discussed here.

8.3.1 Linear Equality Restrictions

If the parameters of the usual linear model (8.1) are subject to the restriction equations

$$C\underline{\beta} = \underline{d} \tag{8.37}$$

where C is a t x p matrix of rank k, the problem is to find, among all vectors $\underline{b}$ which satisfy $C\underline{b} = \underline{d}$, one which minimizes

$$||X\underline{\beta} - \underline{Y}||$$

Obviously, a solution to this problem exists if and only if the system of linear equations (8.37) is consistent, and we shall assume this is the case in this section.

Let us begin by discussing a special case which arises in analysis of variance applications for designed experiments. The design matrix X is a matrix of 0's and 1's, and has rank r (say). Many authors suggest that the "usual restrictions" be placed on the parameters in the model so that a full-rank regression problem results. These restrictions amount to choosing at least p - r linear restrictions to form $C\underline{\beta} = \underline{0}$ such that the rows of C are linearly independent of the rows of X, and C has row rank k = p - r. In other words, $C\underline{\beta}$ is not linearly estimable. Under these conditions the restricted problem has a unique solution obtained by solving for $\underline{b}$ in the full-rank problem

$$\begin{pmatrix} X \\ \hline C \end{pmatrix} \underline{\beta} \cong \begin{pmatrix} \underline{Y} \\ \hline \underline{0} \end{pmatrix} \tag{8.38}$$

or, if normal equations are formed, by solving

$$(X'X + C'C)\underline{\beta} = X'\underline{Y}$$

which has the solution

$$\underline{b} = (X'X + C'C)^{-1}X'\underline{Y} \tag{8.39}$$

Proof that $\underline{b}$ satisfies the restriction equations, and is a least-squares solution, is given in many texts, including Seber (1977). Thus this special restricted least-squares problem can be solved using standard methods given in Section 8.1.

Now let us return to the general restricted problem which we stated initially. Consider the set

$$B = \{\underline{b} \mid C\underline{b} = \underline{d}\}$$

Elements in this set can be characterized as

$$B = \{\underline{b} \mid \underline{b} = C^+\underline{d} + (I_p - C^+C)\underline{z}\}$$

where $\underline{z}$ ranges over p-dimensional vectors. We want to select, from the set B, a solution $\underline{b}$ which minimizes

$$\begin{aligned} ||X\underline{b} - \underline{Y}|| &= ||X[C^+\underline{d} + (I_p - C^+C)\underline{z}] - \underline{Y}|| \\ &= ||X(I_p - C^+C)\underline{z} - (\underline{Y} - XC^+\underline{d})|| \end{aligned}$$

Since there is a unique minimum-length solution given by

$$\hat{\underline{z}} = [X(I - C^+C)]^+(\underline{Y} - XC^+\underline{d}) \tag{8.40}$$

it is a logical choice. Under this choice of $\underline{z}$ it can be shown that

$$\underline{b} = C^+\underline{d} + [X(I - C^+C)]^+(\underline{Y} - XC^+\underline{d}) \tag{8.41}$$

is the *unique minimum-length solution* to the restricted regression problem. Furthermore, when the rank of $(X'|C')$ is p, then the solution (8.41) is unique. (Recall that we are assuming $C\underline{\beta} = \underline{d}$ is consistent.) Proof of these results is given in chapter 20 of Lawson and Hanson (1974).

Equation (8.41) can be written in a more desirable form for computations by utilizing an orthogonal decomposition of C in the form

$$C = H \begin{pmatrix} R_1 & 0 \\ 0 & 0 \end{pmatrix} U'$$
$$= H \begin{pmatrix} R_1 & 0 \\ 0 & 0 \end{pmatrix} \begin{pmatrix} U_1' \\ U_2' \end{pmatrix}$$

where R_1 is k x k nonsingular and U_2' is (p - k) x p. Now we have

$$C^+ = U \begin{pmatrix} R_1^{-1} & 0 \\ 0 & 0 \end{pmatrix} H'$$

and

$$\begin{aligned} I - C^+C &= UU' - UR^+RU' \\ &= UU' - U_1U_1' \\ &= U_2U_2' \end{aligned}$$

Using this result, we rewrite (8.41) as

$$\begin{aligned} \underline{b} &= C^+\underline{d} + (XU_2U_2')^+(\underline{Y} - XC^+\underline{d}) \\ &= C^+\underline{d} + U_2(XU_2)^+(\underline{Y} - XC^+\underline{d}) \end{aligned} \tag{8.42}$$

The fact that $(XU_2U_2')^+ = U_2(XU_2)^+$ can be verified algebraically by checking the four Moore-Penrose conditions.

Lawson and Hanson (1974) suggest the following computational procedure for the important case where C has *full row rank t*. First, transform according to

$$CU = (C_1|0)$$

where C_1 is t x t nonsingular and lower-triangular and U is orthogonal. Partition U as $U = (U_1|U_2)$, where U_1 is p x t. Now transform the augmented matrix using U,

$$\begin{pmatrix} C \\ X \end{pmatrix} U = \begin{pmatrix} C_1 & 0 \\ \tilde{X}_1 & \tilde{X}_2 \end{pmatrix}$$

where $\tilde{X}_1$ consists of the first t columns of XU. The following steps will complete the process:

1. Solve $C_1\underline{z}_1 = \underline{d}$ for $\hat{\underline{z}}_1$. (8.43)
2. Compute

$$\underline{\tilde{Y}} = \underline{Y} - \tilde{X}_1\hat{\underline{z}}_1$$

Note that

$$C^+\underline{d} = U\left(\frac{C_1^{-1}}{0}\right)\underline{d} = U_1C_1^{-1}\underline{d}$$

and

$$XC^+\underline{d} = \tilde{X}_1\hat{\underline{z}}_1$$

3. Compute $\hat{\underline{z}}_2 = \tilde{X}_2^+\underline{\tilde{Y}}$. This quantity is $(XU_2)^+(\underline{Y} - XC^+\underline{d})$ from (8.42).
4. Compute the solution vector

$$\underline{b} = U\left(\frac{\hat{\underline{z}}_1}{\hat{\underline{z}}_2}\right)$$

Note that

$$U\left(\frac{\hat{\underline{z}}_1}{\hat{\underline{z}}_2}\right) = U_1\hat{\underline{z}}_1 + U_2\hat{\underline{z}}_2$$
$$= C^+\underline{d} + U_2(XU_2)^+(\underline{Y} - XC^+\underline{d})$$

as is required in (8.42).

8.3.2 Linear Inequality Restrictions

When the parameters of the linear model (8.1) are subject to inequality restrictions

$$C\underline{\beta} \geq \underline{d} \tag{8.44}$$

the least-squares solutions cannot in general be obtained in closed form. There are several special cases of the least-squares problem with linear inequality restrictions. The most common problem is one in which C is a matrix resulting from a permutation of the columns of $[I_t|0]$ and $\underline{d} = \underline{0}$, that is, some or all of the parameters are restricted to be nonnegative. Other forms of C and $\underline{d}$ yield various types of restrictions. In this subsection we will consider only restrictions in which the t x p matrix C has rank t. We first show that this case is equivalent to the case of nonnegative restrictions, as pointed out by Khuri (1976).

The problem is to find $\underline{b}$ to minimize

$$||X\underline{\beta} - \underline{Y}||$$

subject to

$$C\underline{\beta} \geq \underline{d}$$

where C is t x p and has rank t. Let C be partitioned into $[C_1|C_2]$ where C_1 is t x t and nonsingular. (This many require permutations of the columns of C.) Let X and $\underline{\beta}'$ be partitioned accordingly into $[X_1|X_2]$ and $[\underline{\beta}_1'|\underline{\beta}_2']$. Now let $\underline{\alpha}_1 = C\underline{\beta} - \underline{d}$, $\underline{\alpha}_2 = \underline{\beta}_2$, and $\underline{\alpha}' = [\underline{\alpha}_1'|\underline{\alpha}_2']$. The problem can now be written as: find $\underline{\alpha}$ to minimize

$$||W\underline{\alpha} - \underline{Z}||$$

subject to

$$\underline{\alpha}_1 \geq \underline{0}$$

where $W = [X_1C_1^{-1}|X_2 - X_1C_1^{-1}C_2]$, and $\underline{Z} = \underline{Y} - X_1C_1^{-1}\underline{d}$. Hence, the least-squares problem with linear inequality restrictions of full rank may be solved by using the solution to a least-squares problem with nonnegativity restrictions, and so the subsequent discussion of this subsection will concern only the problem with nonnegativity restrictions.

Solution of the problem with nonnegativity restrictions generally requires the solution of several least-squares problems. One type of algorithm is a search employing branching and bounding. Algorithms of this type were given by Mantel (1969), Gallant and Gerig (1975) and Armstrong and Frome (1976). Mantel's algorithm makes use of the fact that if in the solution to an unrestricted least squares problem the estimates b_{i_1}, b_{i_2}, ..., b_{i_k} are nonpositive while all of the other estimates are positive, at least one of the b_{i_j} (j = 1, 2, ..., k) will be zero in the solution to the same least squares problem with nonnegativity restrictions (see Exercise 8-18). This can be used to reduce the number of least squares problems necessary to be solved since this fact can be used at each stage (see Waterman 1977).

In the algorithm of Armstrong and Frome a binary tree of regression models is set up in which each node corresponds to a model having each parameter either set to zero or left unrestricted, but required to be in the model. After the (unrestricted) least-squares solution is obtained at a given node, two daughter nodes are formed; one by setting one of the free parameters to zero, and the other by leaving that parameter unrestricted but required to be in the model in all descendent nodes. In the root node all parameters are included in the model.

The search is reduced because nodes of the tree can be *fathomed*, in which case no descendent nodes are created from those nodes. A node may be fathomed if the estimates for all free parameters at the node are nonnegative or if the residual sum of squares at the node is larger than the residual sum of squares of the best solution known at that time. (Initially, the residual sum of squares of the best solution is $\underline{Y}'\underline{Y}$, since no solution is known.) The search is further reduced by the fact, proved by Armstrong and Frome (1976), that if fathoming occurs at a node formed by setting $\beta_i = 0$ from a parent node having $b_i < 0$, then the sister node can also be fathomed. We will refer to these three fathoming conditions as rule 1, rule 2, and rule 3; that is:

Rule 1: Estimates for all free parameters are nonnegative.

Rule 2: Residual sum of squares larger than current best bound.

Rule 3: Sister node fathomed.

Example 8.3 will illustrate the method.

Example 8.3

The model $y_i = \beta_0 + \beta_1 X_{1i} + \beta_2 X_{2i} + \beta_3 X_{3i} + e_i$ is to be fitted with the restrictions that $\beta_j \geq 0$ for j = 0, 1, 2, 3. The following data are to be used in the least-squares fit:

X_1	X_2	X_3	y
0.4	94	39	378.4
6.7	77	85	310.0
9.3	8	44	99.0
5.6	41	43	206.0
8.5	33	20	179.0
10.0	71	60	285.2
6.3	88	0	351.9
0.6	54	37	256.7
8.8	63	64	266.9
7.1	66	26	282.6
7.5	63	39	270.3
2.7	65	50	284.2
8.9	17	35	129.3
8.3	31	1	176.0
8.9	39	16	196.4
1.9	66	78	286.1
8.9	77	15	311.6
5.1	89	35	352.9
2.8	57	6	266.1
3.5	22	53	152.9

A fit with no restrictions yields

$$\underline{b}' = [99.00, -2.037, 3.022, -0.1048]$$

which does not satisfy the nonnegativity restrictions. Using the full unrestricted fitted model, we begin the construction of the binary tree. From the root node we create two submodels by selecting one of the variables with a negative coefficient estimate and setting it equal to zero in one case and forcing it to remain in the model in the other case:

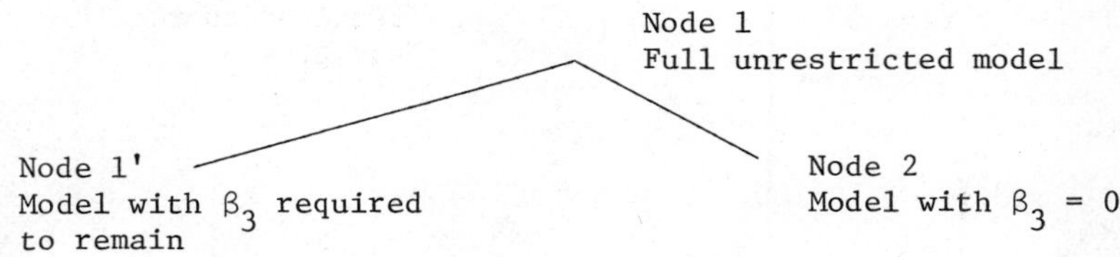

A fit with $\beta_3 = 0$ yields

$$\underline{b}' = [94.70, -1.9256, 3.017, 0]$$

Since this fit does not satisfy the nonnegativity restrictions, we again form two subproblems at node 2:

Node 2
Model with $\beta_3 = 0$

Node 2'
Model with β_1 required to remain

Node 3
Model with $\beta_1 = \beta_3 = 0$

A fit with $\beta_1 = \beta_3 = 0$ yields

$$\underline{b}' = [78.69, 0, 3.093, 0]$$

Node 3 is therefore fathomed by rule 1. At this point we have a new bound for the residual sum of squares to the restricted problem. It is

$$\text{RSS} = 733.5$$

Prior to this step the only known bound was the total sum of squares, $\Sigma\, y_i^2 = 1383500$. Node 2' may be fathomed by rule 3.

We now return to node 1'. There are no computations to be performed here, only the creation of two daughter nodes, using the solution obtained at node 1. Since $b_1 < 0$ and $b_3 < 0$ at node 1', but β_3 is required to remain in the model, we set $\beta_1 = 0$:

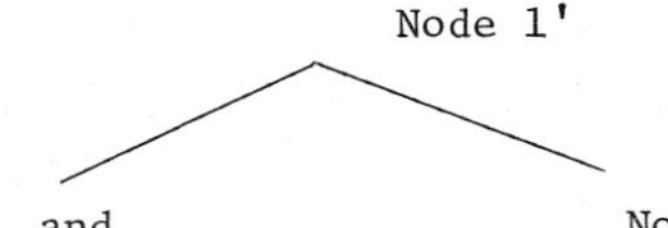

Node 1" Model with β_1 and β_3 required to remain

Node 4 Model with β_3 required to remain and $\beta_1 = 0$

The solution at this point is

$\underline{b}' = [80.99, 0, 3.100, -0.07146]$

and the residual sum of squares is 679.2. Two daughter nodes must be formed from node 4 by setting to zero one of the parameters with an estimate which is positive, since β_3 is required to remain in the model. We choose to set β_2 to zero:

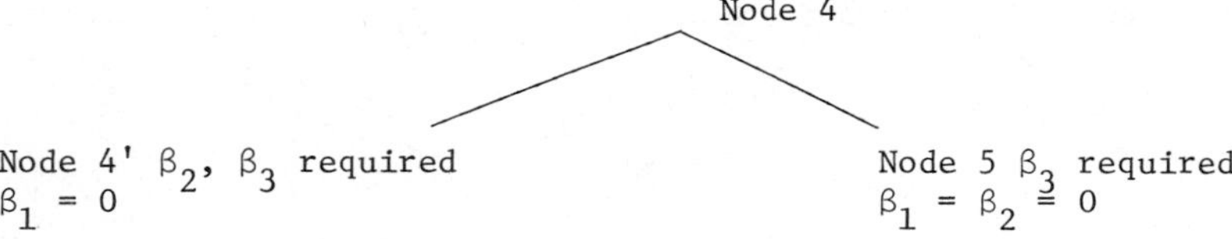

The solution at node 5 is

$\underline{b}' = [243.3, 0, 0, 0.2353]$

This solution satisfies the nonnegativity restrictions, but the residual sum of squares, 112080, is greater than the upper bound. Hence, the best solution currently known is that obtained at node 3 with β_1 and β_3 set to zero.

Node 4' is now fathomed by rule 3, and Node 1" is fathomed by rule 1; hence, the final solution is

$\underline{b}' = [78.69, 0, 3.093, 0]$

as obtained at node 3.

The full binary tree is shown in Fig. 8.2. The solution required regression computations for five different models or submodels. Use of the updating techniques as discussed in Section 8.2 reduces the computations at each node. Armstrong and Frome (1976) suggested choosing the most negative free parameter at each stage when one is to be set to zero (see also Mantel 1969).

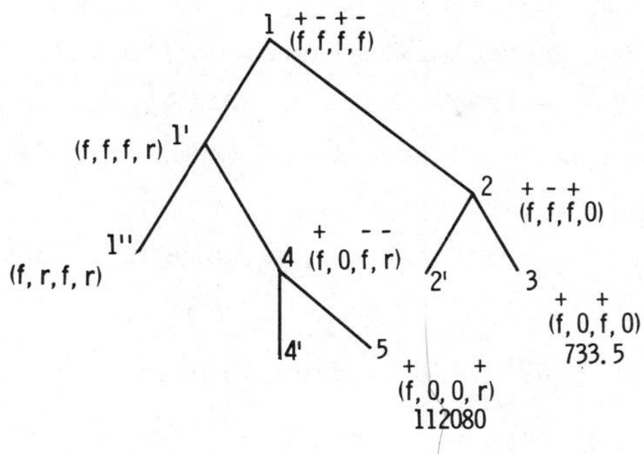

Figure 8.2 Binary tree for the branch and bound algorithm used in Example 8.3. The symbols f, r, or 0 indicate the respective model parameter is allowed to be free, required to be in the model, or set to zero. The sign above the symbol indicates the sign of the estimate. The numbers underneath the symbols are the residual sums of squares for the model indicated.

Another approach to the computational problem for nonnegative least squares uses the Kuhn-Tucker conditions, which are necessary and sufficient conditions for optimality (see, e.g., Sposito 1975). The conditions in the present problem of obtaining least-squares estimates for $\underline{\beta}$ in

$$\underline{Y} = X\underline{\beta} + \underline{\varepsilon}$$

subject to

$$\underline{\beta} \geq \underline{0}$$

are

$$[X'(\underline{Y} - X\underline{b})]_i = 0 \quad \text{if } b_i > 0$$

and

$$[X'(\underline{Y} - X\underline{b})]_i \leq 0 \quad \text{if } b_i = 0$$

Use of these conditions allows for efficient searching through the submodels to find one whose solution is the solution of the restricted full model. Lawson and Hanson (1974) describe and give a FORTRAN program (NNLS) for an algorithm using the Kuhn-Tucker conditions.

Their algorithm begins with $\underline{b}$ set to $\underline{0}$. Next, β_j is allowed to enter the model where j is selected as the index of the maximum element of $[X'(\underline{Y} - X\underline{b})]$, that is, of $X'\underline{Y}$, at the first stage. Parameters are added to or deleted from the model until the Kuhn-Tucker conditions are satisfied.

We will illustrate the method using the data of Example 8.3. At the first stage,

$$\underline{Y}'X = [5041.6, 28954, 318760, 190576]$$

So the first model to be fit is

$$y_i = \beta_2 X_{2i}$$

This yields

$$\underline{b}^{(1)'} = [0, 0, 4.277, 0]$$

and

$$[(\underline{Y} - X\underline{b})'X] = [247, 1740, 0, 7196]$$

Hence, the next model fit is

$$y_i = \beta_2 X_{2i} + \beta_3 X_{3i}$$

This yields

$$\underline{b}^{(2)'} = [0, 0, 3.979, 0.5182]$$

and

$$[(\underline{Y} - X\underline{b})'X] = [194.6, 1394, 0, 0]$$

The next parameter entered is β_1, and

$$\underline{b}^{(3)'} = [0, 4.013, 3.747, 0.3255]$$

with

$$[(\underline{Y} - X\underline{b})'X] = [109.4, 0, 0, 0]$$

When β_0 now enters the model, we get a solution containing negative elements:

$$\underline{b}^{(4)'} = [99.00, -2.037, 3.022, -0.1048]$$

At this point it becomes necessary to remove one of the terms from the model. Lawson and Hanson suggest an efficient method for doing this. When negative elements appear in the solution $\underline{b}^{(i)}$ at the ith stage, and all the elements of $\underline{b}^{(i-1)}$ are positive, determine α so that

$$\alpha = \min_j \left\{ \frac{b_j^{(i-1)}}{b_j^{(i-1)} - b_j^{(i)}} \;\middle|\; b_j^{(i)} \leq 0 \right\}$$

and set

$$\underline{\gamma} = \underline{b}^{(i-1)} + \alpha[\underline{b}^{(i)} - \underline{b}^{(i-1)}]$$

The model terms corresponding to zero elements of $\underline{\gamma}$ are removed and $\underline{b}^{(i+1)}$ is computed from the remaining terms. Should another negative element occur in $\underline{b}^{(i+1)}$, $\underline{\gamma}$ is used instead of $\underline{b}^{(i)}$ to determine a new α and a new $\underline{\gamma}$ to decide on the model for obtaining $\underline{b}^{(i+2)}$. Whenever the negative element occurs following a stage in which all elements had been positive, the previous $\underline{b}$ is used in computing α and $\underline{\gamma}$, however. Continuing with the example, we have

$$\alpha = \frac{4.013}{4.013 + 2.037} = 0.6633$$

and

$$\underline{\gamma}' = \underline{b}^{(3)'} + \alpha[\underline{b}^{(4)'} - \underline{b}^{(3)'}] = [65.67, 0, 3.266, 0.4719]$$

Fitting the model

$$y_i = \beta_0 + \beta_2 X_{2i} + \beta_3 X_{3i}$$

we obtain

$$\underline{b}^{(5)'} = [80.99, 0, 3.100, -0.07146]$$

With

$$\alpha = \frac{0.4719}{0.4719 + 0.07146}$$
$$= 0.8685$$

we update $\underline{\gamma}$:

$$\underline{\gamma}' = \underline{\gamma}^{(old)'} + \alpha[\underline{b}^{(5)'} - \underline{\gamma}^{(old)'}]$$
$$= [78.98,\ 0,\ 3.122,\ 0]$$

Hence, we now fit the model

$$y_i = \beta_0 + \beta_2 X_{2i}$$

obtaining

$$\underline{b}^{(6)'} = [78.69,\ 0,\ 3.093,\ 0]$$

Since all terms are positive, we check the Kuhn-Tucker conditions by computing

$$[(\underline{Y} - X\underline{b})'X] = [0,\ -313.2,\ 0,\ -759.8]$$

The Kuhn-Tucker conditions are satisfied by $\underline{b}^{(6)}$; hence it is the solution. Lawson and Hanson (1974) showed that this procedure will always converge.

In the example considered here, the use of the Kuhn-Tucker conditions to decide which variables to allow to enter the model required solution of six regression problems. The branch-and-bound method required five for this example; however, the regression problems solved by the branch-and-bound technique were larger problems (with more variables). Which method will be more efficient for a given problem will in general depend on the number of positive parameters in the final solution. If all are positive, the branch-and-bound method will obtain the correct solution in one step. If all are zero, the branch-and-bound method will require a number of steps at least as great as the number of parameters; while the method using the Kuhn-Tucker conditions will yield the correct solution in one step.

EXERCISES

8.1 Assume X is n x p of rank p with n > p. Suppose p + 1 Householder reflections are applied, from the left, to the augmented matrix $(X|\underline{Y})$ to produce

$$Q(X|\underline{Y}) = \left(\frac{R}{0}\middle| Q\underline{Y}\right)$$

where the last (n - p - 1) elements of $Q\underline{Y}$ are zero. Does the (p + 1)-th component of $Q\underline{Y}$ have meaning in regression analysis?

8.2 Consider the matrix Q in Eq. (8.4). Partition the columns to form $Q = (Q_1|Q_2)$, where Q_1 consists of the first p columns of Q. Then we have

$$Q'\underline{Y} = \left(\frac{Q_1'\underline{Y}}{Q_2'\underline{Y}}\right) = \left(\frac{\underline{c}_1}{\underline{c}_2}\right)$$

Under the usual assumptions about the random error vector $\underline{e}$ in the model (8.1), find the expectation and the covariance matrix of the (n - p) vector $\underline{c}_2$; hence show that $\underline{c}_2$ is a vector of uncorrelated regression residuals. (See Grossman and Styan 1972 for applications.)

8.3 Using the assumptions and notations of Exercise 8.2, show that

$$Q_2Q_2' = I_n - X(X'X)^{-1}X'$$

8.4 Derive algebraic equations which describe the process of computing R_1^{-1} for the matrix R_1 shown in Eq. (8.9).

8.5 Using the fact that $X(X'X)^-X'X = X$, where $(X'X)^-$ is any g_1 inverse of X'X, show that the function $L'\underline{\beta}$ used in Eq. (8.10) is estimable if and only if $L'(X'X)^-(X'X) = L'$. (Golub and Styan 1973a.)

8.6 Develop a method for determining whether $L'\underline{\beta}$ is estimable. Use the decomposition (8.14) and the result of Exercise 8.5 as a basis for starting. (Golub and Styan 1973a.)

8.7 Verify that $\underline{b}$ in Eq. 8.16 is $X^+\underline{Y}$.

8.8 For any matrix C, rank (C) = rank (C'C) = rank (CC'). Use this fact and the result of Exercise 8.5 to show when $L'\underline{\beta}$ is estimable, then

$$\text{rank } [L'(X'X)^{-}L] = \text{rank } (L)$$

Hint: Note that rank (L') $\leq$ rank $[L'(X'X)^{-}X']$. Does this result hold when L is not full-rank? How does the formula given in Eq. (8.17) change when $L'\underline{\beta}$ is estimable and L is not full-rank?

8.9 Show that a g_3 inverse of X can be used in place of X^+ to obtain $(X'X)^*$ which appears in Eq. (8.19). (See Exercise 7.22a.) Can this fact be used to simplify the method described using Eqs. (8.14) through (8.19)?

8.10 Use the matrix A shown in Eq. (8.18) to derive the analog of Eq. (8.13) for the non-full-rank X matrix case.

8.11 Reduce the formulas (8.20) to the form for computing the mean and sample variance of a single variable and the sample covariance of two variables.

8.12 Show that whenever the first column of the X matrix is nonzero and constant (has the same nonzero value in each row) the sum of residuals $\underline{Y} - X\underline{b}$ is theoretically zero. Hence show that the second check in Section 8.1.3 is valid.

8.13 Prove that the Mullet and Murray procedure (check 5 in Section 8.1.3) is theoretically correct. (*Hint:* First verify the procedure for full-rank X.)

8.14 Use the Garside algorithm with p = 4 and generate the sequence $\{s_j\}$ of integers.

8.15 Show that the $\underline{b}$ given in Eq. (8.41) is the unique minimum-length solution to the constrained problem (8.37).

8.16 Verify that the recurrence relation used to generate polynomials in Section 8.2.4 does indeed give polynomials which are orthogonal over the given x values.

8.17 In order to show how to recover, from the use of *centered* X and observed vector $\underline{Y}$, to the situation in terms of raw values X_R and $\underline{Y}_R$ using Householder transformations as shown in Section 8.1.1, do the following. (Assume X_R is full-rank.)

(a) Define $\underline{1}$ to be the n vector of ones, X_R to be n x (p + 1) with first column $\underline{1}$, X to be the n x p centered matrix, and $\underline{Y}$ to be the centered $\underline{Y}_R$. If

$$C = I_n - \left(\frac{1}{n}\right)\underline{1}\underline{1}'$$

show that C is idempotent and that

$$CX_R = (\underline{0}|X)$$

$$CY_R = \underline{Y}$$

$$CX = X$$

(b) Let Q be an n x n matrix such that

$$Q'X = \left(\frac{R}{0}\right)$$

where R is r x p. Show that the columns of Q are orthogonal to $\underline{1}$, that is, show that $\underline{1}'Q = \underline{0}$.

(c) Partition Q as $Q = (Q_1|Q_2)$, where Q_1 is n x r, and form the (r + 1) x p matrix

$$R_1 = \begin{pmatrix} n^{1/2} & n^{1/2}\bar{\underline{z}}' \\ & \\ 0 & R \end{pmatrix}$$

where $\bar{\underline{z}}'$ is the p-vector of column means of X_R given in Section 8.1.2. Show that

$$U = (n^{-1/2}\underline{1}|Q_1)$$

Also show that U augmented with the columns of Q_2 except for the last column (call this matrix W) is orthogonal and triangularizes X_R to form R_1; that is,

$$W'X_R = \left(\frac{R_1}{0}\right)$$

(d) Find the form of $W'\underline{Y}_R$ as closely in terms of $Q'\underline{Y}$ as you can.

(e) Use the results given in Section 8.1.2 to describe how to recover regression coefficients relative to the regression based on $(X_R|\underline{Y}_R)$ using the coefficients computed from the data $(X|\underline{Y})$.

(f) Using the previous parts, develop a method for recovering $\underline{b}$, R_1, W, the residual sum of squares, the residuals themselves, and $W'\underline{Y}$, where

$$W'X_R = \left(\frac{R_1}{0}\right)$$

using $\hat{\underline{\alpha}}$, R, Q, and other quantities obtained in regression based on $(X|\underline{Y})$.

8.18 Consider the problem of obtaining least squares estimates, $\underline{b}$, for $\underline{\beta}$ in

$$\underline{Y} = X\underline{\beta} + \underline{\varepsilon}$$

subject to

$$\underline{\beta} \geq \underline{0}$$

and with X of full rank. Let $\hat{\underline{\beta}} = (X'X)^{-1}X'\underline{Y}$ (that is, $\hat{\underline{\beta}}$ minimizes $||\underline{Y} - X\underline{\beta}||$ without any restrictions). Suppose that j of the elements of $\hat{\underline{\beta}}$ are nonpositive. Prove that at least one of the j corresponding elements of the restricted estimates $\underline{b}$ is zero. (Waterman 1977.)

REFERENCES AND FURTHER READINGS

Allen, D. M. (1971), Mean Square Error of Prediction as a Criterion for Selecting Variables. *Techno. 13*, 469-476.

Anscombe, F. J., and J. W. Tukey (1963), The Examination and Analysis of Residuals, *Techno. 5*, 141-160.

Armstrong, R. D., and E. L. Frome (1976), Branch and Bound Solution of a Restricted Least Squares Problem, *Techno. 18*, 447-450.

Beale, E. M. L. (1970), Note on Procedures for Variable Selection in Multiple Regression, *Techno. 12*, 909-914.

Beale, E. M. L. (1974), The Scope of Jordan Elimination in Statistical Computing, *J. Inst. Math. Appl. 10*, 138-140.

Beaton, A. E., and J. L. Barone (1977), More on Computational Accuracy: The Error of Measurement. *Proceedings of the Statistical Computing Section, American Statistical Association*, Washington, D. C., 76-81.

Beaton, A. E., D. B. Rubin and J. L. Barone (1976), The Acceptability of Regression Solutions: Another look at Computational Accuracy, *JASA 71*, 158-168.

Berk, K. N. (1976), Tolerance and Condition in Regression Computations, *Proceedings of the Ninth Interface Symposium on Computer Science and Statistics*, edited by D. Hoaglin and R. Welsch, Prindle, Weber & Schmidt, Boston, 202-203.

Bjorck, A. (1967), Solving Linear Least Squares Problems by Gram-Schmidt Orthogonalization, *BIT 7*, 1-21.

Bjorck, A. (1968), Iterative Refinement of Linear Least Squares Solutions, II, *BIT 8*, 8-30.

Bjorck, A., and G. H. Golub (1967), Iterative Refinement of Linear Least Squares Solutions by Householder Transformations, *BIT 7*, 322-377.

Breaux, H. J. (1968), A Modification of Efroymson's Technique for Stepwise Regression Analysis, *CACM 8*, 556-557.

Businger, P., and G. H. Golub (1965), Linear Least Squares Solution by Householder Transformations, *Numer. Math. 7*, 269-276.

Cady, F. B., and W. A. Fuller (1968), Use of Regression Residuals in the Calculation of Orthogonal Polynomial Coefficients, *Agronomy J. 60*, 195-197.

Chambers, J. M. (1971), Regression Updating, *JASA 66*, 744-748.

Chambers, J. M. (1977), *Computational Methods for Data Analysis*, Wiley, New York.

Chun, D. (1968), A Note on a Regression Transformation for Smaller Round-Off Error, *Techno. 10*, 393-396.

Clayton, D. G. (1971), Algorithm AS46: Gram-Schmidt Orthogonalization, *Appl. Stat. 20*, 335-338.

Daniel, C., and F. S. Wood (1971), *Fitting Equations to Data*, Wiley, New York.

Daniel, J. W., W. B. Gragg, L. Kaufman and G. W. Stewart (1976), Reorthogonalization and Stable Algorithms for Updating the Gram-Schmidt QR Factorization, *Math. Comp. 30*, 772-795.

Draper, N. R., and H. Smith (1966), *Applied Regression Analysis*, Wiley, New York.

Driscoll, E., and I. Francis (1977), Some Measures of Regression Package Performance, *Proceedings of the Statistical Computing Section, American Statistical Association*, Washington, D.C., 190-195.

Efroymson, M. A. (1960), Multiple Regression Analysis, in *Mathematical Methods for Digital Computers*, vol. 1, edited by A. Ralston and H. Wilf, Wiley, New York, 191-203.

Enslein, K., A. Ralston, and H. S. Wilf (1977), *Statistical Methods for Digital Computers*, Wiley, New York.

Farebrother, R. W. (1974), Algorithm AS79: Gram-Schmidt Regression, *Appl. Stat. 23*, 477-478.

Fletcher, R. H. (1975), On the Iterative Refinement of Least Squares Solutions, *JASA 70*, 109-112.

Forsythe, A. B., L. Engleman, R. Jennrich and P. May (1973), A Stopping Rule for Variable Selection in Regression, *JASA 68*, 75-77.

Forsythe, G. E. (1957), Generation and Use of Orthogonal Polynomials for Data Fitting on a Digital Computer, *SIAM J. Appl. Math. 5*, 74-88.

Forsythe, G. E., and C. B. Moler (1967), *Computer Solution of Linear Algebraic Systems*, Prentice-Hall, Englewood Cliffs, N.J.

Frane, J. W. (1977a), A Note on Checking Tolerance in Matrix Inversion and Regression, *Techno. 19*, 513-514.

Frane, J. W. (1977b), Recent Developments in BMDP, *Proceedings of the Statistical Computing Section, American Statistical Association*, Washington, D.C., 212-221.

Frane, J. W. (1978), Methods in BMDP for Dealing with Ill-Conditioned Data: Multicollinearity and Multivariate Outliers, *Proceedings of the Computer Science and Statistics: Eleventh Annual Symposium on the Interface*, edited by A. R. Gallant and T. W. Gerig, Institute of Statistics, North Carolina State University, Raleigh, 209-214.

Freund, R. J. (1963), A Warning of Roundoff Error in Regression, *Amer. Stat. 17*, 13-15.

Furnival, G. M. (1971), All Possible Regressions with Less Computation, *Techno. 13*, 403-408.

Furnival, G. M., and R. M. Wilson, Jr. (1974), Regression by Leaps and Bounds, *Techno. 16*, 499-512.

Gallant, A. R., and T. M. Gerig (1975), Computing Methods for Linear Models Subject to Linear Parametric Constraints, *Proceedings of Computer Science and Statistics: Eighth Annual Symposium on the Interface*, edited by J. W. Frane, Health Sciences Computing Facility, UCLA, 233-237.

Garside, M. J. (1965), The Best Subset in Multiple Regression Analysis, *Appl. Stat. 14*, 196-200.

Garside, M. J. (1971), Some Computational Procedures for the Best Subset Selection Problem, *Appl. Stat. 20*, 8-15.

Gentleman, W. M. (1973), Least Squares Computations by Givens Transformations Without Square Roots, *J. Inst. Math. Appl. 12*, 329-336.

Golub, G. H. (1965), Numerical Methods for Solving Linear Least Squares Problems, *Numer. Math. 7*, 206-216.

Golub, G. H. (1966), Iterative Refinement of Least Squares Solution, *Numer. Math. 9*, 139-148.

Golub, G. H. (1969), Matrix Decomposition and Statistical Calculations, in *Statistical Computation*, edited by R. C. Milton and J. A. Nelder, Academic Press, New York, 365-397.

Golub, G., and V. Klema (1978), Software for Rank Degeneracy, *Proceedings of the Computer Science and Statistics: Eleventh Annual Symposium on the Interface*, edited by A. R. Gallant and T. W. Gerig, Institute of Statistics, North Carolina State University, Raleigh, 79-83.

Golub, G. H., and G. P. H. Styan (1973a), Numerical Computations for Univariate Linear Models, *JSCS 2*, 253-274.

Golub, G. H., and G. P. H. Styan (1973b), Some Aspects of Numerical Computations for Linear Models, *Proceedings of Computer Science and Statistics: Seventh Annual Symposium on the Interface*, edited by W. J. Kennedy, Statistical Laboratory, Iowa State University, Ames, 189-192.

Golub, G. H., and J. H. Wilkinson (1966), Note on the Iterative Refinement of Least Squares Solution, *Numer. Math. 7*, 139-148.

Gorman, J. W., and R. J. Toman (1966), Selection of Variables for Fitting Equations to Data, *Techno. 8*, 27-51.

Grossman, S. I., and G. P. H. Styan (1972), Optimality Properties of Theil's BLUS Residuals, *JASA 67*, 672-673.

Hamaker, H. C. (1962), On Multiple Regression Analysis, *Statistica Neerlandica. 16*, 31-56.

Hanson, R. J. (1975), Stably Updating Mean and Standard Deviation of Data, *CACM 18*, 57-58.

Hanson, R. J., and C. L. Lawson (1969), Extensions and Applications of the Householder Algorithm for Solving Linear Least Squares Problems, *Math. Comp. 23,* 787-812.

Hastings, W. K. (1972), Test Data For Statistical Algorithms: Least Squares and ANOVA, *JASA 67*, 874-879.

Heiberger, R. M. (1975), A Procedure for the Review of Statistical Packages and Its Application to the User Interface with Regression Programs, *Proceedings of Computer Science and Statistics: Eighth Annual Symposium on the Interface,* edited by J. W. Frane, Health Sciences Computing Facility, UCLA, 115-121.

Hemmerle, W. J. (1967), *Statistical Computations on a Digital Computer*, Blaisdell, Waltham, Mass.

Herraman, C. (1968), Algorithm AS12: Sums of Squares and Products Matrix, *Appl. Stat. 17*, 289-292.

Hoaglin, D. C., and R. E. Welsch (1978), The Hat Matrix in Regression and ANOVA, *Amer. Stat. 32*, 17-22.

Hocking, R. R. (1972), Criteria for Selection of a Subset Regression: Which One Should Be Used, *Techno. 14*, 967-970.

Hocking, R. R. (1977), Selection of the Best Subset of Regression Variables, in *Statistical Methods for Digital Computers*, edited by K. Enslein, A. Ralston, and H. S. Wilf, Wiley, New York, 39-57.

Hocking, R. R., and R. N. Leslie (1967), Selection of the Best Subset in Regression Analysis, *Techno. 9*, 531-540.

Jennings, L. S., and M. R. Osborne (1974), A Direct Error Analysis for Least Squares, *Numer. Math. 22*, 325-332.

Jennrich, R. I. (1977), Stepwise Regression, in *Statistical Methods for Digital Computers*, edited by K. Enslein, A. Ralston, and H. S. Wilf, Wiley, New York, 58-75.

Kennard, R. W. (1971), A Note on the C_p Statistic, *Techno. 13*, 899-900.

Kennedy, W. J., and T. A. Bancroft (1971), Model Building for Prediction in Regression Based on Repeated Significance Tests, *AMS 42*, 1273-1284.

Kennedy, W. J., J. E. Gentle, and V. A. Sposito (1977), A Computer Oriented Method for Generating Test Problems for L_1 Regression, *Comm. Stat. (B) 6*, 21-27.

Khuri, A. I. (1976), A Constrained Least-Squares Problem, *Comm. Stat. (B) 5*, 82-84.

LaMotte, L. R., and R. R. Hocking (1970), Computational Efficiency in the Selection of Regression Variables, *Techno. 12*, 83-93.

Larson, H. J., and T. A. Bancroft (1963), Sequential Model Building for Prediction in Regression Analyses, *AMS 34*, 462-479.

Lawson, C. L. (1975), Practical Interpretation of Singular Value Analysis, *Proceedings of Computer Science and Statistics: Eighth Annual Symposium on the Interface*, edited by J. W. Frane, Health Sciences Computing Facility, UCLA, 188-190.

Lawson, C. L., and R. J. Hanson (1974), *Solving Least Squares Problems*, Prentice-Hall, Englewood Cliffs, N.J.

Ling, R. F. (1974), Comparison of Several Algorithms for Computing Means and Variances, *JASA 69*, 859-866.

Longley, J. (1967), An Appraisal of Least Squares Problems for the Electronic Computer from the Point of View of the User, *JASA 62*, 819-841.

Longley, J. (1975), Normalized Modified Gram-Schmidt Algorithm for the Solution of Linear Least Squares Equations, *Proceedings of Computer Science and Statistics: Eighth Annual Symposium on the Interface*, edited by J. W. Frane, Health Sciences Computing Facility, UCLA, 158-167.

McCabe, G. P., and M. A. Ross (1975), A Stepwise Algorithm for Selecting Regression Variables Using Cost Criteria, *Proceedings of Computer Science and Statistics: Eighth Annual Symposium on the Interface,* edited by J. W. Frane, Health Sciences Computing Facility, UCLA, 228-232.

Maindonald, J. H. (1977), Least Squares Computation Based on the Cholesky Decomposition of the Correlation Matrix, *JSCS 5,* 247-258.

Mallows, C. L. (1964), Choosing Variables in Linear Regression: A Graphical Aid. Paper presented at the Central Regional Meeting of the Institute of Mathematical Statistics, Manhattan, Kansas, May 7-9.

Mallows, C. L. (1973), Some Comments on C_p, *Techno. 15*, 661-676.

Mantel, N. (1969), Restricted Least Squares Regression and Convex Quadratic Programming, *Techno. 11,* 763-773.

Marquardt, D. W. (1970), Generalized Inverses, Ridge Regression, Biased Linear Estimation, and Nonlinear Estimation, *Techno. 12,* 591-612.

Milton, R. C., and J. A. Nelder (eds.) (1969), *Statistical Computation,* Academic Press, New York.

Moler, C. B. (1967), Iterative Refinement in Floating-Point, *JACM 14,* 316-321.

Mullet, G. M., and T. W. Murray (1971), New Method for Examining Rounding Error in Least Squares Computer Programs, *JASA 66,* 496-498.

Neely, P. M. (1966), Comparison of Several Algorithms for Computation of Means, Standard Deviation, and Correlation Coefficients, *CACM 7,* 496-499.

Newton, R. G., and D. J. Spurrell (1967), A Development of Multiple Regression for the Analysis of Routine Data, *Appl. Stat. 16,* 52-64.

Peters, G., and J. H. Wilkinson (1970), The Least Squares Problem and Pseudo-Inverses, *Comp. J. 13,* 309-316.

Plemmons, R. J. (1974), Linear Least Squares by Elimination and MGS, *JACM 21,* 581-585.

Pringle, R. M., and A. A. Rayner (1971), *Generalized Inverse Matrices with Applications to Statistics,* Griffin and Co., London.

Ralston, A. (1965), *A First Course in Numerical Analysis,* McGraw-Hill, New York.

Ralston, A., and H. S. Wilf (1960), *Mathematical Methods for Digital Computers,* Wiley, New York.

Ralston, A., and H. S. Wilf (1967), *Mathematical Methods for Digital Computers,* vol. 2, Wiley, New York.

Rao, C. R. (1962), A Note on a Generalized Inverse of a Matrix with Applications to Problems in Mathematical Statistics, *JRSS (B) 24,* 152-158.

Rao, C. R., and S. K. Mitra (1971), *Generalized Inverse of Matrices and Its Applications,* Wiley, New York.

Rohde, C. A., and J. R. Harvey (1965), Unified Least Squares Analysis, *JASA 60,* 523-527.

Schatzoff, M.,R. Tsao, and S. Fienberg (1968), Efficient Calculation of all Possible Regressions, *Techno. 10,* 769-779.

Seaman, J. (1977), Use of Computer Resources by Statistical Package Regression Programs, *Proceedings of the Statistical Computing Section, American Statistical Association,* Washington, D.C., 352-357.

Searle, S. R. (1971), *Linear Models,* Wiley, New York.

Seber, G. A. F. (1977), *Linear Regression Analysis,* Wiley, New York.

Shampine, L. F. (1975), Discrete Least Squares Polynomial Fits, *CACM 18*, 179-180.

Snee, R. D. (1977), Validation of Regression Models: Methods and Examples, *Techno. 19,* 415-428.

Sposito, V. A. (1975), *Linear and Nonlinear Programming,* Iowa State University Press, Ames.

Stewart, G. W. (1977), On the Perturbation of Pseudo-Inverse, Projections and Linear Least Squares Problems, *SIAM Review 19,* 634-662.

Stewart, G. W. (1978), Orthogonal Transformations in Regression Calculations, *Proceedings of Computer Science and Statistics: Tenth Annual Symposium on the Interface,* edited by D. Hogben, U.S. Government Printing Office, Washington, D.C., 189-190.

Stoer, J. (1971), On the Numerical Solution of Constrained Least Squares Problems, *SIAM J. Num. Anal. 8,* 382-411.

Swindel, B. F., and D. R. Bower (1972), Rounding Errors in the Independent Variables in a General Linear Model, *Techno. 14,* 215-218.

Van Reeken, A. J. (1968), Dealing with Neely's Algorithm, *CACM 3*, 149-150.

Velleman, P. F., and I. E. Allen (1976), The Performance of Package Regression Routines Under Stress: A Preliminary Trial of a Regression Evaluation Method, *Proceedings of the Statistical Computing Section, American Statistical Association,* Washington, D. C., 297-304.

Velleman, P. F., and I. E. Allen (1977), The Handiness of Package Regression Routines, *Proceedings of the Statistical Computing Section, American Statistical Association,* Washington, D.C. 95-101.

Velleman, P. F., and I. Francis (1975), Measuring Statistical Accuracy of Regression Programs, *Proceedings of Computer Science and Statistics: Eighth Annual Symposium on the Interface,* edited by J. Frane, Health Sciences Computing Facility, UCLA, 122-127.

Velleman, P. F., J. Seaman, and I. E. Allen (1977), Evaluating Package Regression Routines, *Proceedings of the Statistical Computing Section, American Statistical Association,* Washington, D.C., 82-91.

Wampler, R. H. (1970), On the Accuracy of Least Squares Computer Programs, *JASA 65*, 549-565.

Wampler, R. H. (1972), Some Recent Developments in Linear Least Squares Computations, *Proceedings of Computer Science and Statistics: Sixth Annual Symposium on the Interface,* edited by M. E. Tarter, Western Periodicals, North Hollywood, Calif., 94-110.

Wampler, R. H. (1978), Test Problems and Test Procedures for Least Squares Algorithms, *Proceedings of the Computer Science and Statistics: Eleventh Annual Symposium on the Interface,* edited by A. R. Gallant and T. M. Gerig, Institute of Statistics, North Carolina State University, Raleigh, 84-90.

Waterman, M. S. (1977), Least Squares with Nonnegative Regression Coefficients, *JSCS 6*, 67-70.

Welford, B. P. (1962), Note on a Method for Calculating Corrected Sums of Squares and Cross Products, *Techno. 4,* 419-420.

Welsch, R. E., and S. C. Peters (1978), Finding Influential Subsets of Data in Regression Models, *Proceedings of the Computer Science and Statistics: Eleventh Annual Symposium on the Interface,* edited by A. R. Gallant and T. M. Gerig, Institute of Statistics, North Carolina State University, Raleigh, 240-244.

Wilkinson, J. H. (1967), The Solution of Ill-Conditioned Linear Equations, in *Mathematical Methods for Digital Computers,* vol. 2, edited by A. Ralston and H. Wilf, Wiley, New York, 65-93.

Wilkinson, J. H., and C. Reinsch (eds.) (1971), *Handbook for Automatic Computation,* vol. 2, *Linear Algebra,* Springer-Verlag, Berlin.

Youngs, E. A., and E. M. Cramer (1971), Some Results Relevant to Choice of Sum and Sum-of-Products Algorithm, *Techno. 13,* 657-666.

9 / COMPUTATIONAL METHODS FOR CLASSIFICATION MODELS

9.1 INTRODUCTION

This chapter deals with computational aspects of the analysis of data derived from designed experiments which are modeled by linear classification models. These models are of the general type considered in Chapter 8, where the X matrix has elements that are 0's and 1's. Although the methods of Chapter 8 can be applied directly to these models, the special nature of the model and the underlying physical situation dictate some differences in analysis which make extensions of the basic regression computational methods desirable. The multiple-regression computational method when suitably extended and applied to models considered in this chapter is often termed *method for analysis of variance of unbalanced data* or *the method of fitting constants*.

Classification models generally arise from situations wherein one believes that certain *factors* possibly influence the values which are observed for some dependent variable y. The word *factor* is normally used instead of *independent variable* to emphasize that we either cannot or do not choose to measure the factor as we do an independent variable. Rather, we divide the factor into categories or levels and refer to the effect of the various levels of the factor on y.

When one or more factors are assumed to be operating at different levels to affect some observable phenomenon, such as yield of crop or score on an examination, observations are normally taken for selected combinations of these levels. This serves to classify derived data and allows us to imagine cells in an n-way classification of factors by level. These cells are usually termed *subclasses*, and it will be important in subsequent work to distinguish between the cases of *equal* and *unequal* numbers of observations in all subclasses. Data with the same number of observations in each subclass will be referred to as *balanced data*. In other cases, including situations in which there are some empty subclasses, we will call the data *unbalanced*. These are the usual definitions made throughout the literature in statistics. In this section we continue giving some basic definitions to set the stage for discussion of computational methods.

9.1.1 Fixed-effects Models

Consider the model

$$y_{ijk} = \mu + a_i + b_j + ab_{ij} + e_{ijk} \tag{9.1}$$

$$(i = 1, 2, \ldots, I), \quad (j = 1, 2, \ldots, J), \quad (k = 1, 2, \ldots, K)$$

The e_{ijk} in this model are random variables which are usually assumed to be independent, normally distributed, and have zero mean and common variance σ^2. The model term a_i designates the effect of the ith level of the factor A on y. Similarly, b_j signifies the effect of the jth level of factor B on y. The effect of *interaction* between levels of *main* effects A and B appears as ab_{ij} in the model. The overall mean population effect is specified as μ. We assume that the various effects in this model are not random variables; hence we have a *fixed-effects* model (random-effects models will be considered in Section 9.4).

Observations made when factors A and B are each at the first level are modeled by (using I = J = 2)

$$y_{11k} = 1 \cdot \mu + 1 \cdot a_1 + 0 \cdot a_2 + 1 \cdot b_1 + 0 \cdot b_2 + 1 \cdot ab_{11} + \cdots + e_{11k}$$

where the (0,1) coefficients are written to emphasize that dummy regression variables are indeed present. The situation envisioned which gives rise to the model (9.1) is a *completely crossed classification*, since every level of factor B is used with every level of factor A. The cells (i,j) in this case contain an equal number (K) of observations. The data specified by this model will be *balanced*.

As a second example, consider the model

$$y_{ijk\ell} = \mu + a_i + b_{ij} + c_k + ac_{ik} + bc_{ijk} + e_{ijk\ell}$$
$$(i = 1, 2, \ldots, I), \quad (j = 1, 2, \ldots, J),$$
$$(k = 1, 2, \ldots, K), \quad (\ell = 1, 2, \ldots, n_{ijk}) \qquad (9.2)$$

Factors A and C are completely crossed in this case, but factor B is *nested* within A. In nested classification the levels of B occur with each level of A. The effect of factor B and its interaction with factor A are not accounted for separately. The model (9.2) does contain a term for the effect on y of interaction between levels of factors B and C, and such interaction is both meaningful and possible. Note also in model (9.2) that there are possibly unequal numbers in the subclasses as indicated by the limit n_{ijk} of the subscript ℓ.

Assuming I = 2, J = 2, K = 2 in model (9.1), the X matrix of dummy variables and $\underline{Y}$ vector, as used in Chapter 8, are

$$\begin{matrix} (y_{ijk}) & & & (\mu)\ (a_1)\ (a_2)\ (b_1)\ (b_2)\ (ab_{11})\ (ab_{12})\ \ldots \text{ etc.} \end{matrix}$$

$$\underline{Y} = \begin{pmatrix} y_{111} \\ y_{112} \\ y_{121} \\ y_{122} \\ y_{211} \\ y_{212} \\ y_{221} \\ y_{222} \end{pmatrix}, \quad X = \begin{pmatrix} 1 & 1 & 0 & 1 & 0 & 1 & 0 & 0 & 0 \\ 1 & 1 & 0 & 1 & 0 & 1 & 0 & 0 & 0 \\ 1 & 1 & 0 & 0 & 1 & 0 & 1 & 0 & 0 \\ 1 & 1 & 0 & 0 & 1 & 0 & 1 & 0 & 0 \\ 1 & 0 & 1 & 1 & 0 & 0 & 0 & 1 & 0 \\ 1 & 0 & 1 & 1 & 0 & 0 & 0 & 1 & 0 \\ 1 & 0 & 1 & 0 & 1 & 0 & 0 & 0 & 1 \\ 1 & 0 & 1 & 0 & 1 & 0 & 0 & 0 & 1 \end{pmatrix}$$

(9.3)

The X matrix is, of course, rank-deficient. As in previous chapters, we denote the rank of X by r. This is the usual situation when dealing with models of this type. The term *design matrix* is frequently used to describe an X matrix which arises from a classification model.

The rank-deficiency of design matrices means that there is not a unique least-squares solution $\underline{b}$ for the model

$$\underline{Y} = X\underline{\beta} + \underline{e} \tag{9.4}$$

All solutions can be obtained in the form

$$\underline{b} = (X'X)^{-}X'\underline{Y} \tag{9.5}$$

by selecting all possible g_1 inverses, but this is not of much interest. As a random variable, $\underline{b}$ is such that

$$\begin{aligned} E(\underline{b}) &= (X'X)^{-}X'X\underline{\beta} \\ \mathrm{Var}(\underline{b}) &= (X'X)^{-}X'X[(X'X)^{-}]'\sigma^2 \end{aligned} \tag{9.6}$$

The key aspects of what is otherwise an unsatisfactory situation with regard to multiple solutions are that $X\underline{b}$ is *unique* as $\underline{b}$ ranges over solutions (9.5), and the matrix $X(X'X)^{-}X'$ is invariant to $(X'X)^{-}$. Therefore, predicted values $\hat{\underline{Y}}$ and residuals $\underline{Y} - \hat{\underline{Y}}$ are the same for all solutions $\underline{b}$. Also, linear functions of parameters, described as $L'\underline{\beta}$, have a unique estimate $L'\underline{b}$ whenever each row of L' is a linear combination of the rows of X, that is, whenever $L'\underline{\beta}$ consists of functions of $\underline{\beta}$ which are *estimable*. There are obviously infinitely many estimable functions of $\underline{\beta}$, but there can be no more than r in a *linearly independent set* since X has rank r. We will be primarily interested in linearly independent estimable functions, and tests of hypotheses involving such functions, in this chapter. Any hypothesis which is composed entirely of estimable functions is called *testable*. Nontestable hypotheses will not be considered in subsequent discussion.

9.1.2 Restriction Models and Constraints on Solutions

The models given as examples in Section 9.1.1 contained parameters for which no restrictions were specified. Sometimes it is meaningful to include, as part of the model, restrictions on the parameters, in which case the model is called a *restricted model*. An example of restrictions which might be included with the model (9.1) to form a new model are

$$\sum_i a_i = 0, \quad \sum_j b_j = 0, \quad \sum_i ab_{ij} = 0 \quad \text{for all } j$$
$$\sum_j ab_{ij} = 0 \quad \text{for all } i \tag{9.7}$$

In other words, we wish to view the various effects as deviations from mean effects. In this chapter when restrictions on the parameters are included, the restrictions will be nonestimable linear functions of the parameters. It is easily shown that estimable functions with respect to an unrestricted model remain estimable when such nonestimable restrictions are imposed. Although the form of the estimable functions may be altered under the restrictions, the estimates themselves are not changed. It is best to think of restrictions as having very little to do with the computing problems that we will discuss, though they obviously will to some extent dictate the quantities which we wish to compute. Most authors call the imposition of restrictions on a hitherto unrestricted model a *reparametrization* because inclusion of restrictions changes the model, and hence redefines the parameters. Sometimes a restricted model has full rank when the unrestricted model did not. We will not dwell on this point because computationally we are able to handle rank-deficient models in much the same way as full-rank models.

An unrelated idea, that of obtaining a solution subject to given constraints, is also applied in non-full-rank models such as the ones we are considering. Given either a restricted or unrestricted model, it is often desirable to use constraints to simplify the computing problem. It must be emphasized, however, that such constraints are used simply to select a solution $\underline{b}$ from what may be an infinity of solutions. Such constraints do not restrict the model or affect estimable functions, for example. The constraints must be nonestimable functions.

Sometimes, as was mentioned before, a restricted model is in fact full-rank (all parameters estimable). Even in these cases we sometimes use constraints to illuminate the situation and lead us quickly to the solution.

In many instances there is the possibility of confusion when dealing with a restricted model because the constraints imposed to obtain a solution will have the same general algebraic form as the restrictions. In these cases it is easy to lose sight of the fact that the model is indeed restricted, and this leads to confusion about the specific form of estimable functions. We will take care, in the remainder of this chapter, to avoid confusion on this score. Kempthorne (1952), Bancroft (1968), and Searle (1971) discuss restrictions on models and solutions under constraints. These references also provide many worked examples. Hocking and Speed (1975) discuss sources of confusion between restrictions and constraints.

9.1.3 Reductions in Sums of Squares

The models we are working with in this chapter are such that their parameters fall into meaningful and readily identifiable subsets (e.g., the "A effects," "B effects," etc.). We will be interested in processing submodels of the overall model or submodels which contain selected subgroups of the parameters. Description of the various submodels, and test statistics generated through their use, can be simplified by the use of the $R(\cdot)$ notation. In all that follows $R(\cdot)$ will stand for the reduction in total sum of squares due to fitting a submodel. The submodel itself will be designated by quantities shown inside the parentheses. For example, $R(\mu, a)$ will be the reduction in total sum of squares due to fitting

$$y_{ij} = \mu + a_i + e_{ij}$$

Differences in reductions are often needed to provide sums of squares for making desired tests of hypotheses. This leads to a natural extension of the $R(\cdot)$ notation to a form $R(\,|\,)$. The "bar" inside the parentheses will signify that a difference in reduction is being given. For example, $R(a|\mu)$ will describe

$$R(\mu,a) - R(\mu)$$

and

$$R(ab|\mu,a,b) = R(\mu,a,b,ab) - R(\mu,a,b)$$

Thus we can think of $R(a|\mu)$ as the additional reduction in total sum of squares due to fitting μ and a above the reduction obtained by fitting just μ.

9.1.4 An Example

We now consider an example to illustrate aspects of the discussion given previously in this section. The model is

$$y_{ij} = \mu + a_i + e_{ij} \qquad (i = 1, 2, 3), \quad (j = 1, 2) \tag{9.8}$$

For this model the design matrix and $\underline{\beta}$ and $\underline{Y}$ vectors have the form

$$\underline{Y} = \begin{pmatrix} y_{11} \\ y_{12} \\ y_{21} \\ y_{22} \\ y_{31} \\ y_{32} \end{pmatrix} \quad X = \begin{pmatrix} 1 & 1 & 0 & 0 \\ 1 & 1 & 0 & 0 \\ 1 & 0 & 1 & 0 \\ 1 & 0 & 1 & 0 \\ 1 & 0 & 0 & 1 \\ 1 & 0 & 0 & 1 \end{pmatrix} \quad \underline{\beta} = \begin{pmatrix} \mu \\ a_1 \\ a_2 \\ a_3 \end{pmatrix}$$

Normal equations in this case are

$$\begin{pmatrix} 6 & 2 & 2 & 2 \\ 2 & 2 & 0 & 0 \\ 2 & 0 & 2 & 0 \\ 2 & 0 & 0 & 2 \end{pmatrix} \begin{pmatrix} \mu \\ a_1 \\ a_2 \\ a_3 \end{pmatrix} = \begin{pmatrix} Y_{..} \\ Y_{1.} \\ Y_{2.} \\ Y_{3.} \end{pmatrix} \tag{9.9}$$

where $Y_{i\cdot} = \sum_j y_{ij}$ and $Y_{\cdot\cdot} = \sum_i \sum_j y_{ij}$. Suppose we utilize constraints of the form

$$\hat{a}_1 + \hat{a}_2 + \hat{a}_3 = 0$$

Then, for any solution which satisfies the constraints, a set of equations which we can solve in order to obtain an overall solution is

$$\begin{aligned} 6\hat{\mu} \qquad &= Y_{\cdot\cdot} \\ 2\hat{\mu} + 2\hat{a}_1 &= Y_{1\cdot} \\ 2\hat{\mu} + 2\hat{a}_2 &= Y_{2\cdot} \end{aligned} \tag{9.10}$$

and we can now easily select the solution

$$\begin{aligned} \hat{\mu} &= \frac{Y_{\cdot\cdot}}{6} &&= y_{\cdot\cdot} \\ \hat{a}_1 & &&= y_{1\cdot} - y_{\cdot\cdot} \\ \hat{a}_2 & &&= y_{2\cdot} - y_{\cdot\cdot} \\ \hat{a}_3 &= -\hat{a}_1 - \hat{a}_2 &&= y_{3\cdot} - y_{\cdot\cdot} \end{aligned} \tag{9.11}$$

because the linear system (9.10) has a unique solution. (We are denoting means by lower-case letters with dots in place of the appropriate subscripts.)

Thinking in terms of estimable functions it is easily verified, by looking at the design matrix, that $\mu + a_i$ is estimable for each i, but μ by itself is not, nor are the a_i estimable. Since linear combinations of estimable functions are estimable, it follows that $\mu + (\sum_i a_i)/3$ is estimable as is $a_i - a_j$ for $i \neq j$. The best linear unbiased estimate of the latter quantity, for example, is $(y_{i\cdot} - y_{j\cdot})$.

A standard analysis of variance table in this example will appear as

Source	d.f.	Sum of Squares	Mean Square	F
Mean	1	$R(\mu)$	$R(\mu)$	$3R(\mu)/RSS$
A	2	$R(a\|\mu)$	$R(a\|\mu)/2$	$3R(a\|\mu)/(2RSS)$
Error	3	$RSS = TSS - R(\mu,a)$	$RSS/3$	
Total:	6	$TSS = \sum_i \sum_j y_{ij}^2$		

Another useful fact is that the constraints used above can be made to reflect not only in the normal equations but in the design matrix itself. Noting that $\hat{a}_3 = -\hat{a}_1 - \hat{a}_2$, the constraints can be substituted into the model equations and an X matrix in the following form will result:

$$\begin{pmatrix} 1 & 1 & 0 \\ 1 & 1 & 0 \\ 1 & 0 & 1 \\ 1 & 0 & 1 \\ 1 & -1 & -1 \\ 1 & -1 & -1 \end{pmatrix} \quad \text{and} \quad \begin{pmatrix} \mu \\ a_1 \\ a_2 \end{pmatrix} \tag{9.12}$$

We call this transformed X matrix a *reduced X matrix*. The so-called *reduced normal equations* based on this reduced X matrix are

$$\begin{pmatrix} 6 & 0 & 0 \\ 0 & 4 & 2 \\ 0 & 2 & 4 \end{pmatrix} \begin{pmatrix} \mu \\ a_1 \\ a_2 \end{pmatrix} = \begin{pmatrix} Y_{..} \\ Y_{1.} - Y_{3.} \\ Y_{2.} - Y_{3.} \end{pmatrix}$$

Solution of these equations gives the results (9.11). It is obvious from the presentation in Chapter 8 that it is not necessary to form reduced normal equations. The solution can be obtained using orthogonal triangularization of the reduced X matrix if we wish to proceed that way.

The model (9.8) is an unrestricted model. If we add restrictions $a_1 + a_2 + a_3 = 0$ to obtain a restricted model, then there are some changes with regard to estimable functions. First, the new μ becomes estimable, since $\mu + (\sum a_i)/3$ in the unrestricted model reduces to μ with this restriction. Also, it follows from the estimability of $\mu + a_i$ that the a_i themselves will now be estimable. Thus, under the restricted model the solution (9.11) has added significance because it now gives values for estimable functions, and these we know are invariant over solution to the Eqs. (9.9).

Computer programs often output whatever solution is obtained from the normal equations, and the generalized inverse is used to produce the solution. The user must then decide how he wants to interpret these quantities, and such decisions sometimes involve deciding whether the model does include restrictions.

9.2 THE SPECIAL CASE OF BALANCE AND COMPLETENESS FOR FIXED-EFFECT MODELS

In this section we assume that the form of model which is to be utilized reflects completely crossed and/or nested factors and meaningful interactions. Also, we assume that the data on hand are balanced (equal numbers of observations in the subclasses). A general and efficient method for computing the usual quantities needed in analysis of variance will be given. This method stems from the work of many people, including Yates (1934), Hartley (1956, 1962), Hemmerle (1964), Schlater (1965), and Schlater and Hemmerle (1966). It should be emphasized, however, that the assumption is that the data are balanced and complete.

It is normally the case for models of this type that the "usual restrictions" are included with the model and constraints in the same form are employed when obtaining desired solutions. We will adopt this form and describe the "usual restrictions" below. Computations are greatly simplified in this situation. A few simple rules are, as we shall see, all that are needed to have a sound basis for an efficient algorithm. One reason this is the case is because the restricted model is actually full-rank.

The following subsection contains a definition of the "usual restrictions" and provides the basic rules which are followed in computations of many quantities needed for data analysis.

9.2.1 Basic Definitions and Considerations

Further discussion will be facilitated by defining the terms *associated* and *floating* with respect to subscripts on classification model terms. In the models with which we shall deal, one and only one subscript will be *associated* with each main effect symbol and the error symbol. In any given subscripted model term (except e) all subscripts which are present in the model term, and are not associated with a main effect symbol in that term, are called *floating* subscripts. These definitions are made by Hemmerle (1964). For example, in the model

$$y_{ijk\ell m} = \mu + a_i + b_{ij} + c_{ijk} + d_\ell + ad_{i\ell} + bd_{ij\ell} + e_{ijk\ell m} \tag{9.13}$$

the associated and floating subscripts in each model term are shown in Table 9.1.

Table 9.1 Associated and Floating Subscripts in Model (9.13)

Model Term	Associated Subscripts	Floating Subscripts
a_i	i	None
b_{ij}	j	i
c_{ijk}	k	i,j
d_ℓ	ℓ	None
$ad_{i\ell}$	i,ℓ	None
$bd_{ij\ell}$	j,ℓ	i
$e_{ijk\ell m}$	m	i,j,k,ℓ

Using these definitions, and ignoring the error term for now, the *usual restrictions* are obtained by summing each model term over its associated subscripts one at a time and equating each sum to zero. In the model (9.13) the usual restrictions which will be included are

$$\sum_i a_i = \sum_j b_{ij} = \sum_k c_{ijk} = \sum_\ell d_\ell = \sum_i ad_{i\ell} = \sum_\ell ad_{i\ell}$$
$$= \sum_j bd_{ij\ell} = \sum_\ell bd_{ij\ell} = 0 \qquad (9.14)$$

Constraints on the solution vector in a form defined by (9.14) will be the *usual constraints*.

One consequence of using these constraints is that the coefficient matrix in the reduced normal equations is block-diagonal. The blocks correspond to model terms, and the row and column size of each block is the number of degrees of freedom associated with that effect. Thus the columns of the reduced X matrix which correspond to different effects are orthogonal. This is another reason why we can derive especially simple methods for handling the special case considered in this section.

We are now in a position to state the three key rules utilized to obtain estimates of parameters and degrees of freedom and sums of squares in the AOV table. The notation used in stating these rules includes dots in the place of some subscripts on y to indicate an averaging over that subscript. For example

$$y_{\bullet jk} = \frac{1}{I} \sum_i y_{ijk}$$

where I is the upper limit of the i subscript values. These means will be called *partial means*.

Rule 1: For any given model term let F denote the set of floating subscripts and A denote the set of associated subscripts. The *degrees of freedom* for the model term is

$$\left(\prod_{i \varepsilon F} L_i\right)\left[\prod_{j \varepsilon A} (L_j - 1)\right]$$

where L_k denotes the upper limit of the kth subscript in the model.

Rule 2: Estimates of the parameters in a model term are specified as linear combinations of partial means using the following device. Expand symbolically the product

$$\left(\prod_{i \varepsilon F} i\right)\left[\prod_{j \varepsilon A} (j - 1)\right]$$

In each term of this symbolic expansion insert a dot in the location of each subscript that appears in the model but not in the term of the symbolic expansion. (A one in the expansion specifies a dot in every position.) The parameters are now estimated as the linear combination of partial means which has means designated by terms in the symbolic expansion and coefficients defined by corresponding coefficients in the symbolic expansion. (The expansion ij - i yields ij• - i•• which specifies $y_{ij\cdot} - y_{i\cdot\cdot}$ in a three-subscript model.)

Rule 3: The *sum of squares* relative to a model term m is obtained by summing the squares of estimates of the parameters for that term and multiplying this sum of squares by the product p of the limits of all subscripts in the model which do not appear in the model term m.

The sum of squares of rule 3 is, of course, an additional reduction in total sum of squares which might be $R(a|\mu, b, ab)$. However, in view of the orthogonality present in this situation, we have $R(a) = R(a|\mu, b, ab)$. The following example will illustrate use of these rules.

Example 9.1

Consider the following restricted model:

$$y_{ijk\ell} = \mu + a_i + b_j + c_k + ab_{ij} + abc_{ijk} + e_{ijk\ell}$$

$$(i = 1, 2, \ldots, I) \quad (j = 1, 2, \ldots, J)$$

$$(k = 1, 2, \ldots, K) \quad (\ell = 1, 2, \ldots, L)$$

$$\sum_i a_i = \sum_j b_j = \sum_k c_k = \sum_i ab_{ij} = \sum_j ab_{ij} = \sum_i abc_{ijk}$$

$$= \sum_j abc_{ijk} = \sum_k abc_{ijk} = 0 \quad \text{for all } i, j, k$$

Parameters are estimated, using rule 2, by

$$\hat{\mu} = y_{....}$$

$$\hat{a}_i = y_{i...} - y_{....} \qquad (i = 1, 2, \ldots, I)$$

$$\hat{b}_j = y_{.j..} - y_{....} \qquad (j = 1, 2, \ldots, J)$$

$$\hat{c}_k = y_{..k.} - y_{....} \qquad (k = 1, 2, \ldots, K)$$

$$\widehat{ab}_{ij} = y_{ij..} - y_{i...} - y_{.j..} + y_{....}$$

$$\widehat{abc}_{ijk} = y_{ijk.} - y_{ij..} - y_{i.k.} - y_{.jk.} + y_{i...} + y_{.j..} + y_{..k.} - y_{....}$$

Sums of squares and degrees of freedom are specified, using rules 1 and 3, as follows:

Source	d.f.	SS
A	$I - 1$	$JKL \sum_i (\hat{a}_i)^2$
B	$J - 1$	$IKL \sum_j (\hat{b}_j)^2$
C	$K - 1$	$IJL \sum_k (\hat{c}_k)^2$
AB	$(I - 1)(J - 1)$	$KL \sum_i \sum_j (\widehat{ab}_{ij})^2$
ABC	$(I - 1)(J - 1)(K - 1)$	$L \sum_i \sum_j \sum_k (\widehat{abc}_{ijk})^2$
Error	$IJK(L - 1)$	$\sum_i \sum_j \sum_k \sum_\ell (y_{ijk\ell} - y_{ijk.})^2$
Total (corrected):	$IJKL - 1$	$\sum_i \sum_j \sum_k \sum_\ell (y_{ijk\ell} - y_{....})^2$

What has been done thus far in this subsection is to specify, in general terms, the standard form for quantities used in AOV as they appear in most textbooks. It is evident that normal equations are not explicitly formed. All that we needed for computing purposes are the specification of the model and the subarrays of partial means which are computed from the observations.

Thus an efficient and stable algorithm can obviously be developed based on the results given above if adequate attention is given to forming partial means and linear combinations of partial means accurately. The overall algorithm will contain major subalgorithms to perform tasks such as forming arrays of partial means, taking combinations of these to produce estimates of parameters, computing degrees of freedom, and forming sums of squares. Before providing any details on these topics, let us briefly consider the computations specified above, but carried out using matrices.

If we use a matrix approach, instead of the rules, to obtain sums of squares, the matrix expression relative to a given model term m is, using the reduced X matrix,

$$SS_m = \underline{Y}'[X(X'X)^{-1}X' - X_1(X_1'X_1)^{-1}X_1']\underline{Y}$$

$$= \underline{Y}'M_1\underline{Y}$$

where X_1 is obtained from X by deleting all columns which correspond to m. The fact that M_1 is idempotent allows expression of SS_m as

$$SS_m = (M_1\underline{Y})'(M_1\underline{Y})$$

The components of $M_1\underline{Y}$, when squared and summed, give the results obtained in application of rule 3. $M_1\underline{Y}$ is made up of p duplicate sets of estimates of the model term parameters. The number p is the product of limits of subscripts specified in rule 3. Hemmerle (1967) applies the name *residuals for model term m* to the unique elements of $M_1\underline{Y}$. The name *error residuals* is given to the vector

$$M_2\underline{Y} = [I - X(X'X)^{-1}X']\underline{Y}$$

The error sum of squares is $(M_2\underline{Y})'(M_2\underline{Y})$, and M_2 is idempotent.

9.2.2 Computer-related Considerations in the Special Case

A general algorithm for analysis of variance and covariance computations, for the special case considered in this section, is given by Schlater (1965) and Schlater and Hemmerle (1966). This algorithm is fairly extensive and includes such things as formation of variance components and pooling options. We will not attempt to provide details of this algorithm. Rather, in this subsection, a brief discussion of selected aspects of the algorithm will be provided. The algorithm, in its entirety or in part, has been implemented by several different groups of programmers and it does seem to provide consistently accurate results.

Computer software must include the ability to accept user specification of the model. Most programs in use today allow the user to provide some rather natural algebraic specification. The program then deciphers the specification and translates it into numeric coding for subsequent use. There are no established standards for doing this, but many techniques used in compiler construction can be applied to this problem.

Given the problem specification, another task which must be performed is to compute arrays of partial means which are used to produce model term residuals. One approach to this computation is to initially form all possible partial means. This avoids having to reform some partial means later, but it is inefficient in some cases because all partial means are not used in forming model term residuals for every given model. Let us assume that all possible partial means are desired and consider a computer-program-oriented method for obtaining them.

The specified model has (say) t subscripts, so we must form $2^t - 1$ sets of partial means to include with the observations. All of these values will be stored in a one-dimensional array F, with the observations appearing first in the array sorted in the order depicted in Section 9.1.4. Each set of partial means will form a subarray which is stored in consecutive locations in F. The length of each subarray will be stored in a one-dimensional array T. Since the observations appear first in F, T(1) will hold the value $\prod_{i=1}^{t} L_i$, where L_i is the limit of the ith subscript in the model.

Computations will proceed in stages. The first stage stores the observations in F and is called stage 0. Stage 1 deals with the last subscript in the model, stage 2 with next to the last subscript, and so on. Within stage $i > 0$, a total of 2^{i-1} subarrays of partial means will be formed. We will number subarrays of means according to the order in which they are formed. The observations are subarray number 1.

To carry out the computations needed to complete stage $i > 0$ we must form 2^{i-1} subarrays of means and 2^{i-1} values to store in the T array. In the preceding $i - 1$ stages we have created $2^{i-1} - 1$ subarrays of means. Make a 1 to 1 correspondence, in natural order, between the subarrays previously created and those to be created. Having done this, initialize the 2^{i-1} new positions in T by dividing each of the previous 2^{i-1} values in T by L_{t-i+1} which is the limit of the subscript corresponding to stage i. Each subarray of means within stage i will be formed as the sum of M_i elements from its corresponding previously created subarray, that is, element k of subarray $2^{i-1} + j$ $(1 \leq j \leq 2^{i-1})$ is formed by summing appropriate elements of subarray j and then dividing by M_i. The question we must now consider is which elements of subarray j should be used to form the entries in subarray $2^{i-1} + j$.

The techniques used in forming the means which make up subarray $2^{i-1} + j$ proceeds as follows. Conceptually divide subarray j into g_i groups, where g_i is the product of the limits of all subscripts to to the left of that subscript which corresponds to stage i. If $i = t$, then $g_i = 1$. The interval of summation within groups is $I = T(2^{i-1} + j)/g_i$. We now proceed to form a sum by starting with the first element in group 1 and summing M_i elements of subarray j taking elements at interval I. This sum is divided by M_i and becomes the first mean in subarray $2^{i-1} + j$. Next, begin with the second element of subarray j, sum at intervals I until M_i terms are included (this should exhaust the candidate terms in group 1), and divide by M_i to obtain the second element in subarray $2^{i-1} + j$ which is stored in F. We will be able to form I sums for each group, and we treat each group in turn using the same procedure. It is easy to check that $I \cdot g_i$ is $T(2^{i-1} + j)$. Table 9.2 illustrates these computations when the given model contains four subscripts.

Table 9.2 Forming Arrays of Means for the Case of Four Subscripts

Subarray no.	Stage	Means	Value in T array	No. terms in sum	Interval	Subarray to sum over	No. of groups
1	Input	$y_{ijk\ell}$	IJKL				
2	1(ℓ)	$y_{ijk\cdot}$	IJK	L	1	1	IJK
3	2(k)	$y_{ij\cdot\ell}$	IJL	K	L	1	IJ
4		$y_{ij\cdot\cdot}$	IJ	K	1	2	IJ
5	3(j)	$y_{i\cdot k\ell}$	IKL	J	KL	1	I
6		$y_{i\cdot k\cdot}$	IK	J	K	2	I
7		$y_{i\cdot\cdot\ell}$	IL	J	L	3	I
8		$y_{i\cdot\cdot\cdot}$	I	J	1	4	I
9	4(i)	$y_{\cdot jk\ell}$	JKL	I	JKL	1	1
10		$y_{\cdot jk\cdot}$	JK	I	JK	2	1
11		$y_{\cdot j\cdot\ell}$	JL	I	JL	3	1
12		$y_{\cdot j\cdot\cdot}$	J	I	J	4	1
13		$y_{\cdot\cdot k\ell}$	KL	I	KL	5	1
14		$y_{\cdot\cdot k\cdot}$	K	I	K	6	1
15		$y_{\cdot\cdot\cdot\ell}$	L	I	L	7	1
16		$y_{\cdot\cdot\cdot\cdot}$	1	I	1	8	1

This, of course, is an outline of one algorithm which might be included as a component in a major algorithm for AOV. The other tasks, such as forming degrees of freedom and model term residuals, will need to be developed in algorithmic form before a finished program is produced. It is easy to see from this that even moderately general-purpose computer software is not easy or inexpensive to produce.

Additional details relative to an overall algorithm will not be given. There are too many details, and subalgorithms can be structured in several different ways. Instead, we will move back to a lower level of detail and consider how covariates can be processed using the method outlined thus far in AOV.

9.2.3 Analysis of Covariance

In this subsection we consider an extension of the fixed-effects models to ones which include concomitant variables. In matrix notation models with covariates have the form

$$\underline{Y} = X\underline{\beta} + Z\underline{\delta} + \underline{e} \tag{9.15}$$

where $X\underline{\beta}$ is the classification part and $Z\underline{\delta}$ is the so-called continuous part of the model. We will assume that the columns of Z are linearly independent and that the usual restrictions are applied to the classification part of the model. Also, the usual constraints will be utilized and a reduced X matrix considered in computations.

The discussion will concentrate on finding the usual test statistics for testing significance of effects in the model. To this end, consider the reduced normal equations

$$\begin{pmatrix} X'X & X'Z \\ Z'X & Z'Z \end{pmatrix} \begin{pmatrix} \hat{\underline{\beta}} \\ \hat{\underline{\delta}} \end{pmatrix} = \begin{pmatrix} X'\underline{Y} \\ Z'\underline{Y} \end{pmatrix}$$

A test of significance for any selected group of parameters in the classification part of the model has the general form

$$\frac{\left\{\left[\underline{Y}'(X|Z)\begin{pmatrix} \hat{\underline{\beta}} \\ \hat{\underline{\delta}} \end{pmatrix} - \underline{Y}'(X_1|Z)\begin{pmatrix} \hat{\underline{\beta}}_1 \\ \hat{\underline{\delta}}_1 \end{pmatrix}\right] /df_1\right\}}{\{(\underline{Y} - X\hat{\underline{\beta}} - Z\hat{\underline{\delta}})'(\underline{Y} - X\hat{\underline{\beta}} - Z\hat{\underline{\delta}})/df_2\}} \tag{9.16}$$

It will be shown that we can take advantage of the orthogonality in the X matrix and derive an efficient procedure for computing the quantities required to form (9.16). This procedure is described as follows.

Consider two subsets of the normal equations, namely

$$X'X\underline{\hat{\beta}} + X'Z\underline{\hat{\delta}} = X'\underline{Y}$$

and

$$Z'X\underline{\hat{\beta}} + Z'Z\underline{\hat{\delta}} = Z'\underline{Y}$$

The first set of equations yields

$$\underline{\hat{\beta}} = (X'X)^{-1}X'(\underline{Y} - Z\underline{\hat{\delta}})$$

Substituting this solution vector into the second set of equations gives

$$Z'[I - X(X'X)^{-1}X']Z\underline{\hat{\delta}} = Z'[I - X(X'X)^{-1}X']\underline{Y}$$

which is in the form

$$Z'M_2Z\underline{\hat{\delta}} = Z'M_2\underline{Y}$$

or

$$(M_2Z)'(M_2Z)\underline{\hat{\delta}} = (M_2Z)'\underline{Y} \tag{9.17}$$

Note that the matrix M_2 is the same one that operates on $\underline{Y}$ to produce error residuals $M_2\underline{Y}$. Here the operator M_2 will transform the individual columns of the Z matrix using the same operations as those effected on the $\underline{Y}$ vector. For this reason the equations which comprise the linear system (9.17) are called the *error normal equations*.

Using the $\underline{\hat{\delta}}$ derived from Eq. (9.17), we obtain $\underline{\hat{\beta}}$ as

$$\underline{\hat{\beta}} = (X'X)^{-1}X'(\underline{Y} - Z\underline{\hat{\delta}}) \tag{9.18}$$

The error sum of squares for the full model is now expressible as

$$\begin{aligned}(\underline{Y} - X\underline{\hat{\beta}} - Z\underline{\hat{\delta}})'(\underline{Y} - X\underline{\hat{\beta}} - Z\underline{\hat{\delta}}) &= [(I - X(X'X)^{-1}X')(\underline{Y} - Z\underline{\hat{\delta}}]' \\ &\quad [(I - X(X'X)^{-1}X')(\underline{Y} - Z\underline{\hat{\delta}})] \\ &= [M_2(\underline{Y} - Z\underline{\hat{\delta}})]'[M_2(\underline{Y} - Z\underline{\hat{\delta}}] \\ &= (M_2\underline{Y})'(M_2\underline{Y}) - \underline{\hat{\delta}}'Z'M_2\underline{Y} \qquad (9.19)\end{aligned}$$

Now suppose that we omit some parameters in the full model and form the appropriate reduced X matrix, call it X_1. The reduced model is

$$\underline{Y} = X_1\underline{\beta}_1 + Z\underline{\delta}_1 + \underline{e}$$

Proceeding as before, the first subset of the normal equations for the reduced model yields

$$\hat{\underline{\beta}}_1 = (X_1'X_1)^{-1}X_1'(\underline{Y} - Z\hat{\underline{\delta}}_1) \tag{9.20}$$

Substitution into the remaining equations gives

$$Z'[M_2 + M_1]Z\hat{\underline{\delta}}_1 = Z'[M_2 + M_1]\underline{Y}$$

or

$$[(M_2Z)'(M_2Z) + (M_1Z)'(M_1Z)]\hat{\underline{\delta}}_1 = [(M_2Z)' + (M_1Z)']\underline{Y} \tag{9.21}$$

These equations are now solved for $\hat{\underline{\delta}}_1$. Substitution into (9.20) will then give $\hat{\underline{\beta}}_1$. The reader should note in particular how Eq. (9.21) can be obtained by modifying Eqs. (9.17).

We are now able to see how the computations for analysis of covariance can be performed in an efficient manner. Our objective, for the purpose of illustration, is to obtain the test statistic (9.16) which is stated in general terms so that it applies to testing the significance of any effect in the model.

The denominator of (9.16) contains the sum of squares for error. Inspecting Eq. (9.19) we see that the error sum of squares can be formed in the following way:

1. Find the error sum of squares due to analysis of $\underline{Y}$ ignoring the $Z\underline{\delta}$ portion of the model. This sum of squares is $(M_2\underline{Y})'(M_2\underline{Y})$.
2. Solve the error normal equations

$$(M_2Z)'(M_2Z)\hat{\underline{\delta}} = (M_2Z)'\underline{Y} \tag{9.22}$$

   which is the right member of Eq. (9.19)

The error normal equations contain only as many equations as there are concomitant variates in the model, and this number is usually small. Thus, relative to directly solving normal equations, we will save time and effort using this method of finding the error sum of squares so long as $(M_2\underline{Y})'(M_2\underline{Y})$ is easily obtained and the error normal equations are easily formed. Neither of these quantities is difficult to obtain. The error residuals $M_2\underline{Y}$ can be formed using rule 2. This same rule can be applied successively to the columns of Z to produce M_2Z, from which the error normal equations can be formed. Thus the overall error sum of squares is easily obtainable using the procedure described above.

Next consider the numerator of (9.16). The sum of squares in the numerator can be written as

$$\begin{aligned}
\underline{Y}'X\hat{\underline{\beta}} - \underline{Y}'X_1\hat{\underline{\beta}}_1 + \underline{Y}'Z\hat{\underline{\delta}} - \underline{Y}'Z\hat{\underline{\delta}}_1 &= \underline{Y}'X[(X'X)^{-1}X'\underline{Y} - (X'X)^{-1}X'Z\hat{\underline{\delta}}] \\
&\quad - \underline{Y}'X_1[(X_1'X_1)^{-1}X_1'\underline{Y} - (X_1'X_1)^{-1}X_1'Z\hat{\underline{\delta}}_1] \\
&\quad + \underline{Y}'Z\hat{\underline{\delta}} - \underline{Y}'Z\hat{\underline{\delta}}_1 \\
&= (M_1\underline{Y})'(M_1\underline{Y}) + \hat{\underline{\delta}}'Z'M_2\underline{Y} \\
&\quad - \underline{Y}'[I - X_1(X_1'X_1)^{-1}X_1']X\hat{\underline{\delta}}_1 \\
&= (M_1\underline{Y})'(M_1\underline{Y}) + \hat{\underline{\delta}}'Z'M_2\underline{Y} \\
&\quad - \hat{\underline{\delta}}_1'Z'(M_1 + M_2)'\underline{Y} \qquad (9.23)
\end{aligned}$$

Consider the three terms in the right member of Eq. (9.23). The first term is the sum of squares that is used in making the test whenever there are no covariates in the model. The second term is the unadjusted error sum of squares $(M_2\underline{Y})'(M_2\underline{Y})$ minus the adjusted error sum of squares given in the right member of (9.19). The third term is the sum of squares due to regression obtained through solution of the normal type Eqs. (9.21)

Inspection of the linear system (9.21) shows that it can be obtained from the error normal equations by adding the matrix $(M_1Z)'(M_1Z)$ to the coefficient matrix of the error normal equations and the vector $(M_1Z)'\underline{Y}$ to the right-hand-side vector. The matrix M_1 operates on the columns of Z in the same way as it operates on $\underline{Y}$ to form the vector of model term residuals. Thus the matrix $(M_1Z)'(M_1Z)$ is a matrix of sums of squares and cross-products of model term residuals. These residuals can be obtained by treating each column of Z, in turn, as a vector of dependent variate values and using the rules for forming residuals and sums of squares.

To summarize, the previous discussion shows that we can perform analyses of covariance by performing an analysis of variance on the variate y and also on each covariate separately, and then use the model term and error residuals to set up systems in the form of (9.21) for the model terms and (9.17) for error. The reduction in sums of squares obtained through solution of these linear systems are used to adjust the sums of squares obtained in the analysis of $\underline{Y}$ ignoring covariates. This adjustment is accomplished as shown in Eqs. (9.22) and (9.23). The analysis-of-variance problem with covariates is more complicated only in that we must form systems of normal type equations and obtain sums of squares to be used in adjusting the results of the analysis of the dependent variate y ignoring covariates.

Again, we have done nothing more than show the same computational procedure that is used in most textbooks on statistical methods to perform computations in AOC. It is, however, a good one when applied in a computer.

9.3 THE GENERAL PROBLEM FOR FIXED-EFFECTS MODELS

If we admit unbalanced and/or incomplete data, the simple data structure and form of reductions in sums of squares, which were features of the balanced complete situation, no longer apply. When cells are missing it is usually not readily apparent what functions are estimable; hence appropriate testable hypotheses concerning the effects in the model are not easy to derive. Furthermore, the multitude of possible structures and the seemingly special nature of most of these structures makes development of general computing algorithms exceedingly difficult. There are, of course, many useful computer programs which perform analysis-of-variance computations for various classes of data. Notable among these are programs based on algorithms by Harvey (1964), Wilkinson (1970), Bryce (1970), Rubin (1972), Hemmerle (1974), Goodnight (1976a, 1976b), and Dixon and Brown (1977). None are, however, sufficiently general to satisfy the needs of all users. Thus this section contains little more than a discussion of possible approaches to the computing problem in this area. The discussion will be based on procedures suggested by Goodnight (1976a, 1976b) which seem to provide a good starting point for development of a general algorithm for this general area of application.

9.3.1 Estimable Functions

One of the primary objectives in analysis of data of the type under consideration is to test meaningful hypotheses about the effects in the model. This implies that we must specify sets of estimable functions which describe "interesting" relationships among the parameters in the model. We are therefore immediately faced with the problem of somehow specifying the forms of estimable functions for any given model. One approach toward solution of this problem will be discussed in this subsection.

The vector $\underline{\ell}'$ of coefficients of each estimable function $\underline{\ell}'\underline{\beta}$ is in the vector space spanned by the rows of the design matrix X. Thus some subset of r of these rows (r is the rank of X) is a *generating set* for all allowable vectors $\underline{\ell}'$. Using an appropriate subset of the rows of X we can form a linear combination using symbolic coefficients and produce a general form for all estimable functions for the given model. Since this will be helpful in subsequent discussion, let us proceed along this line to see how a general form might be computed.

An automated procedure for producing a generating set, given X, is needed. There are many matrices which have the same row space as X. Included among these are X'X and $(X'X)^*X'X$. This latter matrix has some desirable properties which make it a candidate for use. First, we know from the discussion in Chapter 7 that it has only r nonzero rows and that identity matrix I_r appears as a submatrix. Furthermore, since X has rank r, a reordering of columns to place the independent column vectors first in X (for the purpose of description) leads to a partitioned matrix $Z = (X_1|X_2)$ where $X_2 = X_1K$ for some matrix K, and we have

$$(Z'Z)^*Z'Z = \begin{pmatrix} I_r & K \\ & \\ 0 & 0 \end{pmatrix}$$

The columns of K are coefficients in linear combinations of the columns of X_1 which produce columns in X_2; hence we can expect that the elements in K will generally be 0, 1, or -1, which means that $(X'X)^*X'X$ has an especially simple form. Therefore, using the rows of $(X'X)^*X'X$ as a generating set seems desirable and we shall do this.

Suppose, for example, that we have computed the following matrix M for the model

$$y_{ijk} = \mu + a_i + b_j + ab_{ij} + e_{ijk} \qquad (i = 1, 2), \quad (j = 1, 2, 3), \quad k = (1, \ldots, n_{ij})$$

$$M = (X'X)^*X'X = \begin{pmatrix} 1 & 0 & 1 & 0 & 0 & 1 & 0 & 0 & 0 & 0 & 0 & 1 \\ 0 & 1 & -1 & 0 & 0 & 0 & 0 & 0 & 1 & 0 & 0 & -1 \\ 0 & 0 & 0 & 0 & 0 & 0 & 0 & 0 & 0 & 0 & 0 & 0 \\ 0 & 0 & 0 & 1 & 0 & -1 & 0 & 0 & 0 & 1 & 0 & -1 \\ 0 & 0 & 0 & 0 & 1 & -1 & 0 & 0 & 0 & 0 & 1 & -1 \\ 0 & 0 & 0 & 0 & 0 & 0 & 0 & 0 & 0 & 0 & 0 & 0 \\ 0 & 0 & 0 & 0 & 0 & 0 & 1 & 0 & -1 & -1 & 0 & 1 \\ 0 & 0 & 0 & 0 & 0 & 0 & 0 & 1 & -1 & 0 & -1 & 1 \end{pmatrix} \qquad (9.24)$$

A linear combination of the rows of M, using coefficients t_i, gives the following general form for estimable functions:

Effect	Level		Coefficient
μ			t_1
A	1		t_2
	2		$t_1 - t_2$
B	1		t_4
	2		t_5
	3		$t_1 - t_4 - t_5$
AB	1	1	t_7
	1	2	t_8
	1	3	$t_2 - t_7 - t_8$
	2	1	$t_4 - t_7$
	2	2	$t_5 - t_8$
	2	3	$t_1 - t_2 - t_4 - t_5 + t_7 + t_8$

(9.25)

The t_i's may be chosen arbitrarily. Using this form, we can look at various other general forms relative to specific effects in the model. For example, with respect to A effects, if we set $t_1 = t_4 = t_5 = 0$, the result is as follows:

Effect	Level		Coefficient
μ			0
A	1		t_2
	2		$-t_2$
B	1		0
	2		0
	3		0
AB	1	1	t_7
	1	2	t_8
	1	3	$t_2 - t_7 - t_8$
	2	1	$-t_7$
	2	2	$-t_8$
	2	3	$-t_2 + t_7 + t_8$

From this form it is easily shown, for example, that the function $a_1 - a_2 + ab_{11} + ab_{12} - ab_{13} - ab_{21} - ab_{22} + ab_{23}$ is an estimable function.

The general form of estimable functions also indicates the degrees of freedom associated with each effect. In the previous example there is one degree of freedom for A as indicated by the fact that t_1 and t_2 appear in the A coefficients, but t_1 is the coefficient of μ, so t_2 is the only *free* symbol in the A set. Similarly, t_4 and t_5 are the two free symbols in the B set. The degrees of freedom do, of course, specify the maximum number of linearly independent estimable functions that is associated with each effect.

Specifying general forms of estimable functions provides useful information, but we need some method for selecting suitable linearly independent sets that can be used in testing hypotheses concerning the effects in the model. Given an appropriate set of estimable functions $L'\underline{\beta}$, we know from the discussion in Chapter 8 that the numerator sum of squares for testing H: $L'\underline{\beta} = \underline{0}$ is

$$(L'\underline{b})'[L'(X'X)^{-}L]^{-}(L'\underline{b}) \tag{9.26}$$

where $\underline{b}$ is a solution to the normal equations. Thus production of test criteria is not a problem. We might consider establishing criteria for choice of estimable functions and manipulating the general form to provide a general form of estimable functions for use in testing the significance of each effect in the model under such criteria. Then we could assign numeric values to the symbols involved in each case and produce a maximum rank hypothesis H: $L'\underline{\beta} = \underline{0}$.

The next two subsections deal with selection of "desirable" forms of estimable functions which are used to define hypotheses that are normally tested in AOV applications. In each case the criterion for selection leads to sums of squares that coincide with those recommended in statistical methods texts for use in specified tests of hypotheses.

9.3.2 Selection Criterion 1

A procedure which is sometimes used to obtain a sum of squares for testing significance of a given effect in an unrestricted model is to compute the additional reduction in sums of squares due to the effect in question adjusted for all effects which do not contain the given effect. To illustrate, consider the model

$$y_{ijk} = \mu + a_i + b_j + c_k + ab_{ij} + ac_{ik} + bc_{jk} + e_{ijk\ell} \qquad (9.27)$$

Example reductions for this model are

$$R(a|\mu,b,c,bc), \quad R(b|\mu,a,c,ac), \quad \text{and} \quad R(ab|\mu,a,b,c,ac,bc)$$

As this example illustrates, an effect E_1 is said to *contain* another effect E_2 whenever E_1 has more classification variables than E_2 and all classification variables in E_2 also appear in E_1. Thus ab contains both a and b. The intercept term μ is assumed to have no classification variables and is contained in all other effects. This definition can be extended in an obvious fashion to include containment when there are covariates in the model, but we will assume throughout this subsection that given models do not contain covariates. The reader is referred to Goodnight (1976b) for a more general definition.

The question which will be addressed in this subsection is whether it is possible to establish a criterion for selecting forms of estimable functions so that choices of linearly independent sets will lead to the same sums of squares, using (9.26), as are obtained using the procedure outlined above. We will see that this question can be answered in the affirmative, and a criterion will be given along with an outline of a computer-oriented procedure for performing required computations.

Consider a partitioning (and possible reordering) of the columns of the design matrix X to form

$$X = (X_0|X_1|X_2)$$

where

X_0 consists of the columns of X associated with effects that do not contain an effect E

X_1 consists of the columns of X associated with E

X_2 consists of the columns of X associated with effects which contain E

In this notation we will be interested in the reduction $R(X_1|X_0)$.

Next consider the partitioned matrix

$$V = \begin{pmatrix} X_0'X_0 & X_0'X_1 & X_0'X_2 & X_0'\underline{Y} \\ X_1'X_0 & X_1'X_1 & X_1'X_2 & X_1'\underline{Y} \\ X_2'X_0 & X_2'X_1 & X_2'X_2 & X_2'\underline{Y} \\ \underline{Y}'X_0 & \underline{Y}'X_1 & \underline{Y}'X_2 & \underline{Y}'\underline{Y} \end{pmatrix}$$

We will utilize generalized sweep operations, described in Chapter 7, and perform sweeps on the rows associated with X_0, then X_1, and finally X_2. The specific form of certain submatrices, which we shall need in subsequent discussion, will be given at the end of each stage. We begin by sweeping the rows corresponding to X_0 and obtain

$$V \xrightarrow[\overrightarrow{S}(X_0)]{} \begin{pmatrix} (X_0'X_0)^* & (X_0'X_0)^*X_0'X_1 & (X_0'X_0)^*X_0'X_2 & \hat{\underline{\beta}}_0 \\ -X_1'X_0(X_0'X_0)^* & X_1'M_1X_1 & X_1'M_1X_2 & X_1'M_1\underline{Y} \\ -X_2'X_0(X_0'X_0)^* & X_2'M_1X_1 & X_2'M_1X_2 & X_2'M_1\underline{Y} \\ -\hat{\underline{\beta}}_0' & \underline{Y}'M_1X_1 & \underline{Y}'M_1X_2 & \underline{Y}'M_1\underline{Y} \end{pmatrix}$$

where $M_1 = I - X_0(X_0'X_0)^*X_0'$. Continuing generalized sweep operations gives

$$V \xrightarrow[\overrightarrow{S}(X_0,\overrightarrow{X_1})]{} \begin{pmatrix} A & B & R & \tilde{\underline{\beta}}_0 \\ B' & C & D & \tilde{\underline{\beta}}_1 \\ -R' & -D' & X_2'M_2X_2 & T \\ -\tilde{\underline{\beta}}_0' & -\tilde{\underline{\beta}}_1' & T' & \underline{Y}'M_2\underline{Y} \end{pmatrix}$$

where

$$\underline{\tilde{\beta}}_0 = \underline{\hat{\beta}}_0 - (X_0'X_0)^* X_0'X_1(X_1'M_1X_1)^*X_1'M_1\underline{Y}$$

$$\underline{\tilde{\beta}}_1 = (X_1'M_1X_1)^*X_1'M_1\underline{Y}$$

$$B' = -(X_1'M_1X_1)^*X_1'X_0(X_0'X_0)^*$$

$$C = (X_1'M_1X_1)^*$$

$$D = CX_1'M_1X_2$$

$$M_2 = M_1 - M_1X_1(X_1'M_1X_1)^*X_1'M_1 \qquad (9.28)$$

Finally, completing sweeps on the rows associated with X_2 results in

$$V \xrightarrow{S(X_0,X_1,X_2)} \begin{pmatrix} A_1 & B_1 & R_1 & \underline{b}_0 \\ B_1' & C_1 & D_1 & \underline{b}_1 \\ R_1' & D_1' & S & \underline{b}_2 \\ -\underline{b}_0' & -\underline{b}_1' & -\underline{b}_2' & \underline{Y}'M_3\underline{Y} \end{pmatrix}$$

where

$$S = (X_2'M_2X_2)^*$$

$$D_1 = -(X_1'M_1X_1)^*X_1'M_1X_2(X_2'M_2X_2)^*$$

$$C_1 = C + D(X_2'M_2X_2)^*D'$$

The g_2 inverse of X'X appears in the principal 3 x 3 submatrix of this last matrix, and a least-squares solution to the normal equations $\underline{b}' = (\underline{b}_0', \underline{b}_1', \underline{b}_2')$ also appears.

Suppose that the criterion for choice of form of estimable functions specifies that for the effect E, which defines X_0, X_1 and X_2, the choice is

$$L' = (0|L_1|L_2)$$

where

$$L_1 = (X_1'M_1X_1)^*(X_1'M_1X_1)$$

$$L_2 = (X_1'M_1X_1)^*(X_1'M_1X_2) \qquad (9.29)$$

The question is whether this choice leads to the desired sum of squares.

It can be verified that, for L' given in (9.29),

$$L'(X'X)^*L = C \tag{9.30}$$

and

$$\begin{aligned} L'\underline{b} &= C(X_1'M_1X_1)\underline{b}_1 + C(X_1'M_1X_2)\underline{b}_2 \\ &= (X_1'M_1X_1)^*(X_1'M_1X_1)\underline{\tilde{\beta}}_1 \end{aligned} \tag{9.31}$$

Using (9.29), (9.30), and (9.31) in (9.26) yields the sum of squares

$$(L'\underline{b})'[L'(X'X)^*L]^*(L'\underline{b}) = \underline{\tilde{\beta}}_1'(X_1'M_1X_1)\underline{\tilde{\beta}}_1 \tag{9.32}$$

In the reduction notation the numerator sum of squares used in testing E, in the procedure outlined at the beginning of this subsection, is

$$R(X_1|X_0) = R(X_0,X_1) - R(X_0) \tag{9.33}$$

Using results obtained in sweeping rows of V, it follows that

$$\begin{aligned} R(X_1|X_0) &= (\underline{\tilde{\beta}}_0'|\underline{\tilde{\beta}}_1')\begin{pmatrix} X_0' \\ X_1' \end{pmatrix}\underline{Y} - \underline{\hat{\beta}}_0'X_0'\underline{Y} \\ &= (\underline{\tilde{\beta}}_0' - \underline{\hat{\beta}}_0')X_0'\underline{Y} + \underline{\tilde{\beta}}_1'X_1'\underline{Y} \\ &= [\underline{\tilde{\beta}}_1'X_1' - \underline{\tilde{\beta}}_1'X_1'X_0(X_0'X_0)^*X_0']\underline{Y} \\ &= \underline{\tilde{\beta}}_1'X_1'M_1\underline{Y} \\ &= \underline{\tilde{\beta}}_1'(X_1'M_1X_1)\underline{\tilde{\beta}}_1 \end{aligned} \tag{9.34}$$

Comparing (9.32) and (9.34), we see that the criterion used in selecting L' in (9.29) is desirable in that it leads to the same test of E as would be obtained by the common procedure of computing the additional reduction in sum of squares due to E adjusted for all effects that do not contain E. Furthermore we have available specification of the estimable functions used in this test and these are not readily available if we compute reductions in sums of squares under the usual procedure. The matrix L' as given in (9.29) may contain some zero rows, and these will be deleted before L' is displayed.

The procedure for obtaining reductions in sums of squares for use in testing significance of effects in the model is not the most frequently used procedure. Many data analysts assume a restricted model and compute reductions in a different fashion. We now consider a criterion for selecting estimable functions that leads to the same sums of squares as are produced by this alternative procedure.

9.3.3 Selection Criterion 2

Many data analysts prefer to impose restrictions on the model so that a reduced X matrix having full column rank is obtainable, as was done for the example in Section 9.1.4. The usual restrictions will suffice so long as we have a complete structure, and this is the situation which often exists. For each effect E in the model the additional reduction in sum of squares due to fitting E, adjusted for all other effects in the model, is computed. Obtaining reductions does give sums of squares needed for testing oft-considered hypotheses, but no insight is provided about the form of estimable functions being utilized in the process. As in the preceding subsection, we would like to have a criterion for selecting estimable functions that will lead to the reductions in sums of squares computed under the usual procedure outlined above.

Goodnight (1976b) has developed a criterion which will give sums of squares that are the same as those computed under the usual procedure when the given data are unbalanced but complete and the usual restrictions are imposed. In more complicated cases the criterion selects forms of estimable functions for maximal-rank hypotheses that are sometimes used. Selection is made by operating on the rows of $M = (X'X)^*X'X$ in the following way for each effect E in the model:

Selection Criterion 2. For an effect E,

Step 1: For each column in M whose associated effect does not contain E, look for a nonzero element in the column. For each nonzero element in the column, perform an elementary row operation to zero the element. When all such elements except one have been annihilated, zero the row containing that element.

Step 2: For each column in M which is associated with E do the following. If any nonzero element exists in the column (other than those in a row which has been pivoted on in this step), annihilate the element using an appropriate row operation.

Step 3: Select any nonzero row for which all elements associated with E are zero. Leaving that row unchanged, make all other rows orthogonal to it and then zero the row. Repeat the procedure for each eligible row. At the end of this step the nonzero rows that remain are taken as the rows in the matrix L' in the estimable function $L'\underline{\beta}$ for the E effect.

For a model in the form

$$y_{ijk} = \mu + a_i + b_j + ab_{ij} + e_{ijk}$$
$$(i = 1, 2), \quad (j = 1, 2), \quad (k = 1, \ldots, n_{ij})$$

a set of coefficients in estimable functions for effects in the model are as follows:

	μ	a_1	a_2	b_1	b_2	ab_{11}	ab_{12}	ab_{21}	ab_{22}
L'_A	0	1	-1	0	0	1/2	1/2	-1/2	-1/2
L'_B	0	0	0	1	-1	1/2	-1/2	1/2	-1/2
L'_{AB}	0	0	0	0	0	1	-1	-1	1

Practical experience has shown that in the presence of missing cells or confounding, general forms of estimable functions obtained under the procedure outlined above show considerable imbalance in the coefficients of levels of higher-order effects. This tends to make this selection criterion somewhat undesirable for general use. Nevertheless, it is useful for many applications.

9.3.4 Summary

There are obviously many criteria which can be defined that will select estimable functions having desirable characteristics in some cases. At the present time no single criterion has been proposed which is satisfactory in all situations, and probably none can be found. If this approach, wherein estimable functions are generated as a basis for computations, is to lead to a very general algorithm, a more flexible way to produce functions must be found. Possibly an adaptive-type algorithm which utilizes user specification of desires in form of estimable function will ultimately be developed and become popular.

Future developments in this very important area of statistical computing are impossible to predict. It is an area in need of further research and development.

9.4 COMPUTING EXPECTED MEAN SQUARES AND ESTIMATES OF VARIANCE COMPONENTS

Thus far in this chapter we have considered only fixed-effects models (also called *fixed models*). In this section the more general case, which admits the possibility of random effects, will be considered. For example, if in the model (9.1) we assume that the a_i effects are random and the b_j are fixed effects, then the ab_{ij} are random effects and the model is called a *mixed model* because it contains both fixed and random effects. The usual assumption is that the vectors $\underline{a}$ and $\underline{ab}$ of random effects are mutually independent, have mean $\underline{0}$, and have variance-covariance matrices $\sigma_a^2 I$ and $\sigma_{ab}^2 I$, respectively. The variances σ_a^2 and σ_{ab}^2 and the error variance σ_e^2 are called *variance components* because they are components in the variance of y. When all effects (ignoring the mean effect μ, which is usually fixed) are random, the model is called a *random model*.

Estimation of variance components for the general case of possibly unbalanced data is accomplished using one of several available methods. These methods will be discussed in Section 9.4.2. For a given mixed or random model, a usual procedure is to first calculate desired mean squares in the analysis of variance for the corresponding fixed model. Methods for doing this were described in Sections 9.2 and 9.3. Next, the expected value of each mean square under the given mixed or random model is computed. Finally, the mean squares and their respective expected values may be used to estimate the variance components. Estimates of variance components form the basis for computing test statistics for tests of hypotheses concerning the variance components.

One use for expected mean squares was mentioned above. In addition to their use in estimating variance components, expected mean squares are often of interest in analysis-of-variance applications, whether the assumed model is fixed, mixed, or random, because they suggest which linear combination of mean squares is appropriate for testing an hypothesis under consideration. Thus we will consider the computation of expected mean squares in Section 9.4.1 and then discuss computing estimates of variance components in Section 9.4.2.

9.4.1 Computing Expected Mean Squares

Several methods have been derived for computing expected mean squares. Some are specific to the case of balanced data, and the computations are especially simple in this case (see, e.g., Searle 1971). For the more general case that admits unbalanced data, the "synthesis" method of Hartley (1967) and Rao (1968) as well as other methods by Gaylor, Lucas, and Anderson (1970), Rojas (1972), and Speed and Hocking (1974) are available. The method presented in this subsection is essentially the Speed and Hocking method (see also Goodnight and Speed 1978). As we shall see, it applies to fixed, mixed, or random models.

Consider the mixed model, which we write as

$$\underline{y} = X_0\underline{\beta}_0 + \sum_{i=1}^{k} X_i\underline{\beta}_i + \underline{e} \tag{9.35}$$

where $\underline{\beta}_0$ is a vector containing the fixed effects; the vectors $\underline{\beta}_i$ are random and assumed to be mutually independent and independent of $\underline{e}$ with $\underline{\beta}_i$ assumed to have mean $\underline{0}$ and variance-covariance matrix $\sigma_i^2 I_{m_i}$; $\underline{e}$ is distributed normally with mean $\underline{0}$ and variance-covariance matrix $\sigma_e^2 I$; and m_i is the column dimension of the n x m_i submatrix X_i of the design matrix $X = (X_0, X_1, \ldots, X_k)$. Since mean squares arise from sums of squares, we consider the following general form for a sum of squares:

$$SS = (L'\underline{b})'[L'(X'X)^-L]^{-1}(L'\underline{b})$$

In this form the matrix L' is full row rank and is contained in the row space of X, and $(X'X)^-$ is defined by Eq. (7.22). The sum of squares is expressible as a quadratic form $\underline{y}'A_L\underline{y}$ with symmetric matrix A_L:

$$\begin{aligned} SS &= [L'(X'X)^-X'\underline{y}]'[L'(X'X)^-L]^{-1}[L'(X'X)^-X'\underline{y}] \\ &= \underline{y}'X(X'X)^-L[L'(X'X)^-L]^{-1}L'(X'X)^-X'\underline{y} \\ &= \underline{y}'A_L\underline{y} \end{aligned} \tag{9.36}$$

Note that the A_L matrix is idempotent.

Since sums of squares of interest are expressible as quadratic forms in $\underline{y}$, we next consider the expected value of $\underline{y}'A_L\underline{y}$. It is readily verified that

$$\begin{aligned} E(\underline{y}'A_L\underline{y}) &= \underline{\beta}_0'X_0'A_LX_0\underline{\beta}_0 + \sum_{i=1}^{k} tr(X_i'A_LX_i)\sigma_i^2 + tr(A_L)\sigma_e^2 \\ &= \underline{\beta}_0'X_0'A_LX_0\underline{\beta}_0 + \sum_{i=1}^{k} tr(X_i'A_LX_i)\sigma_i^2 + \sigma_e^2 \cdot \text{rank}\,(A_L) \end{aligned} \tag{9.37}$$

where $tr(A_L)$ denotes the matrix trace [sum of diagonal elements; equal to rank (A_L) when A_L is idempotent], and σ_e^2 is the error variance component. Equation (9.37) shows that the expected mean square which we wish to compute is readily obtained if $X'A_LX$ is available. Let us next consider formation of this matrix.

The matrix $X'A_LX$, with A_L defined as in Eq. (9.36), has a rather complicated form in terms of X and L. However, some simplification is possible when we note that (see Exercise 8.5)

$$L' = L'(X'X)^{-}X'X$$

Using this fact, we obtain

$$X'A_LX = L[L'(X'X)^{-}L]^{-1}L' \tag{9.38}$$

A suggested approach to computation of the matrix $X'A_LX$ given in (9.38) is the following. First, perform the decomposition

$$[L'(X'X)^{-}L]^{-1} = (T'T)^{-1}$$

where T is upper-triangular. Next, form

$$C = (T')^{-1}L' \tag{9.39}$$

Now we have

$$\begin{aligned} X'A_LX &= C'T(T'T)^{-1}T'C \\ &= C'C \end{aligned}$$

If we envision a partitioning of the columns of C into the same-size submatrices as those contained in X as shown in Eq. (9.35), then

$$X'A_LX = \begin{bmatrix} X_0'A_LX_0 & X_0'A_LX_1 & \cdots & X_0'A_LX_k \\ \vdots & \ddots & & \vdots \\ X_k'A_LX_0 & \cdots & & X_k'A_LX_k \end{bmatrix}$$

$$= \begin{bmatrix} C_0'C_0 & C_0'C_1 & \cdots & C_0'C_k \\ \vdots & \ddots & & \vdots \\ C_k'C_0 & \cdots & & C_k'C_k \end{bmatrix} \tag{9.40}$$

Using (9.40), we can write Eq. (9.37) as

$$E(\underline{y}'A_L\underline{y}) = \underline{\beta}_0'C_0'C_0\underline{\beta}_0 + \sum_{i=1}^{k} \mathrm{tr}(C_i'C_i)\sigma_i^2 + r\sigma_e^2$$

where r is the rank of L. Finally, since tr(C'C) is the sum of squares of the elements in C [denoted by SQ(C)], we see that $C_i'C_i$ need not be formed. Thus a desirable computational form for the expected mean square is

$$E(\underline{y}'A_L\underline{y}) = (C_0\underline{\beta}_0)'(C_0\underline{\beta}_0) + \sum_{i=1}^{k} \sigma_i^2\, SQ(C_i) + r\sigma_e^2 \qquad (9.41)$$

where $C = (C_0, C_1, C_2, \ldots, C_k)$ is given by Eq. (9.39). Note that for fixed models σ_e^2 will be the only variance component appearing in Eq. (9.41), and for random models the first term in (9.41) reduces to the single term that consists of a scalar multiple of μ^2.

9.4.2 Variance Component Estimation

One of the most important aspects of the analysis of data from random or mixed models is the estimation of the variance components. An obvious technique for estimation of variance components, which dates back to Daniels (1939), is to take the mean squares in the AOV as estimates of their respective expectations and to estimate the individual components using the obvious linear combinations of the expected mean squares. The computations for this method of estimation, generally called the method of analysis of variance, are quite simple if the form of the expected mean squares is known. The technique is appropriate whether the data are balanced or not. The expected mean squares may be computed as described in Section 9.4.1.

There are certain drawbacks to the analysis-of-variance method for estimating variance components, such as the fact that the estimators may take on negative values. In recent years several other techniques for estimation of variance components have been growing in favor, especially for use with unbalanced data. One of the most widely used methods is that of maximum likelihood (ML). The likelihood function differs considerably depending on the model and only for the very simplest balanced one-way and two-way models do explicit solutions to the likelihood equations exist.

In order to simplify the notation in the following, let us define the matrix H by the expression

$$\sigma^2 H = \sigma_e^2 I + \sum_{i=1}^{k} \sigma_i^2 X_i X_i' \tag{9.42}$$

where the other quantities are as in Eq. (9.35); hence the model (9.35) can be written as

$$\underline{y} = X_0 \underline{\beta}_0 + \underline{\delta} \tag{9.43}$$

where $E(\underline{\delta}) = \underline{0}$ and $\mathrm{Var}(\underline{\delta}) = \sigma^2 H$. The log-likelihood function then is [we are now assuming that the $\underline{\beta}_i$ in Eq. (9.35) have normal distributions]

$$\lambda = 0.5[n \ln(\sigma^2) + \ln(|H|) - (\underline{y} - X_0\underline{\beta}_0)'H^{-1}(\underline{y} - X_0\underline{\beta}_0)/\sigma^2] + c \tag{9.44}$$

where c is a constant. Hence the ML estimates are obtained by maximizing (9.44).

Hartley and Rao (1967) described an iterative technique using a steepest ascent method (see Chapter 10) for the general AOV model. Their technique, as implemented by Hartley and Vaughn (1972), required at each iteration the inversion of the current n x n matrix H (which was actually accomplished by inversion of a smaller matrix but at the expense of several matrix multiplications). Both the first- and second-order partial derivatives of (9.44) (with respect to $\underline{\beta}_0$, σ_e^2, and σ_i^2) involve the expressions $X_i'H^{-1}X_i$, $X_i'H^{-1}X_0$, $X_i'H^{-1}\underline{y}$, and $\underline{y}'H^{-1}\underline{y}$. Hemmerle and Hartley (1973) introduced a transformation, the "W transformation," which considerably reduces the computational effort for obtaining these quantities.

The matrix W is defined as

$$W = \begin{pmatrix} V'H^{-1}V & V'H^{-1}X_0 & V'H^{-1}\underline{y} \\ X_0'H^{-1}V & X_0'H^{-1}X_0 & X_0'H^{-1}\underline{y} \\ \underline{y}'H^{-1}V & \underline{y}'H^{-1}X_0 & \underline{y}'H^{-1}\underline{y} \end{pmatrix} \tag{9.45}$$

where $V = (X_1|X_2|\cdots|X_k)$, and thus W contains the desired quantities involving H^{-1}. The order of W is $(m + k + 1)$ - square, where $m = \sum_{i=1}^{k} m_i$. The important point, however, is the order of the number of computations to obtain W: the number of computations is independent of n. Hemmerle and Hartley (1973) showed that

$$W = W_0 - Z'(D^{-1} + V'V)^{-1}Z \tag{9.46}$$

where

$$W_0 = \begin{pmatrix} V'V & V'X_0 & V'\underline{y} \\ X_0'V & X_0'X_0 & X_0'\underline{y} \\ \underline{y}'V & \underline{y}'X_0 & \underline{y}'\underline{y} \end{pmatrix}$$

$$Z = [V'V|V'X_0|V'\underline{y}]$$

and

$$D = \frac{1}{\sigma^2}\begin{pmatrix} \sigma_1^2 I_{m_1} & & & 0 \\ & \sigma_2^2 I_{m_2} & & \\ & & \ddots & \\ 0 & & & \sigma_k^2 I_{m_k} \end{pmatrix}$$

They also gave a technique for obtaining W. Other, generally more efficient algorithms for computing W have been given by Thompson (1975), Hemmerle and Lorens (1976), and Goodnight and Hemmerle (1978).

The Goodnight-Hemmerle method begins with the matrix

$$\left(\begin{array}{c|c} V'V + D^{-1} & Z \\ \hline Z' & W_0 \end{array}\right) \tag{9.47}$$

and pivots on each diagonal of $V'V + D^{-1}$. The pivots can be of the Gauss-Jordan or sequential Cholesky type. By Eq. (9.46) it is seen that this action leaves W in place of W_0 in (9.47). Storage efficiency may be achieved by noting, as was done by Goodnight and Hemmerle, that all of the quantities needed for the pivots, except the elements of D^{-1}, are contained within the lower-right submatrix (initially W_0). Denoting the initial elements of W_0 as $a_{ij}^{(0)}$ and of $V'V + D^{-1}$ as $d_t^{(0)}$ and denoting the respective elements following the rth pivot as $a_{ij}^{(r)}$ and $d_t^{(r)}$, we have

$$a_{ij}^{(r)} = a_{ij}^{(r-1)} - \frac{a_{ri}^{(r-1)} \cdot a_{rj}^{(r-1)}}{d_r^{(r-1)}} \quad \text{for } i \leq j \tag{9.48}$$

and

$$d_t^{(r)} = d_t^{(r-1)} - \frac{a_{rt}^{(r-1)} \cdot a_{rt}^{(r-1)}}{d_r^{(r-1)}} \quad \text{for } t > r \tag{9.49}$$

Goodnight and Hemmerle then suggest a further simplification by letting i = j = t in (9.48) and subtracting (9.48) from (9.49), yielding

$$d_t^{(r)} - a_{tt}^{(r)} = d_t^{(r-1)} - a_{tt}^{(r-1)} \quad \text{for } t > r$$

Thus,

$$d_t^{(1)} - a_{tt}^{(1)} = (D^{-1})_{tt} \quad \text{for } t > 1$$

and, in general,

$$d_t^{(r)} - a_{tt}^{(r)} = (D^{-1})_{tt} \quad \text{for } t > r$$

or

$$d_r^{(r-1)} = a_{rr}^{(r-1)} + (D^{-1})_{rr} \tag{9.50}$$

Therefore the $d_r^{(r-1)}$ of Eq. (9.48) do not need to be computed at each iteration; that is, Eq. (9.50) requires no updating and may be used in place of (9.49) in the iteration.

Hemmerle and Lorens (1976) discussed an extension of the W transformation for the case in which some of the variance components are zero or near zero. The modification can be implemented in the above procedure by simply not pivoting on any diagonal element in (9.47) which is associated with a zero component.

The problem in ML estimation of variance components, of course, is to maximize (9.44). The role of the W transformation is to reduce the computations and storage required for obtaining various quantities in the derivatives of the log-likelihood function. How to use these derivatives is another matter, and is discussed for the general problem of function optimization in Chapter 10. Hemmerle and Hartley (1973) used a modified Newton-Raphson procedure (see Section 10.2.2) and Jennrich and Sampson (1976) described a pure Newton-Raphson method and Fisher's scoring method, each in conjunction with the W transformation. (See also Jennrich and Sampson 1978 for a discussion of some of the problems to be considered in the development of general-purpose software for variance component estimation.)

The ML estimation as described above involved maximization of the likelihood function simultaneously with respect to a variable representing the fixed effects $\underline{\beta}_0$ and with respect to variables representing the variance components. The fact that this procedure ignores loss of degrees of freedom associated with the estimation of the fixed effects may be considered by some to be a disadvantage of the method. Patterson and Thompson (1971) made a transformation which partitioned the likelihood function into two parts, one of which is entirely free of the fixed effects. Maximizing the part not containing the fixed effects yields the so-called restricted maximum likelihood (REML) estimators. Corbeil and Searle (1976) gave a computational method for the REML estimators that made use of the W transformation and the Newton-Raphson procedure.

Harville (1977) gave a survey of maximum likelihood (ML and REML) techniques, including computational methods, for variance component estimation.

Harville (1977) and Searle (1978) have summarized other methods of estimation of variance components, including MINQUE (minimum norm quadratic unbiased estimation), I-MINQUE (iterative MINQUE), and MIVQUE (minimum variance quadratic unbiased estimation). Liu and Senturia (1977) described a computational method for MINQUE, which also relied on the W transformation. Hartley, Rao, and LaMotte (1978) and Goodnight (1978b) gave computing techniques for types of MIVQUE estimators.

EXERCISES

9.1 Show that the matrices M_1 and M_2 defined in Section 9.2.1 are idempotent.

9.2 Construct the reduced X matrix for the model in Example 9.1 using constraints in the same form as the restrictions, assuming $I = 2$, $J = 3$, $K = 2$, $L = 1$.

9.3 Show that the reduced X matrix obtained in Exercise 9.2 will result if the restrictions are substituted into the model equations and a design-type matrix is formed. In other words, show that a reduced X matrix can be obtained without using constraints.

9.4 Verify that the coefficients (9.25) result as a linear combination of the rows of the matrix M in (9.24).

9.5 Verify the correctness of Eq. (9.30).

9.6 Show that $L'\underline{b}$ reduces to the right member of (9.31) for L' in the form of (9.29).

9.7 Write out the relevant partial derivatives for maximizing (9.44).

9.8 Show that sweeping the upper left submatrix in (9.47) produces W as the lower right submatrix.

REFERENCES AND FURTHER READINGS

Armor, D. J. (1969), A Computer Language for Analysis of Variance, *Proceedings of the Social Statistics Section, American Statistical Association*, Washington, D.C., 91-95.

Bancroft, T. A. (1968), *Topics in Intermediate Statistical Methods*. Iowa State University Press, Ames.

Bargmann, R. E. (1977), Multiple Factor Factorial Experiments, Unbalanced and with Missing Cells: Comparison of Hierarchial Analysis with Other Approaches, *Proceedings of the Statistical Computing Section, American Statistical Association*, Washington, D.C., 27-32.

Bates, C. B., and J. Thomas (1977), Mod-ANOVA: An Analyses of Variance Program for 2^n and 3^n Factorial Experiments, *Proceedings of the Statistical Computing Section, American Statistical Association*, Washington, D.C., 123-126.

Blinn, J. C. (1966), A Statistically-Oriented Computer Language and a System of Statistical Programs, M.S. thesis, Iowa State University, Ames.

Bock, R. D. (1963), Programming Univariate and Multivariate Analysis of Variance, *Techno. 5*, 95-117.

Bock, R. D. (1965), A Computer Program for Univariate and Multivariate Analysis of Variance, *Proceedings of IBM Scientific Computing Symposium on Statistics*, IBM Data Processing Division, White Plains, N.Y., 69-111.

Bowles, R. L. (1950), Accuracy of Certain Approximate Methods in Predicting the Correct Model for Experiments with Unequal Frequencies in the Subclasses, M. S. thesis, Iowa State University, Ames.

Bradley, H. E. (1968), Multiple Classification Analysis for Arbitrary Experimental Arrangements, *Techno. 10*, 13-27.

Bradu, D. (1965), Main Effect Analysis of the General Orthogonal Layout with Any Number of Factors, *AMS 36*, 88-97.

Britton, C., and I. F. Warner (1963), A Computer Program for Analysis of Variance for a Two-Level Factorial Design, *CACM 6*, 308-309.

Bryce, G. R. (1970), A Unified Method for Analysis of Unbalanced Designs, M.S. thesis, Brigham Young University, Provo, Utah.

Bryce, G. R. (1975), The One-Way Model, *Amer. Stat. 29*, 69.

Bryce, G. R., D. T. Scott, and M. W. Carter (1978), Rummage: A Data Analysis System, *Proceedings of Computer Science and Statistics: Eleventh Annual Symposium on the Interface*, edited by A. R. Gallant and T. M. Gerig, Institute of Statistics, North Carolina State University, Raleigh, 402.

Carlson, J. E. (1975), Hypothesis Testing in Linear Models, *Amer. Stat. 29*, 133.

Cramer, S. G. (1967), Programming the Tabular Method of Analysis of Variance for Factorial Experiments, *CACM 10*, 109-110.

Carney, E. J. (1967), Computations of Variances and Covariances of Variance Component Estimates, Ph.D. thesis, Iowa State University, Ames.

Chamberlain, R. L. (1972), The Non-Orthogonal Analysis of Variance, Ph.D. thesis, Iowa State University, Ames.

Claringbold, P. J. (1969), Algorithm 366: Regression Using Certain Direct Product Matrices, *CACM 12*, 688-691.

Cooper, B. E. (1967), A Unifying Computational Method for Analysis of Complete Factorial Experiments, *CACM 10*, 27-34.

Corbeil, R. R., and S. R. Searle (1976), Restricted Maximum Likelihood (REML) Estimation of Variance Components in the Mixed Model, *Techno. 18*, 31-38.

Coulter, B. S. (1964), A Computer Programme for Analysis of Variance in Experiments with Factorial Treatments and One or More Error Terms, *Appl. Stat. 13*, 110-117.

Cramer, E. M. (1976), A Nonorthogonal Analysis of Variance Program, *JASA 71*, 93-95.

Daniels, H. E. (1939), The Estimation of Components of Variance, *JRSS Supplement 6*, 186-197.

Davidson, M. L., and J. Toporek (1977), ANOVA with General Cell Weights, *Proceedings of the Statistical Computing Section, American Statistical Association*, Washington, D.C., 174-179.

Davidson, M. L., and J. D. Toporek (1978), *URWAS: Weighted ANOVA System-User Manual*, Computing Center, University of Rochester, Rochester, N.Y.

Dixon, W. J., and M. B. Brown (eds.) (1977), *BMDP-77: Biomedical Computer Programs, P-Series,* University of California Press, Berkeley.

Eisenhart, C. (1947), The Assumptions Underlying the Analysis of Variance, *Biometrics 3,* 1-21.

Elston, R. C., and N. Bush (1964), The Hypothesis That Can Be Tested When There Are Interactions in an Analysis of Variance Model, *Biometrics 20*, 681-699.

Fisher, R. A. (1935), *The Design of Experiments*, Oliver and Boyd, Edinburgh, Scotland.

Fowlkes, E. (1969), Some Operators of ANOVA Calculations, *Techno. 11*, 511-526.

Francis, I. (1973), Comparison of Several Analysis of Variance Programs, *JASA 68*, 860-865.

Francom, S. F. (1973), A Multiple Linear Regression Approach to the Analysis of Balanced and Unbalanced Designs, M.S. thesis, Brigham Young University, Provo, Utah.

Frane, J. W. (1978), BMD and BMDP Approaches to Unbalanced Data, *Proceedings of the Computer Science and Statistics: Tenth Annual Symposium on the Interface*, edited by D. Hogben, U.S. Government Printing Office, Washington, D.C. 40-47.

Gaylor, D. W., H. L. Lucas, and R. L. Anderson (1970), Calculation of Expected Mean Squares by the Abbreviated Doolittle and Square Root Methods, *Biometrics 26*, 641-655.

Gerig, T. M., and A. R. Gallant (1975), Computing Methods for Linear Models Subject to Linear Parametric Constraints, *JSCS 3*, 283-296.

Gianola, D. (1975), Hypothesis Tests in Linear Models, *Amer. Stat. 29*, 133.

Goodnight, J. H. (1976a), The General Linear Models Procedure, *Proceedings of the First International SAS User's Group*, SAS Institute, Raleigh, N.C.

Goodnight, J. H. (1976b), Computational Methods in General Linear Models, *Proceedings of the Statistical Computing Section, American Statistical Association,* Washington, D. C., 68-72.

Goodnight, J. H. (1978a), Hypothesis Testing in Multi-Way ANOVA Models, *Proceedings of the Computer Science and Statistics: Tenth Annual Symposium on the Interface*, edited by D. Hogben, U.S. Government Printing Office, Washington, D.C. 48-53.

Goodnight, J. H. (1978b), *Computing MIVQUE Estimates of Variance Components*, SAS Institute, Raleigh, N.C.

Goodnight, J. H., and F. M. Speed (1978), *Computing Expected Mean Squares*, SAS Institute, Raleigh, N.C.

Gower, J. C. (1962), The Handling of Multiway Tables on Computers, *Comp. J. 4*, 280-286.

Gower, J. C. (1969), Algorithm AS19: Analysis of Variance for a Factorial Table, *Appl. Stat. 18*, 197-202

Gower, J. C., H. R. Simpson, and A. H. Martin (1967), A Statistical Programming Language, *Appl. Stat. 16*, 89.

Hartley, H. O. (1956), A Plan for Programming Analysis of Variance for General Purpose Computers, *Biometrics 12*, 110-122.

Hartley, H. O. (1962), Analysis of Variance, in *Mathematical Methods for Digital Computers*, vol. 1, edited by A. Ralston and H. Wilf, Wiley, New York.

Hartley, H. O. (1967), Expectation Variances, and Covariances of ANOVA Mean Squares by "Synthesis," *Biometrics 23*, 105-114. Corrigenda: *Biometrics 23*, 853.

Hartley, H. O., and J. N. K. Rao (1967), Maximum Likelihood Estimation for the Mixed Analysis of Variance Model, *Biometrika 54*, 93-108.

Hartley, H. O., J. N. K. Rao, and L. R. LaMotte (1978), A Simple Synthesis-Based Method of Variance Component Estimation, *Biometrics 34*, 233-243.

Hartley, H. O., and W. K. Vaughn (1972), A Computer Program for the Mixed Analysis of Variance Model Based on Maximum Likelihood, in *Statistical Papers in Honor of George W. Snedecor*, edited by T. A. Bancroft, Iowa State University Press, Ames, 129-144.

Harvey, W. R. (1964), *Computing Procedure for a Least Squares Analysis Program,* U.S.D.A. Agricultural Research Services Memorandum ARS-20-8, U.S. Department of Agriculture, Washington, D.C.

Harvey, W. R. (1977), Missing Subclass Problems Using Method 3 of Henderson in Mixed Models, *Proceedings of the Statistical Computing Section, American Statistical Association*, Washington, D.C., 22-26.

Harville, D. A. (1977), Maximum Likelihood Approaches to Variance Component Estimation and to Related Problems, *JASA 72*, 320-340.

Healy, M. J. R., and M. Westmacott (1956), Missing Values in Experiments Analyzed on Automatic Computers, *Appl. Stat. 5,* 203-206.

Heiberger, R. M., and L. L. Laster (1978), Computing Approaches to the Analysis of Variance of Unbalanced Data, *Proceedings of the Computer Science and Statistics: Tenth Annual Symposium on the Interface,* edited by D. Hogben, U.S. Government Printing Office, Washington, D.C., 37-39.

Hemmerle, W. J. (1964), Algebraic Specifications of Statistical Models for Analysis of Variance Computations, *JACM 11,* 234-239.

Hemmerle, W. J. (1967), *Statistical Computations on a Digital Computer*. Blaisdell, Waltham, Mass.

Hemmerle, W. J. (1974), Nonorthogonal Analysis of Variance Using Iterative Improvement and Balanced Residuals, *JASA 69*, 772-779.

Hemmerle, W. J. (1976), Iterative Nonorthogonal Analysis of Covariance, *JASA 71*, 195-199.

Hemmerle, W. J., and E. J. Carney (1969), An Algorithm for Multivariate Analysis of Variance, in *Statistical Computations*, edited by R. C. Milton and J. A. Nelder, Academic Press, Princeton, N.J.

Hemmerle, W. J., and B. W. Downs (1978), Nonhomogeneous Variances in the Mixed AOV Model: Maximum Likelihood Estimation, in *Contributions to Survey Sampling and Applied Statistics: Papers in Honor of H. O. Hartley*, edited by H. A. David, Academic Press, New York, 153-172.

Hemmerle, W. J., and H. O. Hartley (1973), Computing Maximum Likelihood Estimates for the Mixed A.O.V. Model Using the W Transformation, *Techno. 15*, 819-831.

Hemmerle, W. J., and J. A. Lorens (1976), Improved Algorithms for the W-Transform in Variance Component Estimation, *Techno. 18*, 207-211.

Henderson, C. R. (1953), Estimation of Variance and Covariance Components, *Biometrics 9*, 226-252.

Hinkelmann, K. (1975), Hypothesis Testing in Linear Models, *Amer. Stat. 29*, 110.

Hocking, R. R., O. P. Hackney, and T. M. Speed (1978), The Analysis of Linear Models with Unbalanced Data, *Proceedings of the Computer Science and Statistics: Tenth Annual Symposium on the Interface*, edited by D. Hogben, U.S. Government Printing Office, Washington, D.C., 66-70.

Hocking, R. R., and F. M. Speed (1975), A Full Rank Analysis of Some Linear Model Problems, *JASA 70*, 706-712.

Howell, J. R. (1969), Algorithm 359: Factorial Analysis of Variance, *CACM 12*, 631.

James, A. T., and G. N. Wilkinson (1971), Factorization of the Residual Operator and Canonical Decomposition of Nonorthogonal Factors in the Analysis of Variance, *Biometrika 58*, 279-294.

Jennrich, R. I., and P. F. Sampson (1976), Newton-Raphson and Related Algorithms for Maximum Likelihood Variance Component Estimation, *Techno. 18*, 11-18.

Jennrich, R. I., and P. F. Sampson (1978), Some Problems Faced in Making a Variance Component Algorithm into a General Mixed Model Program, *Proceedings of the Computer Science and Statistics: Eleventh Annual Symposium on the Interface,* edited by A. R. Gallant and T. M. Gerig, Institute of Statistics. North Carolina State University, Raleigh, 56-63.

Kempthorne, O. (1952), *The Design and Analysis of Experiments,* Wiley, New York.

Kempthorne, O., G. Zyskind, S. Addelman, T. Throckmorton, and R. White (1961), *Analysis of Variance Procedures*, Aeronautical Research Laboratory, Office of Aerospace Research, ARL 149, U.S. Air Force, Wright-Patterson Air Force Base, Ohio.

Klimbo, L., and P. Nelson (1976), Unbalanced Data, *Amer. Stat. 30,* 205-206.

Kutner, M. H. (1974), Hypothesis Testing in Linear Models (Eisenhart Model I), *Amer. Stat. 29*, 98-100, 133-134.

Leiblein, J. (1959), A General Analysis of Variance Scheme Applicable to a Computer with a Very Large Memory, *JACM 4*, 69.

Liu, L., and J. Senturia (1977), Computation of MINQUE Variance Component Estimates, *JASA 72*, 867-869.

Mexas, A. G. (1969), Some Computational Aspects of Linear Classification Models, Ph.D. thesis, Iowa State University, Ames.

Muller, M. E., and G. N. Wilkinson (1971), Report on ANOVA Workshop, October 19-21, 1970, manuscript, Graduate School, University of Wisconsin, Madison.

Nelder, J. A. (1969), The Description of Statistical Data Structures for Statistical Computing, in *Statistical Computations*, edited by R. C. Milton and J. A. Nelder, Academic Press, New York.

Nelder, J. A. (1974), Analysis of Various Programs, Least Squares, and Two-Way Tables, *Appl. Stat. 23*, 101.

Oliver, I. (1968), Algorithm 330: Factorial Analysis of Variance, *CACM 11*, 431-432.

Overall, J. E., and D. K. Speigel (1969), Concerning Least Squares Analysis of Experimental Data, *Psychological Bulletin 72*, 311-322.

Patterson, H. D., and R. Thompson (1971), Recovery of Inter-Block Information when Block Sizes are Unequal, *Biometrika 58*, 545-554.

Pearce, S. C. (1965), Classification of Non-Orthogonal Designs, *JRSS (A) 126*, 353.

Peng, K. C. (1964), Programming Analysis of Variance by Sequences of Operators and Isomorphic Mappings, *CACM 7*, 415-416.

Preece, D. A. (1971), Iterative Procedures for Missing Values in Experiments, *Techno. 13*, 743-753.

Rao, J. N. K. (1968), On Expectations, Variances and Covariances of ANOVA Mean Squares by "Synthesis," *Biometrics 24*, 963-978.

Rees, D. H. (1966), The Analysis of Variance of Designs with Many Nonorthogonal Classifications, *JRSS (B) 28*, 210.

Rojas, B. (1972), *On Models and Components of Variance*, Technical Report 1970, School of Statistics, University of Minnesota, Minneapolis.

Rubin, D. G. (1972), A Non-Iterative Algorithm for Least Squares Estimation of Missing Values in Any Analysis of Variance Design, *Appl. Stat. 21*, 136-141.

Scheffe, H. (1959), *The Analysis of Variance*, Wiley, New York.

Schlater, J. E. (1965), Analysis of Variance and Covariance Computations on a Digital Computer for Balanced Complete Structures Based on Algebraic Model Specifications, M.S. thesis, Iowa State University, Ames.

Schlater, J. E., and W. J. Hemmerle (1966), Statistical Computations Based Upon Algebraically Specified Models, *CACM 9*, 865-869.

Searle, S. R. (1971), *Linear Models*, Wiley, New York.

Searle, S. R. (1976), Comments on Anova Calculations for Messy Data, *Proceedings of the First International SAS User's Group*, SAS Institute, Raleigh, N.C.

Searle, S. R. (1978), Analysis of Variance of Unbalanced Data from 3-Way and Higher Order Classifications, *Proceedings of the Computer Science and Statistics: Tenth Annual Symposium on the Interface*, edited by D. Hogben, U.S. Government Printing Office, Washington, D.C., 54-57.

Searle, S. R. (1978), A Summary of Recently Developed Methods of Estimating Variance Components, *Proceedings of the Computer Science and Statistics: Eleventh Annual Symposium on the Interface,* edited by A. R. Gallant and T. M. Gerig, Institute of Statistics, North Carolina State University, Raleigh, 64-69.

Speed, F. M. (1969), A New Approach to the Analysis of Linear Models, Ph.D. thesis, Texas A&M University, College Station.

Speed, F. M., and R. R. Hocking (1974), Computation of Expectations, Variances, and Covariances of ANOVA Mean Squares, *Biometrics 30*, 157-169.

Speed, F. M., and R. R. Hocking (1976), The Use of the R( ) Notation with Unbalanced Data, *Amer. Stat. 30*, 30-33.

Speed, F. M., R. R. Hocking, and A. H. Coleman (1977), Three Factor Factorial Analysis with Unequal Number of Observations and Missing Cells, *Proceedings of the Statistical Computing Section, American Statistical Association,* Washington, D.C. 18-21.

Thompson, R. (1975), A Note on the W Transformation, *Techno. 17,* 511-512.

Urquhart, N. S., D. L. Weeks, and C. R. Henderson (1973), Estimation Associated with Linear Models: A Revisitation, *Comm. Stat. (A) 1,* 303-330.

Watson, G. S. (1967), Linear Least Squares Regression, *AMS 38*, 1679-1699.

Wilkinson, G. N. (1970), A General Recursive Procedure for Analysis of Variance, *Biometrika 57*, 19-46.

Wilkinson, G. N. (1978), ANOVA for Non-Orthogonal Data, *Proceedings of the Computer Science and Statistics: Tenth Annual Symposium on the Interface*, edited by D. Hogben, U.S. Government Printing Office, Washington, D.C., 58-65.

Wilkinson, G. N., and C. E. Rogers (1973), Symbolic Description of Factorial Models for Analysis of Variance, *Appl. Stat. 22*, 392-399.

Yates, F. (1934), The Analysis of Multiple Classification with Unequal Numbers in the Subclasses, *JASA 29*, 51-66.

Young, E. H. (1978), A Notation for Specifying Contingent Effects in Analysis of Variance Designs, *Proceedings of the Computer Science and Statistics: Eleventh Annual Symposium on the Interface,* edited by A. R. Gallant and T. M. Gerig, Institute of Statistics, North Carolina State University, Raleigh, 395-399.

Zyskind, G. (1967), On Canonical Forms, Non-negative Covariance Matrices, and Best Single Least Squares Linear Estimators in Linear Models, *AMS 38*, 1092-1109.

Zyskind, G., O. Kempthorne, R. White, E. Dayhoff, and T. Doerfler (1964), *Research on Analysis of Variance and Related Topics*, Aerospace Research Laboratory Office of Aerospace Research, ARL 64-193, U.S. Air Force, Wright-Patterson Air Force Base, Ohio.

Zyskind, G. (1962), On Structure, Relation, Sigma Expectations of Mean Squares, *Sankhya (A) 24*, 117.

10 / UNCONSTRAINED OPTIMIZATION AND NONLINEAR REGRESSION

The statistical estimation of parameters frequently requires optimization (minimization or maximization) of a function. Two of the most common methods of estimation are maximum likelihood and least squares. In maximum likelihood estimation, the estimates of the parameters are taken as those values that maximize the likelihood function given the data. Least-squares estimation yields estimates that minimize the squares of the deviations of the data from the model, summed over all observations. This minimization problem for linear models has been discussed in the preceding two chapters. If the model is nonlinear in the parameters, however, other techniques for minimization are required. A third situation in which the estimation of parameters requires optimization of a function is that in which the criterion of fitting the model is one other than the least-squares criterion (see Chapter 11).

Although in general we may seek the "optimum" of a function (likelihood function, norm of model residuals, etc.), without loss of generality our discussion can be in terms of only the minimum, since if $F(\underline{\hat{\theta}})$ is the minimum of $F(\underline{\theta})$, $-F(\underline{\hat{\theta}})$ is the maximum of $-F(\underline{\theta})$.

We begin the chapter by discussing several methods which are used to compute a vector

$$\underline{\hat{\theta}} = (\hat{\theta}_1, \hat{\theta}_2, \ldots, \hat{\theta}_p)'$$

which minimizes a given scalar valued function $F(\underline{\theta})$. Then we turn to the problems associated with nonlinear least-squares regression computations. These problems include finding $\underline{\hat{\theta}}$, relative to the model

$$y_i = f_i(\underline{\theta}) + e_i \quad (i = 1, 2, \ldots, n)$$

which minimizes

$$F(\underline{\theta}) = \sum_{i=1}^{n} [y_i - f_i(\underline{\theta})]^2$$

The $f_i(\underline{\theta})$ are scalar-valued nonlinear functions of $\underline{\theta}$, and the e_i are random variables assumed to have the same properties as in the multiple linear regression model, which is of course the special model, in the form given above, wherein the $f_i(\underline{\theta})$ are linear functions of $\underline{\theta}$. An example of the type of model we envision in nonlinear regression, with parameters θ_i, is

$$y_i = \theta_1 + \theta_2 \exp(\theta_3 x_i) + e_i \quad (i = 1, \ldots, n)$$

where the x_i are fixed known values of the single independent variable in the model and the y_i are observed values of the dependent variable.

This chapter will not deal with optimization of $F(\underline{\theta})$ subject to constraints on $\underline{\theta}$. This is the problem of *nonlinear programming*, and there is an abundance of literature on this subject. Constrained optimization problems are less often encountered in statistical computing, so we chose to treat the more frequently encountered unconstrained case. Bard (1974) provides a good starting point for the study of estimation under constraints.

10.1 PRELIMINARIES

All computational methods which we will discuss are iterative in nature and are designed to converge to a minimum of the function $F(\underline{\theta})$, provided there is a minimum in the vicinity of the starting point. In general, $F(\underline{\theta})$ may have many local minima in addition to a possible single global minimum. It is unlikely that a numerical method can be found which will always locate an existing global minimum. The best we can hope for generally is a dependable method which converges to a local minimum based on an *initial guess* about a value of $\underline{\theta}$ which minimizes F.

This section is intended to provide definitions and concepts which facilitate the understanding of numerical methods presented in Sections 10.2 and 10.3.

10.1.1 Iteration

All iterative methods which will be discussed proceed, from an initial guess $\underline{\theta}_0$, to produce a sequence $\underline{\theta}_1$, $\underline{\theta}_2$, ... of *iterates* $\underline{\theta}_i$, which we hope converges to the $\hat{\underline{\theta}}$ that minimizes the *objective function* $F(\underline{\theta})$. Computations at the *ith step* consist of determining a vector $\underline{d}_i$, called the *step direction*, and a scalar α_i, called the *step size* (or *step length* when $||\underline{d}_i|| = 1$). These are then used to compute the next iterate

$$\underline{\theta}_{i+1} = \underline{\theta}_i + \alpha_i \underline{d}_i \tag{10.1}$$

Whenever $\underline{\theta}_{i+1}$ is such that $F(\underline{\theta}_{i+1}) < F(\underline{\theta}_i)$, we will say that the step is *acceptable*. An *iterative method is acceptable* if it is designed so that every step that is taken from other than a point where F is minimum is acceptable. We will only consider acceptable methods. These are sometimes called *descent* methods due to the requirement that the objective function be decreased at each step.

The methods which are described in this chapter differ in that they use different criteria for finding directions $\underline{d}_i$ and step sizes α_i. As we shall see in subsequent subsections, there are many ways to construct acceptable methods. Classification and comparison of the various numerical methods for function minimization are, however, based primarily on the way in which $\underline{d}_i$ is determined.

10.1.2 Function Minima

Since $F(\underline{\theta})$ can possibly have relative minima as well as an absolute minimum, we now define what is meant by these terms and consider some characteristics of the function $F(\underline{\theta})$ at such points.

A point $\underline{\theta}^*$ is called a *strong relative minimum* of $F(\underline{\theta})$ if $F(\underline{\theta})$ is defined on a δ neighborhood of $\underline{\theta}^*$ and there exists an ε satisfying $0 < \varepsilon < \delta$ such that

$$F(\underline{\theta}^*) < F(\underline{\theta})$$

for all points such that $0 < ||\underline{\theta}^* - \underline{\theta}|| < \varepsilon$. A *weak relative minimum* exists at $\underline{\theta}^*$ if $\underline{\theta}^*$ is not a strong relative minimum but there exists ε and δ as given above such that

$$F(\underline{\theta}^*) \leq F(\underline{\theta})$$

for all points such that $0 < ||\underline{\theta}^* - \underline{\theta}|| < \varepsilon$. When $\underline{\theta}^*$ is such that $F(\underline{\theta}^*) < F(\underline{\theta})$ for all $\underline{\theta} \neq \underline{\theta}^*$ in a given subspace of p-dimensional euclidean space E^p the point $\underline{\theta}^*$ is said to be a *global or absolute minimum* of $F(\underline{\theta})$ with respect to that subspace. For unconstrained problems the entire space is normally considered. The term *local minimum* is sometimes used instead of relative minimum to describe the points defined above. For functions of a single variable, for example, a weak relative minimum point corresponds to a function value on a "plateau," whereas the function value at a strong relative minimum is at the bottom of a "valley." Fleming (1965) and Murray (1972b) provide a good discussion of these fundamentals.

It will be assumed in all subsequent discussion that $F(\underline{\theta})$ has continuous first- and, as needed, second-order derivatives in the sense that the *gradient vector* of F

$$\underline{g}(\underline{\theta}) = \left[\frac{\partial F(\underline{\theta})}{\partial\theta_1}, \frac{\partial F(\underline{\theta})}{\partial\theta_2}, \ldots, \frac{\partial F(\underline{\theta})}{\partial\theta_p}\right]'$$

and the *Hessian matrix* $G(\underline{\theta}) = (\gamma_{ij})$ with

$$\gamma_{ij} = \frac{\partial^2 F(\underline{\theta})}{\partial\theta_i \partial\theta_j}$$

have continuous elements. The introduction of partial derivatives into the discussion allows definition of a *stationary point* of F to be any point $\underline{\theta}^*$ at which the gradient vector $\underline{g}$ vanishes; in other words

$$\underline{g}^* = \underline{g}(\underline{\theta}^*) = \left[\frac{\partial F(\underline{\theta})}{\partial\theta_1}, \ldots, \frac{\partial F(\underline{\theta})}{\partial\theta_p}\right]'\Bigg|_{\underline{\theta}=\underline{\theta}^*} = \underline{0}$$

A necessary condition that $\underline{\theta}^*$ be a relative minimum of $F(\underline{\theta})$ is that $\underline{\theta}^*$ is a stationary point of $F(\underline{\theta})$. The term *saddle point* of $F(\underline{\theta})$ is applied to any stationary point of $F(\underline{\theta})$ which is not a relative minimum or maximum point.

For the types of functions under consideration, $\underline{\theta}^*$ is a strong relative minimum of F if $\underline{g}^* = \underline{0}$ and $G^* = G|_{\underline{\theta}=\underline{\theta}^*}$ is positive definite. If G^* is only positive semidefinite at a stationary point, then $\underline{\theta}^*$ is a weak relative minimum. The algorithms we will discuss will be designed to seek a strong relative minimum.

10.1.3 Step Direction

Suppose that in the ith iteration the step direction $\underline{d}_i$ has been selected. (Several methods for finding $\underline{d}_i$ are given in Section 10.2.) Let us consider conditions under which the choice of direction leads to an acceptable step.

First note that

$$\rho(\alpha) = F(\underline{\theta}_i + \alpha\underline{d}_i)$$

for given $\underline{\theta}_i$ and $\underline{d}_i \neq \underline{0}$, is solely a function of α. Differentiation with respect to α, and evaluating the resultant function at $\alpha = 0$, gives the *directional derivative* of F at $\underline{\theta}_i$ in the direction $\underline{d}_i$. Carrying out these operations gives

$$\begin{aligned} D &= \frac{d\rho}{d\alpha}\bigg|_{\alpha=0} \\ &= \left(\frac{\partial F}{\partial \underline{\theta}}\right)' \left(\frac{\partial \underline{\theta}}{\partial \alpha}\right)\bigg|_{\alpha=0,\underline{\theta}=\underline{\theta}_i} \\ &= \underline{g}'(\underline{\theta}_i + \alpha\underline{d}_i)\underline{d}_i\big|_{\alpha=0} \\ &= \underline{g}_i'\underline{d}_i \end{aligned} \tag{10.2}$$

where $\underline{g}_i'$ is the transpose of the gradient vector of F evaluated at $\underline{\theta} = \underline{\theta}_i$.

When $\underline{g}_i = \underline{0}$ we are at a stationary point and will not want to make a step. Assume $\underline{g}_i \neq \underline{0}$, in which case a step in the nonzero direction $\underline{d}_i$ will be made if the direction is appropriate. If $D < 0$, it can be shown (e.g., see Ortega and Rheinboldt 1970, page 243) that there is a $\delta > 0$ such that

$$F(\underline{\theta}_i + \alpha \underline{d}_i) < F(\underline{\theta}_i)$$

for all $\alpha \in (0,\delta)$. When D is nonnegative F will increase with α or remain constant. Thus an acceptable step can be made only if $D < 0$, and when such is the case we say that $\underline{d}_i$ is an *acceptable direction*. The following theorem given by Bard (1974) is useful in this regard.

Theorem 10.1

A direction $\underline{d}_i$ is acceptable if and only if there exists a positive definite matrix R such that

$$\underline{d}_i = -R\underline{g}_i$$

Note that whenever $\underline{d}_i = -R\underline{g}_i$ then Eq. (10.2) becomes

$$D = \underline{g}_i'\underline{d}_i = -\underline{g}_i'R\underline{g}_i$$

which is negative whenever $\underline{g}_i \neq \underline{0}$ and R is positive definite. The class of iterative methods which utilize $\underline{d}_i$ in the form given in Theorem 10.1 are called *gradient methods*. The basic iteration (10.1) for these methods becomes

$$\underline{\theta}_{i+1} = \underline{\theta}_i - \alpha_i R_i \underline{g}_i \qquad (\alpha_i \geq 0) \tag{10.3}$$

Gradient methods differ in the way R_i and α_i are chosen.

In Section 10.2 we will describe several computational methods, each method characterized primarily by its choice of step direction. The other major action which must be taken at each step in the iteration is determination of the step size. As will be seen in the next subsection, there are many methods which can be used to obtain the α_i, and selection of step size does not in most cases depend on the way in which $\underline{d}_i$ is determined.

10.1.4 Step Size

Many methods have been proposed for determining step size in the general iterative procedure defined by Eq. (10.1). These are often called *linear search* methods. The basic problem is to find $\alpha^* > 0$ which minimizes

$$\rho(\alpha) = F(\underline{\theta}_i + \alpha\underline{d}_i)$$

We are, of course, assuming that $\underline{d}_i$ is acceptable so that $\rho(\alpha_i) < \rho(0)$ for some $\alpha_i > 0$. In this subsection we will describe some available linear search methods and make references to other methods. The reader may wish to skip the details of this subsection in the first reading of this chapter.

Some of the descent methods described in Section 10.2 are relatively insensitive to the degree to which α^* is approximated, so long as the objective function value is reduced at each step. Other methods seem to need a close approximation $\hat{\alpha}$ of α^* to work well, particularly as the iterates approach a stationary point. It appears to be generally the case that convergence is obtained in fewer iterations when a good approximation of α^* is used at each step, but this does not necessarily imply reduced computing cost because the more exhaustive linear searches require more evaluations of F. Thus the savings that are made by fewer overall iterations may be erased by increased cost of linear search. Efficiency in linear search computations is therefore needed in addition to achievement of suitable accuracy in the approximation of α^*.

There are two basic types of methods for linear search. Methods of the first type use a scheme for systematically reducing the length of the known interval which contains α^* and they utilize comparison of function values as a basis for making decisions at various stages in the process. The method of bisection, mentioned in Chapter 5, is one of this type as is the Illinois method. The other type of methods approximates $\rho(\alpha)$ using a function $\hat{\rho}(\alpha)$ for which a minimum is easily obtained. Second- or third-degree polynomials are sometimes used. The fitting of $\hat{\rho}$ is accomplished using function values $\rho(\alpha_j)$ (and possibly derivatives) for points α_j that span an interval containing α^*. Successive fits are made until a sufficiently good approximate value $\hat{\alpha}$ is obtained. Sometimes a combination of methods of the two types is used. For example, polynomial fits can be made to identify a small interval that contains α^* and then a method of the first type is used to complete the process and find $\hat{\alpha} \cong \alpha^*$.

Regardless of the method used, values of α which bracket α^* are needed initially. These are usually found by starting at $\alpha = 0$ and taking steps through larger α values until an increase in ρ results. Algorithms often utilize a parameter which defines the maximum allowable step size to insure against wildly erratic behavior.

Most linear search methods were derived under the assumption that the function to be used is unimodal over the region of search. This assumption is not always satisfied in practice, but even when it is violated an acceptable step is usually found so that the iteration can continue.

We now consider a linear search method, called the method of *golden section search*, which is of the first type and is a modified form of the so-called Fibonacci search. The golden section search is not quite as efficient as the Fibonacci search (see Keifer 1957), but it is easier to use. The method is designed to reduce the length of the interval of search by a constant factor at each iteration as is shown by the following algorithm.

ALGORITHM GOLDEN ($\varepsilon,\tau,\hat{\alpha}$)

Step	*Description*
	Comment: $\varepsilon > 0$, $\tau > 0$ are user-supplied. Constants v_1 and v_2 are given by $v_1 = (3 - \sqrt{5})/2$ and $v_2 = (\sqrt{5} - 1)/2$. Steps 1 to 6 find the initial bracket.
1.	Compute $\rho(0)$ and $\rho(\tau)$.
2.	If $\rho(\tau) \geq \rho(0)$, set $a_0 = 0$, $b_0 = \tau$, and go to step 7.
3.	Set $i = 0$, $\gamma_0 = 0$.
4.	Set $\gamma_{i+1} = \gamma_i + \tau$.
5.	Compute $\rho(\gamma_{i+1})$.

Algorithm cont.

Step	*Description*
6.	If $\rho(\gamma_{i+1}) \geq \rho(\gamma_i)$, set $a_0 = \gamma_{i-1}$, $b_0 = \gamma_{i+1}$, and go to step 7; otherwise set $i = i + 1$ and go to step 4. Comment: Now $\alpha^* \varepsilon [a_0, b_0]$.
7.	Set $j = 0$.
8.	Set $\ell_j = b_j - a_j$.
9.	If $\ell_j \leq \varepsilon$, go to step 12.
10.	Set $\delta_j = a_j + v_1\ell_j$ and $\sigma_j = a_j + v_2\ell_j$.
11.	If $\rho(\delta_j) < \rho(\sigma_j)$, set $a_{j+1} = a_j$, $b_{j+1} = \sigma_j$, $j = j + 1$, and go to step 8. Otherwise set $a_{j+1} = \delta_j$, $b_{j+1} = b_j$, $j = j + 1$, and go to step 8.
12.	Set $\hat{\alpha} = (a_j + b_j)/2$ and exit.

One desirable feature of this method is that since $v_1 = v_2^2$ and $v_2 = 1 - v_1$, it follows that $\sigma_{j+1} = \delta_j$ when $a_{j+1} = a_j$ and $b_{j+1} = \sigma_j$ at step 11; and $\delta_{j+1} = \sigma_j$ when $a_{j+1} = \delta_j$ and $b_{j+1} = b_j$ at step 11. Therefore, only one function evaluation is needed for each iteration of steps 8 to 11. We also note that the ℓ_j computed in step 8 is such that $\ell_j = v_2^j \ell_0$.

Davidon (1959) suggested a linear search method of the second type for use with the variable metric algorithm (see Section 10.2.3). This search method is a cubic interpolation. The initial requirements are the function values of $F(\underline{\delta}_k)$ and gradient vector values $\underline{g}(\underline{\delta}_k)$, for two points $\underline{\delta}_1$ and $\underline{\delta}_2$ that bracket the minimum, where

$$\underline{\delta}_k = \underline{\theta}_i + \alpha_k \underline{d}_i \qquad (\alpha_1 < \alpha_2)$$

The minimum of the fitted cubic is determined by

$$\hat{\alpha} = \alpha_1 + \tau$$

where

$$\tau = (\alpha_2 - \alpha_1)\left\{1 - \frac{[\underline{g}'(\underline{\delta}_2)\underline{d}_i + a - b]}{[\underline{g}'(\underline{\delta}_2)d_i - \underline{g}'(\underline{\delta}_1)\underline{d}_i + 2a]}\right\}$$

and

$$a = \{b^2 - \underline{g}'(\underline{\delta}_1)\underline{d}_i\underline{g}'(\underline{\delta}_2)\underline{d}_i\}^{1/2}$$

$$b = \frac{F(\underline{\delta}_1) - F(\underline{\delta}_2)}{(\alpha_2 - \alpha_1)/3} + \underline{g}'(\underline{\delta}_1)\underline{d}_i + \underline{g}'(\underline{\delta}_2)\underline{d}_i$$

After an interpolation $\hat{\alpha}$ is taken as the required $\hat{\alpha} \cong \alpha^*$ if $F(\underline{\theta}_i + \hat{\alpha}\underline{d}_i)$ is smaller than each of the values $F(\underline{\delta}_1)$ and $F(\underline{\delta}_2)$. If this is not the case, we can inspect the sign of $\underline{g}'(\underline{\theta}_i + \hat{\alpha}\underline{d}_i)\underline{d}_i$, select a shorter-length interval that contains α^*, and either perform another cubic interpolation or switch to another linear search method.

Cubic interpolation requires evaluation of both the function F and its gradient vector $\underline{g}$. Quadratic interpolation in the linear search does not utilize $\underline{g}$; hence, while less precise, it is useful for minimization methods that do not use $\underline{g}$. Powell (1964) gave a quadratic interpolation which was discussed and extended by Box, Davies, and Swann (1969) and summarized by Jacoby, Kowalik and Pizzo (1972).

Additional linear search methods, and useful discussion of this subject, were given by Bard (1970), Powell (1971), Polak (1971), Murray (1972b), Jacoby, Kowalik, and Pizzo (1972), and Bard (1974).

A somewhat different method which selects α_i to satisfy

$$F(\underline{\theta}_i + \alpha_i\underline{d}_i) \leq F(\underline{\theta}_i) + \tau\alpha_i\underline{g}_i'\underline{d}_i \qquad \tau \in (0,1/2)$$

was suggested in the work of Goldstein (1965) and Armijo (1966). Dennis and More (1977) discussed this method and included criteria for choosing the parameter τ. McCormick (1977) provided an extension to include second derivative information. The Goldstein-Armijo method is helpful in avoiding the situation where very small steps α_i are dictated by the linear search algorithm causing premature termination in the iteration toward $\underline{\hat{\theta}}$. The usual procedure is to choose a τ value and then find α_i as the largest element in the sequence 1, 1/2, 1/4, 1/8, ... which satisfies the inequality given above.

Often, mathematical derivation of a minimization method will assume *exact linear searches* are performed. In other words, it is assumed that the theoretically exact value α^* which minimizes $F(\underline{\theta} + \alpha\underline{d})$ is used. Despite these theoretical considerations, algorithms do not usually attempt to perform linear search to machine precision in an effort to simulate exact linear search.

10.1.5 Convergence of the Iterative Methods

We shall not attempt to deal with mathematical proofs of convergence of iterative nonlinear optimization methods in this chapter. Most proofs presume conditions which are difficult to verify in practical applications, and practitioners frequently do not attempt to do so. Even when a method is supported by a proof of convergence, numerical inaccuracies in application may well result in nonconvergence. Extensive discussion of convergence relative to the iterative methods which are given in Section 10.2 are contained in many publications, such as Ortega and Rheinboldt (1970), Polak (1971), Powell (1972), Broyden, Dennis, and More (1973), and Dennis and More (1977).

The majority of the better methods for nonlinear optimization are designed to easily find the minimum of a convex quadratic function

$$f(\underline{\theta}) = c + \underline{b}'\underline{\theta} + \frac{1}{2}\,\underline{\theta}'A\underline{\theta}$$

where A is a p x p symmetric positive definite matrix. Usually, the method will theoretically converge to the minimum point $\underline{\theta}^*$ in a small finite number of steps. Minimization methods of this kind are said to be *quadratically convergent*. This approach to development of computational methods is desirable because the quadratic is the simplest nontrivial function which has a well-defined minimum, and quadratically convergent methods usually converge rapidly from points near a minimum of a general nonlinear differentiable function. Most methods are sensitive to the initial starting point, and with quadratically convergent methods it may be necessary to select a starting point that is in a region about the minimum point over which the function can be adequately approximated using a quadratic function. Since first- and second-order derivatives of a quadratic function exist continuously, these methods may not perform well for general nonlinear functions that do not have continuous derivatives.

For computational purposes it is not enough to have an algorithm that converges. Consideration must also be given to the rate of convergence, else the user may "break the bank" before coming close to solving a given problem using the algorithm in a computer. The better methods for nonlinear optimization converge at least superlinearly. Loosely speaking, a superlinear convergence means that there is a sequence of scalar values $\{\delta_k\}$ which converges to zero and

$$||\underline{\theta}_{k+1} - \underline{\theta}^*|| \leq \delta_k ||\underline{\theta}_k - \underline{\theta}^*||$$

for each k. (The norm in this case, and throughout this chapter unless stated otherwise, is the usual euclidean norm.) Some methods have a locally quadratic rate of convergence which means that when we are "sufficiently" close to a solution the number of correct digits is approximately doubled at each iteration until machine precision within rounding error is reached.

In the absence of convergence proofs, some degree of confidence can be placed in acceptable methods, that is, methods in which the objective function values form a monotonically decreasing sequence over the iterates. When the objective function does have a minimum, it is not unreasonable to expect that the sequence of iterates $\{\underline{\theta}_i\}$ will converge under this condition.

10.1.6 Termination of Iteration

Regardless of the particular minimization method used, we must have a rule for stopping the iteration process. One natural choice using gradient methods would be to stop when the norm of the gradient vector becomes smaller than some given tolerance ε. This does not, however, always work well in practice because the components of this vector are often difficult to compute to a high degree of accuracy and it is often observed that a minimum is reached while the components of the gradient differ significantly from zero due to rounding error.

One appealing and often used criterion is to monitor the iterates $\underline{\theta}_i$ and stop when the components of this vector cease to change significantly over a selected number of iterations. We must, however, not be too strict in choosing this number because some methods may make very slow progress for several iterations and then suddenly make a comparatively large improvement. Marquardt (1963) suggests a criterion which is essentially the following. Accept $\underline{\theta}_{i+1}$ as the solution whenever the components of $\underline{\theta}_i$, call them $\theta_{i,j}$ $(j = 1, 2, \ldots, p)$, satisfy

$$|\theta_{i,j} - \theta_{i+1,j}| \leq \varepsilon_1(|\theta_{i,j}| + \varepsilon_2) \quad (j = 1, 2, \ldots, p) \tag{10.4}$$

over a given number of iterations, where $\varepsilon_1 = 10^{-4}$ and $\varepsilon_2 = 10^{-3}$ are often used.

There is no known termination criterion that is absolutely satisfactory. Computer programs usually contain input parameters which allow the user some control over the stringency of a given general criterion that is built into the program.

Among the many other termination criteria which have been suggested and which are often used in conjunction with, or in lieu of, the criterion (10.4) are

1. $|F(\underline{\theta}_i) - F(\underline{\theta}_{i+1})| < \varepsilon$
2. $||\underline{g}_i|| < \varepsilon$
3. $\max_j |j\text{th component of } \underline{g}_i| < \varepsilon$

10.2 METHODS FOR UNCONSTRAINED MINIMIZATION

In this section we describe certain selected methods for function minimization. The description in most instances includes variations of a basic method. These variations are usually designed to either eliminate the need for user-supplied derivatives of the objective function, or increase stability of the method, or facilitate the determination of step size. When studying basic methods and their variations, the reader should always consider what quantities are needed in application. A method which requires the user to supply algebraic forms of the gradient vector and Hessian matrix elements will certainly not be as easy to use as a method that requires only first partial derivatives or no derivatives at all.

Several small examples will be given that are solved by hand calculations. These are designed to illustrate the elementary computational steps used to carry out the method. The reader should not, however, be misled by the simplicity of these examples. Good software systems that implement the methods to be described are not at all easy to develop. In fact, the area of unconstrained minimization is very difficult to support with dependable software. In many instances specific devices are built into the more successful programs to help resist numerical difficulties that seem to arise with great frequency in practical applications. The methods given in this section are considered to be the best available, but even with the best computer implementation, no single one will solve the vast majority of problems that arise. Dealing successfully with practical problems in unconstrained minimization often requires "tuning" software for a given difficult problem. By this we mean altering the program to take full advantage of any special features or requirements in the problem. Of course one cannot do this without a thorough understanding of the computational method being employed, and helping the reader to gain such understanding is the primary purpose of this section.

10.2.1 Method of Steepest Descent

Steepest descent is one of the oldest methods for determining the minimum of a function. It is generally accepted that Cauchy (1847) first proposed this method. It is seldom used today because it is often slow to converge. We describe this method mainly for the sake of completeness.

It is easily shown (e.g., see Fleming 1965, page 61) for any given point $\underline{\theta}_i$ where the gradient vector $\underline{g}_i = \underline{g}(\underline{\theta}_i)$ of $F(\underline{\theta})$ is not zero that $\underline{g}_i$ defines the direction of maximum local increase in $F(\underline{\theta})$. Thus if we want to proceed most rapidly downhill on the surface $F(\underline{\theta})$, we should consider $\underline{\theta}$'s in the direction $-\underline{g}_i$ from $\underline{\theta}_i$. This choice of direction defines the method of steepest descent. For this method Eq. (10.1) becomes

$$\underline{\theta}_{i+1} = \underline{\theta}_i - \alpha_i \underline{g}_i \tag{10.5}$$

This is obviously a gradient method for which $R = I_p$ in Theorem 10.1.

Most implementations of this method normalize $\underline{g}_i$ at each step to produce a vector having unit euclidean length. Ortega and Rheinboldt (1970) generalize to other norms by proving that if C is a positive definite symmetric matrix, with usual regularity assumptions about $F(\underline{\theta})$ and its derivatives, the direction of steepest descent in the norm $||\underline{\theta}|| = (\underline{\theta}'C\underline{\theta})^{1/2}$ is given by

$$-C^{-1}\underline{g}(\underline{\theta}) \tag{10.6}$$

One way of viewing the various methods that are presented in this section is to think of them as trying to determine a norm at each iteration which will cause the steepest descent direction to point at the minimum.

An algorithm for steepest descent is as follows, with $||\underline{g}|| = (\underline{g}'\underline{g})^{1/2}$.

ALGORITHM SDSNT$(\varepsilon,\tau,\underline{\theta}_1,\hat{\underline{\theta}})$

Step *Description*

Comment: $\underline{\theta}_1$ is the given initial value and $i = 0$ is assumed on entry.

1. Set $i = i + 1$, compute $\underline{g}(\underline{\theta}_i)$ and $||\underline{g}_i||$.
2. If $||\underline{g}_i|| > \tau$, let $\bar{\underline{g}}_i = \underline{g}_i/||\underline{g}_i||$. Otherwise, exit with $\hat{\underline{\theta}} = \underline{\theta}_i$.
3. Compute step size α_i using one of the methods discussed in Section 10.1.4.

Algorithm cont.

Step	*Description*
4.	Compute $\underline{\theta}_{i+1} = \underline{\theta}_i - \alpha_i \bar{\underline{g}}_i$.
5.	Terminate if appropriate based on the $\underline{\theta}_i$ and tolerance ε. Otherwise return to step 1.

The steepest descent method often tends to make good progress initially and then slows down as the iterates approach a stationary point. Sometimes this method is used for a few iterations to produce a starting value for another algorithm which converges more rapidly given good initial data. Basically, the method is sensitive to small perturbations in direction and step size, so these quantities must be computed to relatively high precision else the iterations will often appear to take a divergent track. This is not to say that the underlying mathematical method is faulty. Some of the most satisfactory convergence theorems, in this area of nonlinear function minimization, have been developed for the steepest descent method. These show that the sequence $\{\underline{g}_i\}$ converges to zero under fairly general regularity conditions (see, e.g., Dennis and More 1977). Account was taken of this fact in step 3 of Algorithm SDSNT where $||\underline{g}_i|| > \tau$ was checked before a desired scaling to producing $\bar{\underline{g}}_i$ was made.

Example 10.1

(Steepest descent for a simple quadratic function)

Consider the following bivariate quadratic function:

$$F(\underline{\theta}) = \theta_1 + \theta_1^2 + \frac{\theta_2^2}{2}$$

For this function we have the gradient vector

$$\underline{g}(\underline{\theta}) = (1 + 2\theta_1, \theta_2)'$$

The method of steepest descent was applied to this function using the starting value $\underline{\theta}_0' = (0,1)$. A summary of the results obtained in the first five iterations is given below with values rounded to five decimal places:

i	θ_1	θ_2	$[\underline{g}(\underline{\theta})]_1$	$[\underline{g}(\underline{\theta})]_2$	$F(\underline{\theta})$
1	-0.35355	0.64645	0.29289	0.64645	-0.01961
2	-0.55990	0.19101	-0.11980	0.19101	-0.22817
3	-0.49348	0.08512	0.01303	0.08512	-0.24634
4	-0.50294	0.02333	-0.00589	0.02333	-0.24972
5	-0.49912	0.00819	0.00176	0.00819	-0.24997

Thus after five iterations we have approximately one significant digit of accuracy in the $\underline{\theta}$ and $\underline{g}$ components, but $F(\underline{\theta})$ shows more accuracy. In this trial an inexact linear search was made at each iteration. Let us compare these results with the following results using hand calculations and exact linear search:

i	θ_1	θ_2	$[g(\underline{\theta})]_1$	$[g(\underline{\theta})]_2$	$F(\underline{\theta})$
1	-2/3	1/3	-1/3	1/3	-1/6
2	-4/9	1/9	1/9	1/9	-13/54
3	-14/27	1/27	-1/27	1/27	-181/729
4	-40/81	1/81	1/81	1/81	-3279/13,122

Note that when the linear search is exact, successive gradient vectors are orthogonal, as they should be for this method. This will, however, not be the case under an inexact linear search. The fact that successive gradient vectors are orthogonal helps to explain why this method tends to make very slow progress when approaching a local minimum.

The objective function used in Example 10.1 is quadratic and fits the general form

$$f(\underline{\theta}) = c + \underline{\theta}'\underline{b} + \frac{1}{2}\,\underline{\theta}'A\underline{\theta}$$

with $c = 0$, $\underline{b}' = (1,0)$, and positive definite matrix

$$A = \begin{pmatrix} 2 & 0 \\ 0 & 1 \end{pmatrix}$$

The gradient vector in this case is easily seen to be

$$\underline{g}(\underline{\theta}) = \underline{b} + A\underline{\theta}$$

so the solution in this quadratic case is easily obtained by solving the system of linear equations

$$A\underline{\theta} + \underline{b} = \underline{0}$$

10.2.2 Newton's Method and Some Modifications

Newton's method (also called the Newton-Raphson method) is a widely used and often-studied method for minimization. The method requires use of both the gradient vector and the Hessian matrix in computations; hence it places more burden on the user to supply derivatives of the objective function than does the steepest descent method. In return for this added inconvenience the method normally displays a quadratic rate of convergence near a stationary point, and when a stationary point is reached the Hessian matrix is available for use in determining the relative strength of the minimum.

Consider a Taylor expansion of F about the point $\underline{\theta}_i$. This takes the form

$$F(\underline{\theta}) = F(\underline{\theta}_i) + (\underline{\theta} - \underline{\theta}_i)'\underline{g}_i + \frac{1}{2}(\underline{\theta} - \underline{\theta}_i)'G_i(\underline{\theta} - \underline{\theta}_i) + E(\underline{\theta} - \underline{\theta}_i) \tag{10.7}$$

where $E(\underline{\theta} - \underline{\theta}_i)$ is the remainder. When $\underline{\theta}$ is close to $\underline{\theta}_i$ we can expect that the quadratic function

$$h_i(\underline{\theta}) = F(\underline{\theta}_i) + (\underline{\theta} - \underline{\theta}_i)'\underline{g}_i + \frac{1}{2}(\underline{\theta} - \underline{\theta}_i)'G_i(\underline{\theta} - \underline{\theta}_i) \tag{10.8}$$

will approximate $F(\underline{\theta})$. The stationary point of $h_i(\underline{\theta})$ is given by the solution to the linear system of equations

$$G_i(\underline{\theta} - \underline{\theta}_i) = -\underline{g}_i \tag{10.9}$$

If G_i is positive definite, this suggests the general iterative scheme

$$\underline{\theta}_{i+1} = \underline{\theta}_i - G_i^{-1}\underline{g}_i \tag{10.10}$$

Equation (10.10) defines *Newton's* iterative method which is in the form of Eq. (10.1) with $\alpha_i = 1$ and $\underline{d}_i = -G_i^{-1}\underline{g}_i$. When $F(\underline{\theta})$ is closely approximated by $h(\underline{\theta})$ in the neighborhood of a strong relative minimum point θ^*, convergence will normally be at a quadratic rate if the Hessian is positive definite at each step. If the initial starting point is "too far" from a minimum, the method will not converge.

One problem with Newton's method is that the Hessian may not be positive definite at each iteration. Thus the method requires modification to insure that the resultant method is acceptable but still retains the desirable characteristics of Newton's method. Several modifications have been proposed. All incorporate a nonconstant step size to help avoid stepping across a minimum and possibly moving outside the neighborhood where a quadratic function gives an adequate fit. We will now describe some of the proposed modifications.

Greenstadt (1967) shows that Newton's method can be significantly more efficient than steepest descent and suggests replacing all negative eigenvalues of G_i with their absolute value and replacing zero eigenvalues with a small positive number. This gives a positive definite matrix $\bar{G}_i$, and (10.10) is used in the modified form

$$\underline{\theta}_{i+1} = \underline{\theta}_i - \alpha_i \bar{G}_i^{-1} \underline{g}_i \tag{10.11}$$

This modification seems to work well in practice. The requirement to compute eigenvalues and vectors of G_i at each iteration (or do a singular-valve decomposition optionally is unfortunate, and generally we would desire a less expensive alternative. Bard (1974), Murray (1972a), and Jacoby, Kowalik, and Pizzo (1972) all discuss Greenstadt's basic method and suggest techniques for implementation in the computer.

One way to improve efficiency becomes apparent when we note that the problem of computing $\underline{d}_i$ at each step requires (1) determining whether G_i is positive definite and modifying G_i if it is not; and (2) solving the system of linear equations

$$\bar{G}_i \underline{d}_i = -\underline{g}_i \tag{10.12}$$

for $\underline{d}_i$ where $\bar{G}_i = G_i$ when G_i is positive definite. This suggests that a method for solving linear systems of equations which flags semidefinite and indefinite G_i will be useful. The Cholesky decomposition method is one that does this. Using the Cholesky method when G_i is positive definite will yield a solution for $\underline{d}_i$ by the usual decomposition and two back-substitution stages, without additional consideration. When the standard Cholesky decomposition of G_i cannot be completed, because G_i is not positive definite, either an eigenvalue-eigenvector decomposition of G_i can be computed or a modified Cholesky decomposition (see Murray 1972a) can be used to compute $\bar{G}_i$. The latter procedure is preferred because the modified Cholesky decomposition is both fast and accurate, and it reduces to the standard Cholesky decomposition when G_i is positive definite. Fiacco and McCormick (1968), Matthews and Davies (1971), and Gill and Murray (1974) also consider use of a Cholesky-type decomposition method in this application. The algorithm given at the end of this subsection is of this type.

An older modification which led to the use of the modified Cholesky decomposition was suggested in the work of Levenberg (1944) and Marquardt (1963). This modification of Newton's method is described by

$$\underline{\theta}_{i+1} = \underline{\theta}_i - \alpha_i(G_i + \tau_i I_p)^{-1}\underline{g}_i \tag{10.13}$$

where $\tau_i \geq 0$ is chosen so that $G_i + \tau_i I_p$ is positive definite. Note that as τ_i approaches infinity the direction in (10.13) approaches that of steepest descent, and when τ_i is small the direction specified is close to the Newton direction. Thus the direction used in (10.13) is an interpolation between the Newton and steepest descent direction. The primary problem with use of (10.13) is that a τ_i must be found at each step which produces an acceptable direction. A small τ_i is preferable so that the computed direction remains close to the Newton direction. Computation of τ_i will require repeated solution of a system of linear equations or, alternatively, computation of eigenvalues and vectors of G_i. (Usually, $\alpha_i \equiv 1$ is used in this method, since τ_i will adjust for step size.) If we elect to perform an eigen-analysis, then, given orthonormal eigenvectors $\underline{v}_j$ and the eigenvalues λ_j of G_i, we select τ_i as the smallest positive value such that $\lambda_j + \tau_i > \varepsilon$ for every j = 1, 2, ..., p. Having done this, $\bar{G}_i^{-1}$ can be computed as

$$(G_i + \tau_i I)^{-1} = \sum_{j=1}^{p} (\lambda_j + \tau_i)^{-1} \underline{v}_j \underline{v}_j'$$

However, as was mentioned previously, computing eigenvalues and vectors at each iteration is expensive. Goldfeldt, Quandt, and Trotter (1966) generalize the form (10.13) and discuss details in application. In some cases a more general form of positive definite diagonal matrix is used instead of $\tau_i I_p$ in Eq. (10.13). This device allows, as in Greenstadt's approach, alteration of the individual eigenvalues of G_i by varying amounts.

We now consider details of an algorithm which uses a modification of Newton's method and employs the Cholesky decomposition of G_i. This algorithm is given by Gill and Murray (1974).

Given that G_i is positive definite, we will begin by considering Cholesky decomposition of $G_i = (\gamma_{jk})$ which can be expressed in the form

$$G_i = LDL'$$

where $L = (\ell_{jk})$ is a p-square unit lower-triangular matrix and $D = (\delta_{jj})$ is diagonal having positive diagonal elements. In this notation, defining equations for the decomposition are, at stage j, restated from Chapter 7 in the form (recall that if j = 1, the sums are defined to be zero)

$$\delta_{jj} = \gamma_{jj} - \sum_{r=1}^{j-1} \ell_{jr}^2 \delta_{rr}$$

$$\ell_{kj} = (\gamma_{kj} - \sum_{r=1}^{j-1} \ell_{jr}\ell_{kr}\delta_{rr})\delta_{jj}^{-1} \qquad k = j + 1, \ldots, p \tag{10.14}$$

Defining $c_{jr} = \ell_{jr}\delta_{rr}$, these equations become

$$\delta_{jj} = \gamma_{jj} - \sum_{r=1}^{j-1} \ell_{jr} c_{jr}$$

$$c_{kj} = \gamma_{kj} - \sum_{r=1}^{j-1} \ell_{jr} c_{kr} \qquad k = j + 1, \ldots, p \tag{10.15}$$

At step j the computations have progressed as summarized by the following array:

$$\begin{array}{llll}
\delta_{11} & & & \\
\ell_{21} & \delta_{22} & & \\
\vdots & \vdots & \ddots & \\
\ell_{j-1,1} & \ell_{j-1,2} & \cdots & \delta_{j-1,j-1} \\
c_{j,1} & c_{j,2} & \cdots & c_{j,j-1} \\
\vdots & & & \vdots \\
c_{n,1} & c_{n,2} & \cdots & c_{n,j-1}
\end{array}$$

The values c_{jr} in the jth row are not required for computation of columns j + 1, ..., p, so the ℓ_{jr} (r = 1, ..., j - 1) can be computed while δ_{jj} is being formed and overwrite the c_{jr} at that time. This ordering of the computations is particularly desirable because when G_i is not positive definite, one or more of the δ_{rr} will be nonpositive, and a modified $\bar{G}_i$ can be found in the course of the decomposition. Even when G_i is positive definite, it may be nearly singular, in which case we will want to improve its condition to help insure that the computations proceed without encountering serious numerical difficulties.

Equations (10.14) show that whenever δ_{jj} is small, some of the ℓ_{jr} can be large. However, since δ_{jj} in (10.14) is positive for positive definite G_i, it follows that

$$\ell_{jr}^2 \delta_{rr} \leq \gamma_{jj}$$

or

$$\ell_{jr}\delta_{rr}^{1/2} \le \gamma_{jj}^{1/2}$$

Hence no element of $LD^{1/2}$ can exceed $\gamma_{jj}^{1/2}$. Thus a large element in $LD^{1/2}$ can occur only if G_i has a large diagonal element. We shall make use of this fact to form a Cholesky decomposition of $\bar{G}_i = \overline{LDL'}$ which is such that (1) all elements of $\overline{LD}^{1/2}$ are bounded above by a chosen value β; (2) all elements of $\bar{D}$ are bounded below by a chosen positive value ξ; and (3) $\bar{G}_i = G_i$ when G_i is positive definite and sufficiently well conditioned. The form of $\bar{G}_i$ will be

$$\bar{G}_i = G_i + M$$

where $M = \text{diag}(m_j)$ will be selected to satisfy the three requirements given above. We will now consider the form of M and suggest a value for each of the parameters β and ξ.

To begin the derivation of M, define

$$\psi_j = \gamma_{jj} - \sum_{r=1}^{j-1} \ell_{jr}c_{jr}$$

Imposing the lower bound ξ as given above $\bar{\delta}_{jj}$ will, in the absence other considerations, be chosen as

$$\bar{\delta}_{jj} = \max\{\xi, |\psi_j|\}$$

where ξ is a small positive tolerance that is only used when δ_{jj} is sufficiently close to zero, and $|\psi_j|$ is specified to account for the possibility of an indefinite G_i. One choice of ξ which has been used in practice is

$$\xi = \max\{m_\varepsilon ||G_i||, m_\varepsilon\}$$

where m_ε is near the smallest positive floating-point number representable by the computer being used.

The value of $\bar{\delta}_{jj}$ given above will be used provided that the resulting column elements in $\overline{LD}^{1/2}$ are less than β. These elements are $\bar{\ell}_{kj}\bar{\delta}_{jj}^{1/2}$, $k = j + 1, \ldots, p$ where $\bar{\ell}_{kj}$ reflects the possible use of modified δ's in (10.14). To determine when $\bar{\delta}_{jj}$ should be further modified define $t = \max(|c_{kj}|,\ k = j + 1, \ldots, p)$. Whenever $t^2 \leq \beta^2\bar{\delta}_{jj}$ we have

$$c_{kj}^2 = (\bar{\ell}_{kj}\bar{\delta}_{jj})^2 \leq \beta^2\bar{\delta}_{jj}$$

Hence

$$|\bar{\ell}_{kj}\bar{\delta}_{jj}^{1/2}| \leq \beta \quad \text{for all } k = j + 1, \ldots, p \tag{10.16}$$

Thus all elements in column j will be bounded above by β whenever $t^2 \leq \beta^2\bar{\delta}_{jj}$; hence we further modify $\bar{\delta}_{jj}$ according to $\bar{\delta}_{jj} = t^2/\beta^2$ when $\bar{\delta}_{jj} < t^2/\beta^2$. In general, $\bar{\delta}_{jj}$ is now given by

$$\max\{\bar{\delta}_{jj},\ t^2/\beta^2\}$$

which will constitute a modification of δ_{jj} only under appropriate conditions.

The elements of M can now be defined. Using the expression

$$\bar{\delta}_{jj} = \psi_j + m_j$$

we have

$$m_j = \begin{cases} \xi - \psi_j & \text{if } \xi \geq \max\{|\psi_j|,\ t^2/\beta^2\} \\ |\psi_j| - \psi_j & \text{if } |\psi_j| \geq \max\{t^2/\beta^2, \xi\} \\ t^2/\beta^2 - \psi_j & \text{if } t^2/\beta^2 \geq \max\{\xi, |\psi_j|\} \end{cases}$$

The only remaining task is to specify β. We need to find a value for β which is large enough to insure that a well-conditioned positive definite G_i will not be modified and small enough to provide a desirable bound in (10.16).

It can be shown that for each $j = 1, 2, \ldots, p$,

$$0 \leq m_j \leq \left[\frac{n_j}{\beta} + (j - 1)\beta\right]^2 + 2[|\gamma_{jj}| + (j - 1)\beta^2] + \xi$$

where

$$n_j = \max\{|\gamma_{kj}|,\ k = j + 1, \ldots, p\}$$

An approximate lower bound for the function of β displayed in the right member of the above inequality, for all j, occurs when

$$\beta^2 = \frac{\max(n_j,\ j = 1, 2, \ldots, p)}{p} = \frac{\tau_1}{p}$$

This is a desirable value for β provided it is sufficiently large so as to satisfy the first condition. To insure that a well-conditioned positive definite matrix is not modified, consider the first of Eqs. (10.14). It can be written as

$$\gamma_{jj} = \sum_{r=1}^{j} \ell_{jr}^2 \delta_{rr} \quad , \ell_{jj} = 1$$

This implies that if

$$\beta^2 \geq \max\{|\gamma_{jj}|,\ j = 1, 2, \ldots, p\} = \tau_2$$

then $\ell_{jr}^2 \delta_{rr} \leq \beta^2$ and no modification in G_i will be made. Thus, a choice of β which will satisfy both of the requirements is given by

$$\beta^2 = \max\{\tau_1/p,\ \tau_2,\ m_\varepsilon\} \tag{10.17}$$

where m_ε is included as an added safety.

The basic elements needed to specify a modified Newton algorithm are now available with one exception. There is the remote possibility that a point $\underline{\theta}_i$ will be encountered where $\underline{g}_i = \underline{0}$ and G_i is not positive definite. In this case we will want to continue the search for a strong local minimum, but special measures must be taken to find a descent direction. This situation will be detected when $||g_i|| < \varepsilon$ and $||M|| > \varepsilon$. The suggested procedure in this case is to solve

$$\bar{L}_i' \underline{y} = \underline{e}_j$$

and set

$$\underline{d}_i = \begin{cases} -\text{sign}(\underline{y}'\underline{g}_i)\underline{y} & \text{if } 0 < ||\underline{g}_i|| < \varepsilon \\ \underline{y} & \text{if } ||\underline{g}_i|| = 0 \end{cases}$$

where $\underline{e}_j$ is a unit vector and j is such that

$$\psi_j - m_j \leq \psi_k - m_k \quad k = 1, 2, \ldots, p$$

ALGORITHM MNA($\underline{\theta}_0$, ε, $\hat{\underline{\theta}}$)

Step *Description*

Comment: $\underline{\theta}_0$ and $\varepsilon > 0$ are inputs. Initialize $i = -1$.

1. Set $i = i + 1$ and compute $\underline{g}_i$ and G_i.
2. Factor G_i as $L_i D_i L_i' = G_i + M$ using the previously described procedure.
3. If $||\underline{g}_i|| < \varepsilon$ and $||E|| = 0$, exit with $\hat{\underline{\theta}} = \underline{\theta}_i$.
4. If $||\underline{g}_i|| > \varepsilon$, solve for $\underline{d}_i$ in

   $$L_i D_i L_i' \underline{d}_i = -\underline{g}_i$$

   If $||\underline{g}_i|| < \varepsilon$ and $||E|| \neq 0$, solve for $\underline{y}$ and $\underline{d}_i$, where

   $$L_i'\underline{y} = \underline{e}_j$$

   $$\underline{d}_i = \begin{cases} -\text{sign}(\underline{y}'\underline{g}_i)\underline{y} & \text{if } 0 < ||\underline{g}_i|| < \varepsilon \\ \underline{y} & \text{if } ||\underline{g}_i|| = 0 \end{cases}$$

5. Compute step size α_i using one of the methods discussed in Section 10.1.4.
6. Set $\underline{\theta}_{i+1} = \underline{\theta}_i + \alpha_i \underline{d}_i$ and return to step 1.

Newton's method and modifications of it are widely used in statistical computing. The method of "scoring" introduced by Fisher (1925) for maximum-likelihood estimation is essentially Newton's method with the Hessian matrix replaced by the matrix of expected values of second derivatives (see, e.g., Kale 1961, 1962, or Harville 1977). Newton's method also forms the basis of some of the better procedures for fitting models under criteria other than least squares, as we shall see in Chapter 11.

10.2.3 Quasi-Newton Methods

In many applications the Hessian matrix G is difficult or impossible to obtain, and it is almost always inconvenient to supply both first and second partial derivatives of the objective function as required by Newton's method. Davidon (1959) considered the possible use of approximations to the Hessian matrix which become closer to the true matrix as the iteration progresses. His original paper stimulated a large amount of subsequent research which has produced exceptionally good gradient methods, called *quasi-Newton* methods (or *variable metric* methods) that are characterized by production of a new approximate Hessian matrix (or its inverse) at each iteration. A possible choice for a starting matrix at the first iteration is the identity which causes the initial direction to be that of steepest descent in the euclidean norm. The basic form of quasi-Newton iterates is

$$\underline{\theta}_{i+1} = \underline{\theta}_i - \alpha_i H_i \underline{g}_i \tag{10.18}$$

where H_i approximates G_i^{-1}.

In order to indicate how H_i might be chosen, consider the quadratic function

$$f(\underline{\theta}) = a + \underline{\theta}'\underline{b} + \frac{1}{2}\underline{\theta}'A\underline{\theta} \tag{10.19}$$

where A is symmetric positive definite. For this function we have

$$\underline{g}_i = \underline{b} + A\underline{\theta}_i$$

at the ith iteration, and the Hessian matrix is the constant A; hence

$$\underline{g}_{i+1} - \underline{g}_i = A(\underline{\theta}_{i+1} - \underline{\theta}_i)$$

Defining $\underline{y}_i = \underline{g}_{i+1} - \underline{g}_i$ and $\underline{\sigma}_i = \underline{\theta}_{i+1} - \underline{\theta}_i$, this equation becomes

$$\underline{y}_i = A\underline{\sigma}_i$$

and

$$A^{-1}\underline{y}_i = \underline{\sigma}_i$$

This suggests that in a general iterative procedure, when H_i approximates A, a choice of H_{i+1} might be made in the form

$$H_{i+1} = H_i + M_i \tag{10.20}$$

for some *correction matrix* M_i such that H_{i+1} satisfies the *quasi-Newton condition*

$$H_{i+1}\underline{y}_i = \underline{\sigma}_i \tag{10.21}$$

where H_i must satisfy

$$H_i\underline{y}_i = \underline{\sigma}_i \tag{10.22}$$

It follows from (10.20) and (10.21) that

$$H_i\underline{y}_i + M_i\underline{y}_i = \sigma_i$$

Under the requirement (10.22), M_i is such that

$$M_i\underline{y}_i = \underline{0}$$

Another requirement which M_i should satisfy is that H_{i+1} be symmetric positive definite whenever H_i is.

At least two different parametric families of matrices have been found which provide desirable candidates for use in updating H_i (see Broyden 1970, Dixon 1972b, and Oren 1976 for more details). The two most often used specific *inverse update* formulas are one by Fletcher and Powell (1963) and another called the Broyden-Fletcher-Goldfarb-Greenstadt-Shanno inverse update. These updates both use rank-two correction matrices M_i. Rank-one corrections have been derived and proposed, but they have certain undesirable characteristics which make them unpopular for general use. The two updates mentioned above are, respectively,

$$H_{i+1} = H_i + \frac{\underline{\sigma}_i\underline{\sigma}_i'}{\underline{\sigma}_i'\underline{y}_i} - \frac{H_i\underline{y}_i\underline{y}_i'H_i}{\underline{y}_i'H_i\underline{y}_i} \tag{10.23}$$

$$H_{i+1} = \left(I - \frac{\underline{\sigma}_i\underline{y}_i'}{\underline{\sigma}_i'\underline{y}_i}\right) H_i \left(I - \frac{\underline{\sigma}_i\underline{y}_i'}{\underline{\sigma}_i'\underline{y}_i}\right) + \frac{\underline{\sigma}_i\underline{\sigma}_i'}{\underline{\sigma}_i'\underline{y}_i} \tag{10.24}$$

We will now consider the update (10.23), which forms the basis for the Davidon-Fletcher-Powell algorithm, and derive certain characteristics of the iterates. Specifically, it will first be shown that the method based on use of (10.18) and (10.23) maintains a positive definite H_{i+1} at each step (if the initial matrix H_0 is positive definite). Then it will be shown that the method is quadratically convergent which, the reader should recall, means convergence in a finite number of iterations when the objective function is strictly convex and quadratic.

Suppose that H_i is positive definite and $\underline{\theta}$ is any nonzero vector. Using (10.23) we have

$$\underline{\theta}'H_{i+1}\underline{\theta} = \underline{\theta}'H_i\underline{\theta} + \frac{\underline{\theta}'\underline{\sigma}_i\underline{\sigma}_i'\underline{\theta}}{\underline{\sigma}_i'\underline{y}_i} - \frac{\underline{\theta}'H_i\underline{y}_i\underline{y}_i'H_i\underline{\theta}}{\underline{y}_i'H_i\underline{y}_i}$$

Let $\underline{q}_1 = H_i^{1/2}\underline{\theta}$ and $\underline{q}_2 = H_i^{1/2}\underline{y}_i$; then

$$\underline{\theta}'H_{i+1}\underline{\theta} = \frac{(\underline{q}_1'\underline{q}_1)(\underline{q}_2'\underline{q}_2) - (\underline{q}_1'\underline{q}_2)^2}{(\underline{q}_2'\underline{q}_2)} + \frac{(\underline{\theta}'\underline{\sigma}_i)^2}{\underline{\sigma}_i'\underline{y}_i}$$

and using the Cauchy-Schwarz inequality gives

$$\underline{\theta}'H_{i+1}\underline{\theta} \geq \frac{(\underline{\theta}'\underline{\sigma}_i)^2}{\underline{\sigma}_i'\underline{y}_i} \tag{10.25}$$

Assuming that exact linear searches are used, it will be the case that $\underline{\sigma}_i'\underline{g}_{i+1} = -\alpha_i\underline{g}_i'H_i\underline{g}_{i+1} = 0$. From this fact it follows that

$$\begin{aligned}\underline{\sigma}_i'\underline{y}_i &= \underline{\sigma}_i'\underline{g}_{i+1} - \underline{\sigma}_i'\underline{g}_i \\ &= -\underline{\sigma}_i'\underline{g}_i \\ &= \alpha_i\underline{g}_i'H_i\underline{g}_i \\ &> 0\end{aligned}$$

since the step size α_i will be positive whenever H_i is positive definite in (10.18). Thus for nonzero $\underline{\theta}$, (10.25) becomes

$$\underline{\theta}'H_{i+1}\underline{\theta} > 0$$

so H_{i+1} is positive definite.

We will now show that the iterative method, based on the use of (10.18) and (10.23) with $H_0 = I_p$, is quadratically convergent. The discussion will generally follow that given by Fletcher and Powell (1963). First note the following, where the given quadratic function is in the form (10.19);

$$\begin{aligned} \underline{y}_i &= \underline{g}_{i+1} - \underline{g}_i \\ &= A\underline{\theta}_{i+1} - A\underline{\theta}_i \\ &= A\underline{\sigma}_i \\ H_{i+1}A\underline{\sigma}_i &= H_{i+1}\underline{y}_i \\ &= H_i\underline{y}_i + \frac{\underline{\sigma}_i\underline{\sigma}_i'\underline{y}_i}{\underline{\sigma}_i'\underline{y}_i} - \frac{H_i\underline{y}_i\underline{y}_i'H_i\underline{y}_i}{\underline{y}_i'H_i\underline{y}_i} \\ &= H_i\underline{y}_i + \underline{\sigma}_i - H_i\underline{y}_i \\ &= \underline{\sigma}_i \end{aligned} \tag{10.26}$$

Also note that since $H_0 = I$, it follows that $\underline{\sigma}_0 = -\alpha_0\underline{g}_0$ and $\underline{\sigma}_1 = -\alpha_1 H_1 \underline{g}_1$, where α_0 and α_1 are both positive. Using these facts, and noting that $\underline{\sigma}_j \neq \underline{0}$ $(0 \leq j \leq p - 1)$ else a stationary point has been located, we have

$$\begin{aligned} \underline{\sigma}_0'A\underline{\sigma}_1 &= \alpha_0\alpha_1\underline{g}_0'AH_1\underline{g}_1 \\ &= -\alpha_1\underline{\sigma}_0'\underline{g}_1 \quad \text{using (10.26)} \\ &= 0 \end{aligned} \tag{10.27}$$

An inductive argument on k will now be given to show that

$$\underline{\sigma}_i'A\underline{\sigma}_j = 0 \qquad 0 \leq i < j < k \tag{10.28}$$

and

$$H_k A\underline{\sigma}_i = \underline{\sigma}_i \qquad 0 \leq i < k \tag{10.29}$$

Since these equations are easily shown to hold for the initial value $k = 2$, assume that they hold for general $k > 2$ so that the problem is to show that (10.28) and (10.29) hold for $k + 1$. Since $f(\underline{\theta})$ is quadratic (as given in Eq. 10.19),

$$\underline{g}_k = \underline{b} + A\underline{\theta}_k$$
$$= \underline{b} + A(\underline{\theta}_{i+1} + \underline{\sigma}_{i+1} + \underline{\sigma}_{i+2} + \cdots + \underline{\sigma}_{k-1})$$
$$= \underline{g}_{i+1} + A \sum_{j=i+1}^{k-1} \underline{\sigma}_j$$

For any $0 \leq i < k$ we have, using the $\underline{g}_k$ as given above,

$$\underline{\sigma}_i'\underline{g}_k = \underline{\sigma}_i'\underline{g}_{i+1} + \sum_{j=i+1}^{k-1} \underline{\sigma}_i'A\underline{\sigma}_j$$
$$= 0 \tag{10.30}$$

Using (10.29) in (10.30) gives

$$\underline{\sigma}_i'AH_k\underline{g}_k = 0 \quad (0 \leq i < k)$$

so, since $\underline{\sigma}_k = -\alpha_k H_k \underline{g}_k$, it follows that

$$\underline{\sigma}_i'A\underline{\sigma}_k = 0 \quad (0 \leq i < k < k + 1) \tag{10.31}$$

As shown above, $\underline{y}_i = A\underline{\sigma}_i$; thus for $0 \leq i < k$,

$$\underline{y}_k'H_kA\underline{\sigma}_i = \underline{y}_k'\underline{\sigma}_i \quad \text{using (10.29)}$$
$$= \underline{\sigma}_k'A\underline{\sigma}_i$$
$$= 0 \quad \text{from (10.31)} \tag{10.32}$$

Now we can complete the inductive argument with

$$H_{k+1}A\underline{\sigma}_i = H_kA\underline{\sigma}_i + \frac{\underline{\sigma}_k\underline{\sigma}_k'A\underline{\sigma}_i}{\underline{\sigma}_k'\underline{y}_k} - \frac{H_k\underline{y}_k\underline{y}_k'H_kA\underline{\sigma}_i}{\underline{y}_k'H_k\underline{y}_k}$$
$$= H_kA\underline{\sigma}_i \quad \text{using (10.31) and (10.32)}$$
$$= \underline{\sigma}_i \quad \text{using (10.29)} \tag{10.33}$$

Equations (10.26), (10.31), and (10.33) prove the induction. Equation (10.28) shows that $\underline{\sigma}_0, \underline{\sigma}_1, \ldots, \underline{\sigma}_{p-1}$ are linearly independent, from which it follows, using (10.29), that $H_p = A^{-1}$

The minimum of $f(\underline{\theta})$ is found in p iterations because (10.30) shows that $\underline{g}_p$ is orthogonal to the linearly independent vectors

$\underline{\sigma}_0, \underline{\sigma}_1, \ldots, \underline{\sigma}_{p-1}$ and this is only possible when $\underline{g}_p = \underline{0}$. Therefore, the quasi-Newton method described above is quadratically convergent.

In application to general nonlinear objective functions, a reasonably accurate linear search method is used to obtain α_i, and (10.18) along with (10.23) or (10.24) provide the iterates. As was the case with Newton's method, the inverse Hessian matrix is available when a stationary point is reached.

A large amount of theoretical and empirical research work has been done to study the rate of convergence of this type of method. Much of this work is summarized by Dennis and More (1977). Indications are that convergence is superlinear when "reasonably good" starting values are used. Accumulated experience with quasi-Newton methods has shown them to be the best among available first-order (only first derivatives required) gradient methods. In fact, quasi-Newton methods are probably the most often used methods for general nonlinear function minimization.

The Davidon-Fletcher-Powell algorithm, as given by Fletcher and Powell (1963), is the following:

ALGORITHM DFP$(\underline{\theta}_0, H_0, \hat{\underline{\theta}}, i=0, \varepsilon)$

Step *Description*

1. Set $i = i + 1$ and compute $\underline{d}_i = -H_i \underline{g}_i$
2. Perform a cubic interpolation linear search for α_i.
3. Set $\underline{\sigma}_i = \alpha_i \underline{d}_i$.
4. Compute $\underline{\theta}_{i+1} = \underline{\theta}_i + \underline{\sigma}_i$ and check for termination.
5. Evaluate $F(\underline{\theta}_{i+1})$ and $\underline{g}(\underline{\theta}_{i+1})$ and check for termination.
6. Set $\underline{y}_i = \underline{g}_{i+1} - \underline{g}_i$.
7. Compute H_{i+1} using (10.23) and return to step 1.

Shanno and Phua (1976) give two algorithms that employ update (10.24) using different starting values and linear search criteria. There is some evidence that (10.24) does provide more stability than (10.23) because it is more robust against departures from exactness in linear search; hence these algorithms are interesting.

Despite the fact that the matrices H_i are guaranteed in theory to be positive definite, in practice it sometimes happens that one of the H_i is semidefinite or indefinite due to accumulated rounding error. Even when the H_i remain positive definite, severe ill-conditioning sometimes causes numerical problems. Various approaches have been taken to handle this problem. Bard (1968) stresses the need for extended-precision computer arithmetic. Jacoby, Kowalik, and Pizzo (1972) suggest periodically resetting H_i to the identity (restarting). Gill and Murray (1972) approach the problem by looking for the most stable procedure at each step, and one which can be used to overcome adverse effects of rounding error by introducing perturbations when necessary. Preferring to deal with an approximate Hessian, rather than the approximate inverse Hessian, they display rank-two updates similar to (10.23) and (10.24) and use a rank-one update (described in Chapter 7) of the Cholesky factorization of H_i^{-1} twice to obtain the Cholesky decomposition of H_{i+1}^{-1}. This update procedure is somewhat similar to the modified Newton method given in the preceding subsection, except that rank-one corrections are incorporated in the present situation and the Cholesky decomposition is updated rather than reformed at each iteration.

Many modifications to the basic quasi-Newton procedure have been suggested. Wolfe (1975) includes a memory term, based on directions previously taken in the iteration, and derives a "quasi-Newton method with memory." This method requires use of a two-dimensional search as well as a linear search. Murtagh and Sargent (1970) studied competing algorithms and recommended procedures for scaling variables and conducting linear search.

Research in this area seems to be progressing on two fronts. First, the requirement for a linear search at each step in the iteration is costly and admits inaccuracies, so attempts are being made to reduce or eliminate this requirement. Davidon (1975), Fletcher (1970), Oren (1973), and Bass (1972) recommend methods that don't require a linear search. Experience with these methods is limited, but results to date are encouraging. The other research frontier is one of eliminating the need for even first derivatives at each iteration. Greenstadt (1972) proposes one method of this type, and Powell (1975) gives a useful survey of algorithms, of various types, that do not require derivatives. Also, Stewart (1967) developed a quasi-Newton modification based on numerically approximating first derivatives using finite differences.

Theoretical work on the general characteristics of quasi-Newton methods abounds. Some publications which the reader may wish to review are Davidon (1968), McCormick and Pearson (1969), Greenstadt (1970), Goldfarb (1970), Broyden (1967), Huang (1970), Dixon (1972a), Broyden, Dennis, and More (1973), Powell (1972), Dennis and More (1974), Oren and Luenberger (1974), Goldfarb (1976), Dennis and More (1977), and Brodlie (1977).

Example 10.2

(Quasi-Newton iteration using Eq. 10.23)

Consider the following quadratic function:

$$f(\underline{\theta}) = a + \underline{b}'\underline{\theta} + \frac{1}{2}\,\underline{\theta}'A\underline{\theta}$$

where

$$a = 0, \quad \underline{b}' = (1,0), \quad A = \begin{pmatrix} 2 & 0 \\ 0 & 1 \end{pmatrix}$$

In this case the objective function has the specific form

$$f(\underline{\theta}) = \theta_1 + \theta_1^2 + \frac{\theta_2^2}{2}$$

and

$$\underline{g}(\theta) = \begin{pmatrix} 1 + 2\theta_1 \\ \theta_2 \end{pmatrix}$$

This is the same function that was used in Example 10.1. We will use starting values

$$H_0 = I, \quad \underline{\theta}_0' = (0,1)$$

Hence

$$\underline{g}_0 = \begin{pmatrix} 1 \\ 1 \end{pmatrix}, \quad \underline{d}_0 = -H_0\underline{g}_0 = -\begin{pmatrix} 1 \\ 1 \end{pmatrix}$$

The first requirement is to find α_0 by linear search. We have

$$f(\underline{\theta}_0 + \alpha_0\underline{d}_0) = \frac{1}{2} - 2\alpha_0 + \frac{3\alpha_0^{\,2}}{2}$$

which reaches the minimum at $\alpha_0 = 2/3$. Using this result, the next iterate $\underline{\theta}_1$ is

$$\underline{\theta}_1 = \underline{\theta}_0 + \alpha_0\underline{d}_0 = \begin{pmatrix} -2/3 \\ 1/3 \end{pmatrix}$$

and

$$\underline{g}_1 = \frac{1}{3}\begin{pmatrix} -1 \\ 1 \end{pmatrix}$$

Note that $\underline{g}_1'\underline{d}_0 = 0$ as should be the case.

To form H_1 we need $\underline{\sigma}_0$ and $\underline{y}_0$ which are found as

$$\underline{\sigma}_0 = \underline{\theta}_1 - \underline{\theta}_0 = -\frac{1}{3}\begin{pmatrix} 2 \\ 2 \end{pmatrix}, \quad \underline{y}_0 = \underline{g}_1 - \underline{g}_0 = -\frac{1}{3}\begin{pmatrix} 4 \\ 2 \end{pmatrix}$$

Now H_1 is computed as

$$H_1 = \begin{pmatrix} 1 & 0 \\ 0 & 1 \end{pmatrix} + \frac{1}{3}\begin{pmatrix} 1 & 1 \\ 1 & 1 \end{pmatrix} - \frac{1}{5}\begin{pmatrix} 4 & 2 \\ 2 & 1 \end{pmatrix}$$

$$= \frac{1}{15}\begin{pmatrix} 8 & -1 \\ -1 & 17 \end{pmatrix}$$

The new direction $\underline{d}_1$ is computed as

$$\underline{d}_1 = -H_1\underline{g}_1 = \frac{1}{5}\begin{pmatrix} 1 \\ -2 \end{pmatrix}$$

Now another linear search must be made to obtain α_1. We have

$$f(\underline{\theta}_1 + \alpha_1\underline{d}_1) = -\frac{1}{6} - \frac{\alpha_1}{5} + \frac{3\alpha_1^2}{25}$$

which is minimized for $\alpha_1 = 5/6$. Hence $\underline{\theta}_2$ has value

$$\underline{\theta}_2 = \underline{\theta}_1 - \alpha_1 H_1 \underline{g}_1 = \frac{1}{2}\begin{pmatrix} -1 \\ 0 \end{pmatrix}$$

and

$$\underline{g}_2 = \begin{pmatrix} 0 \\ 0 \end{pmatrix}$$

The inverse Hessian is H_2, which is computed as

$$H_2 = \frac{1}{15}\begin{pmatrix} 8 & -1 \\ -1 & 17 \end{pmatrix} + \frac{1}{6}\begin{pmatrix} 1 & -2 \\ -2 & 4 \end{pmatrix} - \frac{1}{5}\begin{pmatrix} 1 & -2 \\ -2 & 4 \end{pmatrix}$$
$$= \begin{pmatrix} 1/2 & 0 \\ 0 & 1 \end{pmatrix}$$

10.2.4 Conjugate Gradient Method

While the quasi-Newton methods require only first derivatives, which makes them easier to use than Newton's method, they do use an approximate Hessian matrix which will require a large amount of computer memory for storage when sizable problems are solved. The conjugate gradient method, described in this subsection, is much less demanding in this respect and can be used when solving extremely large problems. It is a gradient method; hence derivatives of the objective function must be available. There is a close and identifiable relationship between quasi-Newton and conjugate gradient methods. This fact is noted by many authors (see, e.g., Myers 1968, Powell 1975, Oren 1976, and Goldfarb 1977), but we shall not attempt to make use of the relationship in this section.

The method of conjugate gradients was first given by Hestenes and Stiefel (1952) and was proposed for solving systems of linear equations. Beckman (1960) gave an algorithm for this application.

One of the first algorithms which used conjugate gradients for solution of the general unconstrained minimization problem was published by Fletcher and Reeves (1964). This algorithm, with some relatively minor modifications, is still in use today.

Basic to the discussion of this method is the definition of conjugate vectors. Vectors (directions) $\underline{d}_0, \underline{d}_1, \ldots, \underline{d}_{p-1}$, with $\underline{d}_j \neq 0$ for each j, are said to be mutually *conjugate* with respect to a given p x p symmetric positive definite matrix A if

$$\underline{d}_j' A \underline{d}_i = 0 \quad \text{for all } i \neq j \tag{10.34}$$

That such a set of vectors exists is apparent because eigenvectors of A have this property. When A = I we have the usual orthogonality condition; thus conjugacy can be thought of as a generalized orthogonality condition.

The theory of the method of conjugate gradients is developed with respect to a convex quadratic objective function, which we again write as

$$f(\underline{\theta}) = a + \underline{\theta}'\underline{b} + \frac{1}{2}\underline{\theta}'A\underline{\theta} \tag{10.35}$$

where A is positive definite. Basically, the method forms linearly independent descent directions $\underline{d}_i$, one at a time, which are mutually conjugate with respect to A, and takes a step according to

$$\underline{\theta}_{i+1} = \underline{\theta}_i + \alpha_i \underline{d}_i$$

for each $\underline{d}_i$. This method can be shown to be quadratically convergent.

The remainder of this subsection will present some details of the method of conjugate gradients. We shall begin by considering the problem of generating conjugate directions. Then the use of gradient directions will be introduced and the conjugate gradient method will be developed for quadratic functions. Finally, ways in which the method is applied to general nonlinear objective functions will be given. The following theorem (see, e.g., Fletcher 1972a) is helpful in gaining an understanding of the method of conjugate gradients as it applies to quadratic functions.

Theorem 10.2

If at each step the linear search is exact, and the search directions $\underline{d}_0, \underline{d}_1, \ldots, \underline{d}_{p-1}$ span a p-dimensional space and are conjugate with respect to the positive definite matrix A in (10.35), then the minimum of $f(\underline{\theta})$ is found in at most p steps; moreover, $\underline{\theta}_{i+1}$ for each $0 \leq i \leq p - 1$ is the minimum point in the subspace generated by the starting point $\underline{\theta}_0$ and the directions $\underline{d}_0, \underline{d}_1, \ldots, \underline{d}_i$.

A proof that no more than p steps will be used to locate the minimum of $f(\underline{\theta})$ is given by Ortega and Rheinboldt (1970, page 271).

To begin the discussion of conjugate directions, suppose that A is known and we have a set of linearly independent directions $\underline{n}_0, \underline{n}_1, \ldots, \underline{n}_{p-1}$ spanning the space of $\underline{\theta}$. Using these data, a set of mutually conjugate directions $\underline{d}_0, \underline{d}_1, \ldots, \underline{d}_{p-1}$ can be obtained. The formula for generating such directions will now be considered. First, let $\underline{d}_0$ have the value

$$\underline{d}_0 = \underline{n}_0$$

For building a set of directions $\underline{d}_0, \underline{d}_1, \ldots, \underline{d}_{p-1}$, the assumptions about the $\underline{n}_j$ insure that the $\underline{d}_i$ are expressible as

$$\underline{d}_i = \underline{n}_i + \sum_{j=0}^{i-1} c_{ij}\underline{d}_j \quad \text{for } 1 \leq i \leq p - 1 \tag{10.36}$$

The coefficients c_{ij} must be chosen so that

$$\underline{d}_i'A\underline{d}_k = 0 \quad (0 \leq k \leq i - 1)$$

By using (10.36), this latter condition can be written as

$$\underline{n}_i'A\underline{d}_k + \sum_{j=0}^{i-1} c_{ij}\underline{d}_j'A\underline{d}_k = 0$$

For mutually conjugate vectors $\underline{d}_0, \underline{d}_1, \ldots, \underline{d}_{i-1}$ this equation shows that the c_{ij} coefficients, for use in (10.36), can take the form

$$c_{ik} = \frac{-\underline{n}_i'A\underline{d}_k}{\underline{d}_k'A\underline{d}_k} \quad (0 \leq k \leq i - 1) \tag{10.37}$$

None of the $\underline{d}_k$ is zero because the $\underline{n}_j$ are linearly independent; hence, since A is positive definite, the denominator in the right member of (10.37) is positive. Equations (10.36) and (10.37) define, in terms of A and $\underline{n}_j$, a set of mutually conjugate directions with respect to A for $\underline{d}_0 = \underline{n}_0$. These equations form a foundation for certain optimization methods which do not use derivatives. The so-called conjugate direction methods are of this type and will be discussed in the next subsection. We will next see how gradient directions are employed to produce the method which is the subject of this subsection.

Suppose that we are willing to consider using the gradient $\underline{g}$ and evaluating it at each point $\underline{\theta}_i$. For the purpose of discussion, consider use of the quadratic function $f(\underline{\theta})$ and assume that exact linear searches are made. From (10.35), $\underline{g}$ takes the form

$$\underline{g} = \underline{b} + A\underline{\theta}$$

and for iterates

$$\underline{\theta}_{i+1} = \underline{\theta}_i + \alpha_i \underline{d}_i \tag{10.38}$$

it follows that

$$\underline{g}_{i+1} - \underline{g}_i = \alpha_i A\underline{d}_i \qquad (0 \le i \le p - 1) \tag{10.39}$$

Define the directions $\underline{n}_i$ in (10.36) to be

$$\underline{n}_i = -\underline{g}_i \qquad (0 \le i \le p - 1) \tag{10.40}$$

(This is allowable, as we shall see in a moment.) Now Eq. (10.37) becomes

$$c_{ik} = \frac{\underline{g}_i'(\underline{g}_{k+1} - \underline{g}_k)}{\underline{d}_k'(\underline{g}_{k+1} - \underline{g}_k)} \qquad (0 \le k \le i - 1) \tag{10.41}$$

The denominator of (10.41) will vanish only when $\alpha_i = 0$ in (10.38) for i = k. This, however, means that the minimum of $f(\underline{\theta})$ has previously been found. These equations, along with (10.36), which are based on gradient directions (10.40), can be used in a p-step procedure to find the minimum of $f(\underline{\theta})$ as stated in Theorem 10.2. Further simplification of (10.36) is possible in this case, and we will now proceed to develop a simpler form of this equation.

One consequence of the fact that during each step a minimum of $f(\underline{\theta})$ is located in a subspace, as Theorem 10.2 states, is that for $(0 \leq j \leq p - 1)$

$$\underline{g}_i' \underline{g}_{j+1} = 0 \quad \text{for each } 0 \leq i \leq j \tag{10.42}$$

To see that this is true, first note that $\underline{g}_{k+1}' \underline{d}_k = 0$ for $0 \leq k \leq p - 1$ because exact linear searches are being made. Now construct the following identity:

$$\underline{g}_{j+1}' \underline{d}_i = \underline{g}_{i+1}' \underline{d}_i + (\underline{y}_{i+1} + \underline{y}_{i+2} + \cdots + \underline{y}_j)' \underline{d}_i \quad (0 \leq i \leq j)$$

where $\underline{y}_k = \underline{g}_{k+1} - \underline{g}_k$. Using (10.39) and substituting for $\underline{y}_k$, we have

$$\underline{g}_{j+1}' \underline{d}_i = \underline{g}_{i+1}' \underline{d}_i + \left(\sum_{k=i+1}^{j} \alpha_k \underline{d}_k' A \right) \underline{d}_i \quad (0 \leq i \leq j)$$

Since $\underline{d}_k' A \underline{d}_i = 0$ for $i < k$, it follows that

$$\underline{g}_{j+1}' \underline{d}_i = 0 \quad (0 \leq i \leq j) \tag{10.43}$$

Using (10.36) with $\underline{n}_i = -\underline{g}_i$ and multiplying by $\underline{g}_{j+1}'$ gives

$$\underline{g}_{j+1}' \underline{d}_i = \underline{g}_i' \underline{g}_{j+1} + \sum_{k=0}^{i-1} c_{ik} \underline{g}_{j+1}' \underline{d}_k \quad (1 \leq i \leq j) \tag{10.44}$$

Using (10.43) in (10.44) and noting that $\underline{d}_0 = -\underline{g}_0$, we see that Eq. (10.42) is correct.

From (10.42) we see that in Eq. (10.41) $c_{ik} = 0$ for $0 \leq k \leq i - 2$, so Eq. (10.36) simplifies to the form

$$\underline{d}_0 = -\underline{g}_0 \tag{10.45}$$

$$\underline{d}_1 = -\underline{g}_1 + c_{10} \underline{d}_0 \tag{10.46}$$

$$\underline{d}_i = -\underline{g}_i + c_{i,i-1} \underline{d}_{i-1} \quad (i > 1) \tag{10.47}$$

where

$$c_{i,i-1} = \frac{\underline{g}_i'(\underline{g}_i - \underline{g}_{i-1})}{\underline{d}_{i-1}'(\underline{g}_i - \underline{g}_{i-1})} \tag{10.48}$$

$$= \frac{\underline{g}_i'\underline{g}_i}{\underline{g}_{i-1}'\underline{g}_{i-1}} \tag{10.49}$$

The form given by (10.49) is obtained using (10.36) and (10.43). Note that if the direction $\underline{d}_0$ is chosen to be other than $\underline{d}_0 = -\underline{g}_0$, and if all other directions $\underline{n}_i$ are chosen as gradient directions, then it is no longer true that $c_{i0} = 0$ for $i > 1$, and $\underline{d}_i$ in (10.47) becomes

$$\underline{d}_i = -\underline{g}_i + c_{i0}\underline{d}_0 + c_{i,i-1}\underline{d}_{i-1} \qquad (\underline{d}_0 \neq -\underline{g}_0,\ i > 1) \tag{10.50}$$

Equations (10.38) and (10.45) to (10.48) define the essential elements in the *method of conjugate gradients* applied to quadratic functions. That it is indeed quadratically convergent follows from the facts that the $\underline{d}_i$ are linearly independent and $\underline{g}_p$ is orthogonal to each of $\underline{d}_0, \ldots, \underline{d}_{p-1}$ when $f(\underline{\theta})$ is the objective function.

Other descriptions of this method are given in many publications, including Kowalik and Osborne (1968), Fletcher (1972a), and Beale (1972). Equations (10.45) to (10.50) are derived by Beale who suggests use of (10.50) in an interesting extension of existing algorithms. Before considering the situation for general nonlinear functions, we will record the steps in the basic p-step procedure outlined above as given by Fletcher and Reeves (1964):

1. Select $\underline{\theta}_0$, compute $\underline{g}_0$, and set $\underline{d}_0 = -\underline{g}_0$.
2. For each $i = 0, 1, \ldots, p-1$,
   (a) Find $\underline{\theta}_{i+1}$ by linear search in the direction $\underline{d}_i$.
   (b) Compute $\underline{g}_{i+1}$.
   (c) Compute $\beta_i = \underline{g}_{i+1}'\underline{g}_{i+1}/\underline{g}_i'\underline{g}_i$.
   (d) Set $\underline{d}_{i+1} = -\underline{g}_{i+1} + \beta_i\underline{d}_i$.

For nonlinear functions $F(\underline{\theta})$ which are not quadratic, the method of conjugate gradients is iterative rather than p-step. The usual procedure is to perform p or p + 1 steps in the form given above, and then *restart* for the next iteration. It is profitable to think of the method as making an infinite sequence of steps with a restart from the usual direction every pth step. More than one way to restart has been suggested. Fletcher and Reeves (1964) restart after every p steps by computing the next direction as the negative gradient (i.e., the form of Eq. 10.45) instead of using (10.47). An algorithm which is essentially that given by Fletcher and Reeves (1964) is given below. Two termination criteria are used. The first checks for zero gradient, and the second compares function values between successive sets of p steps. The algorithm uses Eq. (10.49) to compute $c_{i,i-1}$, but Eq. (10.48) is to be preferred for nonquadratic functions, as is noted by Crowder and Wolfe (1972). Equation (10.48) is sometimes called the Polak-Ribiere form, and most recently developed software systems use this form instead of (10.49).

ALGORITHM FR($\underline{\theta}_0$,ε,tol,$\hat{\underline{\theta}}$)

Step	*Description*
1.	Set k = t = 0 and compute $\underline{g}_0$.
2.	Set $\underline{d}_t = -\underline{g}_t$.
3.	Set k = k + 1 and compute α_{k-1} using cubic interpolation.
4.	Compute $\underline{\theta}_k = \underline{\theta}_{k-1} + \alpha_{k-1}\underline{d}_{k-1}$, $\underline{g}_k$, $c = \|\|\underline{g}_k\|\|^2/\|\|\underline{g}_{k-1}\|\|^2$ and $\underline{d}_k = -\underline{g}_k + c\underline{d}_{k-1}$.
5.	If c < ε, terminate with $\hat{\underline{\theta}} = \underline{\theta}_k$.
6.	If k - t < p, do not restart, go to step 3. Otherwise a restart is needed unless the process has converged. Use the second convergence criterion to compare function values between successive sets of p steps. If they differ by less than tol, exit with the latest value of $\underline{\theta}$. Otherwise set t = k - 1 and go to step 2.

In practice, the linear search is carried out to varying degrees of accuracy by different researchers. Many times, five-figure accuracy in the approximation of α_i is considered to be more than sufficient precision. The check of function values, in step 6, as a termination criterion is needed because the gradient may not be nearly zero at a minimum, due to rounding error; hence the termination criterion used in step 4 is not sufficient by itself. Note the use of t as a pointer to the restart value of k in algorithm FR.

An alternative restart procedure is suggested by Powell (1977b). This alternative procedure uses Eq. (10.48) and Eq. (10.50) for the basic steps and restarts with $\underline{d}_t$ ($t = k - 1$) in place of $\underline{d}_0$ in (10.46), where $\underline{d}_t$ is the current direction when a restart is desired. Here we are using t and k as defined in algorithm FR. Unfortunately, for $k > t + 1$ in this alternative procedure each direction $\underline{d}_k$ must be checked to insure that it is downhill. The suggested criterion is

$$-1.2\|\underline{g}_k\|^2 \leq \underline{d}_k'\underline{g}_k \leq -0.8\|\underline{g}_k\|^2$$

If $\underline{d}_k$ does not satisfy this criterion, a restart is made. Also, a restart is suggested whenever $|\underline{g}_{k-1}'\underline{g}_k| > 0.2\|\underline{g}_k\|^2$. Equation (10.48), instead of (10.49), is used to calculate the c_{ij} coefficients. This modification of the Fletcher-Reeves algorithm has converged more rapidly for some test problems.

The rate of convergence of the conjugate gradient methods described above can sometimes be disappointingly slow, as is shown by Powell (1976). For some problems it now appears that a linear rate of convergence may be the best that can be expected. The work of Crowder and Wolfe (1972) also indicates that this may sometimes be the case.

Several other modifications of the basic conjugate gradient method have been proposed. Notable among these is the Polak-Ribiere algorithm (see Polak 1971, page 52) and PARTAN developed by Shah, Buehler, and Kempthorne (1964) with further work by Papaioannou and Kempthorne (1972). We will not discuss other methods since they are not substantially different from the methods described above. A more recently proposed method by Dixon (1975) is designed to operate without accurate linear search. This method has not, at this time, been extensively tested. It does, however, typify one of the main directions of research that is currently underway, as does the work of Kowalik, Kamgnia, and Boland (1977) and Boland, Kamgnia, and Kowalik (1977) which extends results obtained by Spedicato (1975) on conjugate gradient methods that are scale-invariant.

Example 10.3

(Application of the Conjugate Gradient Method)

Consider the problem in Example 10.2. The objective function is

$$f(\underline{\theta}) = \theta_1 + \theta_1^2 + \frac{\theta_2^2}{2}$$

and the starting point is $\underline{\theta}_0' = (0,1)$. The gradient vector is

$$\underline{g}(\underline{\theta}) = \begin{pmatrix} 1 + 2\theta_1 \\ \theta_2 \end{pmatrix}$$

Using (10.45), we compute

$$\underline{d}_0 = -\underline{g}_0 = -\begin{pmatrix} 1 \\ 1 \end{pmatrix}$$

The first linear search is with respect to

$$f(\underline{\theta}_0 - \alpha_0\underline{g}_0) = \frac{1}{2} - 2\alpha_0 + \frac{3\alpha_0^2}{2}$$

which gives $\alpha_0 = 2/3$. The first point is

$$\begin{aligned} \underline{\theta}_1' &= \underline{\theta}_0' + \alpha_0\underline{d}_0' \\ &= (-2/3,\ 1/3) \end{aligned}$$

and we obtain

$$\underline{g}_1' = (-1/3,\ 1/3)$$

The second direction is obtained using (10.46) and (10.49) as

$$\underline{d}_1 = -\underline{g}_1 + c_{1,0}\underline{d}_0$$

$$= -\underline{g}_1 + \left(\frac{\underline{g}_1'\underline{g}_1}{\underline{g}_0'\underline{g}_0}\right) \underline{d}_0$$

$$= (2/9, -4/9)'$$

(Notice that apart from a scalar multiple, this direction vector $\underline{d}_1$ is the same as that obtained in Example 10.2. Myers 1968 shows that this will be the case for quadratic functions.) The second linear search is carried out using the function

$$f(\underline{\theta}_1 + \alpha_1\underline{d}_1) = -\frac{1}{6} - \frac{6\alpha_1}{27} + \frac{4\alpha_1^2}{27}$$

and $\alpha_1 = 3/4$ is found to yield the minimum of this function. Thus $\underline{\theta}_2$ is found to be

$$\underline{\theta}_2 = \underline{\theta}_1 + \alpha_1\underline{d}_1 = (-1/2, 0)'$$

and

$$\underline{g}_2' = (0,0)$$

10.2.5 Conjugate Direction Method

The methods given in preceding subsections all require use of derivatives of the objective function. It is not difficult to imagine cases where derivatives of the objective function are extremely difficult to obtain. In such cases the method of conjugate directions, which does not use derivatives, can be employed.

Powell (1964) proposed an algorithm based on use of conjugate directions which has performed well in some instances (see Box 1966). This method uses p iterations to form a set of conjugate directions $\underline{d}_0, \underline{d}_1, \ldots, \underline{d}_{p-1}$ using given linearly independent starting directions $\underline{n}_0, \underline{n}_1, \ldots, \underline{n}_{p-1}$ which are usually chosen to be unit vectors. Each iteration produces a new direction $\underline{d}$, which replaces one of the $\underline{n}_i$, and a new starting point $\underline{\theta}_0$ for the next iteration. At the end of p iterations the conjugate directions $\underline{d}_i$ have been formed. To simplify notation we will use $\underline{d}_j$ to denote all directions with the understanding that $\underline{d}_j = \underline{n}_j$ initially. In this notation one iteration of the basic procedure consists of the following steps with respect to the quadratic function (10.35):

1. For each i = 0, 1, ..., p - 1 compute α_i so that $f(\underline{\theta}_i + \alpha_i \underline{d}_i)$ is minimum, and define $\underline{\theta}_{i+1} = \underline{\theta}_i + \alpha_i \underline{d}_i$.
2. For i = 0, 1, ..., p - 2 replace $\underline{d}_i$ by $\underline{d}_{i+1}$ and renumber as $\underline{d}_0, \ldots, \underline{d}_{p-2}$.
3. Replace $\underline{d}_{p-1}$ with $\underline{\theta}_p - \underline{\theta}_0$.
4. Compute α so that $f(\underline{\theta}_p + \alpha(\underline{\theta}_p - \underline{\theta}_0))$ is a minimum and replace $\underline{\theta}_0$ with $\underline{\theta}_0 + \alpha(\underline{\theta}_p - \underline{\theta}_0)$.

Theorem 10.2 indicates that the minimum of a quadratic function will be found in p iterations of this procedure if indeed the final set of $\underline{d}_j$ are mutually conjugate. The fact that they will be mutually conjugate follows from Theorem 10.3, given below.

Theorem 10.3

(Paraphrased from Powell 1964)

Given a space containing vectors $\underline{d}$, $\underline{\theta}_1$, and $\underline{\theta}_2$, if the minimum of the quadratic function (10.35) in the direction $\underline{d}$ from $\underline{\theta}_1$ occurs at $\underline{\theta}_1^*$ and the minimum moving from $\underline{\theta}_2$ in the direction $\underline{d}$ occurs at $\underline{\theta}_2^*$, then $\underline{\theta}_2^* - \underline{\theta}_1^*$ is conjugate to $\underline{d}$.

With Theorem 10.3 in mind, consider iterations of the basic procedure with initial directions being (say) the coordinate directions. The first iteration produces a new point $\underline{\theta}_0^{(1)}$ which minimizes $f(\underline{\theta})$ in the new direction $\underline{d}$ that replaces $\underline{d}_{p-1}$ in step 3. The second iteration begins with $\underline{\theta}_0^{(1)}$ and at the end of step 1 produces a second point $\underline{\theta}_p$ which minimizes $f(\underline{\theta})$ in the direction $\underline{d}_{p-1}$. By Theorem 10.3, $\underline{\theta}_p - \underline{\theta}_0^{(1)}$ is conjugate to $\underline{d}_{p-1}$. Steps 2 and 3 serve to update the set of directions so that the last two in the set are conjugate. Step 4 then finds a point $\underline{\theta}_0^{(2)}$ which minimizes $f(\underline{\theta})$ in the direction found in step 3. This serves to locate the minimum in the space of the conjugate directions. The effect of each iteration is now apparent. One new vector is added to the mutually conjugate set and a new point, which minimizes $f(\underline{\theta})$ in that set, is computed. A second point, needed in Theorem 10.3, will be found in the first step of the next iteration.

One problem with this procedure is that even for a quadratic objective function it is difficult to maintain strong linear independence in the set $\{\underline{d}_i\}$. Powell suggests modifying steps 2 and 3 to replace that one of the candidate vectors which will cause the resulting set of p vectors to be most strongly mutually conjugate in a sense that will now be defined. Theorem 10.4 provides the basis for a criterion of choice.

Theorem 10.4

Let $\underline{\theta}_0, \underline{\theta}_1, \ldots, \underline{\theta}_{p-1}$ be scaled so that $\underline{\theta}_i' A \underline{\theta}_i = 1$ for each $0 \leq i \leq p - 1$. Let X be the p x p matrix having columns $\underline{\theta}_i$. Then the determinant of X is maximum when the $\underline{\theta}_i$ are mutually conjugate with respect to A.

This theorem suggests replacement of a direction $\underline{d}_i$ which will cause the greatest increase in determinant relative to the resulting set of p directions. After some additional derivation, the criterion takes the following form. Replace any one of the $\underline{d}_j$ which is such that $|\gamma_j/\delta| \geq 1$, where

$$\gamma_j = \{2[f(\underline{\theta}_i) - f(\underline{\theta}_{i+1})]\}^{1/2}$$

$$\delta = \{2[f(\underline{\theta}_0) - f_m]\}^{1/2} - \{2[f(\underline{\theta}_p) - f_m]\}^{1/2}$$

$$f_m = \min_{\alpha} f(\underline{\theta}_0 + \alpha(\underline{\theta}_p - \underline{\theta}_0))$$

unless $\underline{\theta}_p + \alpha\underline{d}$ lies between $\underline{\theta}_0$ and $\underline{\theta}_p$. If this is the case, then compute δ as

$$\delta = \{2[f(\underline{\theta}_0) - f_m]\}^{1/2} + \{2[f(\underline{\theta}_p) - f_m]\}^{1/2}$$

If none of the $\underline{d}_j$ satisfies this condition, try one more iteration with the same set of directions. Kowalik and Osborne (1968) present a complete derivation of these equations, and Powell gives an algorithm which performs the p basic iterations using this criterion for replacement of directions.

As was the case with the conjugate gradient method, some restart procedure must be incorporated with the basic conjugate direction iterations when the objective function is not quadratic. An obvious restart procedure is to set all directions back to coordinate directions (unit vectors) after each p iterations. Brent (1973) presents an alternative restart procedure which preserves more of the information available after each p iterations when forming restart directions. We now describe Brent's restart procedure.

At the minimum the conjugate directions satisfy, for quadratic functions,

$$\underline{d}_i' A \underline{d}_i = \lambda_i \qquad (0 \leq i \leq p - 1)$$

In matrix form, defining U to be the matrix having the $\underline{d}_i$ as its columns,

$$U = (\underline{d}_1, \underline{d}_2, \ldots, \underline{d}_{p-1})$$

we have

$$U'AU = D = \mathrm{diag}(\lambda_i)$$

Normalizing the columns of U using $D^{-1/2}$, we obtain

$$V = UD^{-1/2}$$

and V satisfies

$$A^{-1} = VV' \tag{10.51}$$

Since eigenvectors of A and A^{-1} are mutually conjugate with respect to A, and in view of (10.51), an appealing set of restart directions for general use is a set of eigenvectors of VV'. These can be found as the matrix Q in the singular-value decomposition of V, which is

$$V = Q\Sigma R \tag{10.52}$$

where Σ is diagonal. In practical applications a matrix U, formed from the directions available at the end of p iterations, is available but the diagonal matrix D must be computed. A suggested method for approximating the λ_i is based on the facts that for each $0 \leq i \leq p - 1$ within an iteration for the objective function (10.35) we minimize

$$\begin{aligned}\rho_i(\alpha) &= f(\underline{\theta}_i + \alpha\underline{d}_i) \\ &= \frac{\alpha^2}{2}\,\underline{d}_i'A\underline{d}_i + \alpha(\underline{\theta}_i'A\underline{d}_i - \underline{b}'\underline{d}_i) + \left(\frac{1}{2}\,\underline{\theta}_i'A\underline{\theta}_i - \underline{b}'\underline{\theta}_i + c\right)\end{aligned}$$

using linear search, and the second derivative of this function is

$$\frac{d^2\rho(\alpha)}{d\alpha^2} = \underline{d}_i'A\underline{d}_i = \lambda_i \tag{10.53}$$

In application to any nonlinear function $F(\underline{\theta})$ at least three function values are needed in the linear search, so it is always the case that a second finite difference approximation can be used to estimate the derivative λ_i with little additional computation. Thus an approximation to the matrix V of (10.51) is readily formed after each p iterations. Restart directions can then be obtained using a singular-value decomposition to find the matrix Q as given in (10.52).

In addition to the literature previously cited in this subsection the reader may wish to review papers by Zangwill (1967), Powell (1970a), Cohen (1972), Fletcher (1972a), Wolfe (1974), Powell (1975), and Nazareth (1976a) for further discussion of the use of conjugate directions.

10.2.6 Other Derivative-free Methods

Use of conjugate directions is not the only way to avoid taking first derivatives of the objective function. Other possibilities which should be mentioned are

1. to approximate derivatives numerically
2. to use a direct search method.

Some explanation of each of these possibilities is in order.

There are several standard numerical techniques for approximating the gradient vector of a given function $F(\underline{\theta})$ at a point $\underline{\theta}_i$ (see, e.g., Ralston 1965) using only function values. Using the difference-type approximations, the kth component of $\underline{g}_i$, which is denoted by $(\underline{g}_i)_k$, is approximated as follows for forward and central differences, respectively:

$$(g_i)_k \cong \frac{F(\underline{\theta}_i + \delta_k \underline{e}_k) - F(\underline{\theta}_i)}{\delta_k}$$

$$(g_i)_k \cong \frac{F(\underline{\theta}_i + \delta_k \underline{e}_k) - F(\underline{\theta}_i - \delta_k \underline{e}_k)}{2\delta_k}$$

where $\underline{e}_k$ is the kth unit vector and δ_k is small in magnitude. The choice of δ_k requires careful consideration. It must be small enough to give a good approximation to the partial derivative, but if it is too small in a given situation, cancellation of all significant digits will take place in the subtraction indicated above. Stewart (1967) considered this problem and developed an algorithm for choosing δ_k which uses the approximate second derivative information that is available in quasi-Newton methods. Gill and Murray (1972) make a slightly different recommendation as to choice of δ_k. It is not clear, at this time, which of these two algorithms is preferable. In either case approximate derivatives have proven useful in many applications and are recommended by many authors (e.g., Powell 1965, and Brown and Dennis 1972). They cannot, however, provide stability and accuracy equal to that which can be achieved using the actual derivatives. Thus approximating the gradient vector using numerical differentiation techniques is recommended only when the functional form of the gradient is not available.

Numerically approximating derivatives is not limited to first derivatives. Higher-order derivatives can also be approximated. For example, Gill and Murray (1974) suggest a further modification of Newton's method which uses finite differences to approximate the Hessian matrix.

Direct search methods are the second choice for use when derivatives cannot be taken analytically. Methods whose strategy is based only on the comparison of objective function values are called direct search methods. Most of the algorithms for computing step size are, for example, based on direct search methods. The method of conjugate directions, given in the preceding subsection, is usually placed in the class of descent methods rather than being called a direct search method, but it can be viewed as being of this type because the definition of what constitutes a direct search method is rather vague.

None of the direct search methods, applied to general nonlinear functions, appears to be generally competitive with the better gradient descent methods in terms of rate of convergence or overall efficiency. For this reason we will not describe any of these methods. Rather, some relevant literature will be cited to benefit those readers who wish to study such methods. When the objective function does not have continuous derivatives, these direct search methods may be preferable.

Swann (1972) gives an interesting survey of direct search methods. Among direct search methods most often mentioned in the literature are methods by Rosenbrock (1960), Hooke and Jeeves (1961), and Nelder and Mead (1965) (also see O'Neill 1971, and Olsson and Nelson 1975). Jacoby, Kowalik, and Pizzo (1972) provide a good description of these methods, as does the book by Kowalik and Osborne (1968). Bard (1974) gives a brief discussion, but expresses dissatisfaction with the typical performance of methods of this type. Bard recommends using finite difference approximations to derivatives with any good gradient method over use of a direct search method, including Powell's method given in Section 10.2.5.

10.3 COMPUTATIONAL METHODS IN NONLINEAR REGRESSION

The primary problem in nonlinear regression computations is to find a vector $\underline{\hat{\theta}}$ which minimizes the sum of squares [1]

---

[1] The scalar multiplier 1/2 is used for convenience to simplify expressions derived in this section. It has no bearing on the minimization problem.

$$F(\underline{\theta}) = \frac{1}{2} \sum_{t=1}^{n} [y_t - h(\underline{x}_t, \underline{\theta})]^2 \qquad (10.54)$$

The regression model itself is of the form

$$y_t = h(\underline{x}_t, \underline{\theta}^*) + e_t \qquad (t = 1, 2, \ldots, n)$$

where the y_t are observed responses corresponding to fixed known values $\underline{x}_t$ and relative to the response function h; and e_t denotes a random error. The vector $\underline{\theta}^*$ consists of parameters to be estimated. For purpose of discussion in this section we will assume that the e_t are independently and identically distributed normally with mean zero and finite variance σ^2.

The least-squares estimator $\hat{\underline{\theta}}$ of $\underline{\theta}^*$ under fairly general regularity conditions, for large sample size n, is approximately distributed as a multivariate normal random variable having mean vector $\underline{\theta}^*$ and covariance matrix

$$\Sigma = \sigma^2 [J'(\underline{\theta}^*) J(\underline{\theta}^*)]^{-1}$$

The matrix J is the n x p *Jacobian* matrix of the vector-valued function

$$\begin{aligned} \underline{f}'(\underline{\theta}) &= \{[y_1 - h(\underline{x}_1, \underline{\theta})], [y_2 - h(\underline{x}_2, \underline{\theta})], \ldots, [y_n - h(\underline{x}_n, \underline{\theta})]\} \\ &= \{f^{(1)}(\underline{\theta}), f^{(2)}(\underline{\theta}), \ldots, f^{(n)}(\underline{\theta})\} \end{aligned}$$

evaluated at $\underline{\theta}^*$. This matrix has the form

$$J(\underline{\theta}^*) = \left. \begin{pmatrix} \dfrac{\partial f^{(1)}(\underline{\theta})}{\partial \theta_1} & \dfrac{\partial f^{(1)}(\underline{\theta})}{\partial \theta_2} & \cdots & \dfrac{\partial f^{(1)}(\underline{\theta})}{\partial \theta_p} \\ \vdots & & & \\ \dfrac{\partial f^{(n)}(\underline{\theta})}{\partial \theta_1} & \dfrac{\partial f^{(n)}(\underline{\theta})}{\partial \theta_2} & \cdots & \dfrac{\partial f^{(n)}(\underline{\theta})}{\partial \theta_p} \end{pmatrix} \right|_{\underline{\theta}=\underline{\theta}^*} \qquad (10.55)$$

The unknown variance σ^2 is estimated as

$$s^2 = \frac{2F(\underline{\hat{\theta}})}{n - p} \tag{10.56}$$

and the estimate of the covariance matrix is

$$\hat{\Sigma} = s^2[J'(\underline{\hat{\theta}})J(\underline{\hat{\theta}})]^{-1} \tag{10.57}$$

The reader should note that the function $\underline{f}(\underline{\theta})$ and its component functions $f^{(j)}(\underline{\theta})$ now denote general nonlinear functions.

Confidence limits and tests of hypotheses relative to the parameters $\underline{\theta}^*$ in the model are made in the same general fashion as in multiple linear regression, but the computation of $\underline{\hat{\theta}}$ is more difficult for nonlinear models. Also, it must be reemphasized that the stated distributional properties of $\underline{\hat{\theta}}$ are based on the limiting distribution as $n \longrightarrow \infty$. A good introductory discussion of the stochastic properties of nonlinear least-squares estimators is given by Gallant (1975a). Derivation of important properties of $\underline{\hat{\theta}}$ can be found in Jennrich (1969), Hartley (1964), Hartley and Booker (1965), Malinvaud (1970), Gallant (1971), and Chambers and Ertel (1975), for example.

The essential elements to be considered, from the computing standpoint, are $\underline{\hat{\theta}}$ and the matrix (10.57). The major part of this section contains a description of methods for computing these two quantities. The problem of determining $\underline{\hat{\theta}}$ from (10.54) is simply one of minimizing a nonlinear function; hence any of the general methods presented in Section 10.2 might be used to solve this problem. However, Eq. (10.54) is a rather special form of nonlinear function since it is a sum of squares

$$F(\underline{\theta}) = \frac{1}{2} \sum_{t=1}^{n} [f^{(t)}(\underline{\theta})]^2 = \frac{1}{2} \underline{f}'(\underline{\theta})\underline{f}(\underline{\theta}) \tag{10.58}$$

where we write $\underline{\theta}$ to denote the variable in the function to be minimized. It would seem that we should be able to take advantage of the special form of the objective function (10.58) and find specialized computational methods that are more efficient than the general unconstrained minimization methods for this problem. In fact, as we shall see, specialized methods have been derived and are currently in use for nonlinear regression applications. When the specialized methods fail for a given problem, we always have the option of trying the general methods for unconstrained minimization to see if one of these will converge (see, e.g., Vitale and Taylor 1968, Krogh 1974, and McKeown 1975).

Some suggested references for general discussions of the computational problems in nonlinear regression are Bard (1974), Chambers (1977), Ben-Israel (1967), and Davidon (1976b).

10.3.1 Newton's Method for the Nonlinear Regression Problem

Since Newton's method is the most satisfactory among the general minimization methods discussed, in terms of rate of convergence in the vicinity of a minimum, we will begin by considering this method with respect to the objective function (10.58). The gradient vector and Hessian matrix of $F(\underline{\theta})$ have the form

$$\underline{g}(\underline{\theta}) = \left(\frac{\partial F}{\partial \theta_1}, \ldots, \frac{\partial F}{\partial \theta_p}\right)'$$

$$= \begin{pmatrix} \sum_{t=1}^{n} f^{(t)}(\underline{\theta}) \left[\dfrac{\partial f^{(t)}(\underline{\theta})}{\partial \theta_1}\right] \\ \sum_{t=1}^{n} f^{(t)}(\underline{\theta}) \left[\dfrac{\partial f^{(t)}(\underline{\theta})}{\partial \theta_2}\right] \\ \vdots \\ \sum_{t=1}^{n} f^{(t)}(\underline{\theta}) \left[\dfrac{\partial f^{(t)}(\underline{\theta})}{\partial \theta_p}\right] \end{pmatrix}$$

$$= J'(\underline{\theta})\underline{f}(\underline{\theta}) \tag{10.59}$$

$$G(\underline{\theta}) = \begin{pmatrix} \frac{\partial^2 F}{\partial \theta_1^2} & \frac{\partial^2 F}{\partial \theta_1 \partial \theta_2} & \cdots & \frac{\partial^2 F}{\partial \theta_1 \partial \theta_p} \\ & \frac{\partial^2 F}{\partial \theta_2^2} & \cdots & \frac{\partial^2 F}{\partial \theta_2 \partial \theta_p} \\ & & \ddots & \vdots \\ & \text{(sym)} & & \frac{\partial^2 F}{\partial \theta_p^2} \end{pmatrix}$$

where

$$\frac{\partial^2 F}{\partial \theta_i \partial \theta_j} = \sum_{t=1}^{n} \left[f^{(t)}(\underline{\theta}) \frac{\partial^2 f^{(t)}(\underline{\theta})}{\partial \theta_i \partial \theta_j} + \frac{\partial f^{(t)}(\underline{\theta})}{\partial \theta_i} \frac{\partial f^{(t)}(\underline{\theta})}{\partial \theta_j} \right]$$

Thus $G(\underline{\theta})$ is readily seen to have the form

$$G(\underline{\theta}) = J'(\underline{\theta})J(\underline{\theta}) + \sum_{t=1}^{n} f^{(t)}(\underline{\theta})H^{(t)}(\underline{\theta}) \tag{10.60}$$

where $H^{(t)}(\underline{\theta})$ is the Hessian matrix of $f^{(t)}(\underline{\theta})$.

Using these results, the basic Newton iteration has the form

$$\underline{\theta}_{i+1} = \underline{\theta}_i - G_i^{-1}J_i'\underline{f}_i \tag{10.61}$$

where $\underline{f}_i = \underline{f}(\underline{\theta}_i)$ and $J_i = J(\underline{\theta}_i)$. Note that once the Jacobian matrix J_i is obtained, the first term in the sum (10.60) for G_i is essentially available, and one factor in the gradient vector computation (10.59) is also available. In view of these facts, and the expression (10.57), we see that $J(\underline{\theta})$ has an important role in nonlinear regression computations. It should also be noted that there is no assurance that the matrices G_i in the iterative procedure will remain nonsingular, so G_i^{-1} may not always exist.

Newton's method, for $F(\underline{\theta})$ in the form of a sum of squares, has been presented to set the stage for discussion of methods that are normally used in this case. The requirement for second derivatives usually causes practitioners to use other methods that are less demanding in this respect and to try a modified Newton method when other methods fail.

10.3.2 The Modified Gauss-Newton Method

The Gauss-Newton method is applicable to the situation under consideration. This method consists of approximating each of the functions $f^{(k)}(\underline{\theta})$ using only the first two terms in the Taylor series expansion (i.e., a linear approximation). For $F(\underline{\theta})$ in the form (10.58), the Gauss-Newton method selects a direction $\underline{d}_i$ according to

$$J_i'J_i\underline{d}_i = -J_i'\underline{f}_i \tag{10.62}$$

and finds the next iterate as

$$\underline{\theta}_{i+1} = \underline{\theta}_i + \underline{d}_i \tag{10.63}$$

Details of the development of Eqs. (10.62) and (10.63) are as follows. If we expand $f^{(k)}(\underline{\theta})$ in a Taylor series about $\underline{\theta}_j$, the expansion can be seen to have the form

$$f^{(k)}(\underline{\theta}) = f^{(k)}(\underline{\theta}_j) + \left[\frac{\partial f^{(k)}(\underline{\theta})}{\partial\theta_1}, \ldots, \frac{\partial f^{(k)}(\underline{\theta})}{\partial\theta_p}\right]\Bigg|_{\underline{\theta}=\underline{\theta}_j} \cdot (\underline{\theta} - \underline{\theta}_j) + E$$

Dropping the error term and collecting the $f^{(k)}(\underline{\theta})$ as components in $\underline{f}(\theta)$, we have

$$\underline{f}(\theta) \cong \underline{f}(\underline{\theta}_j) + J(\underline{\theta}_j)(\underline{\theta} - \underline{\theta}_j)$$

It follows that

$$\begin{aligned}\frac{1}{2}\underline{f}'(\theta)\underline{f}(\theta) &\cong \frac{1}{2}\underline{f}'(\underline{\theta}_j)\underline{f}(\underline{\theta}_j) + \underline{f}'(\underline{\theta}_j)J(\underline{\theta}_j)(\underline{\theta} - \underline{\theta}_j)\\ &\quad + \frac{1}{2}(\underline{\theta} - \underline{\theta}_j)'J'(\underline{\theta}_j)J(\underline{\theta}_j)(\underline{\theta} - \underline{\theta}_j)\\ \underline{g}(\underline{\theta}) &\cong J'(\underline{\theta}_j)\underline{f}(\underline{\theta}_j) + J'(\underline{\theta}_j)J(\underline{\theta}_j)(\underline{\theta} - \underline{\theta}_j)\end{aligned}$$

This approximate form of $\underline{g}(\underline{\theta})$ suggests iterates in the form defined by Eqs. (10.62) and (10.63).

Notice that this method is essentially Newton's method using an approximation to the Hessian matrix (10.60) which consists of only the first term $J'(\underline{\theta})J(\underline{\theta})$ in that expression. Also note that each iteration in the Gauss-Newton method requires that the system of linear equations (10.62) be solved. These equations are in the form of normal equations, and in this case it is usually better to compute a least-squares solution of

$$J_i\underline{d}_i = -\underline{f}_i \tag{10.64}$$

to preclude introducing excessive amounts of error in the components of the solution $\underline{d}_i$. If there was every assurance that the residuals $f^{(t)}(\underline{\theta})$ were all small near the minimum, we would feel confident that this method would operate like Newton's method due to the form of Eq. (10.60). Unfortunately, large residual problems occur all too frequently and the Gauss-Newton method does not behave as Newton's method.

A modified form of the Gauss-Newton method was suggested by Hartley (1961). This modification consists of introducing a step-shortening parameter so that (10.63) becomes

$$\underline{\theta}_{i+1} = \underline{\theta}_i + \alpha_i\underline{d}_i \qquad (0 \le \alpha_i \le 1)$$

The suggested method for determining α_i is a quadratic polynomial approximation method.

The modified Gauss-Newton method (M.G.N.) was, for several years, the only interesting alternative to use of the steepest descent method. Accumulated experience with M.G.N. has shown that when it does work, the method converges more rapidly than other available methods (see, e.g., Bard 1970, Wolfe 1976, and Ramsin and Wedin 1977), but fortunately it fails too often to be entirely satisfactory. This method seems to be particularly sensitive to starting values, and it often ceases to make satisfactory progress near a point where $J(\underline{\theta})$ is less than full-rank. Also, when some of the residuals $f^{(i)}(\hat{\underline{\theta}})$

are large at the solution, the method usually tends to behave unsatisfactorily. Osborne (1972), McKeown (1975), Dennis and Welsch (1976), and Nazareth (1976b) all discuss this method and cite reasons why it fails for some applications. Basically, the degree to which $J'(\underline{\theta})J(\underline{\theta})$ approximates $G(\underline{\theta})$ seems to serve as a rough index to the expected performance of the method.

An interesting study of various linear search methods used in conjunction with the modified Gauss-Newton method is given by Flanagan, Vitale, and Mendelsohn (1969). Included among the five linear search methods studied was the golden section search and a second-order polynomial interpolation. This latter linear search procedure was the most effective method based on the data used in the numerical tests.

The choice of a reasonable termination criterion must always be made for iterative methods. In the case of M.G.N., a suggestion by Dennis (1977) is appealing. This criterion is derived by noting that at a stationary point $\hat{\underline{\theta}}$ of $F(\underline{\theta})$, as shown by Eq. (10.59), the columns of $J(\hat{\underline{\theta}})$ are orthogonal to $\underline{f}(\hat{\underline{\theta}})$ when the final residuals are not all small. If we denote the jth column of $J(\underline{\theta})$ by $\underline{a}_j(\underline{\theta})$, then the suggested termination criterion is to cease iterating and accept the point $\underline{\theta}_k$ when either

$$||\underline{f}(\underline{\theta}_k)|| \leq \varepsilon_1$$

or

$$\max_{1\leq j\leq p} \frac{|\underline{a}_j'(\underline{\theta}_k)\underline{f}(\underline{\theta}_k)|}{||\underline{a}_j(\underline{\theta}_k)||\cdot||\underline{f}(\underline{\theta}_k)||} \leq \varepsilon_2 \tag{10.65}$$

where ε_1 and ε_2 are small nonnegative numbers. The first inequality is specific to the size of residuals for the current fit, and the second inequality is based on the cosine of the angle between vectors. The important role that the Jacobian matrix has in this method is also emphasized by the above.

Finite-difference approximations to the partial derivatives in $J(\underline{\theta})$ were used by Brown and Dennis (1972) to produce derivative-free methods for M.G.N. and the Levenberg-Marquardt method (described in the following subsection). Tests using both approximate and exact analytical derivatives in these two methods showed little difference in number of successful solutions.

While M.G.N. does provide a desirable alternative to use of other more general methods from a computational standpoint, its lack of dependability in practical applications has caused research workers to seek other methods for computing $\hat{\underline{\theta}}$. One suggested alternative is given in the next subsection.

10.3.3 The Levenberg-Marquardt Modification of Gauss-Newton

When $J(\underline{\theta})$ is rank-deficient, or nearly so, in the Gauss-Newton iteration, the problem of computing the direction in Eq. (10.64) is difficult and may admit multiple solutions. The Levenberg-Marquardt modification, as we have seen previously in Section 10.2.2, transforms $J(\underline{\theta})$ to a better-conditioned full-rank matrix. The usual statement of the modification is to find $\underline{d}_i$ according to

$$(J_i'J_i + \tau^2 I)\underline{d}_i = -J_i'\underline{f}_i \tag{10.66}$$

where τ^2 is a chosen scalar. Solution for $\underline{d}_i$ might better be obtained through a least-squares solution of

$$\begin{pmatrix} J_i \\ \tau I \end{pmatrix} \underline{d}_i = -\begin{pmatrix} \underline{f}_i \\ \underline{0} \end{pmatrix}$$

where $\tau > 0$ must be chosen to condition the problem. Actually, this statement of the method is not strictly accurate. Marquardt (1963) noted that gradient methods are not scale-invariant, so he scaled $J_i'J_i$ to correlation matrix form and normalized $\underline{g}_i$ using the reciprocal of the square root of the corresponding diagonal element in $J_i'J_i$ at each iteration. Then the transformed matrices were used in the form of Eq. (10.66), and the solution $\underline{d}_i^*$ (say) obtained. Using $\underline{d}_i^*$ and the scale factors, $\underline{d}_i$ was then obtained by an inverse scaling.

The Levenberg-Marquardt (L.M.) method seems to be the most-often-used method for nonlinear least-squares estimation problems. It is somewhat more dependable than M.G.N.; nevertheless, it still frequently fails to converge (see, e.g., Meyer and Roth 1972, Meyer 1970, Chambers 1973, and Dennis 1973). Also, the L.M. method sometimes converges very slowly in the neighborhood of a stationary point. Thus while both M.G.N. and L.M. are useful, neither is sufficiently dependable to serve as the sole general nonlinear least-squares method in supporting software systems.

An interesting general comparison of directions generated by the Newton method and the L.M. modification of the Gauss-Newton method is given by Gill and Murray (1976). Suppose that the columns of the $p \times (p-r)$ matrix V span the null space of $J_i'J_i$, and $V'V = I_{p-r}$. Let W be a $(p \times r)$ matrix whose columns span the orthogonal complement of the null space of $J_i'J_i$ such that $W'W = I_r$. Thus we have

$$J_iV = 0 \quad \text{and} \quad W'V = 0$$

Any direction $\underline{d}_i$ in E^p is expressible in the form

$$\underline{d}_i = \underline{d}_i^{(1)} + \underline{d}_i^{(2)}$$

where

$$\underline{d}_i^{(1)} = W\underline{u} \quad \text{and} \quad \underline{d}_i^{(2)} = V\underline{y}$$

for some r vector $\underline{u}$ and (p-r) vector $\underline{y}$. Now Eq. (10.66) becomes

$$J_i'J_iW\underline{u} + \tau^2W\underline{u} + \tau^2V\underline{y} = -J_i'\underline{f}_i \tag{10.67}$$

Multiplying by W' gives

$$(W'J_i'J_iW + \tau^2I)\underline{u} = -W'J_i'\underline{f}_i$$

which uniquely defines $\underline{u}$. Premultiplying (10.67) by V' gives

$$\underline{y} = \underline{0}$$

Thus $\underline{d}_i = \underline{d}_i^{(1)}$ and the search direction is seen to be restricted to the row space of J_i in the L.M. form of the Gauss-Newton method.

If $\underline{d}_i$ denotes a Newton search direction, and the matrix B represents the second term in (10.60) scaled so that $||B|| = 1$; then

$$\underline{\bar{d}}_i = W\underline{\bar{u}} + V\underline{\bar{y}} = \underline{\bar{d}}_i^{(1)} + \underline{\bar{d}}_i^{(2)}$$

and substituting as was done above gives

$$V'BV\underline{\bar{y}} = -V'BW\underline{\bar{u}}$$

Thus, in general, $\underline{\bar{d}}_i$ is not in the space of W and $\underline{\bar{d}}_i$ will not necessarily be similar to $\underline{d}_i$.

Since Newton-type methods are generally more reliable than Gauss-Newton-type methods, one possible avenue toward more dependable methods is to use a closer approximation of

$$\sum_{t=1}^{n} f^{(t)}(\underline{\theta})H^{(t)}(\underline{\theta})$$

than is provided by the term $\tau^2 I$ used in the Levenberg-Marquardt method. We will next discuss several suggestions that have been made in this regard.

10.3.4 Alternative Gradient Methods

The purpose of this subsection is to indicate some of the gradient-type methods which have recently been proposed and are undergoing tests and refinements.

One approach toward generation of alternative search directions is to use a quasi-Newton approach to approximate the Hessian matrix. The update matrix A_k can be selected to satisfy

$$(\underline{\theta}_k - \underline{\theta}_{k-1})'A_k = (J_k'\underline{f}_k - J_{k-1}'\underline{f}_{k-1})$$

using (say) a rank-two update. Nazareth (1976b) utilizes an update matrix of this type and forms a convex combination of approximate Hessian matrices as

$$[(1 - \beta_i)J_i'J_i + \beta_i A_i]\underline{d}_i = -J_i'\underline{f}_i \quad (0 \leq \beta_i \leq 1)$$

to compute a direction $\underline{d}_i$. Adjustment of the scalar parameter β_i allows one to control the extent to which $\underline{d}_i$ is influenced by the Gauss-Newton and quasi-Newton characteristics.

A more direct approach was suggested by Brown and Dennis (1971). This method featured use of a quasi-Newton update to approximate each of the matrices $H^{(k)}$ in Eq. (10.60). The large computer storage requirements make this method undesirable. A more appealing alternative is to approximate the entire second term summation in (10.60) using a single quasi-Newton update form. Betts (1976) derives one method of this type and reports encouraging test results.

Several variations on the above have also been suggested. Welsch and Becker (1975) report good results in robust regression applications obtained by interpolating between directions produced by differing methods, one of which is the direction of steepest descent. Dennis and Welsch (1976) provide more information on this approach and present a good discussion of details needed in application to robust regression models.

Gill and Murray (1976) suggest another approach based on the analysis given for the L.M. modification of the Gauss-Newton method in the preceding subsection. Their approach consists of using a finite-difference-type approximation to the second term in the Hessian matrix (10.60), but restricting the approximation to the orthogonal complement of the row space of J_i. One appealing aspect of this suggestion is that we would not expect J_i to be highly rank-deficient in any iteration, so the extra computation, required for the additional approximation of the Hessian, should generally not be voluminous.

A useful survey of the current situation, insofar as computational methods for nonlinear least-squares applications is concerned, is given by Dennis (1977). The renewed interest in this area of statistical computing research in recent years is expected to continue, and hopefully the marginally satisfactory gradient algorithms that now exist will be replaced by substantially better ones in the near future.

10.3.5 Minimization without Derivatives

We have seen that gradient methods can be implemented, without requiring that the user supply derivatives analytically, by using numerical approximations for derivatives. These are still considered to be derivative-based methods. In this subsection we exclude derivative-based methods and discuss in some detail one of several proposed methods for nonlinear least-squares estimation which does not explicitly depend on the use of derivatives. Mention was made, in Section 10.2, of general minimization methods of this kind, and it was stated that they were not particularly successful compared to gradient methods. However, for the rather special form of objective function under consideration in this section there appears to be more hope for frequent success using such methods.

A method of this type was developed by Ralston (1975) (also see Ralston and Jennrich 1977, 1978). Basically the method consists of the now-familiar device of using hyperplanes to approximate the p-dimensional surface defined by $\underline{h}(\underline{\theta})$ to locate the $\hat{\underline{\theta}}$ which minimizes the euclidean distance between $\underline{h}(\underline{\theta})$ and the observed point $\underline{y}$ in n space. In this method, however, secant hyperplanes are used instead of the tangent hyperplanes which were implied in the work of preceding subsections.

For regression applications the method, at any iteration k, utilizes (p + 1) points

$$\underline{\theta}_k^{(1)}, \underline{\theta}_k^{(2)}, \ldots, \underline{\theta}_k^{(p+1)}$$

of which p are linearly independent. The minimum of $F(\underline{\theta})$ is approximated by considering the hyperplane

$$L_k(\underline{\alpha}) = \sum_{m=1}^{p+1} \alpha_m \underline{h}\left(\underline{\theta}_k^{(m)}\right) \quad , \left(\sum_{m=1}^{p+1} \alpha_m = 1\right) \tag{10.68}$$

and the same relationship between variables

$$\underline{\theta} = \sum_{m=1}^{p+1} \alpha_m \underline{\theta}_k^{(m)} \tag{10.69}$$

where from (10.54)

$$\underline{h}(\underline{\theta}) = [h(\underline{x}_1,\underline{\theta}), h(\underline{x}_2,\underline{\theta}), \ldots, h(\underline{x}_n,\underline{\theta})]'$$
$$= [h_1(\underline{\theta}), h_2(\underline{\theta}), \ldots, h_n(\underline{\theta})]' \qquad (10.70)$$

The major steps at each iteration are to find $\underline{\alpha}^*$, subject to the given constraints, which minimizes the distance between $L_k(\underline{\alpha})$ and $\underline{y} = (y_1, \ldots, y_n)'$; compute $\underline{\theta}^*$ as the linear combination (10.69) using $\underline{\alpha}^*$; and then replace one of the $\underline{\theta}_k^{(j)}$ with $\underline{\theta}^*$.

Note that the response function $\underline{h}(\underline{\theta})$ ranges over a p-dimensional surface in n-dimensional euclidean space, and that the vector $\underline{y}$ does not in general lie on this surface. The value $\hat{\underline{\theta}}$ which produces the smallest distance between $\underline{y}$ and $\underline{h}(\underline{\theta})$ is the minimizer of $F(\underline{\theta})$, as is seen from (we will drop the multiplier 1/2 from Eq. 10.54 in this subsection)

$$F(\hat{\underline{\theta}}) = [\underline{y} - \underline{h}(\hat{\underline{\theta}})]'[\underline{y} - \underline{h}(\hat{\underline{\theta}})]$$

This is somewhat similar to the linear approximation used in the Gauss-Newton method, but as we shall see, the present method utilizes information from previous iterations in a different fashion and does what roughly amounts to numerical differentiation to avoid the need for user-supplied derivatives.

Computationally it is desirable to change the form of Eqs. (10.68) and (10.69) to

$$L_k(\underline{\alpha}) = \underline{h}\left(\underline{\theta}_k^{(p+1)}\right) + \sum_{m=1}^{p} \alpha_m \left[\underline{h}(\underline{\theta}_k^{(m)}) - \underline{h}\left(\underline{\theta}_k^{(p+1)}\right)\right]$$

and

$$\underline{\theta} = \underline{\theta}_k^{(p+1)} + \sum_{m=1}^{p} \alpha_m \left(\underline{\theta}_k^{(m)} - \underline{\theta}_k^{(p+1)}\right)$$

where the point $\underline{\theta}_k^{(p+1)}$ gives the smallest F function value in the set. Let T_k be the (p x p) matrix having columns $\left(\underline{\theta}_k^{(m)} - \underline{\theta}_k^{(p+1)}\right)$

and H_k be the (n x p) matrix having columns $\left[\underline{h}\left(\underline{\theta}_k^{(m)}\right) - \underline{h}\left(\underline{\theta}_k^{(p+1)}\right)\right]$. Then we have from above

$$L_k(\underline{\alpha}) = \underline{h}\left(\underline{\theta}_k^{(p+1)}\right) + H_k\underline{\alpha} \tag{10.71}$$

$$\underline{\theta} = \underline{\theta}_k^{(p+1)} + T_k\underline{\alpha} \tag{10.72}$$

The minimizer of the function

$$F_k(\underline{\alpha}) = [\underline{y} - L_k(\underline{\alpha})]'[\underline{y} - L_k(\underline{\alpha})]$$

is found as

$$\underline{\alpha}^* = (H_k'H_k)^{-1}H_k'\left[\underline{y} - \underline{h}\left(\underline{\theta}_k^{(p+1)}\right)\right] \tag{10.73}$$

and the new point (10.72) is determined as

$$\underline{\theta}^* = \underline{\theta}_k^{(p+1)} + T_k(H_k'H_k)^{-1}H_k'\left[\underline{y} - \underline{h}\left(\underline{\theta}_k^{(p+1)}\right)\right]$$

$$= \underline{\theta}_k^{(p+1)} + T_k\underline{\alpha}^* \tag{10.74}$$

One problem which may arise is that H_k may not retain full column rank for every iteration. The suggested device for dealing with this is to sweep the first p rows of the matrix (selecting the maximum available pivot element for each sweep; e.g., see Jennrich and Sampson 1968)

$$\begin{pmatrix} H_k'H_k & H_k'\left[\underline{y} - \underline{h}\left(\underline{\theta}_k^{(p+1)}\right)\right] \\ \left[\underline{y} - \underline{h}\left(\underline{\theta}_k^{(p+1)}\right)\right]'H_k & 0 \end{pmatrix}$$

and zero any row in which a small pivot element is encountered. This will result in some of the components of $\underline{\alpha}^*$ possibly being zero, but $\underline{\alpha}^*$ will be a least-squares solution, which is all that is required.

In order to avoid having the search collapse into a subspace of the $\underline{\theta}$ space, the following device is used to replace a vector in the $\underline{\theta}_k^{(j)}$ set. Let α_t be the first element in $\underline{\alpha}^*$ such that $|\alpha_t| > \varepsilon$

where ε is a given tolerance. Then for each m = t, ..., p let $\underline{\theta}_{k+1}^{(m)} = \underline{\theta}_k^{(m+1)}$ and $\underline{\theta}_{k+1}^{(p+1)} = \underline{\theta}^*$. Also, when t > 1, let $\underline{\theta}_{k+1}^{(m)} = \left(\underline{\theta}_k^{(m)} + \underline{\theta}_{k+1}^{(p+1)}\right)/2$ for m = 1, 2, ..., t - 1. This latter computation insures that points will not remain in the set indefinitely.

The new vector $\underline{\theta}^*$ may not result in a reduction of the value of the objective function. When this is the case, $\underline{\theta}^*$ is further modified to be the first vector in the sequence

$$\underline{\theta}_j^* = (1 - \delta_j)\underline{\theta}_k^{(p+1)} + \delta_j\underline{\theta}^*$$

for which

$$F(\underline{\theta}_j^*) \leq F\left(\underline{\theta}_k^{(p+1)}\right)$$

where

$$\delta_j = \begin{cases} 1 & j = 0 \\ -(-\frac{1}{2})^j & j = 1, 2, \ldots \end{cases}$$

Normally a small upper limit is placed on the value of j to be considered. If the objective function cannot be reduced, iteration continues anyway since it can be shown that it is possible for the method to converge without such reduction at each step.

Having performed these operations, a possible renumbering of the $\underline{\theta}_{k+1}^{(j)}$ is all that is needed to begin the next iteration. A termination rule based on the successive objective function values is suggested. When the minimizing vector $\hat{\underline{\theta}}$ is found, the estimate of the asymptotic covariance matrix is computed at that point as

$$\hat{\Sigma} = s^2 T(H'H)^{-1}T'$$

where

$$s^2 = \frac{F(\hat{\underline{\theta}})}{n - p}$$

Starting values for this iterative method can be obtained in many ways. One suggestion is to select one vector $\underline{\theta}_1$ and then displace one component at a time according to

$$\underline{\theta}_1^{(i)} = \underline{\theta}_1 + \gamma_i \underline{e}_i$$

where $\underline{e}_i$ is the ith unit vector and $\underline{\gamma} = (\gamma_1, \ldots, \gamma_p)'$ is a vector of user-supplied estimates of displacement needed to reach the minimizing point.

This method, which Ralston named DUD (Doesn't Use Derivatives), has been tested against most of the other available methods for unconstrained minimization. In several instances its performance was remarkably good. It will, however, fail, and it is not known at this time exactly how its success rate compares with the M.G.N. and L.M. methods. Its rate of convergence near a minimum, for some test problems, has matched or exceeded that of the best gradient methods, but this will almost surely not be the case in general. The usual comparative measure between programs in testing is total number of objective function evaluations in a successful trial. With this as a comparative basis, DUD is a competitor with other methods considered in this section.

There are several other methods of this type. Peckham (1970) gives a very similar method, but it requires substantially more computer storage for its operations. Powell (1965) gives another method which is not strictly of this type because it does use numerical differentiation, but it is sufficiently similar to deserve mention. Another method which classifies in this area is the simplex method given by Nelder and Mead (1965).

10.3.6 Summary

As we have seen, there are several different computational methods designed to solve nonlinear least-squares problems. One or more of these methods will often provide satisfactory answers for a given problem. Selection of method and starting values is an art, and experience plays an important role in these activities. Usually, programs will perform a grid search in a user-specified region of the variable space to determine a specific starting value which corresponds to the smallest function value found in the search. The specification of region and grid size is still subject to choice; hence in effect the burden is still on the user to specify starting values.

Despite the availability of several different methods, there is not one that is considered to be universally dependable. If the objective function is indeed special in regression situations, research has not yet uncovered a way to take full advantage of this speciality. Researchers are currently fairly optimistic about the prospects in this area, and several projects are underway which may produce more dependable and efficient algorithms. Meanwhile, as has been pointed out by McKeown (1975), among others, the general unconstrained minimization methods will continue to perform better in some regression applications than the methods that are especially designed for such problems.

Other directions of attack have been taken to help solve certain problems that fall into special classes. One example of this is the work on partially linear nonlinear models. The idea here is that at each iteration the parameters that appear linearly in the model are estimated using linear regression methods and the (hopefully) few remaining parameters that enter the model nonlinearly are estimated using one of the methods given in this section. Harville (1973), Barham and Drane (1972), and Lawton and Sylvester (1971) give descriptions of methodology of this type.

10.4 TEST PROBLEMS

There is a continuing need for problems having known answers to use in testing new software systems and algorithms and reverifying established systems. In this section we will list several test problems which have been used in the past for these purposes. In each case $\underline{\hat{\theta}}$ will denote the computed or true solution and $\underline{\theta}_0$ will denote a starting value.

Problem 10.1 (Greenstadt 1972)

$$F(\underline{\theta}) = (\theta_1 + \theta_2 - 2)^2 + 10^4(\theta_1 - \theta_2)^2$$

$$G(\underline{\theta}) = \begin{bmatrix} 20002 & -19998 \\ -19998 & 20002 \end{bmatrix}$$

$$\hat{\underline{\theta}} = (1,1)'$$

Problem 10.2 (Greenstadt 1972)

$$F(\underline{\theta}) = (\theta_1 + 2\theta_2 + 3\theta_3)^2 + 100(\theta_2 - 1)^2 + (\theta_3 - 2)^2$$

$$G(\underline{\theta}) = \begin{bmatrix} 2 & 4 & 6 \\ 4 & 208 & 12 \\ 6 & 12 & 20 \end{bmatrix}$$

$$\hat{\underline{\theta}} = (-8,1,2)'$$

Problem 10.3 (Rosenbrock 1960)

$$F(\underline{\theta}) = 100(\theta_2 - \theta_1^{\,2})^2 + (1 - \theta_1)^2$$

$$G = \begin{bmatrix} 802 & -400 \\ -400 & 200 \end{bmatrix}$$

$$\hat{\underline{\theta}} = (1,1)' \quad \underline{\theta}_0 = (-1.2,1)'$$

Problem 10.4 (Bass 1972)

$$F(\underline{\theta}) = \sum_{i=1}^{n} \theta_i^{\,2} + \left(\sum_{i=1}^{n} i^{1/2}\theta_i\right)^2 + \left(\sum_{i=1}^{n} i^{1/2}\theta_i\right)^4$$

$$\hat{\underline{\theta}} = \underline{0} \quad \underline{\theta}_0 = (0.1,\ 0.1,\ \ldots,\ 0.1)'$$

Problem 10.5 (Bass 1972)

$$F(\underline{\theta}) = 100[(\theta_3 - 10a)^2 + (b - 1)^2] + \theta_3^{\,2}$$

$$2\pi a = \arctan\left(\frac{\theta_2}{\theta_1}\right) \qquad \theta_1 > 0$$

$$= \arctan\left(\frac{\theta_2}{\theta_1}\right) + \pi \qquad \theta_1 < 0$$

$$b = \left(\theta_1^{\,2} + \theta_2^{\,2}\right)^{1/2}$$

$$\hat{\underline{\theta}} = (1,0,0)' \quad \underline{\theta}_0 = (-1,0,0)'$$

Problem 10.6 (Oren 1973)

$$F(\underline{\theta}) = \sum_{i=1}^{n} \sum_{j=1}^{n} \left[\frac{(\theta_j - 1)(\theta_i - 1)}{j + i - 1}\right]$$

$$\hat{\underline{\theta}} = (1, 1, \ldots, 1)' \quad (\underline{\theta}_0)_i = \frac{4}{i}$$

Problem 10.7 (Fletcher and Powell 1963)

$$F(\underline{\theta}) = (\theta_1 + 10\theta_2)^2 + 5(\theta_3 - \theta_4)^2 + (\theta_2 - 2\theta_3)^4 + 10(\theta_1 - \theta_4)^4$$

$$\underline{\theta}_0 = (3,-1,0,1)'$$

Problem 10.8 (Zangwill 1967)

$$F(\underline{\theta}) = \frac{16\theta_1^{\,2} + 16\theta_2^{\,2} - 8\theta_1\theta_2 - 56\theta_1 - 256\theta_2 + 991}{15}$$

$$\hat{\underline{\theta}} = (4,9)' \quad \underline{\theta}_0 = (3,8)' \quad F(\hat{\underline{\theta}}) = -18.2$$

Problem 10.9 (Box 1966)

$$F(\underline{\theta}) = \sum_{i=1}^{10} \left[\exp(-t_i\theta_1) - \exp(-t_i\theta_2) - \exp(-t_i) + \exp(-10t_i)\right]^2$$

$$t_i = \frac{i}{10} \quad \hat{\underline{\theta}} = (1,10)' \quad \underline{\theta}_0 = (5,0)' \quad F(\hat{\underline{\theta}}) = 0$$

Problem 10.10 (Zangwill 1967)

$$F(\underline{\theta}) = (\theta_1 - \theta_2 + \theta_3)^2 + (\theta_2 - \theta_1 + \theta_3)^2 + (\theta_1 + \theta_2 - \theta_3)^2$$

$$\underline{\theta}_0 = (1/2,1,1/2)'$$

Problem 10.11 (Wolfe 1975)

$$F(\underline{\theta}) = [\exp(\theta_1) - \theta_2]^4 + 100(\theta_2 - \theta_3)^6 + [\tan^{-1}(\theta_3 - \theta_4)]^4 + \theta_1{}^8$$

$\underline{\theta}_0 = (1,2,2,2)'$ $\hat{\underline{\theta}} = (0,1,1,1)'$

Problem 10.12 (Wolfe 1975)

$$F(\underline{\theta}) = (1 - \theta_1)^2 + (1 - \theta_{10})^2 + \sum_{i=1}^{9} (\theta_i{}^2 - \theta_{i+1})^2$$

$\underline{\theta}_0 = (-2, -2, \ldots, -2)'$ $\hat{\underline{\theta}} = (1, 1, \ldots, 1)'$

Problem 10.13 (Goldfeld, Quandt, and Trotter 1966)

$$F(\underline{\theta}) = (2\theta_1{}^2 + 3\theta_2{}^2) \exp(-\theta_1{}^2 - \theta_2{}^2)$$

Problem 10.14 (Kowalik, Kamgnia, and Boland 1977)

$$F(\underline{\theta}) = \left(\frac{\underline{\theta}'A\underline{\theta}}{2} + \underline{b}'\underline{\theta} + 0.25\right)$$

$$A = \begin{pmatrix} 4.7 & 7 & 3.5 & 3 \\ 7 & 14 & 9 & 8 \\ 3.5 & 9 & 8.5 & 5 \\ 3 & 8 & 5 & 7 \end{pmatrix} \quad b = \begin{pmatrix} 0.5 \\ -1 \\ -1.5 \\ 0 \end{pmatrix}$$

$\hat{\underline{\theta}} = (1/2,-5,1/2,0)'$ $\underline{\theta}_0 = (4,4,4,4)'$

Problem 10.15 (Betts 1976)

$$F(\underline{\theta}) = \sum_{i=1}^{15} \left[y_i - \left(\theta_1 + \frac{u_i}{\theta_2 v_i + \theta_3 w_i} \right) \right]^2$$

$\hat{\underline{\theta}} = (0.08241,1.1330,2.3437)'$ $F(\hat{\underline{\theta}}) = 8.214877 \times 10^{-3}$

$\underline{\theta}_0 = (1,1,1)'$

i	y_i	u_i	v_i	w_i
1	0.14	1	15	1
2	0.18	2	14	2
3	0.22	3	13	3
4	0.25	4	12	4
5	0.29	5	11	5
6	0.32	6	10	6
7	0.35	7	9	7
8	0.39	8	8	8
9	0.37	9	7	7
10	0.58	10	6	6
11	0.73	11	5	5
12	0.96	12	4	4
13	1.34	13	3	3
14	2.10	14	2	2
15	4.39	15	1	1

Problem 10.16 (Betts 1976)

$$F(\underline{\theta}) = \sum_{i=1}^{11} \left[y_i - \frac{\theta_1 \left(u_i^2 + \theta_2 u_i \right)}{u_i^2 + \theta_3 u_i + \theta_4} \right]^2$$

$$\hat{\underline{\theta}} = (0.19281, 0.19128, 0.12306, 0.13606)'$$

$$F(\hat{\underline{\theta}}) = 3.070561 \times 10^{-4}$$

$$\underline{\theta}_0 = (0.25, 0.39, 0.415, 0.39)'$$

i	y_i	u_i
1	0.1957	4.
2	0.1947	2.
3	0.1735	1.
4	0.1600	0.5
5	0.0844	0.25
6	0.0627	0.167
7	0.0456	0.125
8	0.0342	0.100
9	0.0323	0.0833
10	0.0235	0.0714
11	0.0246	0.0625

Problem 10.17 (Meyer and Roth 1972)

$$F(\underline{\theta}) = \sum_{i=1}^{5} \left[y_i - \frac{\theta_1 \theta_3 x_{1i}}{1 + \theta_1 x_{1i} + \theta_2 x_{2i}} \right]^2$$

$\underline{\theta}_0 = (10.39,48.83,0.74)'$ $\hat{\underline{\theta}} = (3.13,15.16,0.78)'$

$F(\hat{\underline{\theta}}) = 0.4 \times 10^{-4}$

i	x_1	x_2	y
1	1	1	0.126
2	2	1	0.219
3	1	2	0.076
4	2	2	0.126
5	0.1	0	0.186

Problem 10.18 (Hartley 1961)

$$F(\underline{\theta}) = \sum_{i=1}^{6} [y_i - \theta_1 - \theta_2 \exp(\theta_3 x_i)]^2$$

$\underline{\theta}_0 = (580,-180,-0.160)'$

$\hat{\underline{\theta}} \cong (523.3,-157.0,-0.1994)'$

x_i	y_i	x_i	y_i
-5	127	1	421
-3	151	3	460
-1	379	5	426

EXERCISES

10.1 Provide a proof for Theorem 10.1.

10.2 Show that the directional derivative of $F(\underline{\theta})$ at a point $\underline{\theta}_0$ is maximum in the direction of the gradient at $\underline{\theta}_0$. (Fleming 1965.)

10.3 Devise an algorithm which performs a linear search using bisection.

10.4 In algorithm GOLDEN show that only one function evaluation is needed in each iteration of steps 8 to 11. Also show that $\ell_j = v_2^{\,j}\ell_0$.

10.5 Prove that Newton's method will converge in one iteration to the solution when the objective function is convex and quadratic.

10.6 Derive Eqs. (10.14).

10.7 The derivation of (10.17) used the fact that for each $0 \leq j \leq p$

$$0 \leq m_j \leq \left[\frac{n_j}{\beta} + (j - 1)\beta\right]^2 + 2\left[|\gamma_{jj}| + (j - 1)\beta^2\right] + \xi$$

where $n_j = \max\{|\gamma_{kj}|, k = j + 1, \ldots, p\}$. Derive these bounds for the m_j. (Gill and Murray 1974.)

10.8 Show that in the Gill and Murray modification of Newton's iteration, when G_i is not positive definite then $\psi_j \leq 0$ for at least one $0 \leq j \leq p$. (Gill and Murray 1974.)

10.9 If Eq. (10.23) is expressed in the form

$$H_{i+1} = H_i + W_i - X_i$$

show, for the quadratic objective function (10.19), that

(a) $A^{-1} = \sum_{i=0}^{p-1} W_i$ using Eq. (10.28); and

(b) the form of X_i follows from Eq. (10.29). (Fletcher and Powell 1963.)

10.10 Use the function

$$F(\underline{\theta}) = \theta_1^2 + 2\theta_2^2 + 3\theta_3^2$$

and the starting vectors

$$\underline{\theta}_0 = \begin{pmatrix} 4 \\ 7 \\ 0 \end{pmatrix} \quad \underline{d}_0 = \begin{pmatrix} 1 \\ -2 \\ 1 \end{pmatrix}$$

in the conjugate gradient algorithm of Fletcher and Reeves. (Powell 1976.)

10.11 Give a proof for Theorem 10.2. (Murray 1972.)

10.12 Let A be a p x p positive definite matrix and $\underline{d}_0, \ldots, \underline{d}_{p-1}$ be mutually conjugate with respect to A. Are the vectors $\underline{d}_i$ necessarily linearly independent? Under what conditions will they be linearly independent?

10.13 Consider the function

$$F(\underline{\theta}) = (\theta_1 - \theta_2 + \theta_3)^2 + (\theta_2 - \theta_1 + \theta_3)^2 + (\theta_1 + \theta_2 - \theta_3)^2$$

which is convex and has minimum point (0,0,0)'. Use Powell's basic conjugate direction method with starting point (1/2,1,1/2)' to minimize $F(\underline{\theta})$. (Zangwill 1967.)

10.14 Let $\underline{d}_0, \underline{d}_1, \ldots, \underline{d}_{p-1}$ be given nonzero p vectors, and A be a given p x p positive definite symmetric matrix. Define C to be the p-square matrix having the vectors $\underline{d}_j$ as its columns. Show that if the $\underline{d}_j$ satisfy

$$\underline{d}_j' A \underline{d}_j = 1 \qquad j = 1, 2, \ldots, p$$

then the determinant of C assumes its maximum value if and only if the directions $\underline{d}_j$ are mutually conjugate with respect to A. (Powell 1964.)

10.15 Consider the method of steepest descent and assume an exact linear search is made at each step. What can you say about the relationship between successive directions taken by this method in its search for a minimum?

10.16 Consider the function

$$F(\underline{\theta}) = -\left(\theta_1^2 + \theta_2^2 + 1\right)^{-1}$$

(a) Find the gradient vector and Hessian matrix analytically.

(b) Show that the Hessian matrix is positive definite near the origin.

(c) Analyze the behavior of Newton's basic method ($\alpha_i \equiv 1$) when the starting point $\underline{\theta}_0 = (10,0)'$ is used.

(d) Explain how variable step size ($\alpha_i \not\equiv 1$) can sometimes be beneficial.

10.17 Explain how the Cauchy-Schwarz inequality is used in the derivation of Eq. (10.25).

10.18 Under the conditions of Exercise 10.12, with $\underline{d}_j \neq \underline{0}$ for all j, show that

$$A^{-1} = \sum_{j=0}^{p-1} \frac{\underline{d}_j \underline{d}_j'}{\underline{d}_j' A \underline{d}_j}$$

10.19 Suppose that a quadratic approximation method is desired for the linear search of $F(\underline{\theta}_i + \alpha \underline{d}_i)$. In other words, a quadratic approximation

$$F(\underline{\theta}_i + \alpha \underline{d}_i) \cong a_0 + a_1\alpha + a_2\alpha^2$$

is to be used. Let F be evaluated at the three points α_0, α_1, α_2 and use the notation $v_j = F(\underline{\theta}_i + \alpha_j \underline{d}_i)$. Then a_0, a_1, a_2 can be obtained by solving the linear system

$$v_0 = a_0 + a_1\alpha_0 + a_2\alpha_0^2$$

$$v_1 = a_0 + a_1\alpha_1 + a_2\alpha_1^2$$

$$v_2 = a_0 + a_1\alpha_2 + a_2\alpha_2^2$$

Show that the minimum value α^*, based on this approximation, is given by

$$\alpha^* = -\frac{a_1}{2a_2} = -\frac{1}{2}\,\frac{\begin{vmatrix} 1 & v_0 & \alpha_0^2 \\ 1 & v_1 & \alpha_1^2 \\ 1 & v_2 & \alpha_2^2 \end{vmatrix}}{\begin{vmatrix} 1 & \alpha_0 & v_0 \\ 1 & \alpha_1 & v_1 \\ 1 & \alpha_2 & v_2 \end{vmatrix}}$$

$$= \frac{1}{2}\,\frac{(\alpha_1^2 - \alpha_2^2)v_0 + (\alpha_2^2 - \alpha_0^2)v_1 + (\alpha_0^2 - \alpha_1^2)v_2}{(\alpha_1 - \alpha_2)v_0 + (\alpha_2 - \alpha_0)v_1 + (\alpha_0 - \alpha_1)v_2}$$

whenever $a_2 > 0$. Derive a means for checking to determine whether $a_2 = 0$ in the approximating procedure. Explain why when $v_0 \cong v_1 \cong v_2$, then very poor numerical accuracy can be expected in the last equation.

10.20 Give a proof for Theorem 10.3.

10.21 Show that the update (10.23) satisfies the quasi-Newton condition.

10.22 Show that the update (10.24) satisfies the quasi-Newton condition.

10.23 Does quasi-Newton update (10.24) maintain positive definite iterates H_{i+1} when the starting matrix H_0 is positive definite? Provide a proof to support your conclusion.

10.24 Will the iteration (10.18) have quadratic convergence when update (10.24) is used? [We know that it will when update (10.23) is used.] Provide a proof to support your conclusion. Assume that exact linear search is performed.

10.25 How is the Goldstein-Armijo step size method related to the directional derivative of F at $\underline{\theta}_i$ in the direction $\underline{d}_i$? Indicate how this method avoids generally taking large steps.

10.26 Does the basic Gauss-Newton method produce a descent direction $\underline{d}_i$ at each step? Under what conditions, if any, does it fail to do so?

10.27 Does the Levenberg-Marquardt method produce a descent direction at each iteration? Under what conditions, if any, does it fail to do so? (Dennis 1977.)

10.28 Show how the matrices V and W shown in Eq. (10.67) can be obtained using Householder transformation from the right on the matrix J_i. (Gill and Murray 1976.)

10.29 With respect to the Levenberg-Marquardt method defined through Eq. (10.66), show that $||\underline{d}_i||^2$ approaches zero as $\lambda \longrightarrow \infty$. (Marquardt 1963.)

10.30 Continuing Exercise 10.29, show that as $\lambda \longrightarrow \infty$ the angle which $\underline{d}_i$ makes with $-J_i'\underline{f}$ (the negative gradient) decreases monotonically to zero.

10.31 Discuss the use of Newton's method for maximum likelihood variance component estimation (see Eq. 9.44). What is the role of the W transformation in this application? (Refer to Section 9.4.2.)

REFERENCES AND FURTHER READINGS

Adachi, N. (1971), On Variable Metric Algorithms, *J.O.T.A.* *7*, 391-423.

Aoki, M. (1971), *Introduction to Optimization Techniques*, Macmillan, New York.

Apostol, T. M. (1957), *Mathematical Analysis*, Addison-Wesley, Reading, Mass.

Armijo, L. (1966), Minimization of Functions Having Lipschitz Continuous Partial Derivatives, *Pacific J. Math.* *16*, 1-3.

Bard, Y. (1968), On a Numerical Instability of Davidon-Like Methods, *Math. Comp.* *22*, 665-666.

Bard, Y. (1970), Comparison of Gradient Methods for the Solution of Nonlinear Parameter Estimation Problems, *SIAM J. Num. Anal.* *7*, 157-186.

Bard, Y. (1974), *Nonlinear Parameter Estimation*, Academic Press, New York.

Barham, R. H., and W. Drane (1972), An Algorithm for Least Squares Estimation of Nonlinear Parameters When Some of the Parameters Are Linear, *Techno. 14*, 757-766.

Bass, R. (1972), A Rank Two Algorithm for Unconstrained Minimization, *Math. Comp. 26*, 129-144.

Beale, E. M. L. (1972), A Derivation of Conjugate Gradients, in *Numerical Methods for Nonlinear Optimization*, edited by F. A. Lootsma, Academic Press, New York, 39-44.

Beckley Smith, F., and D. F. Shanno (1971), An Improved Marquardt Procedure for Nonlinear Regressions, *Techno. 13*, 63-74.

Beckman, F. S. (1960), The Solution of Linear Equations by the Conjugate Gradient Method, in *Mathematical Methods for Digital Computers*, vol. 1, edited by A. Ralston and H. Wilf, Wiley, New York, 62-72.

Ben-Israel, A. (1967), On Iterative Methods for Solving Non-Linear Least Squares Problems over Convex Sets, *Israel J. Maths. 5*, 211-224.

Betts, J. T. (1976), Solving the Nonlinear Least Squares Problem: Application of a General Method, *J.O.T.A. 18*, 469-483.

Boland, W. R., E. Kamgnia, and J. S. Kowalik (1979), A Conjugate Gradient Optimization Method Invariant to Nonlinear Scaling, *J.O.T.A. 27*, 221-230.

Box, M. J. (1966), A Comparison of Several Current Optimization Methods, and the Use of Transformations in Unconstrained Problems, *Comp. J. 9*, 67-77.

Box, M. J., D. Davies, and W. H. Swann (1969), *Nonlinear Optimization Techniques*, ICI Monograph no. 5, Oliver and Boyd, Edinburgh, Scotland.

Brent, R. P. (1973), *Algorithms for Minimization without Derivatives*, Prentice-Hall, Englewood Cliffs, N.J.

Brodlie, K. M. (1975), A New Direction Set Method for Unconstrained Minimization without Evaluating Derivatives, *J. Inst. Math. Applics. 15*, 385-396.

Brodlie, K. M. (1977), Unconstrained Minimization, in *The State of the Art in Numerical Analysis*, edited by D. Jacobs, Academic Press, London, 229-268.

Brown, K. M., and J. E. Dennis, Jr. (1971), A New Algorithm for Nonlinear Least Squares Curve Fitting, in *Mathematical Software*, edited by J. Rice, Academic Press, New York, 391-396.

Brown, K. M., and J. E. Dennis, Jr. (1972), Derivative Free Analogues of the Levenberg-Marquardt and Gauss Algorithms for Nonlinear Least Squares Approximation, *Numer. Math. 18*, 289-297.

Broyden, C. G. (1967), Quasi-Newton Methods and Their Application to Function Minimization, *Math. Comp. 21*, 368-381.

Broyden, C. G. (1970), The Convergence of a Class of Double-Rank Minimization Algorithms, parts 1 and 2, *J. Inst. Maths. Applics. 6*, 76-90, 222-231.

Broyden, G. G., J. E. Dennis, Jr., and J. J. More (1973), On the Local and Superlinear Convergence of Quasi-Newton Methods, *J. Inst. Maths. Applics. 12*, 223-246.

Cauchy, A. (1847), Methode Générale Pour la Resolution des Systems d'Equations Simultanes, *C. R. Acad. Sci. Paris 25*, 536-538.

Chambers, J. M. (1973), Fitting Nonlinear Models: Numerical Techniques, *Biometrika 60*, 1-13.

Chambers, J. M. (1977), *Computational Methods for Data Analysis*, Wiley, New York.

Chambers, J. M., and J. E. Ertel (1975), Distribution of Nonlinear Estimates, *Proceedings of Statistical Computing Section, American Statistical Association*, Washington, D.C., 51-56.

Cohen, A. I. (1972), Rate of Convergence of Several Conjugate Gradient Algorithms, *SIAM J. Num. Anal. 9*, 248-259.

Cooper, L., and D. Steinberg (1970), *Introduction to Methods of Optimization*, Saunders, Philadelphia.

Crowder, H. P., and P. Wolfe (1972), Linear Convergence of the Conjugate Gradient Method, *IBM J. Res. and Devel. 16*, 431-433.

Davidon, W. C. (1959), *Variable Metric Method for Minimization*, A.E.C. Research and Development Report ANL-5990, Argonne National Laboratory, Chicago.

Davidon, W. C. (1968), Variance Algorithms for Minimization, *Comp. J. 10*, 406-410.

Davidon, W. C. (1975), Optimally Conditioned Optimization Algorithms without Line Searches, *Mathematical Programming 9*, 1-30.

Davidon, W. C. (1976a), Optimization vs. Quadratic Approximation in Statistics, *Proceedings of Ninth Annual Symposium on Computer Science and Statistics*, edited by D. Hoaglin and R. Welsch, Prindle, Weber and Schmidt, Boston, 158-160.

Davidon, W. C. (1976b), New Least Squares Algorithms, *J.O.T.A. 18*, 185-195.

Dennis, J. E., Jr. (1972), On Some Methods Based on Broyden's Secant Approximation to the Hessian, in *Numerical Methods for Nonlinear Optimization*, edited by F. A. Lootsma, Academic Press, New York, 19-34.

Dennis, J. E., Jr. (1973), Some Computational Techniques for the Nonlinear Least Squares Problem, in *Numerical Solution of Systems of Nonlinear Algebraic Equations*, edited by G. D. Bryne and C. A. Hall, Academic Press, New York, 157-183.

Dennis, J. E., Jr. (1977), Non-Linear Least Squares and Equations, in *The State of the Art in Numerical Analysis*, edited by D. Jacobs, Academic Press, London, 269-312.

Dennis, J. E., Jr., and J. J. More (1974), A Characterization of Superlinear Convergence and Its Application to Quasi-Newton Methods, *Math. Comp. 28,* 549-560.

Dennis, J. E., Jr., and J. J. More (1977), Quasi-Newton Methods, Motivation, and Theory, *SIAM Rev. 19*, 46-89.

Dennis, J. E., Jr., and R. E. Welsch (1976), Techniques for Nonlinear Least Squares and Robust Regression, *Proceedings of the Statistical Computing Section, American Statistical Association*, Washington, D.C., 83-87.

Dixon, L. C. W. (1972a), The Choice of Step Length; A Crucial Factor in Performance of Variable Metric Algorithms, in *Numerical Methods for Nonlinear Optimization*, edited by F. A. Lootsma, Academic Press, New York, 149-170.

Dixon, L. C. W. (1972b), Variable Metric Algorithms: Necessary and Sufficient Conditions for Identical Behavior of Nonquadratic Functions, *J.O.T.A. 10*, 34-40.

Dixon, L. C. W. (1975), Conjugate Gradient Algorithms: Quadratic Termination without Linear Searches, *J. Inst. Maths. Applics. 15*, 9-18.

Fiacco, A. V., and G. P. McCormick (1964), Computational Algorithm for the Sequential Unconstrained Minimization Technique for Nonlinear Programming, *Mgmt. Sci. 10*, 607-617.

Fiacco, A. V., and G. P. McCormick (1968), *Nonlinear Programming: Sequential Unconstrained Minimization Techniques,* Wiley, New York.

Fisher, R. A. (1925), Theory of Statistical Estimation, *Proceedings of Cambridge Philosophical Society 22*, 700-725.

Flanagan, P. D., P. A. Vitale, and J. Mendelsohn (1969), Search Procedures in Nonlinear Regression Problems, *Techno. 11*, 265-284.

Fleming, W. H. (1965), *Functions of Several Variables*, Addison-Wesley, Reading, Mass.

Fletcher, R. (1966), Certification of Algorithm 251: Function Minimization, *CACM 9*, 686.

Fletcher, R. (1970), A New Approach to Variable Metric Algorithms, *Comp. J. 13*, 317-322.

Fletcher, R. (1972a), Conjugate Direction Methods, in *Numerical Methods for Unconstrained Optimization,* edited by W. Murray, Academic Press, New York, 73-86.

Fletcher, R. (1972b), An Algorithm for Solving Linearly Constrained Optimization Problems, *Mathematical Programming 2*, 133-165.

Fletcher, R., and M. J. D. Powell (1963), A Rapidly Convergent Descent Method for Minimization, *Comp. J. 6*, 163-168.

Fletcher, R., and C. M. Reeves (1964), Function Minimization by Conjugate Gradients, *Comp. J. 7*, 149-153.

Gallant, A. R. (1971), Statistical Inference for Nonlinear Regression Models, Ph.D. thesis, Iowa State University, Ames.

Gallant, A. R. (1975a), Nonlinear Regression, *Amer. Stat. 29*, 73-81.

Gallant, A. R. (1975b), The Power of the Likelihood Ratio Test of Location in Nonlinear Regression Models, *JASA 70*, 198-203.

Gill, P. E., G. H. Golub, W. Murray, and M. A. Saunders (1974), Methods for Modifying Matrix Factorizations, *Math. Comp. 28*, 505-536.

Gill, P. E., and W. Murray (1972), Quasi-Newton Methods for Unconstrained Optimization, *J. Inst. Maths. Applics. 9*, 91-108.

Gill, P. E., and W. Murray (1974), Newton-Type Methods for Unconstrained and Linearly Constrained Optimization, *Mathematical Programming 7*, 311-350.

Gill, P. E., and W. Murray (1976), Nonlinear Least Squares and Nonlinearly Constrained Optimization, in *Lecture Notes in Mathematics*, no. 506: *Numerical Analysis*, edited by A. Dold and B. Eckmann, Dundee Conference Proceedings, Springer-Verlag, New York, 134-147.

Gill, P. E., W. Murray, and M. A. Saunders (1975), Methods for Computing and Modifying the LDV Factors of a Matrix, *Math. Comp. 29*, 1051-1077.

Goldfarb, D. (1970), A Family of Variable Metric Methods Derived by Variational Means, *Math. Comp. 24*, 23-26.

Goldfarb, D. (1976), Factorized Variable Metric Methods for Unconstrained Optimization, *Math. Comp. 30*, 796-811.

Goldfarb, D. (1977), Generating Conjugate Directions without Line Searches Using Factorized Variable Metric Updating Formulas, *Mathematical Programming 13*, 94-110.

Goldfeld, S. M., R. E. Quandt, and H. F. Trotter (1966), Maximization by Quadratic Hill Climbing, *Econometrica 34*, 541-551.

Goldstein, A. A. (1962), Cauchy's Method for Minimization, *Numer. Math. 4*, 146-150.

Goldstein, A. A. (1965), On Steepest Descent, *SIAM J. on Control Series A 3*, 147-151.

Goldstein, A. A., and J. F. Price (1971), On Descent from Local Minima, *Math. Comp. 25*, 569-574.

Golub, G. H., and V. Pereyra (1973), The Differentiation of Pseudo-Inverses and Nonlinear Least Squares Problems Whose Variables Separate, *SIAM J. Num. Anal. 10*, 410-423.

Greenstadt, J. (1967), On the Relative Efficiencies of Gradient Methods, *Math. Comp. 21,* 360-367.

Greenstadt, J. (1970), Variations on Variable-Metric Methods, *Math. Comp. 24*, 1-22.

Greenstadt, J. (1972), A Quasi-Newton Method with No Derivatives, *Math. Comp. 26*, 145-166.

Hamilton, P. A., and J. Boothroyd (1969), Certification of Algorithm 251: Function Minimization, *CACM 12*, 512.

Hartley, H. O. (1961), The Modified Gauss-Newton Method for Fitting Nonlinear Regression Functions by Least Squares, *Techno. 3,* 269-280.

Hartley, H. O. (1964), Exact Confidence Regions for the Parameters in Nonlinear Regression Laws, *Biometrika 51*, 347-353.

Hartley, H. O., and A. Booker (1965), Non-Linear Least Squares Estimation, *AMS 36*, 638-650.

Harville, D. A. (1973), Fitting Partially Linear Models by Weighted Least Squares, *Techno. 15*, 509-515.

Harville, D. A. (1977), Variance Component Estimation, *JASA 72,* 320-340.

Hemmerle, W. J. (1975), An Explicit Solution for Generalized Ridge Regression, *Techno. 17*, 309-314.

Hestenes, M. R., and E. Stiefel (1952), Methods of Conjugate Gradients for Solving Linear Systems, *J. Res. Natl. Bureau Standards 49*, 409-436.

Himmelblau, D. M. (1972), A Uniform Evaluation of Unconstrained Optimization Techniques, in *Numerical Methods for Nonlinear Optimization*, edited by F. A. Lootsma, Academic Press, London, 69-97.

Hooke, R., and T. A. Jeeves (1961), Direct Search Solution of Numerical and Statistical Problems, *JACM 8*, 212-229.

Huang, H. Y. (1970), Unified Approach to Quadratically Convergent Algorithms for Function Minimization, *J.O.T.A. 5*, 405-423.

Huang, H. Y., and A. V. Levy (1970), Numerical Experiments on Quadratically Convergent Algorithms for Function Minimization, *J.O.T.A. 6*, 269-282.

Jacoby, S. L. S., J. S. Kowalik, and J. T. Pizzo (1972), *Iterative Methods for Nonlinear Optimization Problems*, Prentice-Hall, Englewood Cliffs, N.J.

Jennrich, R. I. (1969), Asymptotic Properties of Non-Linear Least Squares Estimators, *AMS 40,* 633-643.

Jennrich, R. I., and R. H. Moore (1975), Maximum Likelihood Estimation by Means of Nonlinear Least Squares, *Proceedings of the Statistical Computing Section, American Statistical Association,* Washington, D.C., 57-65.

Jennrich, R. I., and P. F. Sampson (1968), Application of Stepwise Regression to Nonlinear Estimation, *Techno. 10*, 63-72.

Jones, A. P. (1970), Spiral: A New Algorithm for Nonlinear Parameter Estimation Using Least Squares, *Comp. J. 13,* 301-308.

Kale, B. K. (1961), On the Solution of the Likelihood Equation by Iterative Processes, *Biometrika 48,* 452-456.

Kale, B. K. (1962), On the Solution of the Likelihood Equation by Iterative Processes: The Multiparametric Case, *Biometrika 49*, 479-486.

Keifer, J. (1957), Optimal Sequential Search and Approximation Methods under Minimum Regularity Conditions, *SIAM J. Appl. Math. 5,* 105-136.

Kowalik, J. S., E. R. Kamgnia, and W. R. Boland (1977), *A Conjugate Gradient Method Based on an Exponential Model.* Technical Report no. 260, Department of Mathematical Sciences, Clemson University, Clemson, S. Car.

Kowalik, J., and M. R. Osborne (1968), *Methods for Unconstrained Optimization Problems*, American Elsevier, New York.

Krogh, F. T. (1974), Efficient Implementation of a Variable Projection Algorithm for Non-Linear Least Squares Problems, *CACM 17,* 167-169.

Lawton, W. H., and E. A. Sylvester (1971), Elimination of Linear Parameters in Nonlinear Regression, *Techno. 13,* 461-467.

Levenberg, K. (1944), A Method for the Solution of Certain Nonlinear Problems in Least Squares, *Quart, Appl. Math. 2*, 164-168.

McCormick, G. P. (1977), A Modification of Armijo's Step-Size Rule for Negative Curvature, *Mathematical Programming 13*, 111-115.

McCormick, G. P., and J. D. Pearson (1969), Variable Metric Methods and Unconstrained Optimization, in *Optimization*, edited by R. Fletcher, Academic Press, New York, 247-258.

McKeown, J. J. (1975), Specialized versus General Purpose Algorithms for Minimizing Functions that Are Sums of Squared Terms, *Mathematical Programming 9*, 57-68.

Malinvaud, E. (1970), The Consistency of Nonlinear Regressions, *AMS 41,* 956-969.

Mamen, R., and D. Q. Mayne (1972), A Pseudo Newton-Raphson Method for Function Minimization, *J.O.T.A. 10*, 263-276.

Marquardt, D. W. (1963), An Algorithm for Least Squares Estimation of Nonlinear Parameters, *SIAM J. 11*, 431-441.

Matthews, A., and D. Davies (1971), A Comparison of Modified Newton Methods for Unconstrained Optimization, *Comp. J. 14*, 293-294.

Meeter, D. A. (1966), On a Theorem Used in Nonlinear Least Squares, *SIAM J. Appl. Math. 14*, 1176-1179.

Meyer, R. R. (1970), Theoretical and Computational Aspects of Nonlinear Regression, in *Nonlinear Programming*, edited by J. Rosen, O. Mangasarian, and K. Ritter, Academic Press, New York, 465-486.

Meyer, R. R., and P. M. Roth (1972), Modified Damped Least Squares: An Algorithm for Nonlinear Estimation, *J. Inst. Maths. Applics. 9,* 219-233.

Mifflin, R. (1975), A Superlinearly Convergent Algorithm for Minimization Without Calculating Derivatives, *Mathematical Programming 9,* 100-117.

Murray, W. (1972a), Second Derivative Methods, in *Numerical Methods for Unconstrained Optimization*, edited by W. Murray, Academic Press, New York, 57-71.

Murray, W. (ed) (1972b), *Numerical Methods for Unconstrained Optimization*, Academic Press, New York.

Murtagh, B. A., and R. W. H. Sargent (1970), Computational Experience with Quadratically Convergent Minimization Methods, *Comp. J. 13,* 185-194.

Myers, C. E. (1968), Properties of the Conjugate Gradient and Davidon Methods, *J.O.T.A. 2,* 209-219.

Nazareth, L. (1976a), Generation of Conjugate Directions for Unconstrained Minimization, *Math. Comp. 30,* 115-131.

Nazareth, L. (1976b), Some Recent Approaches to Solving Large Residual Non-Linear Least Squares Problems, *Proceedings of the Ninth Interface Symposium on Computer Science and Statistics,* edited by D. Hoaglin and R. Welsch, Prindle, Weber and Schmidt, Boston, 161-167.

Nelder, J. A. and R. Mead (1965), A Simplex Method for Function Minimization, *Comp. J. 7*, 308-313.

Olsson, D. M., and L. S. Nelson (1975), The Nelder-Mead Simplex Procedure for Function Minimization, *Techno. 17*, 45-51.

O'Neill, R. (1971), Algorithm AS47: Function Minimization Using a Simplex Procedure, *Appl. Stat. 20*, 338-345.

Oren, S. S. (1973), Self-Scaling Variable Metric Algorithms without Line Search for Unconstrained Minimization, *Math. Comp. 27,* 873-886.

Oren, S. S. (1976), On Quasi-Newton and Pseudo-Newton Algorithms, *J.O.T.A. 20*, 155-170.

Oren, S. S., and D. G. Luenberger (1974), Self-Scaling Variable Metric (SSVM) Algorithms, I: Criteria and Sufficient Conditions for Scaling a Class of Algorithms, *Management Science 20,* 845-862.

Ortega, J. M., and W. C. Rheinboldt (1970), *Iterative Solution of Nonlinear Equations in Several Variables*, Academic Press, New York.

Osborne, M. R. (1972), Some Aspects of Nonlinear Least Squares Calculations, in *Numerical Methods for Nonlinear Optimization*, edited by F. A. Lootsma, Academic Press, New York, 171-189.

Papaioannou, T., and O. Kempthorne (1972), Parallel Tangents and Steepest Descent Optimization Algorithm: A Computer Implementation with Application to Linear, Partially Linear Models, and Qualitative Data, *JSCS 1*, 349-376.

Peckham, G. (1970), A New Method for Minimizing a Sum of Squares without Calculating Gradients, *Comp. J. 13,* 418-420.

Polak, E. (1971), *Computational Methods in Optimization*, Academic Press, New York.

Powell, M. J. D. (1964), An Efficient Method for Finding the Minimum of a Function of Several Variables without Calculating Derivatives, *Comp. J. 7*, 155-162.

Powell, M. J. D. (1965), A Method for Minimizing a Sum of Squares of Nonlinear Functions without Calculating Derivatives, *Comp. J. 8*, 303-307.

Powell, M. J. D. (1970a), A Survey of Numerical Methods for Unconstrained Optimization, *SIAM Rev. 12*, 79-97.

Powell, M. J. D. (1970b), A New Algorithm for Unconstrained Optimization, *Nonlinear Programming*, edited by J. Rosen, O. Mangasarin, and K. Ritter, Academic Press, New York, 31-65.

Powell, M. J. D. (1971), Recent Advances in Unconstrained Optimization, *Mathematical Programming 1*, 26-57.

Powell, M. J. D. (1972), Quadratic Termination Properties of Minimization Algorithms I and II, *J. Inst. Maths. Applics. 10,* 333-357.

Powell, M. J. D. (1975), A View of Unconstrained Minimization Algorithms that Do Not Require Derivatives, *TOMS 1*, 97-107.

Powell, M. J. D. (1976), Some Convergence Properties of the Conjugate Gradient Method, *Mathematical Programming 11,* 43-49.

Powell, M. J. D. (1977a), Quadratic Termination Properties of Davidon's New Variable Metric Algorithm, *Mathematical Programming 12,* 141-147.

Powell, M. J. D. (1977b), Restart Procedures for the Conjugate Gradient Method, *Mathematical Programming 12*, 241-254.

Ralston, A. (1965), *A First Course in Numerical Analysis,* McGraw-Hill, New York.

Ralston, A., and H. S. Wilf (1960), *Mathematical Methods for Digital Computers,* vol. 1, Wiley, New York.

Ralston, M. L. (1975), DUD: A Derivative-free Algorithm for Nonlinear Regression, Ph.D. thesis, University of California, Los Angeles.

Ralston, M. L., and R. Jennrich (1977), *DUD: A Derivative-free Algorithm for Nonlinear Least Squares,* Technical Report no. 29, Health Sciences Computing Facility, University of California, Los Angeles.

Ralston, M. L. and R. I. Jennrich (1978), Derivative-free Nonlinear Regression, *Proceedings of Computer Science and Statistics: Tenth Annual Symposium on the Interface,* edited by D. Hogben, U.S. Government Printing Office, Washington, D.C., 312-322.

Ramsin, H., and P. Wedin (1977), A Comparison of Some Algorithms for the Nonlinear Least Squares Problem, *BIT 17,* 72-90.

Rosenbrook, H. H. (1960), An Automatic Method for Finding the Greatest or the Least Value of a Function, *Comp. J. 3,* 175-184.

Ross, G. J. S. (1978), MLP: A Maximum Likelihood Program, *Proceedings of Computer Science and Statistics: Tenth Annual Symposium on the Interface,* edited by D. Hogben, U.S. Government Printing Office, Washington, D.C., 87-91.

Shah, B. V., R. J. Buehler, and O. Kempthorne (1964), Some Algorithms for Minimizing a Function of Several Variables, *SIAM J. 12,* 74-92.

Shanno, D. F. (1970), Parameter Selection for Modified Newton Methods for Function Minimization, *SIAM J. Num. Anal. 7,* 366-372.

Shanno, D. F., and P. C. Kettler (1970), Optimal Conditioning of Quasi-Newton Methods, *Math. Comp. 24,* 657-664.

Shanno, D. F., and K. H. Phua (1976), Minimization of Unconstrained Multivariate Functions: Algorithm 500, *TOMS 2,* 87-94.

Smith, F. B., Jr., and D. F. Shanno (1971), An Improved Marquardt Procedure for Nonlinear Regressions, *Techno. 13,* 63-74.

Spedicato, E. (1973), Stability of Huang's Update for the Conjugate Gradient Method, *J.O.T.A. 11,* 459-479.

Spedicato, E. (1975), Recent Developments in the Variable Metric Method for Nonlinear Unconstrained Optimization, in *Towards Global Optimization,* edited by L. Dixon and G. Szego, North-Holland, Amsterdam, 182-195.

Spedicato, E. (1976), A Variable Metric Method for Function Minimization Derived from Invariancy to Nonlinear Scaling, *J.O.T.A. 20,* 314-329.

Stewart, G. W. (1967), A Modification of Davidon's Minimization Method to Accept Difference Approximations to Derivatives, *JACM 14,* 72-83.

Swann, W. H. (1972), Direct Search Methods, in *Numerical Methods for Unconstrained Optimization,* edited by W. Murray, Academic Press, New York, 13-28.

Vitale, P., and D. Taylor (1968), A Note on the Application of Davidon's Method to Nonlinear Regression Problems, *Techno. 10,* 843-849.

Walsh, G. R. (1975), *Methods of Optimization*, Wiley, New York.

Wells, M. (1965), Algorithm 251: Function Minimization, *CACM 8,* 169.

Welsch, R. E. (1978), Nonlinear Statistical Data Analysis, *Proceedings of Computer Science and Statistics: Tenth Annual Symposium on the Interface,* edited by D. Hogben, U.S. Government Printing Office, Washington, D.C., 77-86.

Welsch, R. E., and R. A. Becker (1975), Robust Nonlinear Regression Using DOGLEG Algorithm, *Proceedings of Computer Science and Statistics: Eighth Annual Symposium on the Interface,* edited by J. Frane, Health Sciences Computing Facility, University of California, Los Angeles, 272-279.

Wolfe, M. A. (1975), A Quasi-Newton Method with Memory for Unconstrained Function Minimization, *J. Inst. Maths. Applics. 15,* 85-94.

Wolfe, M. A. (1976), Some Methods for Least Squares Estimation, *J. Inst. Maths. Applics. 18*, 219-236.

Wolfe, P. (1969), Convergence Conditions for Ascent Methods, *SIAM Rev. 11*, 226-235.

Wolfe, P. (1974), Note on a Method of Conjugate Subgradients for Minimizing Non-Differentiable Functions, *Mathematical Programming 7,* 380-383.

Zangwill, W. I. (1967), Minimizing a Function without Calculating Derivatives, *Comp. J. 10*, 293-296.

11 / MODEL FITTING BASED ON CRITERIA OTHER THAN LEAST SQUARES

11.1 INTRODUCTION

While the least-squares criterion of estimation in the linear model remains the most commonly used method, a number of alternative estimation procedures have received widespread attention in recent years. Interest in other methods of estimation has been generated by the unsatisfactory performance of least-squares estimators in certain situations when some model assumptions fail to hold or when large correlations exist among the regressors.

When the distribution of the error vector is not normal, least squares may not yield good estimators. In particular, if the variance of the errors is infinite, the variance of the least-squares estimators is infinite. While intuition may dispel consideration of errors with infinite variance in any but the most pathological of cases, studies have shown that, in fact, certain distributions with infinite variances may be quite appropriate models (see, e.g., Mandelbrot 1963 or Huber 1972).

Even when the majority of the disturbances in the linear model follow a normal distribution, it often occurs that a small number of the observations appear to be from a different distribution; that is, the sample is "contaminated" with "outliers." It is well known that the performance of least-squares estimators is markedly degraded in such situations (see Huber 1972).

The emphasis in this chapter is on alternatives to least-squares estimation in the linear model. However, in Section 11.5 we will briefly consider the analogous situation for nonlinear models.

11.2 MINIMUM L_p NORM ESTIMATORS

We again consider the problem of estimation of the parameters in the linear model

$$\underline{y} = X\underline{\beta} + \underline{\varepsilon} \tag{11.1}$$

where $\underline{y}$ is an n vector of observations,

X is an n x m matrix of observations,

$\underline{\beta}$ is the m vector of parameters,

$\underline{\varepsilon}$ is an n vector of disturbances, and

$n > m$.

Heuristically, the hyperplane corresponding to the estimated parameters should be "close" to the observation points. The closeness of fit is measured by a norm of the vector $\underline{y} - X\underline{\tilde{\beta}}$, where $\underline{\tilde{\beta}}$ is the estimated value of $\underline{\beta}$. The L_p *norms*, defined for the vector $\underline{v}$, with elements v_i, by

$$||\underline{v}||_p = (\sum_i |v_i|^p)^{1/p} \tag{11.2}$$

for $p \geq 1$, appear to be useful measures of fit for the estimation problem. Under the minimum L_p criterion, therefore, the estimator $\underline{\tilde{\beta}}$ is taken so as to solve the minimization problem

$$\min_{\underline{b}} (\sum_i |y_i - \underline{x}_i'\underline{b}|^p)^{1/p} \tag{11.3}$$

where $\underline{x}_i'$ is the ith row of X.

For $p = 2$, the criterion is the familiar least squares, and computational techniques for obtaining the estimators have been discussed in Chapters 8 and 9. Two other values of p that are commonly used are 1 and ∞. For p = 1 the criterion is least absolute values; and for p = ∞, it is the minimax criterion. These two criteria will be discussed separately in Sections 11.2.1 and 11.2.2.

11.2.1 L_1 Estimation

The L_1 criterion of estimation has been referred to in the literature by a variety of names: minimum or least sums of absolute errors (MSAE, LSAE); minimum or least absolute deviations, or errors, or residuals (MAD, MAE, LAD, LAV, LAR); L_1; and so on. The primary virtue of these estimators is perhaps their resistance to outliers in the data and to heavy-tailed error distributions (see, e.g., Rice and White 1964). Gentle (1977) gives a survey and bibliography of work in L_1 estimation.

Since the L_1 norm

$$S(\underline{b}) = \Sigma \left|y_i - \underline{x}_i'\underline{b}\right| \tag{11.4}$$

is a piecewise linear convex surface, any minimum of this norm must lie at one of the vertices of a simplex with $\binom{n}{m}$ points in E^m. There may be more than one minimum point; however, if there are, say, k simplex points at which the minimum occurs, then the (k - 1)-dimensional linear space determined by the k minimal vertices must be a subset of the norm surface (11.4) having a constant value. This nonuniqueness of the estimates may occur even when X is of full rank.

The L_1 estimates and the fitted model have a number of interesting relationships, such as the fact that when the model contains a constant term, the absolute value of the difference in the number of observations with positive deviations and those with negative deviations is no greater than the maximum number of observations on any hyperplane in E^m; and the fact that there is an L_1 model that fits at least m observations exactly. (See, e.g., Appa and Smith 1973 or Gentle, Kennedy, and Sposito 1976, 1977a, 1977b.)

We can distinguish two general types of algorithms for computing the L_1 estimate. One type makes explicit use of the fact that the minimum occurs at one of the simplex points. These algorithms employ linear programming or projection methods to search among the simplex vertices. (It may be noted here that the term *simplex* will be used in this discussion in two different senses. In one sense it refers to a finite set of points, and in the other it refers to a method of linear programming, which uses a finite set of points, but whose name is derived from an early area of application.) The other type of algorithm uses iterative descent methods. The algorithms which work only with the vertices are generally more efficient and more accurate.

An example of an iterative method for L_1 estimation is given in Schlossmacher (1973), using iteratively reweighted least squares. The idea is quite simple and will be considered in Section 11.2.3 for $1 \leq p \leq 2$. Armstrong and Frome (1976a) show that the method is rather inefficient when compared to linear programming techniques we will discuss below.

Singleton (1940) gave a method based on projections that contained the basic idea of some of the later efficient algorithms. It was not until the development of linear programming, however, that L_1 computations became practical in large data sets. The L_1 estimation problem was identified with a primal linear programming problem by Charnes, Cooper, and Ferguson (1955), and a more efficient dual formulation requiring bounded variables was given by Wagner (1959).

In a slight modification of the notation of Charnes et al., the L_1 estimation problem is the linear programming problem

$$\begin{aligned} &\text{minimize} \quad (\underline{1}_n'\underline{e}^+ + \underline{1}_n'\underline{e}^-) \\ &\text{subject to} \quad X\underline{b} + I_n\underline{e}^+ - I_n\underline{e}^- = \underline{y} \\ &\qquad\qquad \underline{e}^+, \underline{e}^- \geq \underline{0} \\ &\qquad\qquad \underline{b} \text{ unrestricted} \end{aligned} \tag{11.5}$$

where $\underline{1}_n$ is an n vector of 1's, and I_n is the nth order identity matrix. (See Sposito 1975, e.g., for a general discussion of linear programming problems and techniques.) The components of $\underline{e}^+$ may be interpreted as vertical deviations above the fitted hyperplane; and the components of $\underline{e}^-$ as vertical deviations below. Clearly $y_i - \underline{x}_i'\underline{b} = e_i^+ - e_i^-$, for i = 1, 2, ..., n. This linear programming problem has n constraints in m + 2n variables. Since in applications the sample size n is likely to be large, this formulation generally requires an unacceptable amount of computation time and storage.

Barrodale and Roberts (1972, 1973) present a modification of the linear programming simplex method that saves storage and gains efficiency in computation by skipping over simplex vertices. The problem can be stated as

$$\begin{aligned} &\text{minimize} && (\underline{1}_n'\underline{e}^+ + \underline{1}_n'\underline{e}^-) \\ &\text{subject to} && X(\underline{r} - \underline{t}) + I_n\underline{e}^+ - I_n\underline{e}^- = \underline{y} \\ &&& \underline{e}^+, \underline{e}^-, \underline{r}, \underline{t} \geq \underline{0} \end{aligned}$$

Because of the relationships among the variables, the computations can be performed using only an (n + 2) by (m + 2) matrix, the additional row and column being needed for labeling. An initial basis is given by the identity multiplier of $\underline{e}^+$ if all y_i are nonnegative. The sign of any row corresponding to a negative y_i is changed, and the unit column from the corresponding element of $\underline{e}^-$ is taken as part of the basis.

The algorithm consists of two stages. In the first stage the pivot column for each of the first m iterations is chosen as that corresponding to the maximum marginal cost associated with r_j or t_j. The pivot element is chosen so that the vector that leaves the basis causes the maximum decrease in the objective function. Thus the pivot element is not necessarily the same as in the usual linear programming simplex algorithm. At the end of the first stage the number of vectors corresponding to an r_j or t_j in the basis will be at least as great as the rank of X. The solution at this stage therefore goes through a number of data points at least as great as the rank of X.

In the second stage of the algorithm, nonbasis columns corresponding to the maximum marginal costs associated with e_i's are interchanged with basic columns corresponding to e_i's that will cause maximum reduction in the objective function. The basis vectors corresponding to r_j's or t_j's are not allowed to leave the basis. At each iteration in the second stage the solution goes through a number of data points at least as great as the rank of X. The algorithm terminates when all the marginal costs are nonpositive.

Example 11.1 The Barrodale-Roberts Modified Simplex Algorithm for L_1 Estimation

Consider using the L_1 criterion to estimate b_1 and b_2 in $y_i = b_1 + b_2x_i + e_i$ using the following data:

x	0	1	1	2	2	3
y	-1	1	2	1	4	5

With $b_j = r_j - t_j$, the full initial tableau is shown in Figure 11.1. As is customary in the linear programming literature, we denote the columns corresponding to the variables r_j, t_j, e_j^+, and e_j^- as $\underline{r}_j$, $\underline{t}_j$, $\underline{e}_j^+$, and $\underline{e}_j^-$. The elements of the vector $\underline{d}$ corresponding to the $\underline{e}_j^+$ or $\underline{e}_j^-$ are the contributions to the objective function from the individual observations. Before any of the $\underline{r}_j$ or $\underline{t}_j$ are brought into the basis, the elements of $\underline{d}$ are merely the absolute values of the y_i's. The corresponding $\underline{e}_j^+$ or $\underline{e}_j^-$ indicate whether the observation lies above or below the line y = 0.

The pivot column is $\underline{r}_2$, since it corresponds to the maximum marginal cost. The usual simplex pivot row would be the fourth one, since it corresponds to the minimum positive d_i/r_{2i}. A Gauss-Jordan pivot on this element in the fourth row of $\underline{r}_2$ would yield a value of 1/2 for r_2. Increasing r_2 even further might decrease the objective function even more; however, it would make e_4^+ negative. Barrodale and Roberts suggest doing this anyway if it will reduce the objective function. This requires that $\underline{e}_4^+$ be replaced in the basis by $\underline{e}_4^-$,

Basis	$\underline{d}$	$\underline{r}_1$	$\underline{r}_2$	$\underline{t}_1$	$\underline{t}_2$	$\underline{e}_1^+$	$\underline{e}_2^+$	$\underline{e}_3^+$	$\underline{e}_4^+$	$\underline{e}_5^+$	$\underline{e}_6^+$	$\underline{e}_1^-$	$\underline{e}_2^-$	$\underline{e}_3^-$	$\underline{e}_4^-$	$\underline{e}_5^-$	$\underline{e}_6^-$
$\underline{e}_1^-$	1	-1	0	1	0	-1	0	0	0	0	0	1	0	0	0	0	0
$\underline{e}_2^+$	1	1	$1^{(2)}$	-1	-1	0	1	0	0	0	0	0	-1	0	0	0	0
$\underline{e}_3^+$	2	1	1	-1	-1	0	0	1	0	0	0	0	0	-1	0	0	0
$\underline{e}_4^+$	1	1	$2^{(1)}$	-1	-2	0	0	0	1	0	0	0	0	0	-1	0	0
$\underline{e}_5^+$	4	1	2	-1	-2	0	0	0	0	1	0	0	0	0	0	-1	0
$\underline{e}_6^+$	5	1	$3^{(3)}$	-1	-3	0	0	0	0	0	1	0	0	0	0	0	-1
Objective Function	14	4	9	-4	-9	-2	0	0	0	0	0	0	-2	-2	-2	-2	-2

Marginal Costs

Figure 11.1 Full Tableau in Example 11.1

which is accomplished by subtracting twice the fourth row from the marginal cost row so as to make the marginal cost of e_4^- zero, and multiplying the fourth row by -1. This would make the marginal cost of r_2 5, which, since it is positive, indicates that the objective function could be reduced still further by a further increase in r_2.

The next smallest positive quotient d_i/r_{2i} is 1, corresponding to the second row. Again we see if a further increase in r_2 is warranted. Replacing $\underline{e}_2^+$ by $\underline{e}_2^-$ in the basis gives a marginal cost of 3 [= 5 - 2(1)] for r_2. Since the marginal cost is positive, we make this substitution and consider the next minimum positive ratio d_i/r_{2i}, which corresponds to the sixth row. Replacing $\underline{e}_6^+$ with $\underline{e}_6^-$ gives a marginal cost of -1 [= 3 - 2(2)] for r_2; hence, we can do no better than to pivot on the sixth row of $\underline{r}_2$. The relevant portion of the tableau prior to pivoting is

Basis	$\underline{d}$	$\underline{r}_1$	$\underline{r}_2$
$\underline{e}_1^-$	1	-1	0
$\underline{e}_2^-$	-1	-1	-1
$\underline{e}_3^+$	2	1	1
$\underline{e}_4^-$	-1	-1	-2
$\underline{e}_5^+$	4	1	2
$\underline{e}_6^+$	5	1	(3)
	10	0	3

etc.

After pivoting on the 3, the condensed tableau is

Basis	$\underline{d}$	$\underline{r}_1$	$\underline{e}_6^+$	$\underline{t}_1$
$\underline{e}_1^-$	1	-1	0	1
$\underline{e}_2^-$	2/3	-2/3	1/3	2/3
$\underline{e}_3^+$	1/3	2/3	-1/3	-2/3
$\underline{e}_4^-$	7/3	-1/3	2/3	1/3
$\underline{e}_5^+$	2/3	1/3	-2/3	-1/3
$\underline{r}_2$	5/3	1/3	1/3	-1/3
	5	-1	-1	1

etc.

Note that at this stage the model is $y_i = 0 + (5/3)x_i$, and the values of $\underline{d}$ corresponding to various $\underline{e}_j^+$ or $\underline{e}_j^-$ are the deviations about this model. The sum of the absolute values of the deviations is 5.

The maximum marginal cost is now 1 corresponding to t_1; hence $\underline{t}_1$ is to be brought into the basis. The minimum positive quotient d_i/t_{1i} corresponds to either the first row or the second row. Checking as before to see if further increase in t_1 would yield greater decrease in the objective function, we obtain a negative marginal cost for t_1; hence, we choose either the first or second row for pivoting. Using the first row yields the condensed tableau

Basis	$\underline{d}$	$\underline{e}_6^+$	$\underline{e}_1^-$
$\underline{t}_1$	1	0	1
$\underline{e}_2^-$	0	1/3	-2/3
$\underline{e}_3^+$	1	-1/3	2/3
$\underline{e}_4^-$	2	2/3	-1/3
$\underline{e}_5^+$	1	-2/3	1/3
$\underline{r}_2$	2	1/3	1/3
	4	-1	-1

The marginal costs of e_6^- and e_1^+ are both -1 (since the sum of the marginal cost of e_i^+ and e_i^- is -2); hence all nonbasic vectors have nonpositive marginal costs. This means there is no need for the second stage. (This is frequently the case.) The estimated values of b_1 and b_2 are $b_1 = r_1^* - t_1^* = -1$ and $b_2 = r_2^* - t_2^* = 2$. The residuals from the model $y_i = -1 + 2x_i$ are given by the elements of $\underline{d}$ corresponding to $\underline{e}_i^+$ or $\underline{e}_i^-$. The other residuals are zero; hence the vector of residuals is (0,0,1,-2,1,0).

A very efficient FORTRAN coding of this algorithm is given by Barrodale and Roberts (1974). The statistician should be forewarned of the notational differences employed in this article (as well as in other articles written by numerical analysts).

There are currently several other algorithms implementing ideas similar to those of Barrodale and Roberts. Gilsinn et al. (1977) reported on empirical studies of some of these algorithms, as well as a novel method of Bartels, Conn, and Sinclair (1976, 1978).

Abdelmalek (1976b) gives a dual simplex algorithm employing an LU decomposition, as described by Bartels and Golub (1969), to attempt to improve the numerical stability in L_1 estimation. Ill-conditioning does not appear to be as much of a problem in L_1 estimation as in least squares, however, and the extra computations for the LU decomposition would not be recommended generally.

Most algorithms for L_1 estimation can be used on linear models not of full rank, without any modification. More efficient procedures may be developed for certain design models, however, as in Armstrong, Elam, and Hultz (1977) or Armstrong and Frome (1977). As with least squares, the L_1 criterion does not lead to unique estimators in non-full-rank models.

Special purpose algorithms for restricted L_1 estimation have been given by Armstrong and Hultz (1977), Barrodale and Roberts (1977a, 1977b), and Bartels and Conn (1978). Barrodale and Roberts (1977c) give a survey of the work on L_1 estimation under linear restrictions (both equality and inequality).

Procedures for efficiently obtaining the best subset of variables of each size have been considered by Roodman (1974), Narula and Wellington (1977b), and Hanson (1977).

A method of obtaining unbiased L_1 estimators, when the error distribution is symmetric and the estimates may not be unique, is given by Sielken and Hartley (1973). The method requires an initial "antisymmetrical" estimator of $\underline{\beta}$. An antisymmetrical estimator $\underline{b}_0$ is one such that for a given realization $\underline{e}$ of the random vector $\underline{\varepsilon}$ in Eq. (11.1),

$$\underline{\beta} - \underline{b}_0(\underline{e}) = -\underline{\beta} + \underline{b}_0(-\underline{e})$$

The initial estimator could be the least-squares estimator, $\hat{\underline{\beta}}$, for example, since

$$\hat{\underline{\beta}}(\underline{e}) = (X'X)^{-1}X'(X\underline{\beta} + \underline{e}) = \underline{\beta} + (X'X)^{-1}X'\underline{e}$$

and

$$\hat{\underline{\beta}}(-\underline{e}) = (X'X)^{-1}X'(X\underline{\beta} - \underline{e}) = \underline{\beta} - (X'X)^{-1}X'\underline{e}$$

(assuming X is of full rank). If $\underline{e}$ is symmetrically distributed about $\underline{0}$, an antisymmetrical estimator is unbiased. Two linear programming problems similar to (11.5) are now defined. (The following forms are equivalent to those of Sielken and Hartley.)

$$\begin{aligned} &\text{minimize} && (\underline{1}_n'\underline{u}^+ + \underline{1}_n'\underline{u}^-) \\ &\text{subject to} && X\underline{b}_1 + I_n\underline{u}^+ - I_n\underline{u}^- = \underline{y} - X\underline{b}_0 \qquad \text{(Problem } P_1\text{)} \\ &&& \underline{u}^+, \underline{u}^- \geq \underline{0} \\ &&& \underline{b}_1 \text{ unrestricted} \end{aligned}$$

and

$$\begin{aligned}
&\text{minimize} && (\underline{1}_n'\underline{u}^+ + \underline{1}_n'\underline{u}^-) \\
&\text{subject to} && -X\underline{b}_1 - I_n\underline{u}^+ + I_n\underline{u}^- = -\underline{y} + X\underline{b}_0 \qquad \text{(Problem } P_2\text{)} \\
& && \underline{u}^+, \underline{u}^- \geq \underline{0} \\
& && \underline{b}_1 \text{ unrestricted}
\end{aligned}$$

The procedure then is as follows

1. Determine $\underline{b}_0$.
2. Select one of the problems P_1 or P_2 with probability 1/2 for each.
3. Obtain an optimal solution $\underline{b}_1^*$ for the problem selected.
4. Estimate $\underline{\beta}$ by $\tilde{\underline{\beta}} = \underline{b}_1^* + \underline{b}_0$.

Since

$$\begin{aligned}
\underline{y}(-\underline{e}) - X\underline{b}_0(-\underline{e}) &= X\underline{\beta} - \underline{e} - X\underline{b}_0(-\underline{e}) \\
&= -\underline{e} + X[\underline{\beta} - \underline{b}_0(-\underline{e})] \\
&= -\underline{e} - X[\underline{\beta} + \underline{b}_0(-\underline{e})] \\
&= -\underline{y}(\underline{e}) + X\underline{b}_0(\underline{e})
\end{aligned}$$

the problems $P_1(\underline{e})$ and $P_2(-\underline{e})$ are identical except for the sign of $\underline{b}_1$, and so are $P_1(-\underline{e})$ and $P_2(\underline{e})$. Hence,

$$\underline{b}_1^* \, [P_1(\underline{e})] = -\underline{b}_1^* \, [P_2(-\underline{e})]$$

and

$$\underline{b}_1^* \, [P_1(-\underline{e})] = -\underline{b}_1^* \, [P_2(\underline{e})]$$

Now,

$$
\begin{aligned}
E(\tilde{\underline{\beta}}) &= E(\underline{b}_0) + E(\underline{b}_1^*) \\
&= \underline{\beta} + E[(\underline{b}_1^*|\underline{e}) + (\underline{b}^*|-\underline{e})] \\
&= \underline{\beta} + E[(\underline{b}_1^*|\underline{e}, P_1) \cdot \frac{1}{2} \\
&\quad + (\underline{b}_1^*|\underline{e}, P_2) \cdot \frac{1}{2} \\
&\quad + (\underline{b}_1^*|-\underline{e}, P_1) \cdot \frac{1}{2} \\
&\quad + (\underline{b}_1^*|-\underline{e}, P_2) \cdot \frac{1}{2}] \\
&= \underline{\beta}
\end{aligned}
$$

that is, $\tilde{\underline{\beta}}$ is unbiased.

11.2.2 L_∞ Estimation

In (11.2) as $p \rightarrow \infty$ the norm becomes max $(|v_i|)$; hence, the *minimum L_∞ criterion* of (11.3) is the *minimax criterion*. It is also frequently referred to as the *Chebyshev criterion*.

The estimates under the minimum L_∞ criterion are very sensitive to outliers; however, they are quite efficient if the distribution of the errors in the model is uniform, or nearly constant with a finite range.

Some computational features of L_∞ estimations are similar to those of L_1 computation. The norm

$$S(\underline{b}) = \max(|y_i - \underline{x}_i'\underline{b}|) \tag{11.6}$$

like (11.4) is a piecewise linear convex surface; and hence there are a finite number of identifiable points at which the minimum must occur. The minimum may not occur at only a single point, even when X is of full rank. The technique of Sielken and Hartley (1973) for obtaining unbiased estimators applies also to L_∞ estimation. The L_∞ estimation problem can also be stated as a linear program, as was noticed by Kelley (1958), who used a dual simplex algorithm.

The (primal) linear programming problem is

$$
\begin{aligned}
\text{minimize} \quad & d \\
\text{subject to} \quad & X\underline{b} + d\underline{1}_n \geq \underline{y} \\
& -X\underline{b} + d\underline{1}_n \geq -\underline{y} \\
& d \geq 0 \\
& \underline{b} \text{ unrestricted}
\end{aligned}
\tag{11.7}
$$

where, as before, $\underline{1}_n$ is an n vector of 1's. Note that the nonnegativity restriction on d (d is a scalar) is merely a consequence of the formulation and not a true restriction.

Barrodale and Phillips (1974) give a very efficient method for obtaining the L_∞ estimates from a dual formulation of the problem:

$$
\begin{aligned}
\text{maximize} \quad & \underline{y}'(\underline{s} - \underline{t}) \\
\text{subject to} \quad & X'(\underline{s} - \underline{t}) = \underline{0} \\
& \underline{1}_m'(\underline{s} + \underline{t}) \leq 1 \\
& \underline{s}, \underline{t} \geq \underline{0}
\end{aligned}
$$

A slack variable is inserted in the inequality constraint, and an artificial variable is inserted in each row of the equality constraints; however, the relationships among the variables are such that only n columns are required. The slack variable and the artificial variables are assigned zero coefficients in the objective function, so the marginal costs of each s_i is $-y_i$. The initial basis consists of vectors associated with the slack and artificial variables.

The algorithm proceeds through three stages. In the first stage, the $\underline{s}_i$ are brought into the basis, replacing the vectors corresponding to the artificial variables. (Note we are using the common notational convention of linear programming of denoting a tableau vector by the name of the variable. The underscore indicates that the symbol represents a vector instead of a variable.) The pivot column is chosen to correspond to the s_i having marginal cost with greatest absolute value, and the row is selected so the pivot has largest absolute value of those corresponding to artificial variables in the basis. Note that the pivot may be negative. This continues until the number of $\underline{s}_i$ in the basis is equal to the rank of X.

Stage 2 consists of a single simplex iteration, which may also require some row interchanges. The pivot column is chosen as before to correspond to the s_i with marginal cost of greatest absolute value. If the marginal cost is positive, however, $\underline{s}_i$ and $\underline{t}_i$ are interchanged. This interchange is done quite simply by changing the sign of the marginal cost and the sign of each element in the pivot column and then adding 2 to the pivot column element opposite the basis vector $\underline{w}$ corresponding to the slack variable. (*Note:* $\underline{s}_i + \underline{t}_i = 2\underline{w}$.) The vector to leave the basis in this stage is $\underline{w}$ (i.e., the last row is the pivot row), which requires that the first m elements of the pivot column be nonpositive to maintain feasibility. If this requirement is not satisfied, row interchanges are performed by simply adding any row, corresponding to a positive element in the pivot column, to the pivot (last) row and then changing the sign of the original row. This obviously also replaces a vector $\underline{s}_i$ in the basis by its corresponding $\underline{t}_i$.

In stage 3 the pivot column is chosen to correspond to the variable s_i or t_i having the most negative marginal cost, and the pivot row is chosen by the usual selection rule of the simplex method. The algorithm terminates when all marginal costs are nonnegative.

Barrodale and Phillips (1975) give an efficient FORTRAN coding of this algorithm.

Abdelmalek (1976a) gives a FORTRAN program implementing the method given by Abdelmalek (1975b), which also proceeds to the solution more rapidly than a straightforward simplex method would. The program is roughly equivalent to that of Barrodale and Phillips in efficiency. Bartels, Conn, and Charalambous (1976) and Hopper and Powell (1977) give other efficient methods for the L_∞ estimation problem.

Special-purpose algorithms for restricted L_∞ estimation have been given by Powell (1974), Lauter (1975), Toutenburg (1975), and Abdelmalek (1977b).

The problem of nonuniqueness of the L_∞ estimator has been considered by Rice (1962), who defined a unique L_∞ solution, called the *strict* solution. Descloux (1963) shows that the strict estimator is the limiting L_p estimator as $p \longrightarrow \infty$ (L_p estimators are unique for full-rank X and $1 < p < \infty$.). An algorithm implementing the underlying concept of Descloux's discussion is given by Duris and Temple (1973) and improved upon by Abdelmalek (1977a).

11.2.3 Other L_p Estimators

In this subsection we will consider the problem (11.3) for $1 < p < \infty$. For this range of p values the exponent 1/p can be ignored and the computing problem becomes one of minimizing the nonlinear function

$$F(\underline{b}) = \sum_{i=1}^{n} |y_i - \underline{x}_i'\underline{b}|^p \tag{11.8}$$

Thus the methods for unconstrained minimization, given in Chapter 10, may be useful in solving this problem. We may have some difficulty because derivatives will not exist continuously at all points, but the fact that $F(\underline{b})$ is convex is certainly advantageous. Let us investigate the use of such methods to see how they might apply.

Using the notation $r_i = y_i - \underline{x}_i'\underline{b}$, we find the gradient vector of $F(\underline{b})$ to be

$$\underline{g}(\underline{b}) = -pX'\underline{w}$$

where the jth component of $\underline{w}$ is

$$w_j = |r_j|^{p-1}\,\text{sign}(r_j)$$

The Hessian matrix of $F(\underline{b})$ is

$$G(\underline{b}) = pX'RX$$

where R is the diagonal matrix

$$R = \text{diag}[(p - 1)|r_j|^{p-2}]$$

Note that for $1 < p < \infty$ a zero residual r_i will cause no problem in the gradient vector definition, but unless $p \geq 2$ the Hessian matrix will not be defined when one of the r_i is zero. Since zero residuals are to be expected in these applications, particularly at the solution, the first-order gradient methods seem to be most appropriate for use when p is in the range $1 < p < 2$. This has proven to be true in practice and, for example, Forsythe (1972) reports satisfactory results using a quasi-Newton method (the DFP method) for problems with $1 < p < 2$.

The potential quadratic rate of convergence of Newton's method has prompted several studies on the use of this method in L_p estimation. Ignoring, for the moment, the fact that R may not be defined in some iterations, the Newton method in this case has the form

$$X'R_kX\underline{d}_k = X'\underline{w}$$

$$\underline{b}_{k+1} = \underline{b}_k + \underline{d}_k$$

The first of these equations can be reformulated by noting that $w_j = |r_j|^{p-2}(y_j - \underline{x}_j'\underline{b})$; hence we have

$$X'R_kX\underline{d}_k = \frac{1}{p - 1} X'R_k\underline{y} - \frac{1}{p - 1} X'R_kX\underline{b}_k$$

Solving for $\underline{d}_k$ gives

$$\underline{d}_k = \frac{1}{p - 1}(X'R_kX)^{-1}X'R_k\underline{y} - \frac{1}{p - 1}\underline{b}_k$$

Thus Newton's method can be specified in the form

$$X'R_k X\underline{\theta}_k = X'R_k \underline{y}$$

$$\underline{b}_{k+1} = \frac{(p - 2)\underline{b}_k + \underline{\theta}_k}{p - 1}$$

This immediately suggests other forms for the iterates which are simpler, more convenient, and are such that they might work well for the convex functions being considered. For example, a natural choice is to iterate according to

$$X'S_k X\underline{b}_{k+1} = X'S_k \underline{y} \tag{11.9}$$

where

$$S_k = \text{diag}(|r_i|^{p-2})$$

$$S_0 = I$$

This is often called *iteratively reweighted least squares*. It is a desirable form because standard multiple linear regression programs usually support weighted least squares, and a simple modification is all that is needed to include an iterative capability.

For $1 < p \leq 2$ Merle and Spath (1974) suggest setting zero residuals r_i to a small $\varepsilon > 0$ and using Newton's method or Eq. (11.9). They investigated these two methods and found that either method converged in the cases they studied empirically. Other studies of the use of Newton's iterates and forms like (11.9) include Gentleman (1965), Fletcher, Grant, and Hebden (1971), Kahng (1972), and Rey (1975). The obvious fact in the methods described above is that some device must be explicitly or implicitly used to insure that a descent direction is obtained for each iteration, else we are departing radically from standard procedures and there is no apparent benefit to be derived from doing this. Equation (11.9) does not necessarily give a descent direction, so we will not expect it, in this form, to converge as often as we might like, and it may make very slow progress even on a convergent track.

A theoretically interesting alternative procedure is suggested by Ekblom (1973), who also proposed the modified Newton method as a desirable choice for L_p problems. Ekblom introduced a perturbation e into the objective function and considered the problem of minimizing

$$F_e(\underline{b}) = \sum_{i=1}^{n}[(y_i - \underline{x}_i'\underline{b})^2 + e^2]^{p/2} \qquad (11.10)$$

For this modified objective function the gradient vector and Hessian are

$$\underline{g}^{(e)}(\underline{b}) = -pX'\underline{w}^{(e)}$$

where

$$w_i^{(e)} = r_i(r_i^2 + e^2)^{(p-2)/2}$$

and

$$G^{(e)}(\underline{b}) = pX'R^{(e)}X$$

where

$$R^{(e)} = \left\{\text{diag}\left(r_i^2 + e^2\right)^{(p-2)/2} + (p-2)r_i^2\left(r_i^2 + e^2\right)^{(p-4)/2}\right\}$$

Whenever $p > 1$ we see that $(p - 2)r_i^2/\left(r_i^2 + e^2\right) > -1$; thus $R^{(e)}$ must have positive elements. Since we have control over the selection of $\underline{d}_k$ at each iteration in Newton's method, we can assume without loss of generality that all of the r_i at each iteration are finite. Thus there are finite positive scalars $0 < \delta_1 < \delta_2$ such that $\delta_1 I \leq R^{(e)} \leq \delta_2 I$, and (assuming a full-rank X matrix)

$$\underline{z}'G^{(e)}\underline{z} = p(X\underline{z})'R^{(e)}(X\underline{z}) \geq p\delta_1(X\underline{z})'(X\underline{z}) > 0 \qquad \text{for all } \underline{z} \neq \underline{0}$$

Hence we are assured that the Hessian matrix will remain positive definite throughout the iterations when $e \neq 0$. Ekblom then proceeds to show that the limiting solution as $e \to 0$ is the solution to the unperturbed problem (11.8). In applications of this procedure a series of minimization problems is envisioned with each successive problem being perturbed in (11.10) by a lesser amount than the preceding problem, and the solution for any given problem provides the starting value for the next problem. Ekblom suggests that the level of accuracy strived for in Newton's iteration need not be high for large e, but it must obviously be increased as e becomes small. Step size in Ekblom's algorithm is obtained using the Goldstein-Armijo method. Thus Ekblom's method can be used for any p in the range $1 < p < \infty$; however, it is not particularly appealing for problems having $p > 2$.

In a comparison of algorithms for L_p estimation, Kennedy and Gentle (1978) found that a quasi-Newton method seems most desirable for $1 < p < 2$, and a modified Newton method works well for $p > 2$.

Finally, an unusual method suggested by Schlossmacher (1973) will be mentioned. It was originally used for L_1 problems, but it might be applied for $2 \geq p > 1$. The method consists of using iteratively reweighted least squares but deleting, at each iteration, the observations and corresponding rows of the X matrix which give rise to zero residuals for that iteration. An observation deleted for one iteration will reenter the computations at the next iteration unless it again gives a zero residual. This iterative method will not always converge, and no proof of convergence under any condition has been derived.

To summarize the above, it is not extremely difficult to obtain a solution to the kind of problem considered in this subsection using any one of several available methods. The primary consideration is one of efficiency.

11.3 OTHER ROBUST ESTIMATORS

The resistance of the L_p estimator to outliers or to heavy-tailed error distributions in general increases as p decreases from 2 to 1, but the efficiency of the estimators for the case of normal errors decreases. The L_1 estimator obtains its strong resistance at considerable sacrifice of efficiency. It is desirable to have estimators that are not only robust to wild observations, but also robust of efficiency. Andrews et al. (1972) proposed and studied a number of such estimators of location for simple models. Computational techniques for these estimators generally are iterative.

One very useful robust estimator of location for the observations $x_1, x_2, \ldots, x_n$ is the M estimate, $\tilde{\mu}$, which is a solution of

$$\sum_i \psi\left(\frac{x_i - \tilde{\mu}}{s}\right) = 0 \tag{11.11}$$

for some function ψ and scale estimate s. Such estimates derive their name from their similarity to maximum-likelihood estimates. (See, e.g., Huber 1972 for a general discussion.)

Analogous estimators have been defined and studied by Relles (1968) and others for the linear regression model (11.1). In this context the estimator $\underline{b}$ is taken as a vector that minimizes

$$\sum_i \phi\left[\frac{y_i - \underline{x}_i'\underline{b}}{s}\right] \tag{11.12}$$

where ϕ is some convex function and $\underline{x}_i'$ is the ith row of X. Equation (11.11) and problem (11.12) are essentially equivalent when ψ is the derivative of ϕ. The function ϕ defines what is called the *influence curve*.

Among the useful functions ψ is the one proposed by Andrews (see Andrews et al 1972):

$$\psi(Z) = \begin{cases} \sin\left(\frac{Z}{c}\right) & \text{if } |Z| < c\pi \\ 0 & \text{otherwise} \end{cases}$$

where c is taken as some appropriate constant. Andrews (1974) gives an iterative procedure to obtain the estimates in the regression model. The procedure, which is similar to (11.9), is essentially an iteratively reweighted least squares using the weights given by the sine function. Using the current estimate $\underline{b}^{(k)}$ and the notation

$$r_i^{(k)} = y_i - \underline{x}_i'\underline{b}^{(k)}$$

weights $w_i^{(k+1)}$ are selected such that

$$[w_i^{(k+1)}]^2 = \begin{cases} [\sin(r_i^{(k)}/s^{(k)})]/r_i^{(k)} & \text{if } |r_i^{(k)}| < c\pi \\ 0 & \text{otherwise} \end{cases}$$

where $s^{(k)}$ is a current scale estimator. The updated estimate $\underline{b}^{(k+1)}$ is then obtained by a least-squares fit of

$$w_i^{(k+1)} y_i = w_i^{(k+1)} \underline{x}_i' \underline{b}^{(k+1)} \qquad (i = 1, 2, \ldots, n)$$

Andrews (1974) defines a median estimator based on a generalized sweep operator for the initial estimates in this iterative process.

As Andrews indicates, it may not be desirable to attempt iteration until convergence. A one-step sine estimator, "SIN1," has been studied by Welsch (1975) and Hill and Holland (1977). They begin with an L_1 estimator (see Section 11.2.1) and do only one iteration using the sine weight.

Huber and Dutter (1974) consider the robust regression estimators based on the ψ function given by

$$\psi(z) = \begin{cases} -c & \text{if } z < -c \\ z & \text{if } |z| < c \\ c & \text{if } z > c \end{cases}$$

Dutter (1975a, 1975b) described three different algorithms for these estimators. Convergence proofs for these methods are given (Dutter 1975a).

Dutter (1977) describes and compares several algorithms for solving the problem of minimizing $F(\underline{b})$ given by

$$F(\underline{b}) = \sum_{i=1}^{n} \phi \left[\frac{y_i - \underline{x}_i'\underline{b}}{s} \right] s + \tau s$$

where s and τ are chosen to provide desirable scaling. The X matrix is assumed to have full rank m. The ϕ function used by Dutter is

$$\phi(t) = \begin{cases} \dfrac{t^2}{2} & \text{if } |t| \le c \\ c|t| - \dfrac{c^2}{2} & \text{if } |t| > c \end{cases}$$

where $c > 0$ is a chosen constant. This ϕ function has a first derivative $\psi(t)$ in the form given above. We will now describe two algorithms which Dutter (1977) found to be somewhat desirable for use in solving robust linear regression problems having rank $(X) = m$. Some additional notation will be needed in the description, and such will now be provided.

Consider a partitioning of the set of indices

$$A = \left\{1, 2, \ldots, n\right\}$$

with respect to the function ψ and the residuals $r_i(\underline{\theta}) = y_i - \underline{x}_i'\underline{\theta}$, $i = 1, 2, \ldots, n$. Let

$$A_0 = \left\{ i \,\middle|\, \frac{r_i}{s} < -c \right\}$$

$$A_1 = \left\{ i \,\middle|\, \left|\frac{r_i}{s}\right| \le c \right\}$$

$$A_2 = \left\{ i \,\middle|\, \frac{r_i}{s} > c \right\}$$

Also let

$$X_0 = X - \sum_{i \notin A_1} \underline{e}_i \underline{x}_i'$$

$$= \sum_{i \in A_1} \underline{e}_i \underline{x}_i'$$

where $\underline{e}_i$ is the ith unit vector. Thus X_0 is constructed from X by zeroing those rows whose index does not belong to the set A_1. Differentiating $F(\underline{\theta})$ gives the gradient vector

$$\underline{g}(\underline{\theta}) = -X'\underline{\psi}[\underline{r}(\underline{\theta})/s]$$

where

$$\underline{\psi}(\underline{t}) = (\psi(t_1), \psi(t_2), \ldots, \psi(t_n))'$$

Partitioning the $\underline{g}(\underline{\theta})$ vector according to the indices defined by the sets A_0, A_1, and A_2, the gradient can be written as

$$\underline{g}(\underline{\theta}) = -\frac{1}{s}X_0'\underline{y} + \frac{1}{s}X_0'X_0\underline{\theta} - c\left[\sum_{i\varepsilon A_2}\underline{x}_i - \sum_{j\varepsilon A_0}\underline{x}_j\right]$$

The Hessian matrix of $F(\underline{\theta})$ is easily seen to be

$$G(\underline{\theta}) = \frac{1}{s}X_0'X_0$$

The first of Dutter's algorithms, which is called algorithm H1, requires that the user supply a starting value $\underline{\theta}_0$, an initial estimate s_0 of the scale factor s, and a tolerance $\varepsilon > 0$ along with the value of c which defines $\psi(t)$. The algorithm computes an updated estimate of the weight s at each iteration. Details are as follows.

ALGORITHM H1$(\underline{\theta}_0, s_0, \varepsilon, c, \tau, X, \underline{y}, \hat{\underline{\theta}})$

Step	*Description*
1.	Set $k = 0$.
2.	Compute $r_i(\underline{\theta}_k) = y_i - \underline{x}_i'\underline{\theta}_k$, $i = 1, \ldots, n$.
3.	Compute a new s value as $\left(s_{k+1}\right)^2 = \frac{1}{2\tau}\sum_{i=1}^{n}\left\{\psi\left[\frac{r_i(\underline{\theta}_k)}{s_k}\right]^2 (s_k)^2\right\}$
4.	Using s_{k+1}, the residuals are transformed (Winsorized) as $(\underline{\delta}_k)_i = \psi\left[\frac{r_i(\underline{\theta}_k)}{s_{k+1}}\right] s_{k+1}$ $i = 1, 2, \ldots, n$
5.	Find a least-squares solution $\underline{d}_k$ in $X\underline{d}_k = \underline{\delta}_k$
6.	Compute $\underline{\theta}_{k+1} = \underline{\theta}_k + \alpha\underline{d}_k$ where a fixed step size α, $0 < \alpha < 2$, is used.

ALGORITHM Cont.

Step	*Description*
7.	If $\|\alpha(\underline{d}_k)_i\| < \varepsilon s_{k+1}(x^{ii})^{1/2}$ for each i = 1, 2, ..., m and $\|s_{k+1} - s_k\| < \varepsilon s_{k+1}$ then go to step 9. Here $x^{ii} = (X'X)^{-1}_{ii}$.
8.	Set k = k + 1 and go to 2.
9.	Set $\hat{\underline{\theta}} = \underline{\theta}^{(k+1)}$ and exit.

The value of α used in steps 6 and 7 is subject to choice. One suggested value is

$$\alpha = \begin{cases} [\Phi(c) - \Phi(-c)]^{-1} & \text{if } 0 < \alpha < 2 \\ 1.9 & \text{otherwise} \end{cases}$$

where Φ is the cumulative distribution function of the standard normal distribution.

Timing comparisons were made between H1 and the Davidon-Fletcher-Powell algorithm for a series of test problems. The average amount of CPU time for these problems using H1 was approximately half that of the DFP algorithm. No data were available which give insight into the relative dependability of these two algorithms so far as simply converging to a solution is concerned.

A second algorithm for robust linear regression differs from H1 only in steps 4 and 5. This algorithm, which is called W1, uses iteratively reweighted least squares to obtain a solution as follows:

Step	*Description*
4.	Compute $(\underline{w}_k)_i = \psi\left[r_i(\underline{\theta}_k)/s_{k+1}\right] / \left[2r_i(\underline{\theta}_k)/s_{k+1}\right]$ if $r_i(\underline{\theta}_k) \neq 0$; otherwise set $(\underline{w}_k)_i = 1/2$ (i = 1, 2, ..., n).
5.	Solve for $\underline{d}_k$ as $\underline{d}_k = (X'W_kX)^{-1}X'W_k\underline{y} - \underline{\theta}_k$ where W_k is an m x m matrix defined by $W_k = \text{diag}\,[(\underline{w}_k)_i]$.

In each of these algorithms there is the requirement, in step 5, to compute a linear least-squares solution. Dutter (1977) used several different standard methods, including singular-value decomposition, to obtain this solution. There were rather sizable differences in computer costs, but no significant difference in the accuracy of the general algorithms H1 and W1 was noted when the method used in step 5 was changed.

Several other algorithms are given by Dutter (1977). Some of these other algorithms were found to be very efficient for some test problems. The primary motivation for considering specialized robust linear regression algorithms is to achieve greater efficiency than that which can be achieved using general methods for function minimization.

Another class of robust estimators is the R estimates, whose name derives from the use of ranks in defining the estimates. Jaeckel (1972) studied the estimates obtained by minimizing

$$\sum_i a[R(y_i - \underline{x}_i'\underline{b})](y_i - \underline{x}_i'\underline{b})$$

where R(•) denotes the rank and a(•) is a score function. Hettmansperger and McKean (1976) considered the computational problems involved in obtaining such an estimate. They gave a two-stage algorithm for the regression model containing an intercept term. Each stage of their algorithm is iterative. We refer the reader to the paper for details.

A simple robust regression estimator was given by Hinich and Talwar (1975) that involved performing separate regressions on different subsets of observations, combining these separate estimates, using the combined estimate to trim the sample, and then performing a regression on the trimmed sample.

Many of the robust and nonparametric procedures involve sorting or ranking. While this is conceptually a very simple operation, the most straightforward methods for sorting are quite inefficient when compared to the better methods available. We will not discuss sorting procedures further, but will refer the reader to the book by Knuth (1973).

11.4 BIASED ESTIMATION

In previous sections we have discussed criteria of estimation that related primarily to the "fit" of the linear model (11.1) to a set of data. Other criteria relate to distributional properties of the estimator, such as expected value, variance, mean-square error, and so forth. The robust estimators generally are designed to have small variance even when model assumptions fail to hold. When the assumption of identical, independent normal distribution for the error term does hold, it is well known that the ordinary least-squares estimator, $\underline{b}$, gives the minimum variance unbiased estimator of $\underline{a}'\underline{\beta}$, for any $\underline{a}$. Even so, this variance may not be small. Consider, for example, the "total variance" of $\underline{b}$ in a full-rank model:

$$\begin{aligned}\sum_j \text{var}\ (b_j) &= \sigma^2\ \text{tr}(X'X)^{-1} \\ &= \sigma^2\ \Sigma\ \lambda_j^{-1} \\ &> \sigma^2\ \lambda_m^{-1}\end{aligned}$$

where λ_j are the eigenvalues of $X'X$ and λ_m is the minimum eigenvalue. Whether or not the assumptions on the error term hold, if $X'X$ is nearly singular, λ_m may be quite small; hence, the minimum variance of the unbiased estimator may be unacceptably large. Equivalently, the expected length (euclidean norm) of $\underline{b}$ may be considered:

$$\begin{aligned}E(\underline{b}'\underline{b}) &= \underline{\beta}'\underline{\beta} + \sigma^2\ \text{tr}(X'X)^{-1} \\ &> \underline{\beta}'\underline{\beta} + \frac{\sigma^2}{\lambda_m}\end{aligned}$$

Thus $\underline{b}$ can be a very long vector when X is ill-conditioned. This is not necessarily caused by numerical inaccuracies or by failure of model assumptions; it is an inherent property of the X matrix.

Hoerl and Kennard (1970a, 1970b) suggested an alternate estimation procedure, ridge regression, to overcome the difficulties noted above. The estimator

$$\underline{b}_k = (X'X + kI)^{-1}X'\underline{y} \qquad k \geq 0 \tag{11.13}$$

is shrunken toward $\underline{0}$, so as to have a shorter length. Hoerl and Kennard show that $\underline{b}_k$, for some $k > 0$, is more desirable than $\underline{b}$ with regard to the minimum mean square error criterion. Obviously, however, $\underline{b}_k$ is biased if $k > 0$.

In applications, k is allowed to vary from zero to some selected upper bound and $\underline{b}_k$ is computed for each k. Individual components of $\underline{b}_k$ are plotted versus k to determine an optimal or near-optimal value of k. The plots of the components generally show a rapid decrease toward zero as k initially moves away from zero, but soon the curve levels off and decreases slowly thereafter. These graphs, which are usually plotted on the same axis, constitute the "ridge trace" defined by Hoerl and Kennard. A "good" value of k is one at which the ridge trace has become fairly stable.

The estimator (11.13) is the least-squares solution to an augmented model (see, e.g., Allen 1974),

$$\begin{pmatrix} \underline{y} \\ 0 \end{pmatrix} = \begin{pmatrix} X \\ \sqrt{k} I_m \end{pmatrix} \underline{\beta} + \underline{\varepsilon} \tag{11.14}$$

Hence computations for the ridge regression estimates can be carried out using the orthogonalization methods of Chapters 7 and 8 without forming the normal equations (11.13). For orthogonal triangularization of (11.14) the row ordering should be as shown, else for small k, row interchange as described in Section 7.2.1 would be necessary.

Ridge regression, however, is one of the few situations in which it may be more desirable to use the normal equations, as in (11.13). These may be solved by orthogonalization methods or sweeps as discussed in Chapter 7.

The fact that the user of ridge regression is responsible for choosing k in Eq. (11.13) detracts from the desirability of generalizing the method in an attempt to extend its utility. Hoerl and Kennard provided some assistance in this regard by defining an iterative procedure for finding "optimal" k values in their so-called generalized ridge regression method, but the procedure is often slow to converge. However, Hemmerle (1975) derived an explicit (non-iterative) solution to the generalized ridge regression problem. His solution is one which would result using optimal k. We will next describe generalized ridge regression and give Hemmerle's procedure for computing an optimal solution. Availability of this explicit solution makes generalized ridge regression an appealing tool for use in data analysis.

Assume that the X matrix in the model (11.1) is such that X'X is in correlation form. If P is an orthogonal matrix of eigenvectors of X'X, and $\Lambda = \text{diag}(\lambda_i)$ contains the m eigenvalues λ_i of X'X, then Λ is expressible as

$$P'X'XP = \Lambda$$

Using P, we now define

$$Z = XP$$

$$\underline{\alpha} = P'\underline{\beta}$$

The model (11.1) can be expressed as

$$\underline{y} = Z\underline{\alpha} + \underline{e}$$

and the *generalized ridge estimate* is defined to be

$$\hat{\underline{\alpha}} = (Z'Z + K)^{-1}Z'\underline{y}$$

where

$$K = \text{diag}(k_i) \qquad (k_i \geq 0, \quad i = 1, 2, \ldots, m)$$

Optimal values of the k_i are taken to be those values that minimize the expectation

$$E[\hat{\underline{\alpha}} - \underline{\alpha})'(\hat{\underline{\alpha}} - \underline{\alpha})]$$

and the associated $\hat{\underline{\alpha}}$ is called the optimal generalized ridge regression solution.

Hemmerle derived the following method for determining an optimal $\hat{\underline{\alpha}}$. First, m scalar values e_i are computed as

$$e_i = \frac{\hat{\sigma}^2}{\lambda_i(\hat{\gamma}_i)^2} \qquad (i = 1, 2, \ldots, m)$$

where

$$\hat{\sigma}^2 = \underline{y}'\underline{y} - \hat{\underline{\gamma}}'Z'\underline{y}$$

and $\hat{\underline{\gamma}}$ is the ordinary least-squares estimate of $\underline{\alpha}$. Using the e_i, we next compute new scalars e_i^* as follows:

$$e_i^* = \frac{1 - 2e_i - (1 - 4e_i)^{1/2}}{2e_i}$$

for each $1 \leq i \leq m$ for which $e_i \leq 1/4$. Finally, the components $\hat{\alpha}_i$ of $\hat{\underline{\alpha}}$ for the optimal K are

$$\hat{\alpha}_i = 0 \qquad \text{if } e_i > 1/4$$

$$\hat{\alpha}_i = \frac{\hat{\gamma}_i}{1 + e_i^*} \qquad \text{if } e_i \leq 1/4$$

Example 11.2 illustrates the use of these equations.

Hemmerle also notes that solution under an optimal K may result in an undesirably large residual sum of squares. He also gives a method for obtaining an explicit solution under a constraint on the size of the residual sum of squares relative to the sum of squared residuals for ordinary least squares. The reader is referred to Hemmerle's paper for a description of this method.

Example 11.2

Suppose that the given regression data are such that

$$X'X = \begin{pmatrix} 1 & 4/5 \\ 4/5 & 1 \end{pmatrix} \qquad \underline{y}'\underline{y} = 53/45$$

and we want an optimal generalized ridge regression solution. The data produced the following quantities:

$$\Lambda = \begin{pmatrix} 9/5 & 0 \\ 0 & 1/5 \end{pmatrix} \qquad \hat{\underline{\gamma}} = \begin{pmatrix} 5/18 \\ 7/9 \end{pmatrix} \qquad Z'\underline{y} = \begin{pmatrix} 9/10 \\ 1 \end{pmatrix}$$

The residual sum of squares is found as

$$\begin{aligned} \hat{\sigma}^2 &= \underline{y}'\underline{y} - \hat{\underline{\gamma}}'Z'\underline{y} \\ &= 3/20 \end{aligned}$$

Now the scalars e_i are determined to be

$$e_1 = \frac{3/20}{(9/10)(5/18)} = \frac{3}{5} > \frac{1}{4}$$

$$e_2 = \frac{3/20}{(1)(7/9)} = \frac{27}{140} < \frac{1}{4}$$

Using e_2, we find e_2^* as

$$\begin{aligned} e_2^* &= \frac{1 - 2e_2 - (1 - 4e_2)^{1/2}}{2e_2} \\ &\cong 0.353096 \end{aligned}$$

Finally, the $\hat{\alpha}_j$ are found to be

$$\begin{aligned} \hat{\alpha}_1 &= 0 \\ \hat{\alpha}_2 &= \frac{\hat{\gamma}_2}{1 + e_2^*} \\ &\cong 0.574813 \end{aligned}$$

Note that $\hat{\alpha}_1$ is zero because we have $e_i > 1/4$ for this regression problem.

Marquardt (1970) discussed ridge regression in a broader context of biased linear estimation procedures. This paper provides an interesting discussion of biased linear estimation and the role that ridge regression has in this area.

11.5 ROBUST NONLINEAR REGRESSION

The preceding sections in this chapter dealt with robust estimation in linear models. The increasing popularity of such estimates in the linear case has led to research and development of analogous methods for robust nonlinear regression. Some computational aspects of this nonlinear problem will be described in this section.

The general situation is one of minimizing a function of the form

$$F(\underline{\theta}) = \sum_{i=1}^{n} \phi\, [y_i - h_i(\underline{\theta})] \tag{11.15}$$

where it is usually the case that $\phi(u) \leq \phi(v)$ whenever $|u| < |v|$. When $\phi(t) = t^2/2$ we have the nonlinear least-squares problem studied in Chapter 10. An example of a robust loss function was previously given as

$$\phi(t) = \begin{cases} \dfrac{t^2}{2} & |t| \leq c \\ c|t| - \dfrac{c^2}{2} & |t| > c \end{cases}$$

where c is a user-supplied parameter. As is typical of loss functions normally employed, this function does not have continous derivatives. It does reduce to the nonlinear regression model for large c.

There are few special software systems for robust nonlinear regression applications. Frequently computer programs for general unconstrained minimization are used to obtain the estimates. This approach does produce answers, but there is usually a need for more flexibility that can only be provided by programs that are tailor-made for robust nonlinear analyses.

An interesting algorithm based on Newton's method is reported by Welsch and Becker (1975) and has been used for robust nonlinear regression applications. We will describe this algorithm using the notation

$$F(\underline{\theta}) = \sum_{i=1}^{n} \phi\,[f_i(\underline{\theta})]$$

$$\underline{\phi}^{(1)}(\underline{t}) = \left(\frac{\partial\phi}{\partial t_1}, \frac{\partial\phi}{\partial t_2}, \ldots, \frac{\partial\phi}{\partial t_n}\right)'$$

$$= [\phi^{(1)}(t_1), \phi^{(1)}(t_2), \ldots, \phi^{(1)}(t_n)]'$$

$$\phi^{(2)}(\underline{t}) = \text{diag}\left(\frac{\partial^2\phi}{\partial t_j^2}\right)$$

The gradient vector and Hessian matrix of F have the form

$$\underline{g}(\underline{\theta}) = J'(\underline{\theta})\underline{\phi}^{(1)}[\underline{f}(\underline{\theta})] \tag{11.16}$$

$$G(\underline{\theta}) = J'(\underline{\theta})\phi^{(2)}[\underline{f}(\underline{\theta})]J(\underline{\theta}) + \sum_{i=1}^{n} \underline{\phi}^{(1)}[f_i(\underline{\theta})]H_i(\underline{\theta}) \tag{11.17}$$

where $H_i(\underline{\theta})$ is the Hessian matrix of $f_i(\underline{\theta})$. The algorithm employed by Welsch and Becker uses a combination of the steepest descent and Newton directions in the course of iteration. The algorithm requires input of an initial value $\underline{\theta}_0$ and an upper bound R on the step size.

Consider the quadratic approximation to $F(\underline{\theta}_i + \underline{d}_i)$

$$A(\underline{d}_i) = F(\underline{\theta}_i) + \underline{g}'(\underline{\theta}_i)\underline{d}_i + \frac{\underline{d}_i' G_i \underline{d}_i}{2}$$

where $\underline{d}_i$ denotes the step (complete with step size) in this discussion. This function is minimized in the gradient direction using step size

$$\alpha_i = \|\underline{g}_i\|^3 / \underline{g}_i' G_i \underline{g}_i$$

If $\alpha_i \geq R$, a step in the negative gradient direction $(\underline{d}_i)_G$ is tried. Using $(\underline{d}_i)_N$ to denote the Newton direction, whenever $\alpha_i < R$ and $\|(\underline{d}_i)_N\| \leq R$, a Newton step is attempted. Finally, when $\alpha_i < R$ and $\|(\underline{d}_i)_N\| > R$, a "dogleg" step is tried. This step, $(\underline{d}_i)_D$ is defined as that point on the line connecting $(\underline{d}_i)_N$ and $(\underline{d}_i)_G$ which is at a distance R from $\underline{\theta}_i$. When any one of the above steps is tried and the objective function is reduced, the step is made as

$$\underline{\theta}_{i+1} = \underline{\theta}_i + \underline{d}_i$$

If none of these attempts proves to be acceptable, R is halved and another iteration is made at the point $\underline{\theta}_i$.

The step bound R is adjusted during the course of computations to help adapt the method to the objective function. If for an acceptable step the reduction in F predicted by the quadratic function, defined as the difference $F(\underline{\theta}_i) - A(\underline{d}_i)$, exceeds a selected multiple of the actual reduction $F(\underline{\theta}_i) - F(\underline{\theta}_i + \underline{d}_i)$, then R is halved before the next iteration. When R is not changed based on the test given above, a further check is made using the derivative

$$S(\delta) = \underline{g}'(\underline{\theta}_i + \delta\underline{d}_i)\underline{d}_i$$

$S(\delta)$ is taken to be the expected change in F for a step in the direction $\underline{d}_i$ from $\underline{\theta}_i + \delta\underline{d}_i$. Naturally, a negative change is desired because F will be decreased. If $S(\delta)$ is assumed to be a linear function, the point δ^* at which $S(\delta) = 0$ is given by

$$\delta^* = \frac{S(0)}{S(0) - S(1)}$$

If $\delta^* < 0$ or $\delta^* \geq 2$, a much larger step could be made if $S(\delta)$ is really linear; so to allow for possibly larger steps, the R value is doubled before the next step is computed. If $0 \leq \delta^* < 2$, the predicted gradient $\underline{g}_i + G_i\underline{d}_i$ is compared with the actual gradient $\underline{g}_{i+1}$ at the point $\underline{\theta}_{i+1}$, and R is doubled if

$$\|\underline{g}_{i+1} - \underline{g}_i - G_i\underline{d}_i\|^2 \leq \|\underline{g}_i\|^2/4$$

because the quality of approximation is good when this inequality is satisfied. In all other cases R is not changed before the next iteration.

Techniques for rescaling variables at each iteration were also developed by Welsch and Becker, and these were incorporated into an algorithm called DOGLEGX. Computational devices for scaling and weighting in various ways made the algorithm particularly well suited to robust regression applications. The authors report some success with this algorithm in practical applications and note that, as is the case for other algorithms, this one also does not work well on some highly nonlinear problems.

Methods other than Newton's method have been employed in these nonlinear applications without great success. Equations (11.16) and (11.17) show that there is still a mildly special structure in these problems, which is greatly accentuated when $\phi(t) = t^2/2$, but the situation for general robust loss functions is more complicated and is generally more difficult to handle. For example, it is obviously easy to generalize the Levenberg-Marquardt method for nonlinear least-squares, but we can expect that the matrix

$$J'(\underline{\theta})\phi^{(2)}[\underline{f}(\underline{\theta})]J(\underline{\theta})$$

will not be positive definite in many cases, and this suggests a higher failure rate in practical applications. Also, in general we can expect robust loss functions to have more weak stationary points which will sometimes complicate matters.

Most of the discussion of nonlinear least-squares regression can be slightly generalized to apply to robust nonlinear regression. Thus it is even more true, for these applications, that much research and development work still needs to be done.

EXERCISES

11.1 Discuss the use of the steepest descent method for finding a solution to the L_1 problem in Eq. 11.4.

11.2 Discuss the possibility of using a quasi-Newton method for solving the L_1 problem defined by Eq. 11.4.

11.3 Give the form of the defining equation for solving the L_1 problem 11.4 using iteratively reweighted least squares. (Armstrong and Frome 1976a).

11.4 Formulate Schlossmacher's algorithm for $2 \geq p \geq 1$. Show that the method will fail for $p > 2$.

11.5 Extend the procedure of Mullett and Murray (Section 8.1.3) to describe a check on the computational accuracy in obtaining an L_p estimator.

11.6 Give the specific form of the gradient vector and Hessian matrix for the function in Eq. (11.15) when

$$\phi(t) = \begin{cases} \frac{c^2}{6} - \frac{c^2}{6}\left[1 - \left(\frac{t}{c}\right)^2\right]^3 & |t| \leq c \\ \frac{c^2}{6} & |t| > c \end{cases}$$

where c is a positive constant.

11.7 Give the defining equations for the generalization of the Levenberg-Marquardt method applied to the function (11.15).

REFERENCES AND FURTHER READINGS

Abdelmalek, N. N. (1971), Linear L_1 Approximation for a Discrete Point Set and L_1 Solution of Overdetermined Linear Equations, *JACM 18*, 41-47.

Abdelmalek, N. N. (1974), On the Discrete Linear L_1 Approximation and L_1 Solutions of Overdetermined Equations, *J. Approx. Theory 11*, 38-53.

Abdelmalek, N. N. (1975a), An Efficient Method for the Discrete Linear L_1 Approximation Problem, *Math. Comp. 29*, 844-850.

Abdelmalek, N. N. (1975b), Chebyshev Solution of Overdetermined Systems of Linear Equations, *BIT 15*, 117-129.

Abdelmalek, N. N. (1976a), A computer Program for the Chebyshev Solution of Overdetermined Systems of Linear Equations, *International J. Numerical Methods in Engineering 10*, 1197-1202.

Abdelmalek, N. N. (1976b), *L_1 Solution of Overdetermined Systems of Linear Equations by a Dual Simplex Method and LU Decomposition*, Technical Report, Division of Electrical Engineering, National Research Council, Ottawa, Canada.

Abdelmalek, N. N. (1977a), Computing the Strict Chebyshev Solution of Overdetermined Linear Equations, *Math. Comp. 31*, 974-983.

Abdelmalek, N. N. (1977b), The Discrete Linear Restricted Chebyshev Approximation, *BIT 17*, 249-261.

Adichie, J. N. (1967), Estimates of Regression Parameters Based on Rank Tests, *AMS 38*, 894-904.

Adichie, J. N. (1974), On Some Robust Properties of Estimates of Regression Based on Rank Tests, *Ann. Institute of Stat. Math. 26*, 223-231.

Allen, D. M. (1974), Relationship between Variable Selection and Data Augmentation and a Method for Prediction, *Techno. 16*, 125-127.

Anderson, D. W. (1965), Linear Programming Time Estimating Equations, *J. Indust. Eng. 16*, 136-138.

Andrews, D. F. (1973), Robust Regression: Practical and Computational Considerations, *Proceedings of Computer Science and Statistics: Seventh Annual Symposium on the Interface*, edited by W. J. Kennedy, Statistical Laboratory, Iowa State University, Ames, 173-175.

Andrews, D. F. (1974), A Robust Method for Linear Regression, *Techno. 16*, 523-531.

Andrews, D. F., P. J. Bickel, F. R. Hampel, P. J. Huber, W. H. Rogers, and J. W. Tukey (1972), *Robust Estimates of Location: Survey and Advances*, Princeton University Press, Princeton, N. J.

Aoki, M. (1965), Successive Generation of Chebyshev Approximate Solution, *J. Basic Engineering 87*, 17-22.

Appa, G., and C. Smith (1973), On L_1 and Chebyshev Estimation, *Math. Programming 5*, 73-87.

Armstrong, R. D., J. J. Elam, and J. W. Hultz (1977), Obtaining Least Absolute Value Estimates for a Two-way Classification Model, *Comm. Stat. (B) 6*, 365-382.

Armstrong, R. D., and E. L. Frome (1976a), A Comparison of Two Algorithms for Absolute Deviation Curve Fitting, *JASA 71*, 328-330.

Armstrong, R. D., and E. L. Frome (1976b), The Calculation of Least Absolute Value Estimators for Two-way Tables, *Proceedings of the Statistical Computing Section, American Statistical Association*, Washington, D.C., 101-106.

Armstrong, R. D., and E. Frome (1977), A Special Purpose Linear Programming Algorithm for Obtaining Least Absolute Value Estimates in a Linear Model with Dummy Variables, *Comm. Stat. (B)*, 383-398.

Armstrong, R. D., and J. W. Hultz (1977), An Algorithm for a Restricted Discrete Approximation Problem in the L_1 Norm, *SIAM J. Num. Anal. 14,* 555-565.

Barrodale, I. (1970), On Computing Best L_1 Approximations, in *Approximation Theory,* edited by A. Talbot, Academic Press, New York, 205-215.

Barrodale, I., and C. Phillips (1974), An Improved Algorithm for Discrete Chebyshev Linear Approximations, *Proceedings of Fourth Manitoba Conference on Numerical Mathematics,* edited by H. C. Williams and B. L. Hartnell, Utilitas Mathematica Publishing, Winnipeg, Canada, 177-190.

Barrodale, I., and C. Phillips (1975), Solution of an Overdetermined System of Linear Equations in the Chebyshev Norm, *TOMS 1,* 264-270.

Barrodale, I., M. J. D. Powell and F. D. K. Roberts (1972), The Differential Correction Algorithm for Rational L_∞ Approximations, *SIAM J. Num. Anal. 9,* 493-503.

Barrodale, I., and F. D. K. Roberts (1970), Applications of Mathematical Programming to ℓ_p Approximation, in *Nonlinear Programming,* edited by J. B. Rosen, O. L. Mangasarian, and K. Ritter, Academic Press, New York, 447-64.

Barrodale, I., and F. D. K. Roberts (1972), *An Improved Algorithm for Discrete L_1 Linear Approximation,* Technical Report TSR 1172, Mathematics Research Center, University of Wisconsin, Madison.

Barrodale I., and F. D. K. Roberts (1973), An Improved Algorithm for Discrete L_1 Linear Approximation, *SIAM J. Num. Anal. 10,* 839-848.

Barrodale, I., and F. D. K. Roberts (1974), Algorithm 478: Solution of an Overdetermined System of Equations in the ℓ_1 Norm, *CACM 17,* 319-320.

Barrodale, I., and F. D. K. Roberts (1977a), An Efficient Algorithm for Discrete ℓ_1 Linear Approximation with Linear Constraints, *SIAM J. Num. Anal.,* (To appear).

Barrodale, I., and F. D. K. Roberts (1977b), *Solution of the Constrained ℓ_1 Linear Approximation Problem,* Technical Report DM-104-IR, Mathematics Department, University of Victoria, Victoria, British Columbia.

Barrodale, I., and F. D. K. Roberts (1977c), Algorithms for Restricted LAV Estimation, *Comm. Stat. (B) 6,* 353-364.

Barrodale, I., F. D. K. Roberts and C. R. Hunt (1970), Computing Best ℓ_p Approximations by Functions Nonlinear in One Parameter, *Comp. J. 13,* 382-386.

Barrodale I., and A. Young (1966), Algorithms for Best L_1 and L_∞ Linear Approximations on a Discrete Set, *Numer. Math. 8,* 295-306.

Bartels, R. (1976), Least Squares Regression as a Subtask in More General Types of Regression, *Proceedings of the Ninth Interface Symposium on Computer Science and Statistics*, edited by D. C. Hoaglin and R. E. Welsch, Prindle, Weber & Schmidt, Boston, 152-157.

Bartels, R. H., and A. R. Conn (1977), LAV Regression: A Special Case of Piecewise Linear Minimization, *Comm. Stat.(B) 6*, 329-340.

Bartels, R. H., and A. R. Conn (1978), Linearly Constrained Discrete L_1 Problems, *TOMS* (To appear).

Bartels, R. H., A. R. Conn, and C. Charalambous (1976), *Minimization Techniques for Piecewise Differentiable Functions: The L_∞ Solution to an Overdetermined Linear System,* Technical Report no. 247, John Hopkins University, Baltimore.

Bartels, R. H., A. R. Conn and J. W. Sinclair (1976), *A FORTRAN Program for Solving Overdetermined Systems of Linear Equations in the L_1 Sense,* Technical Report 236, Department of Mathematical Science, John Hopkins University, Baltimore.

Bartels, R. H., A. R. Conn, and J. W. Sinclair (1978), Minimization Techniques for Piecewise Differentiable Functions: The L_1 Solution to an Overdetermined Linear System, *SIAM J. Num. Anal. 15,* 224-241.

Bartels, R. H., and G. H. Golub (1968a), Stable Numerical Methods for Obtaining the Chebyshev Solution to an Overdetermined System of Equations, *CACM 11,* 401-406.

Bartels, R. H., and G. H. Golub (1968b), Chebyshev Solution to an Overdetermined Linear System, *CACM 11,* 428-430.

Bartels, R. H., and G. H. Golub (1969), The Simplex Method of Linear Programming Using LU Decomposition, *CACM 12,* 266-268.

Bauer, D. F. (1973), On Some Nonparametric Estimates for Regression Parameters, *Comm. Stat. 2,* 225-234.

Boggs, P. T. (1974), A New Algorithm for the Chebyshev Solution of Overdetermined Linear Systems, *Math. Comp. 28,* 203-218.

Burke, O. (1975a), Least Squares Estimators as Robust and Minimax Estimators, *Mathematische Operationsforschung und Statistik 6,* 687-688.

Burke, O. (1975b), Minimax Linear, Ridge, and Shrunken Estimators for Linear Parameters, *Mathematische Operationsforschung und statistik 6,* 697-701.

Chambers, J. M. (1977), *Computational Methods for Data Analysis,* Wiley, New York.

Charnes, A., W. W. Cooper, and R. O. Ferguson (1955), Optimal Estimation of Executive Compensation by Linear Programming, *Management Science 1,* 138-150.

Cheney, E. W. (1974), *A Bibliography for Approximation Theory, 1970-1974,* Technical Report CNA-94, Center for Numerical Analysis, University of Texas, Austin.

Claerbout, J. F., and F. Muir (1973), Robust Modeling with Erratic Data, *Geophysics 38,* 826-844.

Cline, A. K. (1972), Rate of Convergence of Lawson's Algorithm, *Math. Comp. 26,* 167-176.

Conniffe, D., and J. Stone (1973), A Critical View of Ridge Regression, *The American Statistician 22,* 181-187.

Crocker, D. C. (1969), Linear Programming Techniques in Regression Analysis: The Hidden Danger, *AIEE TRANS. 1,* 112-26.

Davies, M. (1967), Linear Approximation Using the Criterion of Least Total Deviations, *JRSS (B) 29,* 101-109.

Dennis, J. E., and R. E. Welsch (1976), Techniques for Nonlinear Least Squares and Robust Regression, *Proceedings of the Statistical Computing Section, American Statistical Association,* Washington, D.C., 83-87.

Descloux, J. (1963), Approximation in L^p and Chebyshev Approximations, *SIAM J. 11,* 1017-1026.

Duris, C. S., and V. P. Sreedharan (1968), Chebyshev and ℓ_1 Solutions of Linear Equations Using Least Squares Solutions, *SIAM J. Num. Anal. 5,* 491-505.

Duris, C. S., and M. G. Temple, (1973), A Finite Step Algorithm for Determining the "Strict" Chebyshev Solution to Ax = b, *SIAM J. Num. Anal. 10,* 690-699.

Dutter, R. (1975a), *Robust Regression: Different Approaches to Numerical Solutions and Algorithms,* Research Report no. 6, Fachgruppe für Statistik, ETH, Zurich.

Dutter, R. (1975b), *Numerical Solution of Robust Regression Problems: Computational Aspects, a Comparison,* Research Report no. 7, Fachgruppe für Statistik, ETH, Zurich.

Dutter, R. (1977), Numerical Solution of Robust Regression Problems: Computational Aspects, a Comparison, *JSCS 5,* 207-238.

Ekblom, H. (1973), Calculation of Linear Best L_p-Approximations, *BIT 13,* 292-300.

Ekblom, H. (1974), L_p-methods for Robust Regression, *BIT 14,* 22-32.

Ekblom, H., and S. Henriksson (1969), L_p-criteria for the Estimation of Location Parameters, *SIAM J. Appl. Math. 17,* 1130-1141.

Fair, R. C. (1974), On the Robust Estimation of Econometric Models, *Ann. Econ. and Soc. Measurement 3,* 667-677.

Fisher, W. D. (1961), A Note on Curve Fitting with Minimum Deviations by Linear Programming, *JASA 56,* 359-362.

Fletcher, R., J. A. Grant, and M. D. Hebden (1971), The Calculation of Linear Best L_p Approximations, *Comp. J. 14,* 276-279.

Fletcher, R., J. A. Grant and M. D. Hebden (1974a), Linear Minimax Approximation as the Limit of Best L_p Approximation, *SIAM J. Num. Anal. 11,* 123-136.

Fletcher, R., J. A. Grant and M. D. Hebden (1974b), The Continuity and Differentiability of the Parameters of Best Linear L_p Approximations, *J. Approximation Theory 10,* 69-73.

Fletcher, R., and M. J. D. Powell (1963), A Rapidly Convergent Descent Method for Minimization, *Comp. J. 6,* 163-168.

Forsythe, A. B. (1972), Robust Estimation of Straight Line Regression Coefficients by Minimizing pth Power Deviations, *Techno. 14,* 159-166.

Gentle, J. E. (1977), Least Absolute Values Estimation: An Introduction, *Comm. Stat.(B) 6,* 313-328.

Gentle, J. E., and T. A. Hanson (1977), Variable Selection Under L_1, *Proceedings of the Statistical Computing Section, American Statistical Association,* Washington, D.C., 228-230.

Gentle, J. E., W. J. Kennedy, and V. A. Sposito (1976), Properties of the L_1-estimate Space, *Proceedings of the Statistical Computing Section, American Statistical Association,* Washington, D.C., 163-164.

Gentle, J. E., W. J. Kennedy and V. A. Sposito (1977a), On Least Absolute Deviations Estimation, *Comm. Stat.(A) 6,* 839-845.

Gentle, J. E., W. J. Kennedy and V. A. Sposito (1977b), On Properties of L_1 Estimators, *Math. Programming 12,* 138-140.

Gentleman, W. M. (1965), *Robust Estimation of Multivariate Location by Minimizing pth Power Deviations,* Ph.D. thesis, Princeton University, Princeton, N.J.

Gilsinn, J., K. Hoffman, R. H. F. Jackson, E. Leyendecker, P. Saunders, and D. Shier (1977), Methodology and Analysis for Comparing Discrete Linear L_1 Approximation Codes, *Comm. Stat. (B),* 399-414.

Glahe, F. R., and J. G. Hunt (1970), The Small Sample Properties of Simultaneous Equation Least Absolute Estimators Vis-a-vis Least Squares Estimators, *Econometrica 38,* 742-753.

Goldstein, A. A., and W. Cheney (1958), A Finite Algorithm for the Solution of Consistent Linear Equations and Inequalities and for the Tchebycheff Approximation of Inconsistent Linear Equations, *Pacific J. Math. 8,* 415-427.

Hanson, T. A. (1977), *Algorithm for Best Subset Selection Under L_1,* Technical Report, Statistics Laboratory, Iowa State University, Ames.

Harter, H. L. (1974a), The Method of Least Squares and Some Alternatives, I., *Int. Statist. Rev. 42,* 147-174.

Harter, H. L. (1974b), The Method of Least Squares and Some Alternatives, II, *Int. Statist. Rev. 42,* 235-264.

Harter, H. L. (1975a), The Method of Least Squares and Some Alternatives, III, *Int. Statist. Rev. 43,* 1-44.

Harter, H. L. (1975b), The Method of Least Squares and Some Alternatives, IV, *Int. Statist. Rev. 43,* 125-190 and 273-278.

Harter, H. L. (1975c), The Method of Least Squares and Some Alternatives, V, *Int. Statist. Rev. 43,* 269-272.

Harter, H. L. (1976), The Method of Least Squares and Some Alternatives, VI, *Int. Statist. Rev. 44,* 113-159.

Harter, H. L. (1977), Nonuniqueness of Least Absolute Values Regression, *Comm. Stat. (A) 6,* 829-838.

Hebden, M. D. (1969), *Topics in Linear Approximation Theory with Special Reference to the Chebyshev Norm,* Ph.D. thesis, University of Leeds, Leeds, England.

Hemmerle, W. J. (1975), An Explicit Solution for Generalized Ridge Regression, *Techno. 17,* 309-314.

Henriksson, S. (1972), *On a Generalization of* L_p*-Approximation and Estimation,* thesis, Lund University, Lund, Sweden.

Hettmansperger, T. P., and J. W. McKean (1976), Computational Problems Involved in Analysis of Linear Models Based on Ranks, *Proceedings of the Statistical Computing Section, American Statistical Association,* Washington, D.C., 88-94.

Hill, R. W., and P. W. Holland (1977), Two Robust Alternatives to Least-Squares Regression, *JASA 72,* 828-833.

Hinich, M. J., and P. P. Talwar (1975), A Simple Method for Robust Regression, *JASA 70,* 113-119.

Hoerl, A. E., and R. W. Kennard (1970a), Ridge Regression: Biased Estimation for Nonorthogonal Problems, *Techno. 12,* 55-67.

Hoerl, A. E., and R. W. Kennard (1970b), Ridge Regression: Applications to Nonorthogonal Problems, *Techno. 12,* 68-82.

Hoerl, A. E., and R. W. Kennard (1976), Ridge Regression: Iterative Estimation of the Biasing Parameter, *Comm. Stat. (A) 5,* 77-88.

Hogg, R. V. (1974), Adaptive Robust Procedures: A Partial Review and Some Suggestions for Future Applications and Theory, *JASA 69,* 909-923.

Holland, P. W. (1976), Robust Test Problems for Robust Regression Problems (with Discussion), *Proceedings of the Ninth Interface Symposium on Computer Science and Statistics,* edited by D. C. Hoaglin and R. E. Welsch, Prindle, Weber & Schmidt, Boston, 99-110.

Holland, P. W., and R. E. Welsch (1977), Robust Regression Using Iteratively Reweighted Least Squares, *Comm. Stat. A6,* 813-828.

Hopper, M. J., and M. J. D. Powell (1977), A Technique that Gains Speed and Accuracy in the Minimax Solution of Overdetermined Linear Equations, in *Mathematical Software,* vol. 3, edited by J. R. Rice, Academic Press, New York,

Housner, G. W., and J. F. Brennan (1948), The Estimation of Linear Trends, *AMS 19,* 380-388.

Huber, P. J. (1972), Robust Statistics: A Review, *AMS 43,* 1041-1067.

Huber, P. J. (1973), Robust Regression: Asymptotics, Conjectures, and Monte Carlo, *Ann. Stat. 1,* 799-821.

Huber, P. J., and R. Dutter (1974), Numerical Solution of Robust Regression Problems, *COMPSTAT 1974: Proceedings in Computational Statistics,* edited by G. Bruckmann, F. Ferschl, and L. Schmetter, Physica Verlag, Vienna, 165-172.

IBM Corporation (1968), *IBM System/360 Scientific Subroutine Package,* IBM Systems Reference Library, White Plains, N.Y.

Jaeckel, L. A. (1972), Estimating Regression Coefficients by Minimizing the Dispersion of the Residuals, *AMS 43,* 1449-1458.

Jureckova, J. (1971), Nonparametric Estimate of Regression Coefficients, *AMS 42,* 1328-1338.

Kahng, S. W. (1972), Best L_p Approximations, *Math. Comp. 26,* 505-508.

Karlovitz, L. A. (1970), An Algorithm for Best L_p Approximation, *J. Approximation Theory 3,* 123-127.

Karst, O. J. (1958), Linear Curve Fitting Using Least Deviations, *JASA 53,* 118-132.

Kelley, J. E. (1958), An Application of Linear Programming to Curve Fitting, *SIAM J. Appl. Math. 6,* 15-22.

Kennedy, W. J., and J. E. Gentle (1977), Examining Rounding Error in LAV Regression Computations, *Comm. Stat. (B) 6,* 415-420.

Kennedy, W. J., and J. E. Gentle (1978), Comparisons of Algorithms for Minimum L_p Norm Linear Regression, *Proceedings of Computer Science and Statistics: Tenth Annual Symposium on the Interface,* edited by D. Hogben, U.S. Government Printing Office, Washington, D.C., 373-378.

Kennedy, W. J., J. E. Gentle and V. A. Sposito (1977), A Computer Oriented Method for Generating Test Problems for L_1 Regression, *Comm. Stat. (B) 6,* 21-27.

Kiountouzis, E. A. (1971), Optimal L_p Approximation Techniques and Data Analysis, *Bull. Soc. Math Greece 12,* 191-206.

Kiountouzis, E. A. (1973), Linear Programming Techniques in Regression Analysis, *Appl. Stat. 22,* 69-73.

Knuth, D. E. (1973), *The Art of Computer Programming*, vol. 3, Addison-Wesley, Reading, Mass.

Kripke, B. R., and T. J. Rivlin (1965), Approximation in the Metric of L^1 (X,u), *Trans. Amer. Math. Soc. 119*, 191-122.

Lauter, H. (1975), A Minimax Linear Estimator for Linear Parameters under Restrictions in the Form of Inequalities, *Mathematische Operationsforschung und Statistik 5*, 689-695.

McCormick, G. F., and V. A. Sposito (1975), A Note on L_1 Estimation Based on the Median Positive Quotient, *Appl. Stat. 24*, 347-350.

McCormick, G. F., and V. A. Sposito (1976), Using the L_2-estimator in L_1-estimation, *SIAM J. Num. Anal. 13*, 337-343.

Mandelbrot, B. (1963), The Variation of Certain Speculative Prices, *J. Business 36*, 394-419.

Marquardt, D. W. (1970), Generalized Inverses, Ridge Regression, Biased Linear Estimation, and Nonlinear Estimation, *Techno. 12*, 591-612.

Marsaglia, G. (1970), One-sided Approximations by Linear Combination of Functions, in *Approximation Theory*, edited by A. Talbot, Academic Press, New York.

Meicler, M. (1968), Chebyshev Solution of an Inconsistent System of n + 1 Linear Equations in n Unknowns in Terms of Its Least Squares Solution, *SIAM Rev. 10*, 373-375.

Meicler, M. (1968), A Steepest Ascent Method for the Chebyshev Problem, *Math. Comp. 23*, 813-817.

Merle, G., and H. Spath (1974), Computational Experiences with Discrete L_p Approximation, *Computing 12*, 315-321.

Micchelli, C. (1974), A Remark on an Algorithm for Best L_p Approximation *J. Approximation Theory 11*, 283-284.

Moursund, D. (1965), Chebyshev Solution of n + 1 Linear Equations in n Unknowns, *JACM 12*, 383-387.

Narula, S. C., and J. F. Wellington (1977a), Algorithm AS108: Multiple Linear Regression with Minimum Sum of Absolute Errors, *Appl. Stat. 26*, 106-111.

Narula, S. C., and J. F. Wellington (1977b), *Subset Selection with Minimum Sum of Weighted Absolute Errors*, Technical Report no. 37-77-P10, School of Management, Rensselaer Polytechnic Institute, Troy, New York.

Narula, S. C., and J. F. Wellington (1977c), An Algorithm for the Minimum Sum of Weighted Absolute Error Regression, *Comm. Stat. (B) 6*, 341-352.

Nather, W. (1972), Minimaxschatzungen in Linearen Regressions-Modell, *Mathmatische Operationsforschung und Statistik 3*, 475-482.

National Bureau of Economics Research (1977), *ROSEPACK,* NBER Computer Research Center, Cambridge, Mass.

Osborne, M. R. (1972), A Class of Methods for Minimizing a Sum of Squares, *Aust. Computer J. 4,* 164-169.

Osborne, M. R., and G. A. Watson (1967), On the Best Linear Chebyshev Approximation, *Comp. J. 10,* 172-177.

Osborne, M. R., and G. A. Watson (1969), An Algorithm for Minimax Approximation in the Nonlinear Case, *Comp. J. 12,* 63-68.

Osborne, M. R., and G. A. Watson (1971), On an Algorithm for Discrete Nonlinear L_1 Approximation, *Comp. J. 14,* 184-188.

Peters, S. C., V. C. Klema, and P. Holland (1978), Software for Iteratively Reweighted Least Squares, *Proceedings of the Computer Science and Statistics: Eleventh Annual Symposium on the Interface,* edited by A. R. Gallant and T. Gerig, Institute of Statistics, North Carolina State University, Raleigh, 380-384.

Pfaffenberger, R. C., and J. J. Dinkel (1978), Absolute Deviations Curve Fitting: An Alternative to Least Squares, in *Contributions to Survey Sampling and Applied Statistics,* edited by H. A. David, Academic Press, New York, 279-294.

Powell, M. J. D. (1974), The Minimax Solution of Linear Equations Subject to Bounds on the Variables, *Proceedings of Fourth Manitoba Conference on Numerical Mathematics,* edited by H. C. Williams and B. L. Hartnell, Utilitas Mathematica Publishing, Winnipeg, Canada.

Rabinowitz, P. (1968), Applications of Linear Programming to Numerical Analysis, *SIAM Rev. 10,* 121-159.

Rao, M. R., and V. Shrinivasan (1972), A Note on Sharpe's Algorithm for Minimizing the Sum of Absolute Deviations in a Simple Regression Problem, *Management Sci. 19,* 222-225.

Relles, D. A. (1968), Robust Regression by Modified Least-Squares, Ph.D. thesis, Yale University, New Haven, Conn.

Rey, W. (1974), Robust Estimates of Quantiles, Location, and Scale in Time Series, *Philips Research Reports 29,* 67-92.

Rey, W. (1975), On Least pth Power Methods in Multiple Regressions and Location Estimations, *BIT 15,* 174-185.

Rice, J. R. (1962), Tchebycheff Approximation in a Compact Metric Space, *Bulletin Amer. Math. Society 68,* 405-410.

Rice, J. R. (1964), *The Approximation of Functions, vol. 1: Linear Theory,* Addison-Wesley, Reading, Mass.

Rice, J. R., and J. S. White (1964), Norms for Smoothing and Estimation, *SIAM Rev. 6,* 243-256.

Robers, P. D., and A. Ben-Israel (1969), An Interval Programming Algorithm for Discrete Linear L_1 Approximation Problems, *J. Approx. Theory 2,* 323-336.

Robers, P. D., and S. S. Robers (1973), Algorithm 458: Discrete Linear L_1 Approximation by Interval Linear Programming, *CACM 16,* 629-631

Roodman, G. (1974), A Procedure for Optimal Stepwise MSAE Regression Analysis, *Operat. Res. 22,* 393-399.

Ross, G. J. S. (1978), MLP: A Maximum Likelihood Program, *Proceedings of Computer Science and Statistics: Tenth Annual Symposium on the Interface,* edited by D. Hogben, U.S. Government Printing Office, Washington, D.C., 87-91.

Sadovski, A. N. (1974), Algorithm AS74: L_1-norm Fit of a Straight Line, *Appl. Stat. 23,* 244-248.

Schlossmacher, E. J. (1973), An Iterative Technique for Absolute Deviations Curve Fitting, *JASA 68,* 857-859.

Sen, P. K., and M. L. Puri (1969), On Robust Nonparametric Estimation in Some Multivariate Linear Models, in *Multivariate Analysis, vol. 2,* edited by P. R. Krishnaian, Academic Press, New York, 33-52.

Sharpe, W. F. (1971), Mean-Absolute-Deviation Characteristic Lines for Securities and Portfolios, *Management Sci. 18,* B1-B13.

Sielken, R. L. and H. O. Hartley (1973), Two Linear Programming Algorithms for Unbiased Estimation of Linear Models, *JASA 68,* 639-641.

Singleton, R. R. (1940), A Method of Minimizing the Sum of Absolute Values of Deviations, *AMS 11,* 301-310.

Smith, A. F. M., and M. Goldstein (1975), Ridge Regression: Some Comments on a Paper of Conniffe and Stone, *Statistician 24,* 61-66.

Sposito, V. A. (1975), *Linear and Nonlinear Programming,* Iowa State University Press, Ames.

Sposito, V. A., W. J. Kennedy and J. E. Gentle (1977), Algorithm AS110: L_p Norm Fit of a Straight Line, *Appl. Stat. 26,* 114-118.

Sposito, V. A., and W. C. Smith (1976), On a Sufficient and a Necessary Condition for L_1 Estimation, *Appl. Stat. 25,* 154-157.

Spyropoulos, K., E. Kiountouzis, and A. Young (1973), Discrete Approximation in the L_1 Norm, *Comp. J. 16,* 180-186.

Stiefel, E. (1960), Note on Jordan Elimination, Linear Programming, and Tchebycheff Approximation, *Numer. Math. 2,* 1-17.

Tewarson, R. P. (1972), On Minimax Solutions of Linear Equations, *Comp. J. 15,* 277-279.

Toutenburg, H. (1975), Minimax-Linear Estimation (MMLE) and Two-phase MMLE in a Restricted Linear Regression Model, *Mathematische Operationsforschung und Statistik 5,* 703-708.

Tukey, J. W. (1962), The Future of Data Analysis, *AMS 33,* 1-67.

Usow, K. H. (1967), On L_1 Approximation II: Computation for Discrete Functions and Discretization Effects, *SIAM J. Num. Anal. 4*, 233-244.

Veidinger, L. (1960), On the Numerical Determination of the Best Approximation in the Chebyshev Sense, *Numer. Math. 2*, 1-17.

Wagner, H. M. (1959), Linear Programming Techniques for Regression Analysis, *JASA 54*, 206-212.

Wagner, H. M. (1962), Nonlinear Regression with Minimal Assumptions, *JASA 57*, 572-578.

Walsh, J. L. (1959), On Extremal Approximations, in *On Numerical Approximation*, edited by R. E. Langer, University of Wisconsin Press, Madison.

Watson, G. A. (1973a), The Calculation of Best Linear One-sided L_p Approximations, *Math. Comp. 27*, 607-620.

Watson, G. A. (1973b), On the Best Linear Chebyshev Approximation, *J. Approximation Theory 7*, 48-58.

Welsch, R. E. (1975), Confidence Intervals for Robust Regression, *Proceedings of the Statistical Computing Section, American Statistical Association*, Washington, D.C., 36-42.

Welsch, R. E. (1978), Nonlinear Statistical Data Analysis, *Proceedings of Computer Science and Statistics: Tenth Annual Symposium on the Interface*, edited by D. Hogben, U.S. Government Printing Office, Washington, D.C., 77-86.

Welsch, R. E., and R. A. Becker (1975), Robust Nonlinear Regression Using the DOGLEG Algorithm, *Proceedings of Computer Science and Statistics: Eighth Annual Symposium on the Interface*, edited by J. W. Frane, Health Sciences Computing Facility, UCLA, Los Angeles, 272-279.

Yohai, V. J. (1974), Robust Estimation in the Linear Model, *Ann. Stat. 2*, 562-567.

12 / SELECTED MULTIVARIATE METHODS

12.1 INTRODUCTION

This chapter is intended to provide a short introduction to computational methods in multivariate analysis. There are many different multivariate methods used in various situations, so an exhaustive treatment of available computational methods would require several chapters of written text. However, a firm foundation for understanding these methods has already been established in preceding chapters. Many of the algorithms employed when solving multivariate computational problems are rather straightforward extensions of well-known algorithms used for other applications. The computational methods discussed in subsequent sections in this chapter are not only important in their own right, but they also serve to provide the flavor of extending and combining univariate-type algorithms to meet the needs in multivariate applications.

12.2 CANONICAL CORRELATIONS

The method of canonical correlations was proposed by Hotelling (1935, 1936) as a technique for studying linear relationships between two sets of variates. To discuss canonical correlations we consider the multivariate normal p vector $\underline{X}$ with nonsingular variance-covariance matrix Σ (sometimes the term *covariance matrix* is used to describe Σ). We partition $\underline{X}$ into the subvectors $\underline{X}_1$ and $\underline{X}_2$ of length t and $q = p - t$, respectively. For convenience we assume $t \leq q$. The variance-covariance matrix Σ is partitioned into the corresponding submatrices Σ_{11}, Σ_{12}, and Σ_{22}. The question of linear relationship between $\underline{X}_1$ and $\underline{X}_2$ may be explored by considering correlations between linear compounds u_i and v_i of $\underline{X}_1$ and $\underline{X}_2$, respectively; that is, $u_i = \underline{\alpha}_i'\underline{X}_1$ and $v_i = \underline{\beta}_i'\underline{X}_2$. The following well-known theorem (see, e.g., Kshirsagar 1972) shows that there is a natural set of linear compounds to consider.

Theorem 12.1

For $\underline{X}_1$ and $\underline{X}_2$ as above, there exists a transformation

$$\underline{u} = A'\underline{X}_1 \quad \text{and} \quad \underline{v} = B'\underline{X}_2$$

from $\underline{X}_1$ and $\underline{X}_2$ to $\underline{u}$ and $\underline{v}$ (A and B are square matrices) such that

$$\text{var}\begin{pmatrix} \underline{u} \\ \underline{v} \end{pmatrix} = \left(\begin{array}{c|c} I_t & P \\ \hline P' & I_q \end{array}\right)$$

where

$$P = [\text{diag}(\rho_1, \rho_2, \ldots, \rho_s, 0, \ldots, 0)|0]$$

$\rho_1 \geq \rho_2 \geq \cdots \rho_s > 0$, and s is the rank of Σ_{12}. Moreover, ρ_1 is the maximum correlation between any two linear compounds of $\underline{X}_1$ and $\underline{X}_2$; ρ_2 is the maximum correlation between any two compounds independent of the first two, that is, $\underline{\alpha}_1'\underline{X}_1$ and $\underline{\beta}_1'\underline{X}_2$; and so on.

The elements of $\underline{u}$ are called the *canonical variables* of the $\underline{X}_1$ space, and those of $\underline{v}$ are called the canonical variables of the $\underline{X}_2$ space. The columns of A and B are called *canonical vectors*. The quantity ρ_i is the ith *canonical correlation*.

When $t = q = 1$, the canonical correlation is the ordinary correlation between $\underline{X}_1$ and $\underline{X}_2$; and when $t = 1$ and $q > 1$, the (single) canonical correlation is the multiple correlation coefficient of $\underline{X}_1$ regressed on $\underline{X}_2$.

For each i $(i = 1, 2, \ldots, s)$, the quantities of Theorem 12.1 can be shown to satisfy the system

$$\begin{pmatrix} -\rho_i\Sigma_{11} & \Sigma_{12} \\ \Sigma'_{12} & -\rho_i\Sigma_{22} \end{pmatrix} \begin{pmatrix} \underline{\alpha}_i \\ \underline{\beta}_i \end{pmatrix} = \underline{0} \qquad (12.1)$$

where $\underline{\alpha}_i$ and $\underline{\beta}_i$ are the ith columns of A and B, respectively. Equation (12.1) may equivalently be written as

$$\Sigma_{12}\Sigma_{22}^{-1}\Sigma'_{12}\underline{\alpha}_i = \rho_i^{\ 2}\Sigma_{11}\underline{\alpha}_i$$

and

$$\underline{\beta}_i = \frac{1}{\rho_i}\Sigma_{22}^{-1}\Sigma'_{12}\underline{\alpha}_i \qquad (12.2)$$

By substituting the sample corrected sums of squares and cross-products matrices S_{11}, S_{12}, and S_{22} for Σ_{11}, Σ_{12}, and Σ_{22} in the discussion above, we can define the sample canonical variables and canonical correlations. Letting $\underline{X}_{11}, \underline{X}_{12}, \ldots, \underline{X}_{1n}$ and $\underline{X}_{21}, \underline{X}_{22}, \ldots, \underline{X}_{2n}$ be random samples of the random vectors $\underline{X}_1$ and $\underline{X}_2$, of length t and q, respectively, with n > t, q, and

$$\overline{\underline{X}}_1 = \frac{1}{n}\sum_{i=1}^{n}\underline{X}_{1i} \qquad \overline{\underline{X}}_2 = \frac{1}{n}\sum_{i=1}^{n}\underline{X}_{2i}$$

$$Z_1 = \begin{pmatrix} \underline{X}'_{11} - \overline{\underline{X}}'_1 \\ \underline{X}'_{12} - \overline{\underline{X}}'_1 \\ \vdots \\ \underline{X}'_{1n} - \overline{\underline{X}}'_1 \end{pmatrix} \qquad Z_2 = \begin{pmatrix} \underline{X}'_{21} - \overline{\underline{X}}'_2 \\ \underline{X}'_{22} - \overline{\underline{X}}'_2 \\ \vdots \\ \underline{X}'_{2n} - \overline{\underline{X}}'_2 \end{pmatrix}$$

we form the matrices of corrected sums of squares and cross-products

$$S_{11} = \frac{Z'_1 Z_1}{n - 1}$$

$$S_{12} = \frac{Z'_1 Z_2}{n - 1}$$

$$S_{22} = \frac{Z'_2 Z_2}{n - 1}$$

The *sample canonical vectors* and *correlations* are defined to be solutions to equations similar to (12.2),

$$S_{12}S_{22}^{-1}S'_{12}\underline{a}_i = r_i^2 S_{11}\underline{a}_i$$

$$\underline{b}_i = \frac{1}{r_i} S_{22}^{-1}S'_{12}\underline{a}_i \tag{12.3}$$

with $r_1 \geq r_2 \geq \cdots \geq r_t$. If Σ_{11} and Σ_{22} are of full rank, with probability one S_{11} and S_{22} are of full rank and each $r_i > 0$. For the present, let us assume that S_{11} and S_{22} are of full rank (i.e., Z_1 and Z_2 are of full rank). The system (12.3) is equivalent to the system

$$\begin{pmatrix} 0 & S_{12} \\ S'_{12} & 0 \end{pmatrix} \begin{pmatrix} \underline{a}_i \\ \underline{b}_i \end{pmatrix} = r_i \begin{pmatrix} S_{11} & 0 \\ 0 & S_{22} \end{pmatrix} \begin{pmatrix} \underline{a}_i \\ \underline{b}_i \end{pmatrix} \tag{12.4}$$

Olkin (1951) and Golub (1969), using orthogonal factorizations of Z_1 and Z_2,

$$Z_1 = Q_1R_1$$

$$Z_2 = Q_2R_2$$

where $Q_1'Q_1 = I_t$, $Q_2'Q_2 = I_q$, and R_1 and R_2 are upper-triangular (see Section 7.2), showed (12.4) can be written as

$$\begin{pmatrix} 0 & \Omega \\ \Omega' & 0 \end{pmatrix} \begin{pmatrix} \underline{\xi}_i \\ \underline{\eta}_i \end{pmatrix} = r_i \begin{pmatrix} \underline{\xi}_i \\ \underline{\eta}_i \end{pmatrix} \tag{12.5}$$

where $\Omega = (R_1')^{-1}S_{12}R_2^{-1}$, $\underline{\xi}_i = R_1\underline{a}_i$, and $\underline{\eta}_i = R_2\underline{b}_i$. The matrix Ω can be written more simply as $\Omega = Q_1'Q_2$. Equation (12.5) shows that the r_i are just the singular values of Ω; and $\underline{\xi}_i$ and $\underline{\eta}_i$ are the columns of U and V in the singular-value decomposition of Ω (see Section 7.2.4),

$$\Omega = U \Lambda V'$$

The vectors $\underline{a}_i$ and $\underline{b}_i$ are easily recovered from $\underline{\xi}_i$ and $\underline{\eta}_i$, since R_1 and R_2 are triangular matrices.

In summary, the procedure for obtaining sample canonical vectors and correlations from samples $\underline{X}_{11}, \underline{X}_{12}, \ldots, \underline{X}_{1n}$ and $\underline{X}_{21}, \underline{X}_{22}, \ldots, \underline{X}_{2n}$ is as follows:

1. Form the adjusted data matrices Z_1 and Z_2.
2. Factor Z_1 into Q_1R_1 and Z_2 into Q_2R_2.
3. Form $Q_1'Q_2$ and obtain its singular-value decomposition $U \Lambda V'$.
4. Obtain $\underline{a}_i$ and $\underline{b}_i$, using R_1^{-1}, R_2^{-1}, U, and V.

An alternate computational procedure may be formulated directly from Eq. (12.3) and the Cholesky decomposition (see Section 7.4) of S_{11}^{-1} and S_{22}^{-1}. Horst (1961) described such an algorithm for the canonical analysis between two random vectors as discussed in this section, and also considered the extension of canonical analysis to several random vectors simultaneously.

If Z_1 or Z_2 is not of full rank, the matrices R_1 and S_{11} or R_2 and S_{22} used above would have to be modified slightly. The modification consists simply of reducing the dimensionality. In the case of R_1 or R_2, this is done by redefining R_1 or R_2 to be the largest upper-triangular matrix of full rank in the factorization $Z_1 = Q_1R_1$ or $Z_2 = Q_2R_2$. The Cholesky decomposition of S_{11}^{-1} or S_{22}^{-1} in the second method mentioned above is, of course, actually performed on S_{11} and S_{22}, and the inversion is then made. If S_{11} or S_{22} is singular, the Cholesky decomposition will uncover this fact, and the procedure is modified so that the matrix actually decomposed is a nonsingular submatrix of largest order (see Section 7.4 and Cohen 1969).

In computing applications the question of singularity of matrices is not a simple one. In the applications considered here, small values relative to the precision of the computer arithmetic may be taken as indications of singularity, and a number of the canonical correlations may be set to zero.

If sample correlation matrices are used in Eq. (12.3), instead of the sample covariance matrices S_{11}, S_{12}, and S_{22}, the same canonical correlations, R_i, will be obtained. The canonical vectors, however, will be different; and the canonical variables must be computed after standardizing the original data with the reciprocal of the sample standard deviations.

12.3 PRINCIPAL COMPONENTS

The principal components of a random vector $\underline{X}$ are the elements of an orthogonal transformation of $\underline{X}$ which have zero correlations. Such transformations were originally considered by Galton, Edgeworth, and Karl Pearson around the turn of the century; and the concept was developed essentially in its present-day form by Hotelling (1933).

The population principal components are defined as follows. Let $\underline{X}$ be a random p vector with variance-covariance matrix Σ. Let Γ be a p x p orthogonal matrix such that $\Sigma = \Gamma\Delta\Gamma'$, where $\Delta = \text{diag}(\delta_1, \delta_2, \ldots, \delta_p)$ and $\delta_1 \geq \delta_2 \geq \cdots \geq \delta_p > 0$ are the eigenvalues of Σ. Let $\underline{V} = \Gamma'\underline{X}$. Then the elements $v_1, v_2, \ldots, v_p$ of $\underline{V}$ are called, respectively, the first principal component of $\underline{X}$, the second principal component, and so on. If any of the eigenvalues of Σ are repeated, Γ is not uniquely determined; also it may be the case that some of the eigenvalues are zero. Neither of these situations presents any special problems. (In the case of, say, p - q zero eigenvalues, instead of the p x p orthogonal matrix Γ, we form a p x p matrix Γ^* consisting of a q x q orthogonal matrix in the upper left and p - q zero rows and columns.)

It follows immediately from the relation $\underline{V} = \Gamma'\underline{X}$ that the variances of the principal components, the v_i's, are the δ_i's. Also, it is clear that both the trace and the determinant of the variance-covariance matrix of $\underline{V}$ are the same as the trace and determinant of Σ; so that by either of the usual measures of "total variation" of a random vector, both $\underline{V}$ and $\underline{X}$ have the same overall variability.

Principal component analysis involves the use of the sample analogue of Σ in the discussion above. For a random sample of size n, $\underline{X}_1, \underline{X}_2, \ldots, \underline{X}_n$, with

$$\overline{\underline{X}} = \frac{1}{n} \sum_{i=1}^{n} \underline{X}_i$$

and

$$Z = \begin{pmatrix} \underline{X}_1' & - \overline{\underline{X}}' \\ \underline{X}_2' & - \overline{\underline{X}}' \\ & \vdots \\ \underline{X}_n' & - \overline{\underline{X}}' \end{pmatrix}$$

the analogue of Σ is S, given by

$$S = \frac{Z'Z}{n - 1}$$

It is clear, however, from the discussion of canonical correlations in Section 12.2 and the discussion of Chapter 7, that it is not necessary to form S; the appropriate method for computing the principal components is by the singular-value decomposition of Z,

$$Z = U \Lambda V'$$

The first principal component is the first column of V; the second principal component is the second column; and so on. The sample variances are the squares of the diagonal elements of Λ divided by $n - 1$. This representation of principal components and the implied computational procedure were discussed by Whittle (1952) and later authors.

12.4 FACTOR ANALYSIS

Factor analysis is a frequently used technique in multivariate data analysis. One of the primary uses of this technique is to study the covariances or correlations among a large set of variables and define groups or clusters of variables as factors such that the dominant variables in each factor are highly correlated. In this section we will describe several algorithms that have been developed to support factor-analytic applications.

The usual model for factor analysis is

$$\underline{Y} = \Lambda\underline{X} + \underline{\varepsilon} \tag{12.6}$$

where $\underline{Y}$ is a random p vector of observations, $\underline{X}$ is a k vector of unobserved random *factors,* $\underline{\varepsilon}$ is a random error vector, and Λ is a p x k matrix of parameters, called *factor loadings,* to be estimated. It is assumed that $\underline{X}$ and $\underline{\varepsilon}$ are independent and that their expected values are zero,

$$E(\underline{X}) = E(\underline{\varepsilon}) = \underline{0}$$

Furthermore, the variance-covariance matrices of $\underline{Y}$, $\underline{X}$, and $\underline{\varepsilon}$ are assumed to be $V(\underline{Y}) = \Sigma$, $V(\underline{X}) = I$, and $V(\underline{\varepsilon}) = \Gamma^2$, where Γ^2 is diagonal.

Under these assumptions it follows that Σ is expressible in the form

$$\Sigma = \Lambda\Lambda' + \Gamma^2 \tag{12.7}$$

The ith diagonal element of Γ^2, denoted by γ_i^2, is called the *specific variance* of the ith response, and the diagonal elements of $\Lambda\Lambda'$ are called *communalities* of the responses.

The data that are used in factor analysis are p vector observations of $\underline{Y}$, which we denote by $\underline{Y}_1, \underline{Y}_2, \ldots, \underline{Y}_n$. The vectors $\underline{Y}_j$ can be used to form an n x p matrix $Z = (z_{ij})$ whose rows are the $\underline{Y}_i'$. Using Z, the sample variance-covariance matrix of the n responses can be computed as described in Chapter 8. Let $S = (s_{ij})$ denote this matrix of sample variances and covariances.

The main computational problem relative to factor analysis is (for given k) to estimate Λ and Γ^2 using the sample information available in the S matrix. This amounts to fitting Σ to S in such a way that Λ and Γ^2 are estimated. Normally, the choice of k is made by the research worker based on the experimental situation, and often several different values of k (i.e., numbers of factors) are tried in an effort to best explain the observed phenomenon in terms of unobserved underlying factors. Although sometimes the associated correlation matrix is used instead of S, for purposes of subsequent discussion we shall assume that S is used and that the inverse matrix has elements denoted by $S^{-1} = (s^{ij})$.

Several different criteria have been proposed for fitting Σ to S. The three most popular criteria are the following (see Timm 1975 and Joreskog 1977).

Unweighted least squares

The criterion is to minimize the nonlinear function

$$F_u(\Lambda,\Gamma) = \frac{\text{trace}(S-\Sigma)^2}{2} \tag{12.8}$$

with respect to Λ and Γ.

Generalized least squares

The criterion is to minimize

$$F_g(\Lambda,\Gamma) = \frac{\text{trace}(I-S^{-1}\Sigma)^2}{2} \tag{12.9}$$

Maximum likelihood

This criterion is to minimize

$$F_m(\Lambda,\Gamma) = \text{trace}(\Sigma^{-1}S) - \ln|\Sigma^{-1}S| - p \tag{12.10}$$

where $|\Sigma^{-1}S|$ denotes the determinant of the given matrix.

Discussion and derivations relative to these criteria are given by Joreskog (1967, 1972), Timm (1975), Lawley and Maxwell (1971), and Joreskog and Goldberger (1972). One problem which is immediately obvious is that Λ is not uniquely defined in (12.7) since, for example, an orthogonal transformation T can be applied to obtain $\Lambda_1 = \Lambda T$, and Σ is not altered when Λ_1 replaces Λ in (12.7). In order to obtain a formulation of the problem which includes specification of one from the set of candidate Λ matrices, it is usually assumed that $\Lambda'\Lambda$ is diagonal when criterion (12.8) is used and that $\Lambda'\Gamma^{-2}\Lambda$ is diagonal when either criterion (12.9) or (12.10) is selected. Once a solution matrix $\hat{\Lambda}$ is obtained, it is normally subjected to right transformations (called *rotations*) in an effort to obtain a form $\hat{\Lambda}_1 = \hat{\Lambda}T$ which is as "simple" as possible. Exactly what constitutes a simple form for $\hat{\Lambda}_1$ is a difficult question. Thurstone (1947) provides very reasonable criteria which are often used. We shall not consider the various computational methods for rotation; Cattell and Khanna (1977) give an extensive discussion of this subject.

The main computing problem under any one of the three criteria (12.8), (12.9) or (12.10) is to minimize a nonlinear function F of variables which are the elements of Λ and the γ_i^2 on the diagonal of Γ^2. The best among available algorithms utilize a two-stage iterative approach to solving this problem. The first stage consists of minimizing F with respect to Λ for some given Γ value. Letting $\tilde{\Lambda}$ denote the estimate of Λ which minimizes F given Γ, the function $F(\tilde{\Lambda},\Gamma)$ is then minimized with respect to Γ (this is the second-stage computation). Using the $\tilde{\Gamma}$ obtained as the solution to the second-stage minimization problem, stage 1 is repeated to produce a new $\tilde{\Lambda}$, and the two-stage iterative procedure continues thus until successive $\tilde{\Gamma}$'s agree within prescribed tolerance. The final solutions $\hat{\Lambda}$ and $\hat{\Gamma}$ are taken to be the $\tilde{\Lambda}$ and $\tilde{\Gamma}$ produced in the last iteration. The reason that minimization proceeds in two stages for each iteration is that the first-stage computation reduces to computing eigenvalues and vectors of a given symmetric matrix (a relatively straightforward numerical problem), so the second-stage minimization involves a function of a substantially smaller number of variables. Normally, Newton's method (see Section 10.2.2) is used in the second stage to find the $\tilde{\Gamma}$ which minimizes $F(\tilde{\Lambda},\Gamma)$.

Derivation of specific computationally oriented algebraic equations requires a substantial amount of mathematical manipulation. We will not give these details. The reader is referred to Joreskog (1977), Dixon and Brown (1977), and Joreskog and Van Thillo (1971) for details of the differentiation and associated simplification of resulting expressions that produces such equations. The following summary of these results is given relative to each of the criteria (12.8), (12.9), and (12.10). In each case, α_j (j = 1, ..., p) will denote the eigenvalues of the indicated matrix and $\underline{v}_j$ will denote corresponding orthonormal eigenvectors having components

$$\underline{v}_j' = (v_{1j}, v_{2j}, \ldots, v_{pj})$$

Unweighted Least Squares

$$F_u(\Lambda,\Gamma) = \frac{\text{trace}(S - \Sigma)^2}{2}$$

$\Lambda'\Lambda$ assumed to be diagonal

Stage 1 matrix: $A = S - \Gamma^2$

Eigenvalues of A: $\alpha_1 \geq \alpha_2 \geq \cdots \geq \alpha_p$

Orthonormal eigenvectors of A: $\underline{v}_1, \underline{v}_2, \ldots, \underline{v}_p$

Value of $\tilde{\Lambda}$ given Γ: $\tilde{\Lambda} = VD^{1/2}$

where $D = \text{diag}(\alpha_1, \alpha_2, \ldots, \alpha_k)$

$V = (\underline{v}_1, \underline{v}_2, \ldots, \underline{v}_k)$

Stage 2 function to be minimized: $F_u(\tilde{\Lambda},\Gamma) = \sum_{r=k+1}^{p} \frac{\alpha_r^2}{2}$

Derivatives needed for gradient vector and Hessian matrix elements:

$$\frac{\partial F_u(\tilde{\Lambda},\Gamma)}{\partial\gamma_i} = -2\gamma_i \sum_{r=k+1}^{p} \alpha_r v_{ir}^2$$

$$\frac{\partial^2 F_u(\tilde{\Lambda},\Gamma)}{\partial\gamma_i \partial\gamma_j} = 4\left[\gamma_i\gamma_j \sum_{r=k+1}^{p} v_{ir}v_{jr} \sum_{s=1}^{k} \frac{\alpha_r + \alpha_s}{\alpha_r - \alpha_s} v_{is}v_{js} + \delta_{ij} \sum_{r=k+1}^{p} \left(\gamma_i^2 - \frac{\alpha_r}{2}\right) v_{ir}^2\right]$$

(Note: δ_{ij} is the Kronecker delta.)

Starting values: $\gamma_i = \left[\dfrac{1 - k/2p}{s^{ii}}\right]^{1/2}$

when $S^{-1} = (s^{ij})$ exists. Otherwise

$$\gamma_i = \frac{s_{ii}}{2}$$

when S is not positive definite.

Generalized Least Squares

$$F_g(\Lambda,\Gamma) = \text{trace}(S^{-1}\Sigma - I)^2$$

$\Lambda'\Gamma^{-2}\Lambda$ assumed to be diagonal (S must be positive definite)

Stage 1 matrix: $A = \Gamma S^{-1}\Gamma \qquad A^{-1} = (a^{ij})$

Eigenvalues of A: $\alpha_1 \leq \alpha_2 \leq \cdots \leq \alpha_p$

Orthonormal eigenvectors of A: $\underline{v}_1, \underline{v}_2, \ldots, \underline{v}_p$

Value of $\tilde{\Lambda}$ given Γ: $\tilde{\Lambda} = \Gamma V(D^{-1} - I)^{1/2}$

where $D = \text{diag}(\alpha_1, \ldots, \alpha_k)$

$V = (\underline{v}_1, \underline{v}_2, \ldots, \underline{v}_k)$

Stabilizing transformation: $\theta_i = \ln(\gamma_i^2)$

Stage 2 function to be minimized:

$$F_g(\tilde{\Lambda},\underline{\theta}) = \sum_{j=k+1}^{p} \left[\frac{(\alpha_j - 1)^2}{2}\right]$$

Derivatives needed for gradient vector and Hessian matrix elements:

$$\frac{\partial F_g(\tilde{\Lambda},\underline{\theta})}{\partial\theta_i} = \sum_{j=k+1}^{p} \left(\alpha_j^2 - \alpha_j\right) v_{ij}^2$$

$$\frac{\partial^2 F_g(\tilde{\Lambda},\underline{\theta})}{\partial\theta_i \partial\theta_j} = \frac{\delta_{ij}\partial F_g(\tilde{\Lambda},\underline{\theta})}{\partial\theta_i} + \sum_{r=k+1}^{p} \alpha_r v_{ir} v_{jr}$$

$$\times \left[\sum_{s=1}^{k} \alpha_s \frac{\alpha_r + \alpha_s - 2}{\alpha_r - \alpha_s} v_{ir} v_{is} + a^{ij} \gamma_i \gamma_j \right]$$

Starting values: $\theta_i = \ln \left(\dfrac{1 - k/2p}{s^{ii}} \right)$

where $S^{-1} = (s^{ij})$

Maximum Likelihood

$F_m(\Lambda,\Gamma) = \text{trace}(\Sigma^{-1}S) - \ln|\Sigma^{-1}S| - p$

$\Lambda'\Gamma^{-2}\Lambda$ assumed to be diagonal (S must be positive definite)

Stage 1 matrix: $A = \Gamma S^{-1}\Gamma$

Eigenvalues of A: $\alpha_1 \leq \alpha_2 \leq \cdots \leq \alpha_p$

Orthonormal eigenvectors of A: $\underline{v}_1, \underline{v}_2, \ldots, \underline{v}_p$

Value of $\tilde{\Lambda}$ given Γ: $\tilde{\Lambda} = \Gamma V(D^{-1} - I)^{1/2}$

where $D = \text{diag}(\alpha_1, \ldots, \alpha_k)$

$V = (\underline{v}_1, \underline{v}_2, \ldots, \underline{v}_k)$

Stabilizing transformation: $\theta_i = \ln(\gamma_i^2)$

Stage 2 function to be minimized:

$$F_m(\tilde{\Lambda},\underline{\theta}) = \sum_{j=k+1}^{p} \left[\ln(\alpha_j) + \frac{1}{\alpha_j} - 1 \right]$$

Derivatives needed for gradient vector and Hessian matrix elements:

$$\frac{\partial F_m(\tilde{\Lambda},\underline{\theta})}{\partial\theta_i} = \sum_{j=k+1}^{p} \left(1 - \frac{1}{\alpha_j}\right) v_{ij}^2$$

$$\frac{\partial^2 F_m(\tilde{\Lambda},\underline{\theta})}{\partial\theta_i \partial\theta_j} = -\delta_{ij} \left(\frac{\partial F_m(\tilde{\Lambda},\underline{\theta})}{\partial\theta_i}\right)$$

$$+ \sum_{r=k+1}^{p} v_{ir}v_{jr} \left[\sum_{s=1}^{k} \frac{\alpha_r + \alpha_s - 2}{\alpha_r - \alpha_s} v_{ir}v_{js} + \delta_{ij}\right]$$

Starting values: $\theta_j = \ln \left[\frac{1 - k/2p}{s^{ii}}\right]$

where $S^{-1} = (s^{ij})$

There are a few additional details which must be considered before these methods are sufficiently dependable to include in general software systems. One such detail concerns the so-called Heywood case where Γ is singular or very nearly so. When one of the γ_i is nearly zero, the corresponding estimate $\hat{\gamma}_i^2$ will cause numerical problems beginning with determination of the θ_i in the generalized least-squares and maximum-likelihood methods. Joreskog (1977) describes a way to circumvent this and other potential difficulties in use of the three basic methods given previously in this section, and the reader may refer to this work for needed details.

12.5 MULTIVARIATE ANALYSIS OF VARIANCE

In Chapter 9 we considered analysis of variance computations relative to certain univariate classification models. The handling of multivariate models proceeds in a similar fashion although there are some additional considerations which we shall describe in this section. Since there is a close relationship between the computing problems in the univariate and multivariate cases, the presentation in this section will be brief.

The usual model for multivariate analysis of variance has the general form

$$Y = XB + F \tag{12.11}$$

where Y is an n x q matrix of observed values, X is the n x p rank r design matrix, B is a p x q matrix of parameters, and F is an n x q matrix of random errors. Thus instead of a single-dimensional response (i.e., $q \equiv 1$), which was the case considered in Chapter 9, we now have a q-dimensional situation with the same design matrix X in each dimension. If the same assumptions are made about each column of F, as were made about the vector $\underline{e}$ in the univariate model, then the estimability condition and minimum variance estimates of linear functions of the elements of any particular column of the B matrix follows as in the univariate case.

Tests of hypotheses in multivariate analysis of variance are usually based on the assumption that the rows of F (call them $\underline{f}_i'$) are independent and identically distributed according to the multivariate normal distribution

$$\underline{f}_i \sim MVN(\underline{0},\Sigma)$$

The form of a linear hypothesis is

$$H_0: \quad L'BW = \Gamma$$

The matrix L' is g x p and has rank $g \leq r$. The W matrix is q x u and has rank $u \leq q$. Test statistics are scalar-valued functions of the matrices

$$H = (L'\hat{B}W - \Gamma)'[L'(X'X)^{-}L]^{-1}(L'\hat{B}W - \Gamma)$$

$$E = W'Y'[I - X(X'X)^{-}X']YW$$

where

$$\hat{B} = (X'X)^{-}X'Y \tag{12.12}$$

In most cases W is specified to be the identity matrix and Γ is the zero matrix. This is the form we shall use in subsequent discussion. Under this choice of W and Γ the simplified expressions for the quantities given above are

$$H_0: \quad L'B = 0 \tag{12.13}$$

$$H = (L'\hat{B})'[L'(X'X)^{-} L]^{-1}(L'\hat{B}) \tag{12.14}$$

$$E = Y'[I - X(X'X)^{-}X']Y \tag{12.15}$$

Note that when q = 1 these are the expressions utilized in Chapters 8 and 9.

Several different test statistics have been proposed for testing hypotheses in the form of H_0. The statistics are functions of the eigenvalues of the matrix HE^{-1}. If $\lambda_1 \geq \lambda_2 \geq \cdots \geq \lambda_q$ denote the eigenvalues of HE^{-1}, then three frequently used test statistics are the following:

S1. Wilk's Likelihood Ratio

$$\Lambda = \frac{\det(E)}{\det(E + H)}$$

$$= \prod_{j=1}^{t} \frac{1}{1 + \lambda_j}$$

where rank (H) = t. For large n, the statistic

$$-[n - r - 0.5(u - t + 1)] \ln(\Lambda)$$

is approximately distributed as chi-square with qt degrees of freedom.

S2. Hotelling-Lawley Trace

$$\text{trace}(HE^{-1}) = \sum_{i=1}^{t} \lambda_i$$

is such that the statistic

$$\frac{2(sn + 1)\ \text{trace}(HE^{-1})}{s^2(2m + s + 1)}$$

is, for large n, approximately distributed as

$$F(z|s(2m + s + 1), 2(sn + 1))$$

where

$$s = \min(g,t)$$

$$m = 0.5(|g - t| - 1)$$

S3. A Test based on the largest eigenvalue λ_1 rejects H_0 if

$$\lambda_1 > d(\alpha)$$

where $d(\alpha)$ is the 100α percentage point of the greatest root distribution. Heck (1960) has prepared charts which give approximate percentage points for this distribution.

Timm (1975) and Morrison (1967) are two recommended references for an extensive description of estimation and testing in this multivariate situation.

Computational methods for determining $\hat{B}$ were given in Chapter 8. The same considerations apply here in that either X or X'X can be used as a basis for proceeding. With regard to estimable functions, again the same considerations apply. The matrix function L'B is estimable if each element of the matrix is an estimable function as defined in Chapter 8.

The need to compute eigenvalues of HE^{-1} is a major additional requirement over univariate computations. A usual procedure is to first decompose the symmetric positive definite matrix E as

$$E = QDQ' \tag{12.16}$$

where Q is an orthogonal matrix of eigenvectors of E and D is the diagonal matrix containing the eigenvalues of E. Having done this, eigenvalues of HE^{-1} are obtained as the eigenvalues of the matrix

$$D^{-1/2}Q'HQD^{-1/2} \tag{12.17}$$

Most standard statistical software packages contain routines which perform computations needed in multivariate analysis of variance and covariance. Specification of the matrix L in the hypothesis (12.13) is as much a problem in the multivariate case as it is in univariate computer analyses. Some programs approach the problem by forming the classical reductions in sums of squares, hence bypassing explicit representation of L. As we have seen in Chapter 9, this leaves the specific form of the hypothesis in doubt in some cases. Other programs either generate L according to preselected criteria or require that the user supply this matrix.

In addition to the general purpose software packages, there are more specialized packages designed to perform multivariate analyses. Finn (1977) gives a good discussion of available software and describes a general algorithm for multivariate regression and analysis of variance. Bock (1963, 1965, 1975) has developed one of the most popular algorithms for this application, and it is used as a basis for Finn's discussion. Hemmerle and Carney (1969) describe an efficient algorithm for handling balanced complete data in multivariate analysis. This algorithm uses an extension of the method described in Chapter 9. Finally, Hohwald and Heiberger (1977) give a comparative evaluation of the capabilities and limitations of the multivariate components in the four packages BMDP (Dixon and Brown), MANOVA (Clyde 1969), MULTIVARIANCE (Finn 1972), and SAS76 (Barr et al. 1976).

EXERCISES

12.1 Verify the equivalence of Eqs. (12.4) and (12.5) under the given transformation.

12.2 Show that Eq. (12.7) is a valid expression of Σ.

12.3 Explain why Λ is not uniquely defined by Eq. (12.7) and discuss the actions which are usually taken to handle the problem of nonunique Λ.

12.4 Why do algorithms for factor analysis usually proceed in two stages rather than taking a direct approach to minimize a nonlinear function of the variables in Λ and Γ?

12.5 Verify that the partial derivatives given for the unweighted least-squares method are indeed correctly stated.

12.6 What actions may be taken to increase stability in the given factor analysis algorithms for cases where Γ is singular?

12.7 Show that eigenvalues of HE^{-1} can be determined by the procedure outlined in Eqs. (12.16) and (12.17). Why is this approach considered instead of directly determining eigenvalues of the matrix product HE^{-1}?

REFERENCES AND FURTHER READINGS

Anderson, T. W. (1958), *An Introduction to Multivariate Statistical Analysis,* Wiley, New York.

Barr, A. J., J. H. Goodnight, J. P. Sall and J. T. Helwig (1976), *A User's Guide to SAS76,* SAS Institute, Raleigh, N.C.

Bock, R. D. (1963), Programming Univariate and Multivariate Analyses of Variance, *Techno. 5,* 95-117.

Bock, R. D. (1965), A Computer Program for Univariate and Multivariate Analysis of Variance, *Proceedings of Scientific Symposium on Statistics,* chaired by L. Robinson, 69-111.

Bock, R. D. (1975), *Multivariate Statistical Methods in Behavioral Research,* McGraw-Hill, New York.

Cattell, R. B., and D. K. Khanna (1977), Principles and Procedures for Unique Rotation in Factor Analysis, in *Statistical Methods for Digital Computers,* edited by K. Enslein, A. Ralston, and H. S. Wilf, Wiley, New York, 166-202.

Clarke, M. R. B. (1970), A Rapidly Convergent Method for Maximum-likelihood Factor Analysis, *British Journal of Mathematical and Statistical Psychology 23,* 43-52.

Clyde, D. J. (1969), *MANOVA: Multivariate Analysis of Variance on Large Computers,* Clyde Computing Service, Miami, Florida.

Cohen, C. (1969), An Investigation of the Geometry of Subspace for Some Multivariate Statistical Models, Ph.D. Thesis, University of Illinois, Urbana.

Cooley, W. W., and P. R. Lohnes (1971), *Multivariate Data Analysis,* Wiley, New York.

Cramer, C. M., and R. D. Bock (1966), Multivariate Analysis, *Review of Educational Research 36,* 604-617.

Dempster, A. P. (1969), *Elements of Continuous Multivariate Analysis,* Addison-Wesley, Reading, Mass.

Dixon, W. J., and M. B. Brown (eds.)(1977), *BMDP Biomedical Computer Programs P-Series,* University of California Press, Los Angeles.

Engelman, L., J. Frane, and R. Jennrich (1976), BMD and BMDP Programs for Analysis of Variance, *Proceedings of the Statistical Computing Section, American Statistical Association,* Washington, D.C., 148-153.

Finn, J. D. (1972), *MULTIVARIANCE: Univariate and Multivariate Analysis of Variance and Covariance and Regression,* National Educational Resources, Chicago.

Finn, J. D. (1974), *A General Model for Multivariate Statistical Analysis,* Holt, Rinehart and Winston, New York.

Finn, J. D. (1977), Multivariate Analysis of Variance and Covariance, in *Statistical Methods for Digital Computers,* vol. 3, edited by K. Enslein, A. Ralston, and H. S. Wilf, Wiley, New York, 203-264.

Frane, J. W., and M. Hill (1976), Factor Analysis as a Tool for Data Analysis, *Comm. Stat. (A) 5,* 487-506

Golub, G. H. (1969), Matrix Decomposition and Statistical Calculations, in *Statistical Computation,* edited by R. C. Milton and J. A. Nelder, Academic Press, New York, 365-398.

Good, I. J. (1969), Some Applications of the Singular Decomposition of a Matrix, *Techno. 11,* 823-831.

Harman, H. H. (1967), *Modern Factor Analysis,* University of Chicago Press, Chicago.

Harman, H. H. (1977), Minres Method of Factor Analysis, in *Statistical Methods for Digital Computers,* edited by K. Enslein, A. Ralston, and H. S. Wilf, Wiley, New York, 154-165.

Harris, R. J. (1975), *A Primer of Multivariate Statistics,* Academic Press, New York.

Heck, D. L. (1960), Charts of Some Upper Percentage Points of the Distribution of the largest Characteristic Root, *AMS 31,* 625-692.

Hemmerle, W. J., and E. J. Carney (1969), An Algorithm for Multivariate Analysis of Covariance, in *Statistical Computation,* edited by R. C. Milton and J. A. Nelder, Academic Press, New York, 413-426.

Hohwald, J., and R. M. Heiberger (1977), An Evaluation of MANOVA Programs, *Proceedings of the Statistical Computing Section, American Statistical Association,* Washington, D.C., 250-255.

Horst, P. (1961), Relations among m Sets of Measures, *Psychometrika 26,* 129-149.

Hotelling, H. (1933), Analysis of a Complex of Statistical Variables into Principal Components, *J. Educational Psychology 24,* 417-441, 498-520.

Hotelling, H. (1935), The Most Predictable Criterion, *J. Educational Psychology 26,* 139-142.

Hotelling, H. (1936), Relations between Two Sets of Variates, *Biometrika 28,* 321-377.

Jennrich, R. I., and S. M. Robinson (1969), A Newton-Raphson Algorithm for Maximum Likelihood Factor Analysis, *Psychometrika 34,* 111-123.

Jennrich, R. I., and P. F. Sampson (1966), Rotation for Simple Loadings, *Psychometrika 31,* 313-323.

Joreskog, K. G. (1967), Some Contributions to Maximum Likelihood Factor Analysis, *Psychometrika 32,* 443-482.

Joreskog, K. G. (1972), Factor Analysis By Generalized Least Squares, *Psychometrika 37,* 243-260.

Joreskog, K. G. (1977), Factor Analysis by Least-squares and Maximum-likelihood Methods, in *Statistical Methods for Digital Computers,* edited by K. Enslein, A. Ralston, and H. S. Wilf, Wiley, New York, 125-153.

Joreskog, K. G., and A. S. Goldberger (1972), Factor Analysis by Generalized Least Squares, *Psychometrika 37,* 243-259.

Joreskog, K. G., and M. Van Thillo (1971), *New Rapid Algorithms for Factor Analysis by Unweighted Least Squares, Generalized Least Squares, and Maximum Likelihood,* Research Memorandum 71-5. Educational Testing Service, Princeton, N.J.

Kaiser, H. F. (1958), The VARIMAX Criterion for Analytic Rotation in Factor Analysis, *Psychometrika 23,* 187-260.

Kaiser, H. F. (1970), A Second Generation Little Jiffy, *Psychometrika 35,* 401-415.

Kaiser, H. F., and J. Caffrey (1965), Alpha Factor Analysis, *Psychometrika 30,* 1-14.

Kshirsagar, A. M. (1972), *Multivariate Analysis,* Marcel Dekker, New York.

Lawley, D. N. and A. E. Maxwell (1971), *Factor Analysis as a Statistical Method,* 2nd ed., Butterworth, London.

Morrison, D. F. (1967), *Multivariate Statistical Methods,* McGraw-Hill, New York.

Olkin, I. (1951), On Distribution Problems in Multivariate Analysis, Ph.D. thesis, University of North Carolina, Chapel Hill.

Pillai, K. C. (1960), *Statistical Tables for Tests of Multivariate Hypotheses,* Manila, Statistical Center, University of the Philippines, Manila.

Pruzek, R. M., J. R. Sherry, N. F. Walker, and R. A. Pfeiffer (1976), Interactive Multivariate Analysis Generalized for Explorations of Data-IMAGE, *Proceedings of the Ninth Interface Symposium on computer Science and Statistics,* edited by D. Hoaglin and R. Welsch, Prindle,Weber and Schmidt, Boston, 275-279.

Rao, C. R. (1955), Estimation and Tests of Significance in Factor Analysis, *Psychometrika 20,* 93-111.

Rao, C. R. (1973), *Linear Statistical Inference and Its Applications,* 2nd ed. Wiley, New York.

Roy, J. (1958), Step-down Procedure in Multivariate Analysis, *AMS 29,* 1177-1187.

Roy, S. N. (1957), *Some Aspects of Multivariate Analysis,* Wiley, New York.

Schatzoff, M. (1966), Exact Distributions of Wilk's Likelihood Ratio Criterion, *Biometrika 53,* 347-358.

Thurstone, L. L. (1940), Experimental Study of Simple Structure, *Psychometrika 5,* 153-168.

Thurstone, L. L. (1947), *Multiple Factor Analysis,* University of Chicago Press, Chicago.

Timm, N. H. (1975), *Multivariate Analysis with Applications in Education and Psychology,* Brooks/Cole, Monterey, Calif.

Tucker, L. R., and C. Lewis (1973), A Reliability Coefficient for Maximum Likelihood Factor Analysis, *Psychometrika 38,* 1-10.

Whittle, P. (1952), On Principal Components and Least Square Methods of Factor Analysis, *Skandinavian Aktuar 35,* 223-239.

INDEX